Texts and
Monographs
in Physics

Nail R. Sibgatullin

Oscillations and Waves

in Strong Gravitational and Electromagnetic Fields

With 26 Figures

Springer-Verlag
Berlin Heidelberg New York
London Paris Tokyo
Hong Kong Barcelona

Professor Dr. Nail R. Sibgatullin
Faculty of Mechanics and Mathematics, Moscow State University
SU-119899 Moscow, USSR

Translator:
Dr. Nathaniel M. Queen
Department of Mathematics, The University of Birmingham
P.O. Box 363, Birmingham B15 2TT, United Kingdom

Editors

Wolf Beiglböck
Institut für Angewandte Mathematik
Universität Heidelberg
Im Neuenheimer Feld 294
D-6900 Heidelberg 1
Fed. Rep. of Germany

Joseph L. Birman
Department of Physics, The City College
of the City University of New York
New York, NY 10031, USA

Robert P. Geroch
Enrico Fermi Institute
University of Chicago
5640 Ellis Ave.
Chicago, IL 60637, USA

Elliott H. Lieb
Department of Physics
Joseph Henry Laboratories
Princeton University
Princeton, NJ 08540, USA

Tullio Regge
Istituto di Fisca Teorica
Università di Torino, C. so M. d'Azeglio, 46
I-10125 Torino, Italy

Walter Thirring
Institut für Theoretische Physik
der Universität Wien, Boltzmanngasse 5
A-1090 Wien, Austria

Title of the original Russian edition:
Kolebaniya i volny v silnykh gravitatsionnykh i ėlektromagnitnykh polyakh
© Nauka, Moscow 1984

ISBN-13: 978-3-642-83529-2 e-ISBN-13: 978-3-642-83527-8
DOI: 10.1007/978-3-642-83527-8

Library of Congress Cataloging-in-Publication Data. Sibgatullin, N. R. (Nail ' Rakhimovich) [Kolebaniia i volny v sil 'nykh gravitatsionnykh i ėlektromagnitnykh poliakh. English] Oscillations and waves in strong gravitational and electromagnetic fields / Nail R. Sibgatullin. p. cm.–(Texts and monographs in physics) Translation of: Kolebaniia i volny sil 'nykh gravitatsionnykh i ėlektromagnitnykh poliakh. Includes bibliographical references. ISBN 0-387-19461-4 (U.S.) 1. Gravitational fields. 2. Electromagnetic fields. 3. Perturbation (Quantum dynamics) I. Title. II. Series. QB283.S5513 1990 530.1'4–dc20 89-11542

Softcover reprint of the hardcover 1st edition 1991

57/3140-543210 – Printed on acid-free paper

Dedicated to the memory of my father,
Rakhim Khaĭrulovich Sibgatullin

Preface to the English Edition

This book is an updated and modified translation of the Russian edition of 1984. In the present edition, certain sections have been abridged (in particular, Sects. 6.1 and 8.3) and the bibliography has been expanded. There are more detailed discussions of the group properties of integrable systems of equations of mathematical physics (Sect. 3.4) and of the Riemannian problem in the context of the infinite-dimensional internal symmetry groups of these systems of equations. There is an extended discussion of the reasons for the acceleration and retardation of pulsars in connection with more recent achievements of X-ray astronomy. Part of the material of Chap. 8 of the Russian edition has been included in Chap. 7; thus the number of chapters has been reduced to seven.

S. Chandrasekhar set for me an example of brilliant analytical penetration into the essence of physical problems, and my book touches on his work in many instances. The results of modern quantum theories of strong fields are not presented, but they can be found in the fundamental monographs *Quantum Electrodynamics of Strong Fields* by W. Greiner, B. Müller, J. Rafelski (Springer-Verlag, Berlin, Heidelberg, New York 1985) and *Quantum Effects in Intense External Fields* [in Russian] by A. Grib, S. Mamaev, W. Mostepanenko (Energoatomizdat, Moscow 1988).

This book was translated by Dr. N. M. Queen; I am very grateful to him.

I thank sincerely H. Latta, C.–D. Bachem, V. Rehman, S. von Kalckreuth for preparing of the english manuscript.

Moscow, August 1990 *Nail R. Sibgatullin*

Preface

> "The current of life which flows day and night in my veins flows in the universe and dances a measured dance."
>
> *Rabindranat Tagor*

This book is devoted to the investigation of waves and oscillations in the presence of strong gravitational and electromagnetic fields.

A detailed study is made of the propagation of gravitational and electromagnetic waves in the pseudo-Riemannian manifolds of the general theory of relativity. Much attention is given to the classical problems of the theory of black holes and waves in their vicinity. Rigorous results of the mathematical theory of black holes as well as of stationary axially symmetric fields are expounded, and the properties of electromagnetic and gravitational waves propagating in the gravitational fields of charged and neutral black holes are analyzed in detail.

An account is given of the principles of relativistic hydrodynamics, magnetohydrodynamics, and the acoustics of a relativistic gas. Scale-invariant motions of an ultrarelativistic gas are analyzed in detail in the framework of the special and general theories of relativity. An outline is given of the theory of the equations of state of an ideal gas under strong compression, and also at high temperatures. The development of nonhomogeneities in models of the Universe with a cosmological magnetic field is investigated. The reader who is interested only in problems of relativistic hydrodynamics can confine himself to Chaps. 5 and 6 of the book, where a brief introduction to cosmology is given at the same time. Chapter 2 represents an introduction to the classical theory of black holes. Chapters 1 and 4 contain an account of the principles of the wave dynamics of gravitational and electromagnetic fields in general relativity. Some acoustic phenomena in strong gravitational fields and manifestations of weak nonlinearity for oscillations and waves in restricted systems in external electromagnetic and gravitational fields are considered in Chap. 7.

As a guide for the reader, each chapter is prefaced with an elementary introduction to the physical problems.

A knowledge of the elements of tensor analysis is sufficient for reading the book. From the material, the novice reader can, if he wishes, master the various mathematical methods of the theory of waves in Newtonian mechanics and in the special and general theories of relativity, and can gain an idea of the new results in this field.

The author has made use of material from lectures given by him in the Faculty of Mechanics and Mathematics at Moscow University. A bibliography is provided to enable the reader to make a detailed study of problems related to the subjects of the book but treated here with insufficient completeness.

The author is deeply grateful to Academician L. I. Sedov, who suggested that this book be written, for numerous fruitful discussions of the problems considered, and to Professors V. I. Arnol'd, A. A. Starobinsky and R. A. Sunyaev for valuable remarks on the manuscript.

The author expresses sincere gratitude to the editor, V. V. Rozantseva, for her work on improving the manuscript, and to Drs. G. A. Alekseev and Alberto Garsia for a number of helpful remarks.

Moscow, December 1983 *Nail R. Sibgatullin*

Contents

1. Gravitational Waves in Strong Gravitational Fields

Gravitational fields cannot propagate instantaneously. The problem of detecting gravitational waves experimentally is an unsolved fundamental problem of modern experimental physics. The solution of this problem requires corresponding instrumentation with greater accuracy (the measured effect must exceed the level of the errors in the measurements), as well as the search for an experimental arrangement in which the expected effect is most apparent for the existing level of accuracy of the experimental facilities.

Near the Earth, space-time is asymptotically flat and gravitational waves are weak. Therefore it is comparatively easy to perform a theoretical analysis of the properties of gravitational waves on the flat background in any of the existing theories of the gravitational field. In particular, in the general theory of relativity it reduces to the determination of solutions of the classical wave equation.

The pseudo-Riemannian manifolds near objects with strong gravitational fields (neutron stars, pulsars, quasars, and black holes) are essentially curved. Waves propagate very differently in pseudo-Riemannian and flat manifolds. This can be seen if only from the fact that weak waves in curved spaces do not in general satisfy Huygens's principle in the narrow sense, i.e., there is no backward front for perturbations with compact support [1.1–3].

In this chapter we study some general properties of wave fields on the background of pseudo-Riemannian manifolds in general relativity, the conditions on surfaces of discontinuity in generalized theories of the gravitational field, and the interference of plane waves in general relativity. At the beginning of the chapter, we explain the Newman-Penrose formalism, which is a very convenient theoretical tool for investigating problems having degenerate algebraic properties.

1.1 Formalism of Complex Null Tetrads and Petrov Classification of Algebraic Types of the Weyl Tensor

A set of space-time points M_4 in which physical processes take place possesses in itself certain properties in general relativity. We shall assume that this set is a smooth four-dimensional manifold. By the definition of a manifold, each point of M_4 is contained in some open neighborhood Q, which can be mapped in a one-to-one manner onto some open simply connected domain q in the Euclidean space R^4. The mapping of φ together with the neighborhood Q is called a *chart* or a *local coordinate system.* If two neighborhoods Q and Q' have a nonempty intersection $Q \cap Q'$, then the maps (φ, Q) and (φ', Q') must be compatible, i.e., the transition from the coordinates $\varphi(Q \cap Q')$ to the coordinates $\varphi'(Q \cap Q')$ and vice versa has at least the smoothness class C^2. The manifold M_4 is assumed to be *Hausdorff*, i.e., for any two distinct points in it, it is possible to find nonintersecting neighborhoods containing them. The set of neighborhoods $\{Q\}$ forms an open covering of M_4: $M_4 = \cup Q$. The totality of charts forms the *atlas of the manifold* M_4.

With each point x of the manifold M_4 we can associate a linear space TM_x tangent to the manifold at the point x. As a basis in TM_x we can take four vectors $\vec{\mathcal{E}}_i$, $i = 1,2,3,4$; the vector $\vec{\mathcal{E}}_i$ is a tangent vector to the coordinate line x^i: $\vec{\mathcal{E}}_i = \partial/\partial x^i$.

The geometry of a pseudo-Riemannian manifold is completely characterized by specifying on the manifold M_4 a symmetric second-rank tensor g_{ij}, which at each point can be reduced to the form $g_{ij} = \eta_{ij}$, $\eta_{\alpha\alpha} = -1$ $(\alpha = 1,2,3)$, $\eta_{44} = 1$, $\eta_{ij} = 0$ $(i \neq j)$ by means of affine transformations. Thus, locally a pseudo-Riemannian space has the structure of Minkowski space. The quadratic form $g_{ij}dx^i dx^j$ determines the interval ds^2 between two neighboring points with coordinates x^1, x^2, x^3, x^4 and $x^1 + dx^1$, $x^2 + dx^2$, $x^3 + dx^3$, $x^4 + dx^4$. The interval on time-like curves is characterized by the proper time: $ds^2 = c^2 d\tau^2$, where c is the speed of light. On space-like curves, $ds^2 = -dl^2$, where dl is the distance between two neighboring points on the curve. The interval is equal to zero between two neighboring points on isotropic curves (light rays).

The space-time manifold is assumed to be *orientable*, i.e., the Jacobian of the coordinate transformation for compatibility of all charts can always be chosen to be positive.

In order to formulate differential equations which do not depend on the choice of the coordinate system on each neighborhood of the manifold M_4, it is necessary to specify on it a *connection*, i.e., a set of functions Γ^i_{jk} $(i,j,k = 1,2,3,4)$, called the *Christoffel symbols*, which under a coordinate transformation $y^i = f^i(x^k)$ change according to the law

$$\Gamma'^p_{qr}(y) = \frac{\partial y^p}{\partial x^i}\left[\Gamma^i_{jk}(x)\frac{\partial x^j}{\partial y^q}\frac{\partial x^k}{\partial y^r} + \frac{\partial^2 x^i}{\partial y^q\,\partial y^r}\right] .$$

The connection in Riemannian spaces is consistent with the metric, i.e., the covariant derivatives of the metric tensor are identically equal to zero, $\nabla_i\, g_{jk} = 0$, and it is symmetric with respect to the lower indices. Hence the Christoffel symbols can be expressed uniquely in terms of the metric as

$$2\Gamma^i_{jk} = g^{im}(g_{jm,k} + g_{km,j} - g_{jk,m}) .$$

Here and in what follows, a comma indicates a derivative with respect to the coordinates whose index appears after the comma.

The degree of curvature of a Riemannian manifold is characterized by the Riemann-Christoffel curvature tensor

$$2R_{iklm} = (g_{im,kl} + g_{kl,im} - g_{il,km} - g_{km,il}) + 2g_{np}\left(\Gamma^n_{kl}\Gamma^p_{im} - \Gamma^n_{km}\Gamma^p_{il}\right) \quad . \tag{1.1.1}$$

The Ricci tensor is formed from the Riemann tensor by means of a contraction with respect to the first and third indices: $R_{ik} = R^s_{isk}$. In order to characterize the intrinsic properties of the gravitational field which do not depend on the choice of the scale factor $a^2(x)$ in the class of metrics conformal to a given metric, $g'_{ij} = a^2(x)g_{ij}$, we introduce the concept of the *Weyl conformal-curvature tensor* W_{iklm}:

$$R_{iklm} = W_{iklm} + \tfrac{1}{2}(g_{il}R_{km} + g_{km}R_{il} - g_{im}R_{kl} - g_{kl}R_{im}) - \frac{R}{6}(g_{il}g_{km} - g_{im}g_{kl}) \quad . \tag{1.1.1'}$$

The Weyl tensors for a given metric g_{ij} and for the metric $g'_{ij} = a^2 g_{ij}$, where a is an arbitrary function of x^i, are equal when one of the indices is contravariant and the remaining ones are covariant: $W'^i_{jkl} = W^i_{jkl}$. By definition, the Weyl tensor satisfies the identity $W^s_{isk} = 0$.

In addition to the fields of the basis vectors $\mathcal{E}_i$ on the neighborhood Q, for various reasons it is convenient to introduce the fields of the orthonormalized vectors $\boldsymbol{h}_a$ $(a = 1, 2, 3, 4)$ satisfying, by definition, the relations

$$g_{ij}h^i_a h^j_b = \eta_{ab} \quad ,$$

where η_{ab} are the components of the metric tensor in Minkowski space and h^i_a $(i = 1, 2, 3, 4)$ are the components of the vector $\boldsymbol{h}_a$ in the local system of coordinates x^i. Because of the orientability of the manifold, we can impose on h^i_a the requirement of positivity of the determinant $\det(h^i_a)$, where $\boldsymbol{h}_4$ is a time-like vector directed to the future. For the matrices (h^i_a), we introduce the inverse matrices (ξ^a_i). The coefficients of the metric g_{ij} can be represented in the form

$$g_{ij} = \xi^a_i \xi^b_j \eta_{ab} \quad .$$

In particular cases, the coefficients h^i_a can be holonomic, which means that they can be represented in the form of partial derivatives of the functions $x^i(y^a)$, i.e., $h^i_a = \partial x^i/\partial y^a$. In these cases, the metric tensor can be reduced to the principal axes directly for all points of the neighborhood Q by means of a single coordinate transformation. This means that the neighborhood Q is, in fact, a flat (more precisely, a pseudo-Euclidean) piece of the manifold M_4.

By definition, the *tetrad components* of an arbitrary tensor $T^{i\cdots}_{j\cdots}$ are the expressions $T^{b\cdots}_{a\cdots} \equiv T^{i\cdots}_{j\cdots}\xi^b_i \ldots h^j_a \ldots$.

In many cases, it is very convenient to introduce complex null tetrads, which have been widely applied since the work of *Newman* and *Penrose* [1.4] and *Sachs* [1.5]. In contrast to the original work of *Newman* and *Penrose* [1.4], who used the spinor representation (see also the reviews of [1.6,7]), in what follows we shall confine ourselves to a purely vector treatment.

We introduce, instead of the fields of the four vectors $\boldsymbol{h}_a$, the fields of the following four vectors, two of which are complex:

$$\begin{aligned} \sqrt{2}\boldsymbol{n} &= \boldsymbol{h}_1 + \boldsymbol{h}_4, \quad \sqrt{2}\boldsymbol{l} = \boldsymbol{h}_4 - \boldsymbol{h}_1, \\ \sqrt{2}\boldsymbol{m} &= \boldsymbol{h}_2 + \mathrm{i}\boldsymbol{h}_3, \quad \sqrt{2}\boldsymbol{m}^* = \boldsymbol{h}_2 - \mathrm{i}\boldsymbol{h}_3 \ . \end{aligned} \tag{1.1.2}$$

The world lines of particles moving with the speed of light are integral curves of the vectors $\boldsymbol{l}$ and $\boldsymbol{n}$. The length of each of the vectors $\boldsymbol{l}$, $\boldsymbol{n}$, $\boldsymbol{m}$, $\boldsymbol{m}^*$ is equal to zero. Therefore these vectors are isotropic, and the tetrads introduced by the method indicated above are said to be *null*. It follows from the definition (1.1.2) that the vectors $\boldsymbol{l}$, $\boldsymbol{n}$, $\boldsymbol{m}$, $\boldsymbol{m}^*$ satisfy the normalization conditions

$$\begin{aligned} &\boldsymbol{l}\cdot\boldsymbol{l} = \boldsymbol{m}\cdot\boldsymbol{m} = \boldsymbol{n}\cdot\boldsymbol{n} = \boldsymbol{m}^*\cdot\boldsymbol{m}^* = \boldsymbol{l}\cdot\boldsymbol{m} = \boldsymbol{n}\cdot\boldsymbol{m} = 0, \\ &\boldsymbol{l}\cdot\boldsymbol{n} = -\boldsymbol{m}\cdot\boldsymbol{m}^* = 1 \ , \end{aligned} \tag{1.1.3}$$

where, by definition, the scalar product $\boldsymbol{a}\cdot\boldsymbol{b}$ is equal to the number $g^{ij}a_ib_j$ in the local system of coordinates x^i.

It follows from the relations (1.1.2) that the metric tensor in the system of coordinates x^i can be expressed in terms of the components of the vectors $\boldsymbol{l}$, $\boldsymbol{n}$, $\boldsymbol{m}$, $\boldsymbol{m}^*$ in this same system:

$$g_{ij} = l_in_j + l_jn_i - m_im_j^* - m_i^*m_j \ .$$

From the complex null basis, it is very convenient to construct second-rank antisymmetric tensors (bivectors F_{ij}). We introduce the tensor F_{ij}^* *dual* to the tensor F_{ij} according to the rule[1]

$$2F_{ij}^* = \varepsilon_{ijkl}F^{kl} \ ,$$

where $\varepsilon_{ijkl} \equiv \sqrt{-g}\, e_{ijkl}$, and the component e_{ijkl} is equal to +1 or −1 if (i,j,k,l) is formed from $(1,2,3,4)$ by means of an even or an odd number of transpositions of the indices, respectively, and is equal to zero if two or more of the numbers i,j,k,l are equal. The contravariant components of the pseudotensor ε^{ijkl} are $-e_{ijkl}/\sqrt{-g}$.

Let us form the bivector $(F_{ij} - \mathrm{i}F_{ij}^*)/2$ and expand it in terms of the vectors of the null basis $\boldsymbol{l}$, $\boldsymbol{n}$, $\boldsymbol{m}$, $\boldsymbol{m}^*$. It follows from the definition of a null tetrad that the determinant formed from the contravariant components of the vectors $\boldsymbol{l}$, $\boldsymbol{n}$, $\boldsymbol{m}$, $\boldsymbol{m}^*$ in the local coordinate system is equal to $\mathrm{i}/\sqrt{-g}$. Therefore we have the relations

$$\mathrm{i}\varepsilon_{ijkl}l^kn^l = -m_jm_i^* + m_im_j^*, \quad \mathrm{i}\varepsilon_{ijkl}m^km^{*l} = -l_jn_i + l_in_j \ . \tag{1.1.4}$$

[1] The duality relations have been analyzed, for example, in [1.8].

We denote by M_{ij} the bivector $(l_j n_i - l_i n_j + m_i m_j^* - m_j m_i^*)$. Then it follows from (1.1.4) that

$$2M_{ij} = -\mathrm{i}\varepsilon_{ijkl}M^{kl} \ . \tag{1.1.5}$$

Similarly, we can establish the relations

$$\mathrm{i}\varepsilon_{ijkl}l^k m^l = -(l_i m_j - l_j m_i), \quad \mathrm{i}\varepsilon_{ijkl}n^k m^{*l} = -(n_i m_j^* - n_j m_i^*) \ . \tag{1.1.6}$$

Definitions. A bivector A_{ij} is said to be *simple* if it can be represented in the form $a_i b_j - a_j b_i$, where $\boldsymbol{a}$ and $\boldsymbol{b}$ are vectors.

A bivector G_{ij} is said to be *self-dual* if it satisfies the relation

$$2\mathrm{i}G_{ij} = \varepsilon_{ijkl}G^{kl} \ .$$

The simple bivectors $V_{ij} = l_i m_j - l_j m_i$ and $U_{ij} = n_j m_i^* - n_i m_j^*$, like M_{ij}, are self-dual, since they satisfy the equations (1.1.6) analogous to (1.1.5).

As is readily verified, the complex bivector $(F_{ij} - \mathrm{i}F_{ij}^*)/2$ is a self-dual bivector and must therefore be expanded in terms of self-dual basis bivectors. Besides the basis bivectors $\boldsymbol{V}$, $\boldsymbol{M}$, and $\boldsymbol{U}$, a complete basis in terms of which an arbitrary bivector is expanded includes also their complex-conjugate bivectors $\boldsymbol{V}^*$, $\boldsymbol{M}^*$, and $\boldsymbol{U}^*$. However, these bivectors are not self-dual, since they satisfy the equations

$$2V_{ij}^* = \mathrm{i}\varepsilon_{ijkl}V^{*kl}, \quad 2M_{ij}^* = \mathrm{i}\varepsilon_{ijkl}M^{*kl}, \quad 2U_{ij}^* = \mathrm{i}\varepsilon_{ijkl}U^{*kl} \ .$$

Therefore the self-dual bivector $2F_{ij}^+ = F_{ij} - \mathrm{i}\varepsilon_{ijkl}F^{kl}/2$ has the following expansion with respect to simple bivectors

$$F_{ij}^+ = \Phi_0 U_{ij} + \Phi_1 M_{ij} + \Phi_2 V_{ij} \ , \tag{1.1.7}$$

where the complex numbers Φ_0, Φ_1, Φ_2 are the tetrad components of the bivector $\boldsymbol{F}^+$ in the three-dimensional complex space with the basis bivectors $\boldsymbol{V}$, $\boldsymbol{M}$, $\boldsymbol{U}$. The components Φ_0, Φ_1, Φ_2 of the complex bivector F_{ij}^+ can also be obtained directly from the components F_{ij} of a real bivector (for example, the bivector of the electromagnetic field) as follows

$$\Phi_0 = F_{ik}l^i m^k, \quad \Phi_1 = \tfrac{1}{2}F_{ik}(l^i n^k - m^i m^{*k}), \quad \Phi_2 = -F_{ik}n^i m^{*k} \ .$$

Each transformation of the six-parameter group of proper Lorentz transformations, which is the group of orthogonal rotations of Minkowski space, can be written explicitly as a transition at a point from the null tetrad $\boldsymbol{l}$, $\boldsymbol{n}$, $\boldsymbol{m}$, $\boldsymbol{m}^*$ to the tetrad $\boldsymbol{l}'$, $\boldsymbol{n}'$, $\boldsymbol{m}'$, $\boldsymbol{m}^{*\prime}$ by means of the direct product of the following transformations:

I) A rotation in the plane of the space-like vectors $\boldsymbol{h}_2$ and $\boldsymbol{h}_3$ through an angle χ:

$$\boldsymbol{l}' = \boldsymbol{l}, \quad \boldsymbol{n}' = \boldsymbol{n}, \quad \boldsymbol{m}' = \boldsymbol{m}\exp(\mathrm{i}\chi) \ .$$

II) A Lorentz rotation in the plane of the vectors $\boldsymbol{h}_1$ and $\boldsymbol{h}_4$ (a transition to a coordinate system moving with respect to the original one): $\boldsymbol{l}' = A\boldsymbol{l}$, $\boldsymbol{n}' = \boldsymbol{n}/A$, $\boldsymbol{m}' = \boldsymbol{m}$, where A is an arbitrary real number.

III) A rotation of the vector $\boldsymbol{n}$ with $\boldsymbol{l}$ fixed:

$$\boldsymbol{l}' = \boldsymbol{l}, \quad \boldsymbol{n}' = aa^*\boldsymbol{l} + \boldsymbol{n} + a\boldsymbol{m} + a^*\boldsymbol{m}^*, \quad \boldsymbol{m}' = a^*\boldsymbol{l} + \boldsymbol{m} \ ,$$

where a is an arbitrary complex number.

IV) A rotation of the vector $\boldsymbol{l}$ with $\boldsymbol{n}$ fixed:

$$\boldsymbol{l}' = bb^*\boldsymbol{n} + \boldsymbol{l} + b^*\boldsymbol{m} + b\boldsymbol{m}^*, \quad \boldsymbol{n}' = \boldsymbol{n}, \quad \boldsymbol{m}' = b\boldsymbol{n} + \boldsymbol{m} \ ,$$

where b is an arbitrary complex number.

We emphasize that the transformed basis $\boldsymbol{l}'$, $\boldsymbol{n}'$, $\boldsymbol{m}'$, $\boldsymbol{m}^{*\prime}$ also satisfies the normalization conditions (1.1.3).

Under these transformations, the bivectors $\boldsymbol{U}$, $\boldsymbol{M}$, and $\boldsymbol{V}$ of the complex basis and the components Φ_0, Φ_1, Φ_2 transform as follows:

I) $$\boldsymbol{U}' = \mathrm{e}^{-\mathrm{i}\chi}\boldsymbol{U}, \quad \boldsymbol{M}' = \boldsymbol{M}, \quad \boldsymbol{V}' = \mathrm{e}^{\mathrm{i}\chi}\boldsymbol{V},$$
$$\Phi_0' = \mathrm{e}^{\mathrm{i}\chi}\Phi_0, \quad \Phi_1' = \Phi_1, \quad \Phi_2 = \mathrm{e}^{-\mathrm{i}\chi}\Phi_2;$$

II) $$\boldsymbol{U}' = \frac{1}{A}\boldsymbol{U}, \quad \boldsymbol{M}' = \boldsymbol{M}, \quad \boldsymbol{V}' = A\boldsymbol{V},$$
$$\Phi_0' = A\Phi_0, \quad \Phi_1' = \Phi_1, \quad \Phi_2' = \frac{1}{A}\Phi_2;$$

III) $$\boldsymbol{U}' = a^2\boldsymbol{V} - a\boldsymbol{M} + \boldsymbol{U}, \quad \boldsymbol{M}' = \boldsymbol{M} - 2a\boldsymbol{V}, \quad \boldsymbol{V}' = \boldsymbol{V};$$
$$\Phi_0' = \Phi_0, \quad \Phi_1' = \Phi_1 + a\Phi_0, \quad \Phi_2' = a^2\Phi_0 + 2a\Phi_1 + \Phi_2;$$

IV) $$\boldsymbol{U}' = \boldsymbol{U}, \quad \boldsymbol{M}' = \boldsymbol{M} - 2b\boldsymbol{U}, \quad \boldsymbol{V}' = b^2\boldsymbol{U} - b\boldsymbol{M} + \boldsymbol{V};$$
$$\Phi_0' = \Phi_0 + 2b\Phi_1 + b^2\Phi_2, \quad \Phi_1' = \Phi_1 + b\Phi_2, \quad \Phi_2' = \Phi_2 \ .$$

The transformations (I), (II), (III), and (IV) constitute Abelian subgroups of the group of Lorentz transformations. Thus, we can say that the formalism of null tetrads provides a representation of the Lorentz group in a three-dimensional complex space [1.9].

Suppose that the quadratic equation $a^2\Phi_0 + 2a\Phi_1 + \Phi_2 = 0$ is nondegenerate, i.e., for $\Phi_0 \neq 0$ it has distinct roots. In the case $\Phi_0 = 0$ we shall assume that it is nondegenerate if $\Phi_1 \neq 0$. Applying the transformation (III) with a equal to one of the roots of this equation, in the new coordinate system we obtain $\Phi_2' = 0$ and $\Phi_1' \neq 0$, in view of the assumption of nondegeneracy. If $\Phi_0 \neq 0$, we can, after this, apply the transformation (IV) in order to reduce the component Φ_0 to zero. The corresponding value of b satisfies a linear equation and is therefore determined uniquely.

Thus, in the general case the bivector F_{ij} at a fixed point with a perfectly definite null tretrad has the form

$$F_{ij} = \Phi_1 M_{ij} + \Phi_1^* M_{ij}^* \quad .$$

The vacuum electromagnetic field at each point is completely described by the antisymmetric second-rank tensor F_{ij}. As is well known, the energy-momentum tensor of the vacuum electromagnetic field has the form

$$4\pi T_{ij} = -\left[F_{ik}F_j^k - \tfrac{1}{4}g_{ij}F_{lm}F^{lm}\right] \quad . \tag{1.1.8}$$

Using the expressions (1.1.7), it can be represented in the form

$$\begin{aligned}4\pi T_{ij} = {} & \Phi_0\Phi_0^* n_i n_j + \Phi_1\Phi_1^*(n_i l_j + n_j l_i + m_i m_j^* + m_i^* m_j) \\ & + \Phi_2\Phi_2^* l_i l_j - 2(m_i^* n_j + n_i m_j^*)\Phi_0\Phi_1^* \\ & + 2\Phi_0\Phi_2^* m_i^* m_j^* - 2(l_i m_j + m_i l_j)\Phi_2\Phi_1^* + \text{c.c.} \quad ,\end{aligned}$$

where c.c. denotes the complex conjugate of the preceding expression.

In the canonical basis, only the second term in the expansion (1.1.7) of F_{ij}^+ is nonzero. If the quadratic equation written above is degenerate, i.e., has multiple roots, then the application of the transformation (III) with the parameter a equal to the root of this equation simultaneously reduces both Φ_2 and Φ_1 to zero. The last remaining component Φ_0 can be made equal to unity by applying the transformations (I) and (II). Therefore the energy-momentum tensor of the electromagnetic field in the degenerate case can be represented in the form

$$4\pi T_{ij} = n_i n_j \quad .$$

The degenerate case is realized in plane electromagnetic waves when both invariants of the electromagnetic field reduce to zero.

It follows from (1.1.7) that in the general case the invariants of the electromagnetic field can be expressed in terms of the components Φ_0, Φ_1, and Φ_2 as follows:

$$\begin{aligned}&\boldsymbol{H}^2 - \boldsymbol{E}^2 = \tfrac{1}{2}F_{lm}F^{lm} = 4\,\mathrm{Re}\{\Phi_0\Phi_2 - \Phi_1^2\}, \\ &2\boldsymbol{E}\cdot\boldsymbol{H} = \tfrac{1}{4}F_{ij}F_{lm}\varepsilon^{ijlm} = 4\,\mathrm{Im}\{\Phi_0\Phi_2 - \Phi_1^2\}, \\ &\boldsymbol{H}^2 - \boldsymbol{E}^2 + 2\mathrm{i}\boldsymbol{E}\cdot\boldsymbol{H} = 4(\Phi_0\Phi_2 - \Phi_1^2) \quad .\end{aligned}$$

Let us consider now the algebra of a fourth-rank tensor R_{ijkl} which is antisymmetric in the first and second indices, and also in the third and fourth indices, and symmetric with respect to transpositions of the first and second pairs of indices. Contracting the components of R_{ijkl} with respect to the first and third indices, we obtain a symmetric second-rank tensor R_{jl}. Contraction of the components of the tensor R_{jl} gives a scalar R. An example of the tensor R_{ijkl} is the Riemann curvature tensor. The algebraic properties of the tensor R_{ijkl} also apply to the tensors

$$p_{ijkl} = R_{ik}g_{jl} + R_{jl}g_{ik} - R_{il}g_{jk} - R_{jk}g_{il}, \quad q_{ijkl} = (g_{ik}g_{jl} - g_{il}g_{jk})R \ ,$$

whose components are linear in R_{ik} and R.

The tensor $W_{ijkl} = R_{ijkl} + ap_{ijkl} + bq_{ijkl}$ also possesses antisymmetry within the first and second pairs of indices and symmetry with respect to transposition of these pairs. It is easy to choose the coefficients a and b in such a way that the tensor W_{ijkl} gives zero after contraction with respect to the first and third indices. The corresponding values of a and b are $-1/2$ and $1/6$. An example of the tensor W_{ijkl} is the Weyl tensor introduced above.

In the same way that this was done for the electromagnetic field tensor, we form the complex combination

$$2W^{+}_{ijkl} \equiv W_{ijkl} - \frac{\mathrm{i}}{2}\varepsilon_{ijst}W^{st}_{kl} \ . \tag{1.1.9}$$

If the first pair of indices i and j is fixed, the tensor W^{+}_{ijkl} is a self-dual bivector, and it can therefore be represented in the form

$$W^{+}_{ijkl} = a_{ij}U_{kl} + a_{kl}U_{ij} + b_{ij}M_{kl} + b_{kl}M_{ij} + c_{ij}V_{kl} + c_{kl}V_{ij} \ . \tag{1.1.10}$$

The bivectors a_{ij}, b_{ij}, and c_{ij} can be expanded with respect to a complete basis of bivectors U_{ij}, M_{ij}, V_{ij}, U^{*}_{ij}, M^{*}_{ij}, V^{*}_{ij}. In (1.1.10) we now use the condition $W^{s}_{ksm} = 0$. It is easy to verify that only five linearly independent terms in (1.1.10) give zero after contraction, namely,

$$\begin{aligned} &V_{ij}V_{kl}, \quad U_{ij}U_{kl}, \quad M_{ij}M_{kl} + U_{ij}V_{kl} + V_{ij}U_{kl}, \\ &U_{ij}M_{kl} + U_{kl}M_{ij}, \quad V_{ij}M_{kl} + V_{kl}M_{ij} \ . \end{aligned}$$

All the remaining terms of (1.1.10) give linearly independent symmetric second-rank tensors after contraction, and therefore their coefficients must vanish.

Thus, in the general case, an arbitrary self-dual tensor having the algebraic properties of the Weyl conformal curvature tensor can be reduced to the form

$$\begin{aligned} -W^{+}_{ijkl} = {} & \psi_0 U_{ij}U_{kl} + \psi_1(U_{ij}M_{kl} + U_{kl}M_{ij}) \\ & + \psi_2(M_{ij}M_{kl} + V_{ij}U_{kl} + V_{kl}U_{ij}) \\ & + \psi_3(V_{ij}M_{kl} + V_{kl}M_{ij}) + \psi_4 V_{ij}V_{kl} \ . \end{aligned} \tag{1.1.11}$$

The factor -1 is introduced here in order to conform to the notation of *Newman* and *Penrose* [1.4]. It is readily seen from this that the Newman-Penrose scalars ψ_0, ψ_1, ψ_2, ψ_3, ψ_4 of the tensor W_{ijkl} can be calculated directly from the components of the real tensor W_{ijkl}:

$$\begin{aligned} &\psi_0 = -W_{iklm}l^i m^k l^l m^m, \quad \psi_1 = -W_{iklm}l^i n^k l^l m^m, \\ &\psi_2 = -W_{iklm}l^i m^k m^{*l} n^m, \quad \psi_3 = -W_{iklm}n^i l^k n^l m^{*m}, \\ &\psi_4 = -W_{iklm}n^i m^{*k} n^l m^{*m} \ . \end{aligned} \tag{1.1.11$'$}$$

The complex tensors W^{+}_{ijkl} represent symmetric second-rank matrices W_{AB} in a three-dimensional complex space:

$$
\begin{aligned}
&-W^{+}_{UU} = \psi_0, \quad -W^{+}_{MU} = \psi_1, \quad -W_{MM} = -W_{UV} = \psi_2, \\
&-W_{VM} = \psi_3, \quad -W_{VV} = \psi_4 \ .
\end{aligned} \tag{1.1.12}
$$

Under the transformations (I), the components of tensors with the algebraic properties of the Weyl tensor in a complex null basis transform as follows:

$$
\psi_0' = e^{2i\chi}\psi_0, \quad \psi_1' = e^{i\chi}\psi_1, \quad \psi_2' = \psi_2, \psi_3' = e^{-i\chi}\psi_3, \quad \psi_4' = e^{-2i\chi}\psi_4 \ ; \tag{1.1.13}
$$

under the transformations (II),

$$
\psi_0' = A^2\psi_0, \quad \psi_1' = A\psi_1, \quad \psi_2' = \psi_2, \quad \psi_3' = \frac{1}{A}\psi_3, \quad \psi_4' = \frac{1}{A^2}\psi_4 \ ; \tag{1.1.14}
$$

under the transformations (III),

$$
\begin{aligned}
\psi_0' &= \psi_0, \quad \psi_1' = \psi_1 + a\psi_0, \\
\psi_2' &= \psi_2' + 2a\psi_1 + a^2\psi_0, \\
\psi_3' &= \psi_3 + 3a\psi_2 + 3a^2\psi_1 + a^3\psi_0, \\
\psi_4' &= \psi_4 + 4a\psi_3 + 6a^2\psi_2 + 4a^3\psi_1 + a^4\psi_0 \ ;
\end{aligned} \tag{1.1.15}
$$

under the transformations (IV),

$$
\begin{aligned}
\psi_0' &= \psi_0 + 4b\psi_1 + 6b^2\psi_2 + 4b^3\psi_3 + b^4\psi_4, \\
\psi_1' &= \psi_1 + 3b\psi_2 + 3b^2\psi_3 + b^3\psi_4, \\
\psi_2' &= \psi_2 + 2b\psi_3 + b^2\psi_4, \\
\psi_3' &= \psi_3 + b\psi_4, \quad \psi_4' = \psi_4 \ .
\end{aligned} \tag{1.1.16}
$$

It is worthwhile to note that for the transformations (III) the expressions ψ_3', ψ_2', ψ_1', ψ_0' are obtained by successive differentiation of the expression for ψ_4' with respect to a and multiplication by a numerical factor. Similarly, for the transformations (IV) the expressions for ψ_1', ψ_2', ψ_3', ψ_4' are obtained from ψ_0' by successive differentiation with respect to b and multiplication by a numerical factor; for example, $\psi_1' = 0.25\partial\psi_0'/\partial b$, $\psi_2' = 12^{-1}\partial^2\psi_0'/\partial b^2$, etc.

Petrov Type I [1.10]. In general, the equation $\psi_4' = 0$ [see (1.1.15)] has four distinct roots a_1, a_2, a_3, a_4. By applying the transformation (III) with one of these roots a_i, we reduce the component ψ_4' to zero, with $\psi_3' \neq 0$. Next, we turn to the equation $\psi_0' = 0$. This equation will have three distinct roots $b_1 = (a_i - a_{i-1})^{-1}$, $b_2 = (a_i - a_{i+1})^{-1}$, $b_3 = (a_i - a_{i+2})^{-1}$. By applying the transformation (IV) with one of these roots b, we can make the component ψ_0' vanish, if it was not already equal to zero, with $\psi_1' \neq 0$.

By means of the transformations (I) and (II), we can ensure that $\psi_1 = 1$.

As a result, we find that a tensor W^{+}_{AB} of Petrov type I can be transformed to the form

$$-W^{+}_{AB} = \begin{pmatrix} 0 & 1 & \psi_2 \\ 1 & \psi_2 & \psi_3 \\ \psi_2 & \psi_3 & 0 \end{pmatrix} . \tag{1.1.17}$$

Thus, a Weyl tensor of Petrov type I has four independent real invariants.

Petrov Type II. If the equation $\psi'_4 = 0$ has one pair of multiple roots, then by applying the transformation (III) with the multiple root a we make the components ψ_4 and ψ_3 vanish simultaneously. Next, we require that ψ'_0 vanish after the transformation (IV). Two different values of b are then determined, and it becomes possible to reduce the component ψ_0 to zero by means of two different transformations (IV). The component ψ_1 can be made equal to unity by means of the transformations (I) and (II). Thus, a Weyl tensor of Petrov type II in a canonical basis has the form

$$-W^{+}_{AB} = \begin{pmatrix} 0 & 1 & \psi_2 \\ 1 & \psi_2 & 0 \\ \psi_2 & 0 & 0 \end{pmatrix} \quad \text{or} \quad \begin{pmatrix} 0 & 0 & \psi_2 \\ 0 & \psi_2 & 1 \\ \psi_2 & 1 & 0 \end{pmatrix} . \tag{1.1.18}$$

Clearly, a Weyl tensor of type II is characterized by one complex or two real invariants.

Petrov Type D. Suppose now that the equation $\psi'_4 = 0$ has pairs of equal roots. By applying the transformation (III) with one of the roots, we make the components ψ_4 and ψ_3 vanish simultaneously. After this, the quadratic equation $\psi'_0 = 0$ will have a multiple nonzero root b. By applying the transformation (IV) with this root b, we make the components ψ_0 and ψ_1 vanish simultaneously. Therefore in a canonical basis a Weyl tensor of type D will have the form

$$-W^{+}_{AB} = \begin{pmatrix} 0 & 0 & \psi_2 \\ 0 & \psi_2 & 0 \\ \psi_2 & 0 & 0 \end{pmatrix}$$

and will possess two real invariants.

Petrov Type III. Suppose that the equation $\psi'_4 = 0$ has a root a of multiplicity three. Then the transformation (III) with this root a makes it possible to reduce ψ_2, ψ_3, and ψ_4 to zero. After this, a unique transformation (IV) will make it possible to reduce the component ψ_0 to zero. The component ψ_1 can be made equal to unity by means of the transformations (I) and (II). In a canonical basis, a Weyl tensor of Petrov type III has the form

$$-W^{+}_{AB} = \begin{pmatrix} 0 & 0 & 0 \\ 0 & 0 & 1 \\ 0 & 1 & 0 \end{pmatrix} \quad \text{or} \quad \begin{pmatrix} 0 & 1 & 0 \\ 1 & 0 & 0 \\ 0 & 0 & 0 \end{pmatrix} .$$

Petrov Type N. In this case, the equation $\psi'_4 = 0$ has a root of multiplicity four. Therefore, after the transformation (III) with the parameter a equal to this root there remains only the single nonzero component ψ_0, which can be made equal to unity by means of the transformations (I) and (II). A Weyl tensor of type N in a canonical basis has the form

$$-W^{+}_{AB} = \begin{pmatrix} 0 & 0 & 0 \\ 0 & 0 & 0 \\ 0 & 0 & 1 \end{pmatrix} \quad \text{or} \quad \begin{pmatrix} 1 & 0 & 0 \\ 0 & 0 & 0 \\ 0 & 0 & 0 \end{pmatrix} .$$

Algebraic Invariants of the Weyl Tensor. The invariants I_1 and I_2 of the Weyl tensor given by

$$I_1 = W_{iklm}W^{iklm}, \quad 2I_2 = \varepsilon_{iklm}W^{ikpq}W^{lm}_{pq} ,$$

which are quadratic in the components, are related to the Newman-Penrose scalars $\psi_0, \ldots, \psi_4$ as follows:

$$I_1 + \mathrm{i}I_2 = 2(W^{+}_{iklm}W^{+iklm}) = 16(3\psi_2^2 + \psi_0\psi_4 - 4\psi_1\psi_3) . \tag{1.1.19}$$

Since in a canonical basis the Weyl tensor in the general case is characterized by four real scalars (Petrov type I), it is also useful to calculate the invariants I_3 and I_4 of the Weyl tensor, cubic in the components, given by

$$\begin{aligned} &I_3 = W_{iklm}W^{lmpq}W^{ik}_{pq}, \quad 2I_4 = \varepsilon_{pqst}W_{iklm}W^{lmpq}W^{stik}, \\ &I_3 + \mathrm{i}I_4 = 96(\psi_0\psi_2\psi_4 + 2\psi_1\psi_2\psi_3 - \psi_2^3 - \psi_0\psi_3^2 - \psi_1^2\psi_4) . \end{aligned} \tag{1.1.20}$$

We note that if the quadratic invariants I_1 and I_2 of the Weyl tensor are equal to zero, then the fourth-rank tensor $W^{+}_{ikst}W^{+st}_{lm}$ possesses the algebraic properties of the Weyl tensor. For Petrov types N and III, the invariants I_1, I_2, I_3, and I_4 of the Weyl tensor vanish, but the Weyl tensor itself is nonzero.

Maxwell's Equations and Rotation Coefficients of the Null Tetrad. The eight Maxwell equations $-\nabla_j F^{ij} = 4\pi j^i/c$ and $\nabla_{[i} F_{kl]} = 0$ (or $\nabla_i F^{*ij} = 0$) can be written as the four complex equations

$$-\nabla_i F^{+ki} = \frac{2\pi j^k}{c} . \tag{1.1.21}$$

If in these equations we substitute the expansion (1.1.7) of the bivector F^{+ij} with respect to simple bivectors and consider the projections of the equations onto the basis vectors $\boldsymbol{l}$, $\boldsymbol{n}$, $\boldsymbol{m}$, $\boldsymbol{m}^*$ of the null tetrad, we obtain a form of Maxwell's equations using the formalism of null tetrads.

Before writing down the results, we note that in the general case a basis of isotropic vectors $\boldsymbol{l}$, $\boldsymbol{n}$, $\boldsymbol{m}$, $\boldsymbol{m}^*$ rotates continuously in a local chart in going from point to point.

Following Newman and Penrose, we introduce a notation for the operators of covariant differentiation along the directions of the basis vectors,

$$\delta \equiv m^i\nabla_i\,, \quad \delta^* \equiv m^{*i}\nabla_i\,, \quad \Delta \equiv n^i\nabla_i\,, \quad D \equiv l^i\nabla_i \quad , \tag{1.1.22}$$

and for the rotation coefficients of the null tetrad,

$$\varrho = m^i\delta^* l_i, \quad \sigma = m^i\delta l_i, \quad \tau = m^i\Delta l_i, \quad k = m^i D l_i \quad , \tag{1.1.23}$$

$$\begin{aligned} &\lambda = -m^{*i}\delta^* n_i, \quad \mu = -m^{*i}\delta n_i, \\ &\nu = -m^{*i}\Delta n_i, \quad \pi = -m^{*i} D n_i \quad , \end{aligned} \tag{1.1.24}$$

$$\begin{aligned} &2\alpha = n^i\delta^* l_i - m^{*i}\delta^* m_i, \quad 2\beta = n^i\delta l_i - m^{*i}\delta m_i \quad , \\ &2\gamma = n^i\Delta l_i - m^{*i}\Delta m_i, \quad 2\varepsilon = n^i D l_i - m^{*i} D m_i \quad . \end{aligned} \tag{1.1.25}$$

The complex conjugates of these scalars are obtained by replacing the operator δ by δ^* and the vector $\boldsymbol{m}$ by $\boldsymbol{m}^*$, and vice versa.

The derivatives of the basis bivectors $\boldsymbol{U}$, $\boldsymbol{M}$, $\boldsymbol{V}$ along the directions of the vectors of the null tetrad $\boldsymbol{l}$, $\boldsymbol{n}$, $\boldsymbol{m}$, $\boldsymbol{m}^*$ are self-dual bivectors and can therefore be expanded with respect to these bivectors. The coefficients of these expansions are, in fact, the quantities ϱ, σ, τ, etc., determined by (1.1.23–25):

$$\begin{aligned} \delta^*\boldsymbol{U} &= -2\alpha\boldsymbol{U} - \lambda\boldsymbol{M}, & \delta^*\boldsymbol{M}/2 &= \varrho\boldsymbol{U} - \lambda\boldsymbol{V}, \\ \delta\boldsymbol{U} &= -2\beta\boldsymbol{U} - \mu\boldsymbol{M}, & \delta\boldsymbol{M}/2 &= \sigma\boldsymbol{U} - \mu\boldsymbol{V}, \\ \Delta\boldsymbol{U} &= -2\gamma\boldsymbol{U} - \nu\boldsymbol{M}, & \Delta\boldsymbol{M}/2 &= \tau\boldsymbol{U} - \nu\boldsymbol{V}, \\ D\boldsymbol{U} &= -2\varepsilon\boldsymbol{U} - \pi\boldsymbol{M}, & D\boldsymbol{M}/2 &= k\boldsymbol{U} - \pi\boldsymbol{V}, \\ \delta^*\boldsymbol{V} &= 2\alpha\boldsymbol{V} - \varrho\boldsymbol{M}, & \Delta\boldsymbol{V} &= 2\gamma\boldsymbol{V} + \tau\boldsymbol{M}, \\ \delta\boldsymbol{V} &= 2\beta\boldsymbol{V} - \sigma\boldsymbol{M}, & D\boldsymbol{V} &= 2\varepsilon\boldsymbol{V} + k\boldsymbol{M}, \end{aligned} \tag{1.1.26}$$

It follows from (1.1.26) that the coefficients (1.1.23–25) are the "connection coefficients" in the three-dimensional complex vector space of $\boldsymbol{U}$, $\boldsymbol{M}$, and $\boldsymbol{V}$.

We now calculate the divergence of the bivectors $\boldsymbol{U}$, $\boldsymbol{M}$, $\boldsymbol{V}$ and write the result in terms of the projections onto the basis vectors of the null tetrad:

$$\begin{aligned} &m_j^*\nabla_i U^{ij} = -\lambda, && m_j\nabla_i U^{ij} = -2\gamma + \mu, && n_j\nabla_i U^{ij} = -\nu, \\ &m_j^*\nabla_i M^{ij} = 2\pi, && m_j\nabla_i M^{ij} = 2\tau, && n_j\nabla_i M^{ij} = 2\mu, \\ &m_j^*\nabla_i V^{ij} = \varrho - 2\varepsilon, && m_j\nabla_i V^{ij} = -\sigma, && n_j\nabla_i V^{ij} = \tau - 2\beta, \\ &l_j\nabla_i U^{ij} = \pi - 2\alpha, && l_j\nabla_i M^{ij} = 2\varrho, && l_j\nabla_i V^{ij} = -k \quad . \end{aligned} \tag{1.1.27}$$

By means of these equations and the notation (1.1.22–25), we can now write down Maxwell's equations in the formalism of null tetrads:

$$-\lambda\Phi_0 + (\delta^* + 2\pi)\Phi_1 + (-D + \varrho - 2\varepsilon)\Phi_2 = 2\pi j_i m^{*i}/c \quad , \tag{1.1.28}$$

$$(\Delta + \mu - 2\gamma)\Phi_0 - (\delta - 2\tau)\Phi_1 - \sigma\Phi_2 = 2\pi j_i m^i/c \quad , \tag{1.1.29}$$

$$-\nu\Phi_0 + (\Delta + 2\mu)\Phi_1 - (\delta - \tau + 2\beta)\Phi_2 = 2\pi j_i n^i/c \quad , \tag{1.1.30}$$

$$(\delta^* + \pi - 2\alpha)\Phi_0 - (D - 2\varrho)\Phi_1 - k\Phi_2 = 2\pi j_i l^i/c \quad . \tag{1.1.31}$$

Bianchi Identities in the Formalism of Null Tetrads. In contrast to the case of flat spaces, in Riemannian spaces repeated covariant derivatives of the components of vectors and tensors cannot be permuted. For example, for the covariant derivatives of the components of an arbitrary vector $\boldsymbol{u}$ we have the identities

$$(\nabla_i \nabla_j - \nabla_j \nabla_i) u_k = -R^s_{kij} u_s \ , \tag{1.1.32}$$

$$(\nabla_i \nabla_j - \nabla_j \nabla_i)\nabla_l u_k = -R^s_{kij}\nabla_l u_s - R^s_{lij}\nabla_k u_s \ , \tag{1.1.33}$$

where the tensor R_{ijkl} is defined by (1.1.1).

Differentiating (1.1.32) covariantly with respect to x^l and antisymmetrizing with respect to i, j, l, we eliminate the third derivatives of u_k from (1.1.32, 33) by means of an antisymmetrization of (1.1.33) with respect to i, j, l. Then, in view of the arbitrariness of the vector $\boldsymbol{u}$, by using the algebraic properties of the Riemann tensor we obtain the Bianchi identities

$$\nabla_i R_{jlks} + \nabla_j R_{liks} + \nabla_l R_{ijks} = 0 \ . \tag{1.1.34}$$

From (1.1.34) there follow the identities

$$\nabla_i R^i_{jkl} = \nabla_k R_{jl} - \nabla_l R_{jk} \ , \tag{1.1.35}$$

$$\nabla_i R^{*i}_{jkl} = 0 \ , \tag{1.1.36}$$

where $2R^*_{iklm} = \varepsilon_{ikst} R^{st}_{lm}$.

Multiplying (1.1.36) by $-\mathrm{i}$ (where i is the imaginary unit), combining this equation with (1.1.35), and using (1.1.1′, 9), we have

$$2\nabla_i W^{+i}_{jkl} = -\sigma^{\ i}_{jkq}\nabla_i R^q_l - \sigma^{\ i}_{jql}\nabla_i R^q_k + \tfrac{1}{3}\sigma^{\ i}_{jkl}\nabla_i R + \nabla_k R_{jl} - \nabla_l R_{jk} \ , \tag{1.1.37}$$

$$2\sigma_{ijkl} = g_{ik}g_{jl} - g_{il}g_{jk} - \mathrm{i}\varepsilon_{ijkl} \ .$$

In terms of the tetrad components, the Ricci tensor can be represented in a form analogous to the form of the energy-momentum tensor (1.1.8) of the electromagnetic field[2]:

$$\begin{aligned} R_{ik} - g_{ik}R/4 = 2[&\Phi_{00}n_i n_j + \Phi_{22}l_i l_j + 2\Phi_{11}(n_{(i}l_{j)} + m_{(i}m^*_{j)}) \\ &+ m_i m_j \Phi_{20} + m^*_i m^*_j \Phi_{02} - 2m^*_{(i}m_{j)}\Phi_{01} \\ &- 2m_{(i}n_{j)}\Phi_{10} - 2l_{(i}m_{j)}\Phi_{21} - 2l_{(i}m^*_{j)}\Phi_{12}] \ . \end{aligned} \tag{1.1.39}$$

In view of the equalities (1.1.26, 27), it is not difficult to take the divergence of the Weyl tensor in the form (1.1.11), but the calculation of the right-hand

[2] The convenience of the notation Φ_{AB} for the tetrad components of the tensor $R_{ij} - g_{ij}R/4$ is clear if only from the fact that Einstein's equations $R_{ik} = \kappa T_{ik}$ for the gravitational field created by an electromagnetic field (see Sect. 1.2) in the formalism of null tetrads can be written in the elegant form

$$\Phi_{AB} = \kappa\Phi_A\Phi^*_B/(4\pi), \quad A, B = 0, 1, 2 \ . \tag{1.1.38}$$

side of (1.1.37) presents major difficulties. By rewriting (1.1.37) in terms of the projections onto the basis vectors $\boldsymbol{l}$, $\boldsymbol{n}$, $\boldsymbol{m}$, $\boldsymbol{m}^*$ and dividing both sides by -2, we obtain the form of the Bianchi identities in the formalism of null tetrads. This form is given in the following equations, where before each identity we indicate the method of obtaining it by means of a convolution of both sides of (1.1.37) with the vector components in the square brackets [we recall that the tetrad components of the Weyl tensor $\psi_0, \psi_1, \psi_2, \psi_3, \psi_4$ are determined by (1.1.11), and the tetrad components of the Ricci tensor by (1.1.39)]:

$$\begin{aligned}[l^j l^k m^l];\ &(\delta^* + \pi - 4\alpha)\psi_0 - (D - 4\varrho - 2\varepsilon)\psi_1 - 3k\psi_2 \\ &= (\delta - 2\alpha^* - 2\beta + \pi^*)\Phi_{00} - (D - 2\varrho^* - 2\varepsilon)\Phi_{01} \\ &\quad - k^*\Phi_{02} + 2\sigma\Phi_{10} - 2k\Phi_{11}\ ,\end{aligned} \tag{1.1.40}$$

$$\begin{aligned}[m^j l^k m^l];\ &(\Delta + \mu - 4\gamma)\psi_0 - (\delta - 4\tau - 2\beta)\psi_1 - 3\sigma\psi_2 \\ &= -\lambda^*\Phi_{00} + (\delta + 2\pi^* - 2\beta)\Phi_{01} \\ &\quad - (D - \varrho^* + 2\varepsilon^* - 2\varepsilon)\Phi_{02} + 2\sigma\Phi_{11} - 2k\Phi_{12}\ ,\end{aligned} \tag{1.1.41}$$

$$\begin{aligned}[l^j(l^k n^l - m^k m^{*l})];\ &-\lambda\psi_0 + (\delta^* + 2\pi - 2\alpha)\psi_1 - (D - 3\varrho)\psi_2 - 2k\psi_3 \\ &= \tfrac{1}{3}[(\Delta + \mu^* - 2\mu - 2\gamma - 2\gamma^*)\Phi_{00} - (\delta^* - 2\tau^* - 2\alpha - 2\pi)\Phi_{01} \\ &\quad - \sigma^*\Phi_{02} + 2(\delta + \pi^* - 2\alpha^* + \tau)\Phi_{10} - 2(D - 2\varrho^* + \varrho)\Phi_{11} \\ &\quad - 2k^*\Phi_{12} + 2\sigma\Phi_{20} - 2k\Phi_{21}] - \tfrac{1}{8}DR\ ,\end{aligned} \tag{1.1.42}$$

$$\begin{aligned}[n^j l^k m^l];\ &-\nu\psi_0 + (\Delta + 2\mu - 2\gamma)\psi_1 - (\delta - 3\tau)\psi_2 - 2\sigma\psi_3 \\ &= \tfrac{1}{3}[-\nu^*\Phi_{00} + (\Delta + 2\mu^* - 2\gamma - 2\mu)\Phi_{01} \\ &\quad - (\delta^* - \tau + 2\beta^* - 2\alpha - 2\pi)\Phi_{02} \\ &\quad - 2\lambda^*\Phi_{10} + 2(\delta + 2\pi^* + \tau)\Phi_{11} - 2(D - \varrho^* + 2\varepsilon^* + \varrho)\Phi_{12} \\ &\quad + 2\sigma\Phi_{21} - 2k\Phi_{22}] + \tfrac{1}{8}\delta R\ ,\end{aligned} \tag{1.1.43}$$

$$\begin{aligned}[l^j m^{*k} n^l];\ &-2\lambda\psi_1 + (\delta^* + 3\pi)\psi_2 - (D - 2\varrho + 2\varepsilon)\psi_3 - k\psi_4 \\ &= \tfrac{1}{3}[-2\nu\Phi_{00} + 2\lambda\Phi_{01} + 2(\Delta + \mu^* - 2\gamma^* - \mu)\Phi_{10} \\ &\quad - 2(\delta^* - 2\tau^* - \pi)\Phi_{11} - \sigma^*\Phi_{12} + (\delta + \pi^* - 2\alpha^* + 2\tau + 2\beta)\Phi_{20} \\ &\quad - (D - 2\varrho^* + 2\varrho + 2\varepsilon)\Phi_{21} - k^*\Phi_{22}] - \tfrac{1}{8}\delta^* R\ ,\end{aligned} \tag{1.1.44}$$

$$\begin{aligned}[n^j(l^k n^l - m^k m^{*l})];\ &-2\nu\psi_1 + (\Delta + 3\mu)\psi_2 - (\delta - 2\tau + 2\beta)\psi_3 - \sigma\psi_4 \\ &= \tfrac{1}{3}[-2\nu\Phi_{01} + 2\lambda\Phi_{02} - 2\nu^*\Phi_{10} + 2(\Delta + 2\mu^* - \mu)\Phi_{11} \\ &\quad - 2(\delta^* - \tau^* + 2\beta - \pi)\Phi_{12} - \lambda^*\Phi_{20} + (\delta + 2\pi^* + 2\beta + 2\tau)\Phi_{21} \\ &\quad - (D - \varrho^* + 2\varepsilon^* + 2\varepsilon + 2\varrho)\Phi_{22}] + \tfrac{1}{8}\Delta R\ ,\end{aligned} \tag{1.1.45}$$

$$\begin{aligned}[m^{*j} m^{*k} n^l];\ &-3\lambda\psi_2 + (\delta^* + 2\alpha + 4\pi)\psi_3 - (D - \varrho + 4\varepsilon)\psi_4 \\ &= -2\nu\Phi_{10} + 2\lambda\Phi_{11} + (\Delta - 2\gamma^* + 2\gamma + \mu^*)\Phi_{20} \\ &\quad - (\delta^* + 2\alpha - 2\tau^*)\Phi_{21} - \sigma^*\Phi_{22}\ ,\end{aligned} \tag{1.1.46}$$

$$\begin{aligned}[n^j m^{*k} n^l];\ &-3\nu\psi_2 + (\Delta + 2\gamma + 4\mu)\psi_3 - (\delta - \tau + 4\beta)\psi_4 \\ &= -2\nu\Phi_{11} + 2\lambda\Phi_{12} - \nu^*\Phi_{20} + (\Delta + 2\gamma + 2\mu^*)\Phi_{21} \\ &\quad - (\delta^* - \tau^* + 2\beta^* + 2\alpha)\Phi_{22}\ .\end{aligned} \tag{1.1.47}$$

Connection of the Tetrad Components of the Weyl and Ricci Tensors with the Rotation Coefficients of the Null Tetrad. In order to express the tetrad components of the Weyl tensor and the Ricci tensor in terms of the rotation coefficients of the null tetrad, we can make use of the identity (1.1.32), writing it in terms of the projections onto the vectors of the null basis $\boldsymbol{l}$, $\boldsymbol{n}$, $\boldsymbol{m}$, $\boldsymbol{m}^*$. It is sufficient to replace the vector $\boldsymbol{u}$ in (1.1.32) by one of the vectors of the null basis and make use of the relation (1.1.1′) between the components of the Riemann tensor and the Weyl tensor.

Since all the calculations follow the same pattern, we give the derivation of only the single component ψ_1, indicating for the remaining tetrad components of W_{iklm} and R_{ik} the choice of the vector $\boldsymbol{u}$ and the method of projection.

It follows from (1.1.32), in particular, that

$$\begin{aligned} 2\psi_1 &= (\nabla_i \nabla_j l_k - \nabla_j \nabla_i l_k) m^k (l^i n^j - m^i m^{*j}) \\ &= -W^s_{kij} l_s m^k (l^i n^j - m^i m^{*j}) \quad . \end{aligned} \tag{1.1.48}$$

Using the notation (1.1.22), from (1.1.48) we have

$$\begin{aligned} 2\psi_1 = m^k \{ &(D\Delta - \Delta D - \delta\delta^* + \delta^*\delta) l_k \\ &+ \nabla_i l_k (-Dn^i + \Delta l^i + \delta m^{*i} - \delta^* m^i) \} \quad . \end{aligned} \tag{1.1.49}$$

It is easy to verify that

$$\begin{aligned} \Delta l_k &= l_k (n_i \Delta l^i) - m_k (m_i^* \Delta l^i) - m_k^* (m_i \Delta l^i), \\ m^k \nabla_j l_k &= l_j (m^k \Delta l_k) + n_j (m^k D l_k) - m_j (m^k \delta^* l_k) - m_j^* (m^k \delta l_k) \quad . \end{aligned}$$

Therefore it follows from (1.1.49) that

$$\begin{aligned} 2\psi_1 = m^k \{ &D[l_k (n_i \Delta l^i) - m_k^* (m_i \Delta l^i)] \\ &- \Delta [l_k (n_i D l^i) - m_k^* (m_i D l^i)] - \delta [l_k n_i \delta^* l^i - m_k^* m_i \delta^* l^i] \\ &+ \delta^* [l_k n_i \delta l^i - m_k^* m_i \delta l^i] \} - (l_i D n^i)(m_j \Delta l^j) \\ &+ (m_i D n^i)(m_j \delta^* l^j) + (m_i^* D n^i)(m_j \delta l^j) + (n_i \Delta l^i)(m_i D l^j) \\ &- (m_i \Delta l^i)(m_j \delta^* l^j) - (m_i^* \Delta l^i)(m_j \delta l^j) - (m_i D l^i)(m_j^* \delta n^j) \\ &- (m_i \delta m^{*i})(m_j \delta^* l^j) - (m_i \Delta l^i)(m_j^* \delta l^j) - (l_i \delta^* m^i)(l_j \Delta m^j) \\ &+ (n_i \delta^* m^i)(l_i D m^j) + (m_i^* \delta^* m^i)(m_j \delta l^j) \quad . \end{aligned} \tag{1.1.50}$$

Substituting the expressions (1.1.23–25) into (1.1.50), we finally obtain

$$\begin{aligned} 2\psi_1 = &(D + \varepsilon^* - \varepsilon - \varrho^*)\tau - (\delta - \beta - \alpha^* + \pi^*)\varrho \\ &+ (\delta^* - 3\alpha + \beta^* - \tau^* - \pi)\sigma - (\Delta + \mu^* - \mu - 3\gamma - \gamma^*)k \quad . \end{aligned} \tag{1.1.51}$$

Thus, the tetrad components of the Weyl tensor can be expressed in terms of the coefficients (1.1.23–25) by means of the projection (1.1.47) as follows:

$$\psi_0 = 2(\nabla_{[i}\nabla_{j]} l_k) m^k l^i m^j$$
$$= (D - 3\varepsilon + \varepsilon^* - \varrho - \varrho^*)\sigma - (\delta - 3\beta - \tau - \alpha^* + \pi^*)k \ , \qquad (1.1.52)$$
$$\psi_1 = (n^k \nabla_{[i}\nabla_{j]} l_k - m^{*k}\nabla_{[i}\nabla_{j]} m_k) l^i m^j$$
$$= (D + \varepsilon^* - \varrho^*)\beta - \sigma(\alpha + \pi) - (\delta - \alpha^* + \pi^*)\varepsilon + k(\mu + \gamma) \ , \qquad (1.1.53)$$
$$\psi_2 = 2 l^i m^j n^k \nabla_{[i}\nabla_{j]} m^*_k$$
$$= (D + \varepsilon + \varepsilon^* - \varrho^*)\mu - (\delta - \alpha^* + \pi^* + \beta)\pi$$
$$+ k\nu - \sigma\lambda + R/12 \ , \qquad (1.1.54)$$
$$\psi_2 = 2 m^{*i} n^j m^k \nabla_{[i}\nabla_{j]} l_k = -(\Delta - \gamma - \gamma^* + \mu^*)\varrho$$
$$+ (\delta^* + \beta^* - \tau^* - \alpha)\tau + k\nu - \delta\lambda + R/12 \ , \qquad (1.1.55)$$
$$2\psi_2 = (l^i n^j - m^i m^{*j})(n^k \nabla_{[i}\nabla_{j]} l_k - m^{*k}\nabla_{[i}\nabla_{j]} m_k)$$
$$= (D + 2\varepsilon + \varepsilon^* - \varrho^* + \varrho)\gamma + \varrho\mu - (\Delta - \mu + \mu^* - \gamma^*)\varepsilon$$
$$- (\delta + 2\beta + \tau + \pi^* - \alpha^*)\alpha + (\delta^* - \pi - \tau^* + \beta^*)\beta$$
$$- \tau\pi + k\nu - \alpha\lambda - R/12 \ , \qquad (1.1.56)$$
$$\psi_3 = 2(l^i n^j - m^i m^{*j}) n^k \nabla_{[i}\nabla_{j]} m^*_k$$
$$= (D + 3\varepsilon + \varepsilon^* + \varrho - \varrho^*)\nu - (\Delta + \gamma - \gamma^*)\pi$$
$$- (\delta + 3\beta + \tau - \alpha^* + \pi^*)\lambda + (\delta^* + \alpha + \beta^* - \tau^*)\mu \ , \qquad (1.1.57)$$
$$\psi_3 = m^{*i} n^j (n^k \nabla_{[i}\nabla_{j]} l_k - m^{*k}\nabla_{[i}\nabla_{j]} m_k)$$
$$= (\delta^* + \beta^* - \tau^*)\gamma - (\Delta - \gamma^* + \mu^*)\alpha + \nu(\varrho + \varepsilon) - \lambda(\tau + \beta) \ , \qquad (1.1.58)$$
$$\psi_4 = 2 n^i m^{*j} m^{*k} \nabla_{[i}\nabla_{j]} n_k$$
$$= (\delta^* + \pi - \tau^* + 3\alpha + \beta^*)\nu - (\Delta + 3\gamma - \gamma^* + \mu + \mu^*)\lambda \ . \qquad (1.1.59)$$

Similarly, we can obtain the tetrad components of the Ricci tensor:

$$\Phi_{00} = 2 l^i m^{*j} m^k \nabla_{[i}\nabla_{j]} l_k$$
$$= (D - \varepsilon - \varepsilon^* - \varrho)\varrho - (\delta^* - 3\alpha + \pi - \beta^*)k + \tau k^* - \sigma\sigma^* \ , \qquad (1.1.60)$$
$$2\Phi_{01} = 2(l^i n^j + m^i m^{*j}) m^k \nabla_{[i}\nabla_{j]} l_k$$
$$= (D - \varepsilon - 2\varrho + \varepsilon^* + \varrho^*)\tau - (\Delta - 3\gamma + \mu - \gamma^* - \mu^*)k$$
$$+ (\delta - \beta - \alpha^* - \pi^*)\varrho - (\delta^* - 3\alpha + \pi + \tau^* + \beta^*)\sigma \ , \qquad (1.1.61)$$
$$\Phi_{10} = l^i m^{*j}(n^k \nabla_{[i}\nabla_{j]} l_k - m^{*k}\nabla_{[i}\nabla_{j]} m_k)$$
$$= (D + 2\varepsilon - \varrho - \varepsilon^*)\alpha - \pi\varrho - \beta\sigma^*$$
$$- (\delta^* - \beta^* + \pi)\varepsilon + \gamma k^* + k\lambda \ , \qquad (1.1.62)$$
$$\Phi_{20} = 2 l^i m^{*j} n^k \nabla_{[i}\nabla_{j]} m^*_k = (D + 3\varepsilon - \varrho - \varepsilon^*)\lambda$$
$$- (\delta^* + \alpha + \pi - \beta^*)\pi + \nu k^* - \mu\sigma^* \ , \qquad (1.1.63)$$
$$\Phi_{02} = 2 m^i n^j n^k \nabla_{[i}\nabla_{j]} l_k$$
$$= (\delta - \beta - \tau + \alpha^*)\tau - (\Delta - 3\gamma + \gamma^* + \mu)\sigma + k\nu^* - \varrho\lambda^* \ , \qquad (1.1.64)$$
$$2\Phi_{11} = 2(l^i n^j + m^i m^{*j})(n^k \nabla_{[i}\nabla_{j]} l_k - m^{*k}\nabla_{[i}\nabla_{j]} m_k)$$
$$= (D - \varrho + \varrho^* + \varepsilon^*)\gamma - (\Delta - 2\gamma - \gamma^* + \mu - \mu^*)\varepsilon$$
$$+ (\delta + 2\beta - \tau - \alpha^* - \pi^*)\alpha - (\delta^* + \pi + \tau^* + \beta^*)\beta \ , \qquad (1.1.65)$$

$$\begin{aligned}
2\Phi_{21} &= 2(l^i n^j + m^i m^{*j}) n^k \nabla_{[i} \nabla_{j]} m^*_k \\
&= (D + \varrho^* + \varepsilon^* - \varrho + 3\varepsilon)\nu - (\Delta + \gamma - \gamma^* + 2\mu - \mu^*)\pi \\
&\quad + (\delta + 3\beta - \alpha^* - \tau - \pi^*)\lambda - (\delta^* + \alpha + \beta^* + \tau^*)\mu \ , \qquad (1.1.66)\\
\Phi_{12} &= m^i n^j (n^k \nabla_{[i} \nabla_{j]} l_k - m^{*k} \nabla_{[i} \nabla_{j]} m_k) \\
&= (\delta - \tau + \alpha^*)\gamma - (\Delta - 2\gamma + \mu + \gamma^*)\beta \\
&\quad - \mu\tau - \alpha\lambda^* + \varepsilon\nu^* + \nu\sigma \ , \qquad (1.1.67)\\
\Phi_{22} &= 2 n^i m^j m^{*k} \nabla_{[i} \nabla_{j]} \\
&= (\delta + 3\beta - \tau + \alpha^*)\nu - (\Delta + \mu + \gamma + \gamma^*)\mu + \pi\nu^* - \lambda\lambda^* \ . \qquad (1.1.68)
\end{aligned}$$

We give the form of the Weyl equations for the neutrino spinor field with the components Φ and Ψ in the Newman-Penrose formalism:

$$\begin{aligned}
D\Phi + \delta^*\Psi &= (\varrho - \varepsilon)\Phi + (\alpha - \pi)\Psi, \\
\delta\Phi + \Delta\Psi &= (\tau - \beta)\Phi + (\gamma - \mu)\Psi \ .
\end{aligned}$$

The energy-momentum tensor of the neutrino field has the form

$$\begin{aligned}
\Phi_{00} &= \mathrm{i}\kappa[\Psi D\Psi^* - \Psi^* D\Psi + k\Phi\Psi^* - k^*\Phi^*\Psi^* + (\varepsilon - \varepsilon^*)\Psi\Psi^*], \\
2\Phi_{01} &= \mathrm{i}\kappa[\Psi\delta\Psi^* - \Psi^*\delta\Psi - \Psi D\Phi^* + \Phi^* D\Psi \\
&\quad + \sigma\Phi\Psi^* - (\varrho^* + \varepsilon + \varepsilon^*)\Psi\Phi^* + (\beta - \alpha^* - \pi^*)\Psi\Psi^*], \\
\Phi_{02} &= -\mathrm{i}\kappa[\Psi\delta\Phi^* - \Phi^*\delta\Psi + \sigma\Phi\Phi^* + (\alpha^* + \beta)\Psi\Phi^* + \lambda^*\Psi\Psi^*], \\
2\Phi_{11} &= \mathrm{i}\kappa[\Phi D\Phi^* - \Phi^* D\Phi + \Psi\Delta\Psi^* - \Psi^*\Delta\Psi + (\varepsilon^* - \varepsilon)\Phi\Phi^* \\
&\quad + (\tau + \pi^*)\Phi\Psi^* - (\tau^* + \pi)\Psi\Phi^* + (\gamma - \gamma^*)\Psi\Psi^*], \\
2\Phi_{12} &= \mathrm{i}\kappa[\Phi\delta\Phi^* - \Phi^*\delta\Phi - \Psi\Delta\Phi^* + \Phi^*\Delta\Psi \\
&\quad + (\alpha^* - \beta - \tau)\Phi\Phi^* + \lambda^*\Phi\Psi^* - (\mu + \gamma + \gamma^*)\Psi\Phi^* - \nu^*\Psi\Psi^*], \\
\Phi_{22} &= \mathrm{i}\kappa[\Phi\Delta\Phi^* - \Phi^*\Delta\Phi + (\gamma^* - \gamma)\Phi\Phi^* + \nu^*\Phi\Psi^* - \nu\Psi\Phi^*] \ .
\end{aligned}$$

1.2 Gravitational Waves and Generalized Solutions of the Equations of the Electrovacuum

In the general theory of relativity, gravitational fields are characterized by ten unknown functions, which endow space-time in a local coordinate system with a pseudo-Riemannian metric g_{ij} $(i, j = 1, \ldots, 4)$. All test particles in gravitational fields move according to "inertia" — along trajectories with zero 4-acceleration, provided that no forces of nongravitational type act on them. The equations of general relativity relate the geometrical characteristics of the curvature of space-time to the energy and momentum of the matter embedded in it. These equations have the form

$$R_{ij} - \tfrac{1}{2} g_{ij} R = \kappa T_{ik} \quad , \tag{1.2.1}$$

where R_{ik} is the Ricci tensor, T_{ik} is the energy-momentum tensor of the matter, $\kappa = 8\pi G/c^4$, G is the gravitational constant, and c is the speed of light in vacuum. A necessary condition for the integrability of (1.2.1) is that the divergence of T_{ik} be equal to zero, since, by virtue of the Bianchi identities, the divergence of the left-hand side of (1.2.1) is identically equal to zero.

The part of the operator R_{ij} containing second derivatives of the components of the metric tensor (the *principal part* of R_{ik}) takes the simplest form in the so-called *harmonic coordinate system* of Lanczos, determined by the condition $(\sqrt{-g}\, g^{ij})_{,j} = 0$, where it reduces to the operator $g^{ij}\partial^2_{ij}$ for each component of the metric tensor:

$$\partial^2_{ij}\varphi \equiv \varphi_{,ij} \equiv \partial^2\varphi/\partial x^i \partial x^j \quad .$$

The characteristics of Einstein's equations (1.2.1) are the isotropic surfaces determined by the equations $g^{ij}u_{,i}u_{,j} = 0$. On the isotropic surfaces $u = \text{const}$, the field of directions $u_{,i}$ forms a congruence of bicharacteristics, which are isotropic geodesics[3].

As was noted by *Petrov* [1.11] in his review of studies of the problems of gravitation, at the present time "there exists a rather motley collection of attempts to describe gravitational waves". After the classification of the types of Weyl tensor given by Petrov, there appeared many studies in which the definition of gravitational waves was based on algebraic criteria which distinguish the classes of solutions of Einstein's equations (see the references in [1.11, 12]).

A stimulus for attempts at an invariant definition of a gravitational radiation field was provided by a paper of *Pirani* [1.13], in which it was asserted that only algebraically special gravitational fields (of Petrov type II or III) are wave-like. A gravitational radiation field is joined onto a non-wave-like field of algebraically general type along the wave front, at which there is a break of the Riemann tensor, which has a degenerate structure (of Petrov type N). The mechanism of transport of energy and momentum by means of gravitational waves [1.14] remains unclear in this approach, as before, and the various algebraically invariant definitions of gravitational waves actually provide only a means of distinguishing exact solutions of the nonlinear hyperbolic system.

It is necessary to adopt a differential approach as a basis for solving the problem. The nonlinear hyperbolic system of Einstein equations is characterized by many distinctive features of hyperbolic systems: there is no effect of inversion of plane waves in it, but there is a characteristic tendency for focusing of any normal congruence of bicharacteristics with a nonzero shear (see Sect. 1.2.1).

Wave solutions of the equations are characterized by different scales. It is natural to call the region with a smooth variation of the solution the *background*,

[3] Isotropic geodesics are also called null geodesics and light rays. The bicharacteristics of the equations of the electromagnetic field in vacuum are also isotropic geodesics.

and the region with a strong variation the *wave*. The nonlinear hyperbolic system of gravitational equations obeys a generalized Huygens principle [1.2]. This principle asserts that the features of the solution at a point P depend asymptotically only on the features of the initial data localized near the intersection of the surface of the initial Cauchy data with the conoid of bicharacteristics emanating from the point P into the past. The dependence of the solution on only the initial data belonging to the intersection of the conoid of rays with vertex at the point P with the Cauchy hypersurface (the Huygens principle in the narrow sense) holds for the wave equation and for Maxwell's equations [1.1, 3] only for a particular class of pseudo-Riemannian spaces, namely, those with the interval

$$ds^2 = 2\,du\,dv - g_{ab}(u)\,dx^a dx^b, \quad a, b = 1, 2 \ .$$

Therefore initially sharp variations of gravitational fields in general relativity remain sharp in the future as they propagate, although they may be smeared slightly.

In the case of a uniform background, the process of propagation of waves does not depend on their wavelength. It is to this case that the majority of exact solutions interpreted as waves refer. However, in the general case of a background which varies in space and time, only solutions which vary sharply in comparison with the background can work their way through it along isotropic geodesics. Such generalized solutions are well known for linear hyperbolic systems under the name of *running waves* [1.2]. In this case, the characteristic scale of variation of the background is the same as the characteristic scale of variation of the coefficients of the principal part of the differential operator.

For the nonlinear equations of general relativity, a characteristic analog of solutions of the running wave type is provided by solutions describing rapidly oscillating trains of waves or breaks in the derivatives of the components of the metric tensor. In view of the localization of such solutions along characteristic surfaces, the transport equations for solutions of the running-wave type can be interpreted as equations for conservation of the flux of "energy" of a wave along tubes formed by isotropic geodesics.

We note that the results given below may be invalid for generalized (non-Einstein) models of the gravitational field. Breaks in the field in non-Einstein theories of gravitation were studied in a paper by *Sedov* [1.15] (see Sect. 1.4), who derived algebraic conditions on nonisotropic surfaces with a strong break in the gravitational field.

The problem of experimental detection of gravitational waves was discussed, for example, in [1.16, 17]. Problems concerning the generation of gravitational waves by celestial bodies were considered in [1.18, 19].

1.2.1 Properties of Families of Isotropic Geodesics in General Relativity

For an arbitrary continuous vector field $\boldsymbol{l}$ on a manifold M_4, there exists an associated system of ordinary differential equations

$$dx^i/d\tau = l^i(x^1, x^2, x^3, x^4), \quad i = 1,2,3,4 \ .$$

The solutions of this system determine integral curves of the vector field l^i. We denote by D the covariant derivative along the direction of the vector field l^i: $D \equiv l^i \nabla_i$. The integral curves are geodesics on the manifold if there exists a scalar $p(x)$ such that $Dl^i = pl^i$. In this case, instead of the field $\boldsymbol{l}$ we can introduce the field $\boldsymbol{l}' = g\boldsymbol{l}$, with $g^{-1}Dg + p = 0$, for which $D\boldsymbol{l}' = 0$. For geodesics, this corresponds to the choice of an affine parameter on them.

We shall call the pseudovector ω with $2\omega^i \equiv \varepsilon^{iklm} l_k \nabla_l l_m$ the *rotation of the vector field* $\boldsymbol{l}$. The rotation of a vector field $\boldsymbol{l}$ is equal to zero if it has an integrating factor: $l_i = p\,\partial u/\partial x^i$. The converse is also true. If the rotation ω of a vector field is equal to zero, then there exist functions p and u such that $l_i = p\,\partial u/\partial x^i$, and the family of surfaces $\{u\}$ stratifies some open region in M_4.

Lemma 1. For a vector field $\boldsymbol{l}$ of smoothness class C^2 for which $Dl_i = 0$ with $l^k l_k = \text{const}$, the divergence of the rotation vector is equal to zero.

Indeed, let us choose, at a given point in the tangent space, four linearly independent vectors $A^{(i)}$ $(i = 1,2,3,4)$:

$$\det A_j^{(i)} = e^{iklm} A_i^{(1)} A_k^{(2)} A_l^{(3)} A_m^{(4)} \neq 0, \quad A^{(1)} = \boldsymbol{l} \ .$$

It is easy to show that $2\nabla_i \omega^i = \varepsilon^{iklm}[l_k \nabla_i \nabla_l l_m + \nabla_i l_k \nabla_l l_m]$.

The first group of terms in $\nabla_i \omega^i$ vanishes because of the algebraic properties of the Riemann-Christoffel tensor. We represent the second group of terms in $\nabla_i \omega^i$ in the form

$$I_2 = \left|\det A_j^{(i)}\right|^{-1} e_{i_1 i_2 i_3 i_4} A_i^{(i_1)} A_j^{(i_2)} A_k^{(i_3)} A_e^{(i_4)} \nabla^i l^j \nabla^k l^e \ .$$

In each term of this sum, there is either a factor of the form $\nabla_k l_m l^m$ or a factor of the form Dl_k. Therefore I_2 is equal to zero, in view of the conditions of the lemma.

For an isotropic vector field for which $Dl_i = \alpha l_i$, the assertion of the lemma can also be obtained from the vanishing of the imaginary part of (1.1.60).

We shall consider below the family of isotropic geodesics with zero rotation, in terms of which we shall construct the gradient field of the tangent vectors $l_i = u_{,i}$.

Let $\mathcal{D}$ be the intersection of the family with some space-like hypersurface intersecting each geodesic only once. For the surface $\mathcal{D}$, the unit normal vector q is time-like, and the vector $\boldsymbol{l} - q(\boldsymbol{l}\cdot q)$ is tangential to $\mathcal{D}$. Let $\boldsymbol{u}$ and $\boldsymbol{v}$ be fields of space-like unit vectors tangential to $\mathcal{D}$ and orthogonal to the vector $\boldsymbol{l} - q(\boldsymbol{l}\cdot q)$.

The vector $\boldsymbol{n} \equiv [2(\boldsymbol{l}\cdot\boldsymbol{q})^2]^{-1}[2\boldsymbol{q}(\boldsymbol{l}\cdot\boldsymbol{q}) - \boldsymbol{l}]$ is isotropic and orthogonal to $\boldsymbol{u}$ and $\boldsymbol{v}$. Moreover, $\boldsymbol{l}\cdot\boldsymbol{n} = 1$. We introduce the notation $\boldsymbol{m} \equiv (\boldsymbol{u} + \mathrm{i}\boldsymbol{v})/\sqrt{2}$.

We now define a field of null tetrads in a world tube of the manifold M_4 by means of parallel transport of the tetrads along the isotropic geodesics $\boldsymbol{l}$ which form this tube. For such a tetrad field, $D\boldsymbol{n} = D\boldsymbol{m} = D\boldsymbol{l} = 0$. Therefore the rotation coefficients of the null tetrads have the values $k = \pi = \varepsilon = 0$; $\alpha^* + \beta = \tau$, $\mathrm{Im}\{\varrho\} = 0$, $\mathrm{Re}\{\varrho\} = -\nabla_i l^i/2$. Equation (1.1.60) takes the form

$$\Phi_{00} = (D - \varrho)\varrho - \sigma\sigma^* \quad . \tag{1.2.2}$$

The derivative D is equal to the derivative with respect to the affine parameter. We shall show that $\nabla_i l^i$ has the meaning of the rate of relative change of the area of the wave front. In fact, if we go over to a system of coordinates u, α, ξ^1, ξ^2, where $u = \text{const}$ is the equation of an isotropic surface, α is an affine parameter of a geodesic on this surface, and ξ^1 and ξ^2 are Lagrangian coordinates of the geodesic, then the interval takes the form

$$ds^2 = 2\,du\,d\alpha + g_0\,du^2 + g_A du\,d\xi^A - g_{AB}\,d\xi^A d\xi^B; \quad A, B = 1, 2 \quad . \tag{1.2.3}$$

In this isotropic-geodesic coordinate system, the components of the vector $\boldsymbol{l}$ have the values $l_u = 1$, $l_\alpha = l_A = 0$. Therefore $\nabla_i l^i = \partial \ln\sqrt{g}/\partial\alpha$, where $g = g_{11}g_{22} - g_{12}^2$.

The area Δ of an element of surface cut out on a two-dimensional front $u = \text{const}$, $\alpha = \text{const}$ by neighboring rays is $\sqrt{g}\,d\xi^1 d\xi^2$. Consequently, (1.2.2) can be written in the form

$$\frac{d^2}{d\alpha^2}\sqrt{\Delta} = -(\sigma\sigma^* + \Phi_{00})\sqrt{\Delta} \quad . \tag{1.2.4}$$

The coefficient $\Phi_{00} = \kappa T_{ij} l^i l^j/2$ for an electromagnetic field and an ideal gas is always greater than zero. For the case of an arbitrary continuous medium with energy-momentum tensor T^{ij}, the condition $T_{ij} l^i l^j \geq 0$ is a plausible and reasonable restriction; it is called the *weak energy condition* [1.20]. Thus, the coefficient $\sigma\sigma^* + \Phi_{00}$ cannot be negative. Therefore the second derivative of the square root of Δ with respect to the affine parameter α cannot be positive.

Equation (1.2.4) can be regarded as a linear equation for $\sqrt{\Delta}$. If a ray does not leave the region occupied by matter as the affine parameter α increases (for example, in closed models of the universe), then for $\sigma\sigma^* + \Phi_{00} \geq \text{const} > 0$ the solution of (1.2.4) has a set of zeros corresponding to focal points of the isotropic geodesics, where $\Delta \to 0$ [1.21].

Property A. If the rays were focused at the initial instant, i.e., the area of the corresponding element of the wave front was decreasing ($\partial\Delta/\partial\alpha < 0$) at $\alpha = 0$, then according to (1.2.4) this area will decrease monotonically with increasing α until it reaches zero.

Property B. On an isotropic geodesic $\lambda(\alpha)$, extended to sufficiently large values of the affine parameter α, there exist two points $q = \lambda(\alpha_q)$ and $r = \lambda(\alpha_r)$ through which another isotropic geodesic passes, provided that at some point $P \in \lambda(\alpha)$ the Newman-Penrose scalar has a value $\psi_0 \neq 0$ in a null tetrad for which the vector $\boldsymbol{l}$ is directed along the tangent to $\lambda(\alpha)$. These two points q and r are said to be conjugate to each other. The set of points $\tilde{\Sigma}$ at which the rays emitted orthogonally to a two-dimensional space-like surface Σ are focused is called the set *conjugate* to the surface Σ, and the points of the set $\tilde{\Sigma}$ themselves are called the *focal points*.

In order to prove Property B, we consider in a sufficiently small neighborhood of the point p the set of two-dimensional elementary areas for which $\varrho > 0$. Because of Property A, the rays emitted orthogonally to these surfaces will certainly intersect the geodesic $\lambda(\alpha)$ for finite α. We shall show that even if $\varrho = 0$ on an area, the rays emitted from this area will still intersect $\lambda(\alpha)$ for a finite value $\alpha = \alpha_0$. Indeed, it follows from (1.1.52) that

$$\psi_0 = D\sigma - 2\varrho\sigma \quad .$$

If at the point p we have $\psi_0 \neq 0$, then for rays for which $\varrho = 0$ at the point p, the quantity σ becomes nonzero with increasing α and, according to (1.2.2), the scalar ϱ becomes positive with increasing α. It follows from (1.2.4) that the area of the wave front then contracts to zero after a finite interval of variation of α.

Let us now choose a point $r = \lambda(\alpha_r)$, where $\alpha_r > \alpha_0$, and emit light rays from it into the "past". Suppose that not one of these rays intersects the geodesic $\lambda(\alpha)$ in the interval of variation (α_p, α_0) (otherwise, Property B would already be proved). Then for those rays which pass near $\lambda(\alpha)$ in a sufficiently small neighborhood of the point p, the scalar ϱ cannot be positive, for otherwise these rays would be focused to $\lambda(\alpha)$ without reaching the point α_0. Therefore near the point p we have parameter values $\varrho < 0$ for rays emitted from the point $r = \lambda(\alpha_r)$. But then it follows from Property A that when a ray is extended into the past on the geodesic $\lambda(\alpha)$ a focal point $q = \lambda(\alpha_q)$ is certainly formed. The points q and r are conjugate.

Definition. For a given point p of the manifold M_4, we denote by $\mathcal{I}^+(p)$ [respectively, $\mathcal{I}^-(p)$] the set of points belonging to M_4 which can be joined to the point p by time-like or isotropic curves directed to the future (respectively, to the past) with respect to the point p. We call $\mathcal{I}^+(p)$ the *future set* for p. We shall denote the boundary of $\mathcal{I}^+(p)$ by $\partial\mathcal{I}^+(p)$.

Property C. Isotropic geodesics emitted orthogonally to a two-dimensional space-like orientable compact surface Σ belong to the boundary of the future set $\partial\mathcal{I}^+(\Sigma)$ only until the set conjugate to Σ is reached, and after this the rays enter the interior part $\mathcal{I}^+(\Sigma) - \partial\mathcal{I}^+(\Sigma)$ of the future set for Σ.

Accordingly, if an isotropic geodesic between the points p and r contains a point conjugate to the point p, it can be deformed into a time-like curve passing through the points p and r.

Property C was proved in [1.20], where an explicit construction was presented.

1.2.2 Propagation of Breaks in the Gravitational Field and Their Algebraic Classification

We define a *break in the field of order* k on an isotropic hypersurface $u(x^i) = 0$ to be a generalized solution of the field equations in which the metric g_{ij} together with all its derivatives up to the order $k-1$ inclusive are continuous on the hypersurface $u(x^i) = 0$, while some of the derivatives of g_{ij} of order k and above have a break in passing through $u = 0$, having finite limits as $u \to 0$ from the "left" and "right" of $u = 0$. The class of admissible coordinate transformations $x^{i'} = f^i(x^j)$ must belong to the class C^k, and by means of breaks in the derivatives of $f^i(x^j)$ of order $k+1$ it is not possible to eliminate the breaks in all the derivatives of g_{ij} of order k.

The components of the Riemann tensor can be discontinuous only for breaks in the field of order 1 or 2. In the case of breaks of order 1, there must be additional algebraic relations between the strengths of the discontinuities in the first derivatives. Since by a "break" we mean here a weak limit of smooth solutions which vary "sharply" across $u = 0$ and "smoothly" along $u = 0$, by virtue of the field equations this concept becomes meaningless near focal points, where solutions which vary sharply across the hypersurface $u = 0$ will also vary sharply along the bicharacteristics.

The theory of Hadamard breaks was first applied to Einstein's equations by *Stellmacher* [1.22]. His ideas were elaborated by others [1.23–25]. In what follows, we outline the results of [1.23].

A convenient working tool for the analysis of breaks is the formalism of null tetrads (see Sect. 1.1). Suppose that on the surface $u(x^i) = 0$ the field l^i is equal to the field of tangents to the bicharacteristics $l_i = u_{,i}$, and let u_i and v_i be two space-like unit vectors orthogonal to l_i and to each other, and tangential to the hypersurface $u = 0$. We have $u_i l^i = v_i l^i = 0$, $u_i u^i = v_i v^i = -1$. We define a complex isotropic vector m_i and an outgoing vector n_i on the hypersurface $u = 0$ by the relations $\sqrt{2}\, m_i \equiv u_i + \mathrm{i} v_i$; $n_i l^i = 1$, $n_i m^i = 0$. The field of tetrads introduced in this way is not defined uniquely. It is possible to go from one field to another by means of the transformations (I–III) given in Sect. 1.1. We define fields of null tetrads in the neighborhood of the hypersurface $u = 0$ by means of parallel transport along isotropic geodesics starting from the surface $u = 0$ along the direction of the vectors n_i. Therefore in this neighborhood of the hypersurface of the break, $u = 0$, the rotation coefficients ν, τ, and γ are equal to zero.

By the definition of a break of order 1, the metric and its interior derivatives are continuous at $u = 0$, i.e., in the notation (1.1.22) we have

$$[g_{ij}] = [\delta g_{ij}] = [D g_{ij}] = 0 \ ;$$

there are breaks only in the derivatives of g_{ij} along the outgoing direction, $[\Delta g_{ij}] = \zeta_{ij}$. Functions which define coordinate transformations $x'^i = f^i(x^j)$ can have breaks in the second derivatives along the direction $\boldsymbol{n}$. Such functions satisfy the Hadamard conditions

$$\left[\frac{\partial^2}{\partial x^i \, \partial x^k} f^m\right] = \gamma^{m'} l_i l_k \ .$$

We find from these conditions that the discontinuity of ζ_{ij} in going over to the system of coordinates $x^{i'}$ transforms as follows

$$\frac{\partial f^k}{\partial x^j}\frac{\partial f^l}{\partial x^i}\zeta'_{kl} = \zeta_{ij} + l_i\gamma_j + l_j\gamma_i \ . \tag{1.2.5}$$

The algebraic conditions imposed on the discontinuity of ζ_{ij} can be obtained either [1.25] as a result of integration, over a layer of small thickness ε around the hypersurface $u = 0$, of continuous solutions which vary sharply across the hypersurface $u = 0$ and a subsequent transition to the limit $\varepsilon \to 0$ or [1.15], since the scalar curvature R is the Lagrangian for Einstein's gravitational equations, by means of the variational conditions for the continuity of the generalized momentum at $u = 0$:

$$\left[\frac{\partial R}{\partial(\partial g_{ij}/\partial x^k)}\right] l_k = 0 \ .$$

Writing out these conditions, we obtain

$$(\zeta_{ij} - g_{ij}\zeta^k_k/2)l^i = 0 \ . \tag{1.2.6}$$

From ζ_{ij} satisfying these conditions it is possible to form an invariant quantity which does not change under any of the possible transformations (I–III),

$$2|P^2| = \zeta_{ij}\zeta^{ij} - \zeta^2/2 \ ,$$

where P is a complex scalar function whose modulus is an invariant characteristic of the break. By expanding ζ_{ij} in terms of the basis vectors of the tetrad constructed above and substituting this expansion into (1.2.6), we readily obtain the general form of ζ_{ij}:

$$\zeta_{ij} = P m_i m_j + P^* m_i^* m_j^* + l_i\gamma_j + l_j\gamma_i \ . \tag{1.2.7}$$

Owing to (1.2.5), the terms $l_i\gamma_j + l_j\gamma_i$ in (1.2.7) can be eliminated by means of an appropriate choice of the coordinates.

The rotation coefficients ϱ, σ, and k do not contain derivatives of the metric along the outgoing direction and are therefore continuous at $u = 0$. The coefficients π, μ, ε, α, and β are continuous as a consequence of the conditions (1.2.6) on the surface of the break:

$$2[\pi] = -\zeta_{ij} l^i m^{*j} = 0, \quad 2[\mu] = \zeta_{ij} m^i m^{*j} = 0,$$
$$2[\varepsilon] = -\zeta_{ij} l^i l^j = 0, \quad 4[\alpha] = -\zeta_{ij} m^{*i} l^j = 0,$$
$$4[\beta] = \zeta_{ij} m^i l^j = 0 \ .$$

From the definition (1.1.24) we have $2[\lambda] = -P$. By the construction of the field of null tetrads in the neighborhood of the surface of the break, $u = 0$, the rotation coefficients τ, γ, and ν are equal to zero.

On the surface of the break itself, $k = 0$ (in general, $k \neq 0$ for $u \neq 0$). At $u = 0$, we can also make the coefficients π and ε vanish if the tetrad field on the surface of the break is obtained by parallel transport along the geodesics l^i.

The tetrad components of the Ricci tensor as second-order differential operators of the components of the metric tensor can be divided into three groups as follows:

a) The breaks in the components Φ_{00} and Φ_{10} are equal to zero, since according to (1.1.60, 62), Φ_{00} and Φ_{10} can be expressed in terms of rotation coefficients which are continuous on the surface of the break. The component Φ_{01} is equal to $(\Phi_{10})^*$ (the asterisk indicates complex conjugation) and is therefore also continuous. By means of Einstein's equations, we obtain from this the conditions for conservation of the flux of energy and momentum at the break, $[n_i T^i_j] = 0$.

b) The break in the components $\Phi_{20} = \Phi^*_{02}$ is determined by the function P, and according to (1.1.63) we have

$$[\Phi_{20}] = [D\lambda] - [\lambda]\varrho = -\tfrac{1}{2}(DP + \tfrac{1}{2}P\nabla_i l^i) \ . \tag{1.2.8}$$

c) The breaks in the components Φ_{11}, Φ_{12}, Φ_{21}, and Φ_{22} cannot be expressed in terms of the function P. By means of Einstein's equations, the breaks in these components can be expressed in terms of the breaks in the corresponding components of the energy-momentum tensor.

If the characteristic surfaces of the nongravitational fields are isotropic (for example, neutrino and electromagnetic fields), as for the gravitational field, then the discontinuity in the component Φ_{20} is nonzero. As is shown below, breaks in the gravitational field produce breaks in these fields, and vice versa.

We shall first analyze the relations at the breaks which follow from Maxwell's equations for the electromagnetic field. Maxwell's equations (1.1.28–31) imply algebraic conditions on the breaks of the tetrad components F_{ij}: $[\Phi_0] = [\Phi_1] = 0$. Therefore $[F^+_{ij}] = fV_{ij}$, where $f = [\Phi_2]$. Taking the difference of the equations (1.1.28) written on both sides of the surface of the break, we obtain

$$Df + \tfrac{1}{2}f\nabla_i l^i - \tfrac{1}{2}\Phi_0 P = 0 \ , \tag{1.2.9}$$

where Φ_0 is the tetrad component of the electromagnetic field tensor, continuous at the break. The discontinuity of the component $\Phi^{(\mathrm{em})}_{20}$ of the electromagnetic field at the break is readily calculated:

$$[\Phi^{(\mathrm{em})}_{20}] = \frac{\kappa}{4\pi}[\Phi_2\Phi^*_0] = \frac{\kappa}{4\pi} f\Phi^*_0 \ . \tag{1.2.10}$$

We turn now to the Weyl equations (see the end of Sect. 1.1). It follows from them that there can be a break in only the component Φ of the spinor of the neutrino field, and the component Ψ must remain continuous. However, a break in the component Φ is not related to a break in the gravitational field, and we shall assume in what follows that this break is equal to zero, owing to the initial conditions.

The break in the component $\Phi_{20}^{(\nu)}$ of the energy-momentum tensor of the neutrino field can be readily calculated from the relations of Sect. 1.1:

$$[\Phi_{20}^{(\nu)}] = -\mathrm{i}\kappa P\Psi\Psi^*/2 \ . \tag{1.2.10'}$$

Therefore from (1.2.8, 10, 10′) we finally obtain

$$0 = DP + \frac{1}{2}\nabla_i l^i P + \frac{4G}{c^4} f\Phi_0^* + \frac{8\pi G}{c^4}\mathrm{i}P\Psi\Psi^* \ , \tag{1.2.11}$$

where G is the gravitational constant. The system of equations (1.2.9, 11) is a closed system of ordinary differential equations for f and P, describing the time evolution of the break in the gravitational and electromagnetic fields.

It is interesting that the presence of the neutrino field has no influence on the total strength of the breaks in the gravitational and electromagnetic fields. Indeed, it follows from (1.2.9, 11) that

$$\nabla_i \left[l^i \left(\frac{8G}{c^4}|f^2| + |P^2| \right) \right] = 0 \ .$$

Let us assume that there is no external electromagnetic field. Then from (1.2.11) we obtain

$$P\sqrt[4]{-g} = \mathrm{const} \cdot \exp\left(-\mathrm{i}\frac{8\pi G}{c^4} \int\limits_{\alpha_0}^{\alpha} \Psi\Psi^* \, d\alpha \right) \ , \tag{1.2.11'}$$

where g is the determinant of the metric tensor in the isotropic-geodesic coordinate system. It follows from this equation that in the presence of a neutrino field the polarization vector of a gravitational wave undergoes a rotation. In fact, the argument of the complex function P characterizes the angle between the polarization vector of the gravitational wave and the initial polarization vector under parallel transport along the ray, and the greater the intensity of the neutrino field, the greater the rotation, according to (1.2.11′), experienced by the polarization vector of the gravitational wave. In the presence of external electromagnetic fields, this effect is superimposed on the effect of successive mutual conversion of the breaks in the electromagnetic and gravitational fields, discovered in [1.26].

Let $\Phi_0 = a\exp(\mathrm{i}b)$, where a and b are real numbers. In order to determine the effect of a variation of b along the rays, we go over to an isotropic-geodesic coordinate system and the new variables $\mathcal{P} = P\sqrt[4]{-g}\exp(\mathrm{i}b)$, $\mathcal{F} = 2\sqrt{G}f\sqrt[4]{-g}/c^2$, $2\sqrt{G}\,a/c^2 \equiv \Phi$. Then the system of equations (1.2.9, 11) can be rewritten in the form

$$\frac{d}{d\alpha}\mathcal{P}+\Phi\mathcal{F}+\mathrm{i}\left(\frac{8\pi G}{c^4}\Psi\Psi^*-\frac{\partial b}{\partial\alpha}\right)\mathcal{P}=0,\quad \frac{d}{d\alpha}\mathcal{F}-\frac{1}{2}\mathcal{P}=0\ .$$

It follows from this system that the influence of the neutrino field on the evolution of the breaks is equivalent to the influence of a certain effective external electromagnetic field $\tilde{\Phi}_0$ with the same intensity a as Φ but with a different law of variation of the argument along the ray: $\tilde{\Phi}_0 = a\exp(\mathrm{i}\tilde{b})$, where

$$\tilde{b} = b - \frac{8\pi G}{c^4}\int\limits_{\alpha_0}^{\alpha}\Psi\Psi^*\, d\alpha\ .$$

Therefore the successive mutual conversion of the gravitational and electromagnetic breaks in the neutrino-electromagnetic vacuum is accompanied by a rotation of the polarization vector of these waves, i.e., we have the same situation here as in an arbitrary electrovacuum wave [1.26].

The foregoing conclusions hold also for the case of short waves propagating in neutrino and electromagnetic fields.

Let us establish the geometrical meaning of the invariant P for a congruence of isotropic geodesics intersecting the surface of a break. A congruence can be characterized [1.5, 27] by the Sachs optical scalars μ and λ [$\mathrm{Re}\{\mu\}$ has the meaning of the rate of convergence of the null geodesics, $\mathrm{Im}\{\mu\}$ measures their rotation, and λ characterizes the shear (distortion) of the congruence]. By means of the conditions (1.2.7), it can be shown that the expansion parameter μ is continuous at the break, while the shear parameter has a discontinuity of magnitude $[\lambda] = -P/2$, as we have noted above.

In order to describe the algebraic structure of a break in the Weyl tensor, we expand this tensor in terms of the basis bivectors U_{ij}, M_{ij}, and V_{ij} according to (1.1.11). Calculating the breaks in the tetrad components ψ_0 and ψ_1 of the Weyl tensor according to (1.1.52, 53), we find that they are equal to zero.

It follows from this that a first-order break in the Weyl tensor is a tensor with special algebraic properties, i.e., a Weyl tensor of Petrov type II, and the tangent to a bicharacteristic of the surface of the break is its principal isotropic direction (Debever vector).

Using (1.1.54) and the fact that $R = 0$ for the electromagnetic field, we obtain for the break in the component ψ_2 the expression

$$[\psi_2] = -\sigma[\lambda] = \sigma P/2\ .$$

Making use of (1.1.57, 66), we have

$$[\psi_3] = -[\Delta\pi] - (\delta + 3\beta - \alpha^*)[\lambda] = 2[\Phi_{21}^{(\mathrm{em})}] + (\delta + 3\beta - \alpha^*)P\ .$$

Therefore the break in the component ψ_3 is also determined by the break in the component of the electromagnetic field, $[\Phi_2] = f$:

$$[\Phi_{21}^{(\mathrm{em})}] = \kappa f\Phi_1^*/(4\pi)\ .$$

The tetrad component ψ_4 contains second derivatives of the metric along the vector $\boldsymbol{n}$, which give rise to a Dirac δ function after passing to the limit of a discontinuous solution from a sequence of continuous solutions. According to (1.1.59), near the surface of the break we can separate from $[\psi_4]$ the principal singular part in the form

$$\psi_4 = \delta(u)P/2 + \tilde{\psi}_4, \quad \lim_{\varepsilon\to 0} \int_{-\varepsilon}^{\varepsilon} \tilde{\psi}_4 \, du = 0 \ ,$$

where the quantity $\tilde{\psi}_4$ can be discontinuous at $u = 0$. To find the law of variation of the discontinuity $[\tilde{\psi}_4]$ along the bicharacteristics of the surface of the break, we use the Bianchi identity (1.1.45). We represent $\Delta\Phi_{20}$ near $u = 0$ in the form

$$\Delta\Phi_{20}^{(\mathrm{em})} = \kappa\Delta(\Phi_2\Phi_0^*/(4\pi)) = f\delta(u)\Phi_0^*\kappa/(4\pi) + \tilde{\Delta}\Phi_{20}^{(\mathrm{em})},$$
$$\lim_{\varepsilon\to 0} \int_{-\varepsilon}^{\varepsilon} (\tilde{\Delta}\Phi_{20}^{(\mathrm{em})}) \, du = 0 \ .$$

Then (1.1.46) determines the law of evolution of $[\tilde{\psi}_4]$ along the surface of the break [all the δ functions drop out of (1.1.46), since their coefficients satisfy (1.2.8)]:

$$\begin{aligned} &-(D-\varrho)[\tilde{\psi}_4] + (\delta^* + 2\alpha)[\psi_3] + 3P\psi_2 \\ &\quad = -P\Phi_{11}^{(\mathrm{em})} + [\tilde{\Delta}\Phi_{20}^{(\mathrm{em})}] + \{\mu^*\Phi_0^* f - (\delta^* + 2\alpha)(f\Phi_1^*)\}\kappa/(4\pi) \\ &\quad\quad - \sigma^*[\Phi_{22}^{(\mathrm{em})}] \ . \end{aligned}$$

Thus, near normal points of the surface of a first-order break, the Weyl tensor has the structure

$$W_{iklm} = \delta(u)N_{iklm} + \theta(u)II_{iklm} + I_{iklm} \ ,$$

where $\delta(u)$ and $\theta(u)$ are, respectively, the Dirac and Heaviside functions of the characteristic function u, the quantities N_{iklm}, II_{iklm}, and I_{iklm} are bitensors of the corresponding Petrov types, and the principal directions of degenerate Weyl tensors coincide with the tangents to the isotropic geodesics of the surface of the break. In general, a first-order break in the gravitational field is always accompanied by breaks in the invariants of the Weyl tensor, as can be seen from the relations (1.1.19, 20) between these invariants and the Newman-Penrose scalars.

If a congruence of isotropic geodesics on the surface of a break has a zero shear $\sigma = 0$, then $\psi_0 = 0$ and $[\psi_2] = 0$. In this case, the discontinuity of the Weyl tensor has the Petrov type-III structure, and according to (1.1.19, 20) the breaks in the invariants of the Weyl tensor are completely determined by the breaks in only the first derivatives of g_{ij}. In this case, it follows from Einstein's vacuum equations that $\psi_1 = 0$, and the fields of the Weyl tensor must have an algebraically special form on both sides of the break in some neighborhood of the surface $u = 0$ (the Goldberg-Sachs theorem [1.4]). If the tensor describing the break in the Weyl tensor is of Petrov type III, then a field of type II can be joined at the discontinuity only onto another field of type II or onto one of type D, while a field of type III can be joined onto a field of type III or, in special cases, onto one of type N or empty space. An interesting example in which breaks of this kind can occur is provided by a solution found by *Robinson* and *Trautman* [1.28], for which $\sigma \equiv 0$.

We shall study the way in which the discontinuity in the rotation of the principal isotropic directions of the Weyl tensor different from $\boldsymbol{l}$ is related to the breaks in the components of the Weyl tensor in the case of second-order breaks in the gravitational field. In this case $[\psi_0] = [\psi_1] = [\psi_2] = [\psi_3] = 0$, and according to (1.1.45) the discontinuity $[\psi_4]$ satisfies the equation [1.29]

$$D[\psi_4] + \tfrac{1}{2}(\nabla_i\, l^i)[\psi_4] = 0 \ .$$

In order to make the directions $\boldsymbol{n}$ coincide with the principal isotropic directions of the Weyl tensor "before" the break by means of the transformation (III) given in Sect. 1.1, the complex number a_- must satisfy the quartic equation [see (1.1.15)]

$$\psi_4 + 4a_-\psi_3 + 6(a_-)^2\psi_2 + 4(a_-)^3\psi_1 + (a_-)^4\psi_0 = \psi_4' = 0 \ . \tag{1.2.12}$$

Here and in what follows, primes are used to denote the Newman-Penrose scalars in the new field of tetrads. The values of the other scalars ψ_A' $(A = 0, 1, 2, 3)$ can be obtained by successive differentiation of the expression for ψ_4' with respect to a_- and division of the result by the coefficient obtained for the corresponding ψ_A'. After the surface of the break, the complex number a_+ must satisfy the equation

$$\psi_4 + [\psi_4] + 4a_+\psi_3 + 6(a_+)^2\psi_2 + 4(a_+)^3\psi_1 + (a_+)^4\psi_0 = 0 \ . \tag{1.2.13}$$

Let us subtract (1.2.13) from (1.2.12) and put $[a] = a_+ - a_-$. This gives

$$-[\psi_4] = 4[a]\psi_3' + 6[a]^2\psi_2' + 4[a]^3\psi_1' + [a]^4\psi_0' \ . \tag{1.2.14}$$

The rotation of the principal isotropic direction at $u = 0$ is due to the break $[a]$. From (1.2.14) we obtain the following results for weak discontinuities:

$[a] \sim [\psi_4]$ if $\psi_3' \neq 0$, $\psi_4' = 0$ before the break (a nondegenerate Debever vector);

$$\left.\begin{array}{l} [a] \sim \sqrt{[\psi_4]} \\ [a] \sim ([\psi_4])^{1/3} \\ [a] \sim ([\psi_4])^{1/4} \end{array}\right\} \text{ if } \left\{\begin{array}{l} \psi_4' = \psi_3' = 0,\ \psi_2' \neq 0 \\ \psi_4' = \psi_3' = \psi_2' = 0,\ \psi_1' \neq 0 \\ \psi_4' = \psi_3' = \psi_2' = \psi_1' = 0,\ \psi_0' = 0 \end{array}\right\} \text{ before the break.}$$

This means that the higher the degree of degeneracy of the Weyl tensor, the stronger the discontinuity in the rotation of its principal isotropic directions at the surface of the break. We emphasize that this analysis has a local character.

1.2.3 Decay of an Arbitrary Break in the Vacuum Gravitational Field

Suppose that on a space-like hypersurface T we are given the first and second quadratic forms with coefficients $\mathcal{G}_{\alpha\beta}$ and $b_{\alpha\beta}$, respectively, satisfying in empty space the four constraint equations (consequences of the Gauss and Peterson-Codazzi equations)

$$\nabla_\beta b^\beta_\alpha - \nabla_\alpha b = 0, \quad -\tfrac{1}{4} b^\beta_\alpha b^\alpha_\beta + \tfrac{1}{2}\tilde{R} + \tfrac{1}{4} b^2 = 0 \ , \tag{1.2.15}$$

where $b = b^\alpha_\alpha$, $\tilde{R}$ is the scalar curvature of the surface, and the covariant differentiation is expressed in terms of $\mathcal{G}_{\alpha\beta}$. We assume that on the hypersurface T there exists a two-dimensional surface Σ with no boundary on which the coefficients of the second quadratic form and the derivatives of $\mathcal{G}_{\alpha\beta}$ along the normal to Σ have breaks. We introduce on Σ a field of orthogonal tetrads $\boldsymbol{m}$, $\boldsymbol{m}^*$, $\boldsymbol{\sigma}$, where $\boldsymbol{\sigma}$ is the normal to Σ, $\sqrt{2}\,\boldsymbol{m} = \boldsymbol{u} + \mathrm{i}\boldsymbol{v}$, and $\boldsymbol{u}$ and $\boldsymbol{v}$ define an arbitrary continuous field of orthogonal unit vectors on Σ. (If such a field does not exist globally on Σ because of its topological properties, we must cover Σ by means of charts with such vector fields.) If we go over to a different parameterization of T, the breaks in the derivatives of $\mathcal{G}_{\alpha\beta}$ on Σ acquire additional "fictitious" contributions from the transformation of the coordinates $\sigma_{(\alpha}A_{\beta)}$, where A_β is an arbitrary vector on Σ.

We shall assume that the initial break on Σ is the limit as $\varepsilon \to 0$ of a sequence of smooth initial data with a sharp variation in a layer of thickness ε in the neighborhood of Σ, where the smooth data also satisfy the constraint equations. Then the functions $\mathcal{G}_{\alpha\beta}$, $\partial_\gamma \mathcal{G}_{\alpha\beta}$, $b_{\alpha\beta}$ and their interior derivatives along Σ remain finite, while the second derivatives of $\mathcal{G}_{\alpha\beta}$ and the first derivatives of $b_{\alpha\beta}$ along the normal $\boldsymbol{\sigma}$ tend to infinity as $\varepsilon \to 0$.

Integrating the constraint equations (1.2.15) over the ε layer, in the limit $\varepsilon \to 0$ we obtain

$$[b_{\alpha\beta}]\sigma^\alpha - [b]\sigma_\beta = 0, \quad \mathcal{G}^{\alpha\beta}[\partial_{(\sigma)}\mathcal{G}_{\alpha\beta}] + [\partial_{(\sigma)}\mathcal{G}_{\alpha\beta}]\sigma^\alpha\sigma^\beta = 0 \ . \tag{1.2.16}$$

Projecting $[\partial_{(\sigma)}\mathcal{G}_{\alpha\beta}]$ and $b_{\alpha\beta}$ onto the vectors of the chosen tetrad, by means of these equations we obtain

$$\begin{aligned} [b_{\alpha\beta}] &= a\sigma_\alpha\sigma_\beta + P_2 m_\alpha m_\beta + P_2^* m^*_\alpha m^*_\beta, \\ [\partial_{(\sigma)}\mathcal{G}_{\alpha\beta}] &= P_1 m_\alpha m_\beta + P_1^* m^*_\alpha m^*_\beta + \sigma_{(\alpha}A_{\beta)} \ , \end{aligned}$$

where a and P_2 are functions characterizing the breaks in the tetrad components of the second quadratic form, and the function P_1 characterizes the breaks in the first derivatives of the metric $\mathcal{G}_{\alpha\beta}$.

We construct two isotropic surfaces u_+ and u_- passing through Σ. The initial break on Σ is divided into two breaks. With an appropriate choice of the affine parameter, the initial breaks in the first derivatives of the metric tensor on u_+ and u_- will be determined by the expressions

$$P_+(\alpha = 0) = P_1 + P_2, \quad P_-(\alpha = 0) = P_1 - P_2$$

respectively.

The Cauchy problem for initial data which are smooth off the surface Σ but have breaks on Σ can be divided into three problems, for each of which it is easy to prove the local existence of an analytic solution in the case of analytic initial data.

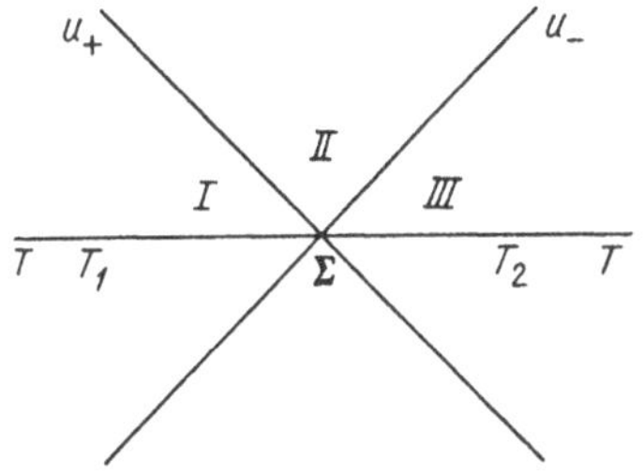

Fig. 1.1. Regions for the Cauchy problem with discontinuous initial data

The surface of the break Σ divides the hypersurface T into two parts T_1 and T_2, as shown in Fig. 1.1.

We complete the definition of the vector fields $\boldsymbol{m}$ and $\boldsymbol{\sigma}$ on T_1 and T_2, for example, by means of parallel transport of the triad given on Σ along the geodesics on T emanating from Σ in the direction of the normal. Then on T we can specify arbitrarily the first quadratic form and the tetrad component of the second quadratic form $b_{\alpha\beta}m^\alpha m^\beta$ (with a smooth dependence off the surface Σ). The tetrad components of $b_{\alpha\beta}m^\alpha m^{*\beta}$ and $b_{\alpha\beta}\sigma^\alpha m^\beta$, which are continuous on Σ, need be specified only on Σ. In fact, the constraint equations $\nabla_\alpha b^\alpha_\beta - \nabla_\beta b = 0$ contain derivatives along the vector $\boldsymbol{\sigma}$ of only these tetrad components. The scalar $b_{\alpha\beta}\sigma^\alpha\sigma^\beta$ can be expressed in terms of $b_{\alpha\beta}m^\alpha m^{*\beta}$, $b_{\alpha\beta}m^\alpha\sigma^\beta$, $\mathcal{G}_{\alpha\beta}$, $b_{\alpha\beta}m^\alpha m^\beta$ as a consequence of the second equation in (1.2.15), which is linear in $b_{\alpha\beta}\sigma^\alpha\sigma^\beta$.

An analytic solution of Einstein's equations in the regions I and III shown in Fig. 1.1 can be obtained in the form of series in the affine parameter of the normal congruences of isotropic geodesics by using the isotropic-geodesic coordinate system defined by (1.2.3). After solving Einstein's equations in regions I and III, for the region II shown in the figure we obtain a Goursat problem with initial data on the characteristics u_+ and u_-. The problem with data on the characteristics for Einstein's equations was analyzed by *Dautcourt* [1.30] and *Sachs* [1.31]. The solution of the problem in region II requires a knowledge of only the parameter describing the shear of the isotropic geodesics on u_+ and u_-, whose values can be obtained by solving the problems in regions I and III. In addition, on

Σ it is necessary to specify the initial value of the expansion parameter of the bicharacteristics for each of the isotropic geodesics of u_+ and u_- and the exterior rotation of the field m on Σ, i.e., $M^{ij}\nabla_i m_j$. The interaction of plane gravitational and electromagnetic waves in region II is described in Sect. 1.3.

1.2.4 The Interaction of Short Gravitational and Electromagnetic Waves in Arbitrary External Electromagnetic Fields

Owing to the nonlinearity of the system of Einstein's and Maxwell's equations, there is always an interaction between the gravitational and electromagnetic fields in a vacuum[4]. The interaction of wave-like fields is of special interest.

We have already mentioned the wave properties inherent in solutions possessing different scales of variation in different regions of space-time. In this case, it is natural to call the region with a smooth variation of the solution the *background* and the region with a "sharp" variation the "wave".

Most of the known exact wave solutions with distinguished algebraic properties have a uniform background (see, for example, [1.5, 32, 33]). A single-stage process of conversion of electromagnetic waves into gravitational waves and vice versa in the presence of a constant transverse magnetic or electric field was considered in [1.34–36].

In the general case, only solutions which vary sharply in comparison with the background can work their way through nonhomogeneities of the background along isotropic geodesics without undergoing scattering.

Such solutions (of the running-wave type) in the absence of external electromagnetic fields are characterized by propagation according to the laws of geometrical optics along isotropic geodesics in curved space-time [1.37]. It follows from the transport equation for the amplitudes of the waves that the intensity (luminosity) of the radiation is inversely proportional to the elementary area of the wave front cut out by these same geodesics — the light rays. The polarization vector of a wave undergoes parallel transport along the rays which carry the wave. When the small amplitude of the gravitational waves becomes comparable in order of magnitude with the wavelength (in a system of units in which the speed of light in a vacuum is equal to unity), trains of short gravitational waves begin to distort the background, and the process of propagation of the waves must be considered together with the problem of determining the background itself.

We shall show in what follows that essentially new effects occur when short waves propagate in arbitrary electromagnetic fields. Either kind of wave — gravitational or electromagnetic — gives rise to the appearance of the other, and they propagate with a mutual modulation of the amplitudes of the waves.

The total "energy" of the gravitational and electromagnetic waves obeys an equation of "continuity" with an isotropic velocity vector along the rays which

[4] The gravitational and electromagnetic fields described by the system of Einstein's and Maxwell's equations in empty space are also called the *electrovacuum*.

carry the wave. In the quasiclassical approximation, developed below, we can speak of a gas of photons and gravitons which are converted into each other. The total distribution function of these particles will satisfy the Liouville equation. Einstein's equations for the background will contain the total energy-momentum tensor of the photons and gravitons as the root-mean-square value of the "noise" in the waves. The reverse effect of the waves on the background manifests itself in exactly the same way as in the absence of external electromagnetic fields [1.37, 38].

Besides the mutual conversion of the waves, another interesting effect which takes place in the general case is the fact that the plane of polarization of waves that were initially linearly polarized rotates with respect to the tetrad subjected to parallel transport along a bicharacteristic — a light ray.

When waves propagate in the field of a rotating black hole, the plane of polarization at the "exit" of a wave from the gravitational field is found to be rotated with respect to its direction at the "entrance". The rotation of the plane of polarization can be determined by means of parallel transport with respect to the pseudo-Euclidean infinity [1.39, 40]. This fact is connected with the nature of parallel transport in a curved Riemannian space. The present author [1.26] pointed out an effect which has an essentially different character, namely, an additional rotation of the polarization vector with respect to a tetrad undergoing parallel transport when there is an electromagnetic background. We note that, independently of the contributions [1.23, 26] by the present author, in 1973–1975 there appeared several publications in which the effect of mutual conversion of electromagnetic and gravitational waves was investigated in various special cases, namely, in a uniform magnetic field [1.41], in a dipole magnetic field [1.42], and in static electrovacuum fields [1.43, 44]. In an interesting paper, *Choquet-Bruhat* [1.45] obtained transport equations for the amplitudes of waves, and the effect of distortion of the background was demonstrated by means of the Nordström-Reissner-Vaidya solution.

In what follows, we discuss the results of [1.23, 26], where the most general case was studied. Thus, we shall seek an asymptotic solution $\tilde{g}_{ij}, \tilde{F}_{ij}$ of the electrovacuum equations in the form

$$\begin{aligned}\tilde{g}_{ij} &= \dot{g}_{ij}(x,\omega) + \varepsilon h_{ij}(\omega u, x, \omega),\\ \tilde{F}_{ij} &= \dot{F}_{ij}(x,\omega) + \varepsilon\omega f_{ij}(\omega u, x, \omega)\quad ,\end{aligned} \tag{1.2.17}$$

where ω is a large parameter and ε is a small parameter, with $\varepsilon\omega \lesssim 1$, and from among the variables x we have distinguished the *fast variable* ωu. The background components $\dot{g}_{ij}$ and $\dot{F}_{ij}$ do not depend on the fast variable. We shall assume that the functions h_{ij} and f_{ij} have periodic dependences on $\omega u \equiv \xi$, so that these functions can be expanded in Fourier series in the harmonics $\cos(n\omega u)$ and $\sin(n\omega u)$. The term "short wave" means that the characteristic scale of variation of the background metric and of the electromagnetic field is much larger than the wavelength of a monochromatic wave, since their ratio is of order $\omega \gg 1$.

We shall give an example of a solution for which the expansions (1.2.17) hold. Suppose that at some instant of time (i.e., on some space-like hypersurface) a wave was initially sinusoidal with a small amplitude of order $\varepsilon \sim \omega^{-1}$. Then, because of the nonlinearity of the electrovacuum equations, higher harmonics $\cos(n\omega u)$ and $\sin(n\omega u)$ $(n = 2, 3, \ldots)$ appear in the subsequent evolution of the wave. It can be seen from the system of equations for the electrovacuum that the coefficients of the higher harmonics will then be of order ω^{-n}, and the background metric $\mathring{g}_{ij}$ and the field $\mathring{F}_{ij}$ are asymptotic expansions in powers of $1/\omega^2$.

The expansions (1.2.17) correspond to running waves of the electrovacuum, and therefore the perturbations h_{ij} cannot be reduced to terms of order $1/\omega$ by means of a coordinate transformation of the form

$$x^{i\prime} = x^i + \varepsilon\xi^i(\omega u, x, \omega)/\omega \quad , \tag{1.2.18}$$

for which

$$h'_{ij} = h_{ij} + \dot{\xi}_i u_{,j} + \dot{\xi}_j u_{,i} + \mathrm{O}(1/\omega), \quad \mathring{g}'_{ij} = \mathring{g}_{ij} + \mathrm{O}(\varepsilon/\omega) \quad .$$

Here the dots denote derivatives with respect to the "fast variable", and the indices are lowered or raised by means of the background metric.

Using the metric (1.2.17), the components of the Ricci tensor can be represented in the form of an expansion in powers of the small parameter:

$$\begin{aligned}
\tilde{R}_{ij} &= \mathring{R}_{ij} + \varepsilon\omega^2 \overset{1}{R}_{ij} + \varepsilon\omega \overset{2}{R}_{ij} + \varepsilon^2\omega^2 \overset{3}{R}_{ij} + \mathrm{O}(\varepsilon\omega) + \mathrm{O}(\varepsilon^2\omega^2) \\
-2\overset{1}{R}_{ij} &= \ddot{h}_{ij} l^k l_k - l_i \ddot{L}_j - l_j \ddot{L}_i \quad , \\
-2\overset{2}{R}_{ij} &= 2D\dot{h}_{ij} + h_{ij}\nabla_k l^k - \nabla_j \dot{L}_i - \nabla_i \dot{L}_j \\
&\quad - l_j(\nabla_k \dot{h}^k_i - \nabla_i \dot{h}/2) - l_i(\nabla_k \dot{h}^k_j - \nabla_j \dot{h}/2) \quad , \\
2\overset{3}{R}_{ij} &= -[L_k(\dot{h}^k_j l_i + \dot{h}^k_i l_j - l^k h_{ij}) + \tfrac{1}{2}h(l_i \dot{L}_j + l_j \dot{L}_i)] \\
&\quad - \dot{L}_i \dot{L}_j + l_m l^m(\dot{h}^l_i \dot{h}_{lj} + h\ddot{h}_{ij}/2) \\
&\quad + l_i l_j[\tfrac{1}{2}(h_{km}h^{km} - h^2/2)^{\cdot\cdot} - \tfrac{1}{2}\dot{h}^l_m \dot{h}^m_l + \tfrac{1}{4}\dot{h}^2] \quad ;
\end{aligned} \tag{1.2.19}$$

here $l_i \equiv u_{,i}$; $D \equiv l^k\nabla_k$, $L_k \equiv h^j_k l_j - hl_k/2$, $h \equiv h^i_i$.

Let us assume that $\varepsilon\omega = 1$ and equate the terms of order ω in Einstein's equations. Then we obtain $\overset{1}{R}_{ij} = 0$.

If the scalar $l_k l^k \equiv g^{ij}u_{,i}u_{,j}$ were not equal to zero, it would follow from the equation $\overset{1}{R}_{ij} = 0$ that

$$\ddot{h}_{ij} l_k l^k = l_i \ddot{L}_j + l_j \ddot{L}_i \quad .$$

Therefore by choosing the transformation

$$x^{i\prime} = x^i - \omega^{-2} \int L^i(\xi, x)\, d\xi, \quad \xi \equiv \omega u$$

it would be possible to make h_{ij} equal to zero (apart from terms of order $1/\omega$) in the system of coordinates $x^{i\prime}$, which contradicts the assumption that the leading terms in h_{ij} cannot be eliminated by means of a coordinate transformation.

It follows that $l^k l_k = 0$. This means that the surfaces $u = \text{const}$ are isotropic. The isotropic surfaces $u = \text{const}$ determine a family of isotropic geodesics with no rotation in the space of the background.

We shall assume that in the irradiation region (the region filled with the light rays of the wave) the field of null tetrads is obtained by parallel transport along the rays $\boldsymbol{l}$ of a tetrad field given on some hypersurface which each ray intersects only once.

From the conditions that $\overset{1}{R}_{ij}$ vanish we also obtain $l_i \ddot{L}_j + l_j \ddot{L}_i = 0$, from which it follows that $L_i = 0$ or

$$h_i^j l_j - \tfrac{1}{2} h l_i = 0 \ .$$

These four linear algebraic relations impose constraints on the ten components h_{ij}, so that only six of the quantities h_{ij} are independent.

It is possible to represent h_{ij} explicitly in terms of the null tetrad $\boldsymbol{l}$, $\boldsymbol{n}$, $\boldsymbol{m}$, $\boldsymbol{m}^*$ in the form

$$h_{ij} = P m_i m_j + P^* m_i^* m_j^* + l_i A_j + l_j A_i \ . \tag{1.2.20}$$

Here the complex scalar P characterizes the actual gravitational wave (its modulus describes the amplitude of the wave, while the argument of P describes the direction of the polarization vector). The term $l_i A_j + l_j A_i$ in h_{ij} can be made equal to zero by means of an appropriate coordinate transformation (1.2.18). Similarly, it follows from Maxwell's equations $\nabla_i F^{ij} = 0$ that $l_i f^{ij} = 0$, from which we obtain

$$2 f_{ij} = f V_{ij} + f^* V_{ij}^*, \quad V_{ij} \equiv 2(l_i m_j - l_j m_i) \ .$$

By averaging Einstein's equations $\tilde{R}_{ij} = \kappa \tilde{T}_{ij}$ (where $\tilde{T}_{ij}$ is the energy-momentum tensor of the electromagnetic field) with respect to the fast variable ωu, we obtain, apart from terms of order $\mathrm{O}(1/\omega^2)$,

$$\dot{R}_{ij} = \kappa \dot{T}_{ij} + \frac{u_{,i} u_{,j}}{4\pi} \int_0^{2\pi} \left\{ |\dot{P}^2| + \frac{8G|f^2|}{c^4} \right\} d\xi \ . \tag{1.2.21}$$

For an initially sinusoidal wave, the higher harmonics will have small amplitudes in comparison with the first harmonic, so that (1.2.21) can be rewritten in the form

$$\dot{R}_{ij} = \kappa \dot{T}_{ij} + u_{,i} u_{,j} [|\dot{P}^2|/4 + 2G|f^2|/c^4] \ . \tag{1.2.21$'$}$$

This simple expression has been obtained on the basis of the fact that $\overset{3}{R}_{ij}$ is greatly simplified by using $L_k = 0$ $(k = 1,2,3,4)$ and $l_m l^m = 0$, and also the fact that terms which are total derivatives with respect to ξ drop out from $\overset{3}{R}_{ij}$ after this quantity is averaged with respect to ξ.

Thus, when Einstein's equations are averaged, the system of background equations contains the squares of the amplitudes of the gravitational and electromagnetic waves, which tend to produce an additional curvature of the background in relation to a given background solution of Einstein's and Maxwell's equations.

In Einstein's equations, we now equate the terms containing the zeroth power of ω for the first harmonic. Then it follows from Einstein's equations that $2\pi \overset{2}{R}_{ij} = \kappa f_{l(i} \dot{F}^l_{j)}$. Multiplying both sides of these equations by $m^{*i} m^{*j}$ and summing over i and j, we obtain

$$D\dot{P} + \frac{1}{2}\dot{P}\nabla_k l^k + \frac{4G}{c^4}\Phi_0^* f = 0 \ . \tag{1.2.22}$$

Equation (1.2.22) could have been obtained directly from the equation $4\pi\Phi_{20} = \kappa\Phi_2\Phi_0^*$, where the expression for Φ_{20} is given by (1.1.63).

It follows from Maxwell's equation (1.1.28) that

$$Df + \tfrac{1}{2} f\nabla_k l^k - \tfrac{1}{2}\Phi_0 \dot{P} = 0 \ . \tag{1.2.23}$$

From (1.2.22, 23) it is easy to obtain the following "continuity" equation for the total intensity of the electromagnetic and gravitational waves:

$$\nabla_i \left\{ l^i \left[|f^2| + \frac{c^4}{8G} |\dot{P}^2| \right] \right\} = 0 \ . \tag{1.2.24}$$

In an isotropic-geodesic coordinate system constructed on the rays l_i [see (1.2.3)], it follows from this equation that

$$\sqrt{g}\left[|f^2| + \frac{c^4}{8G} |\dot{P}^2| \right] = \Omega(u, \xi^1, \xi^2) \ .$$

Therefore along fixed rays (u = const, ξ^1 = const, ξ^2 = const) we have conservation of the quantity

$$\left[|f^2| + \frac{c^4}{8G} |P^2| \right] \sqrt{g} \ .$$

The quantity $\sqrt{g}\, d\xi^1 d\xi^2$ has the meaning of the elementary area on the surface of the wave front u = const, α = const cut out by these same rays. The total intensity of the waves is inversely proportional to the elementary area on the wave front.

From the "continuity" equation for the total intensity of a wave and the equation (1.2.21) for the geometry of the background it follows that mutually coupled trains of electromagnetic and gravitational waves act on the background in exactly the same way as purely gravitational waves in the absence of external electromagnetic fields [1.37]. It has been shown [1.38] that in this case it is possible to introduce a distribution function for massless particles which obeys the Liouville equation.

The tetrad component Φ_0 in the expansion of the bivector of the electromagnetic field with respect to the tetrad $\boldsymbol{l}$, $\boldsymbol{n}$, $\boldsymbol{m}$, $\boldsymbol{m}^*$ characterizes the radiation of the external field along the rays which carry short waves, since the square of its modulus is proportional to the flux of energy and momentum of the external field along the isotropic direction:

$$|\Phi_0^2| = 4\pi \dot{T}_{ij} l^i l^j \quad ,$$

where $\dot{T}_{ij}$ is the energy-momentum tensor of the background. According to (1.2.22, 23), all the special properties of short waves are related to the component Φ_0.

We shall show [1.26] that the variation of the component Φ_0 along the rays can be expressed in terms of the rotation coefficients of the canonical tetrad for the bivector of the background electromagnetic field. Let $\tilde{\boldsymbol{l}}$, $\tilde{\boldsymbol{n}}$, $\tilde{\boldsymbol{m}}$, $\tilde{\boldsymbol{m}}^*$ be a tetrad in which the bivector of the background field takes the canonical form in the nondegenerate case: $F_{ij} = \tilde{\Phi}_1 \tilde{M}_{ij} + \tilde{\Phi}_1^* \tilde{M}_{ij}^*$ (see Sect. 1.1).

We first define a new orientation of the vector $\boldsymbol{l}$ by means of the transformation (IV) in such a way that it becomes tangential to a given congruence of isotropic geodesics with no rotation. We then stretch it by means of the transformation (II), so that it coincides with the vector $l_i \equiv u_{,i}$, where $\{u\}$ is a given family of isotropic surfaces.

The parameters A and b of the transformations (II) and (IV) will then satisfy the following ordinary differential equations along the isotropic geodesics:

$$\begin{aligned} D(bA) &= A[\tilde{m}_i D\tilde{l}^i + b\tilde{m}_i D\tilde{m}^{*i} + |b^2|\tilde{m}_i D\tilde{n}^i], \\ -DA &= A[\tilde{n}_i D\tilde{l}^i + b\tilde{n}_i D\tilde{m}^{*i} + b^*\tilde{n}_i D\tilde{m}^i] \quad . \end{aligned} \tag{1.2.25}$$

By means of the transformations (III) and (I) with the parameter a and the angle θ, we rotate the vectors $\tilde{\boldsymbol{n}}$ and $\tilde{\boldsymbol{m}}$ so that $\boldsymbol{n}$ and $\boldsymbol{m}$ satisfy the condition of parallel transport along the congruence. Then the parameter θ satisfies the equation

$$\mathrm{i}D\theta = \tilde{m}_i^* D\tilde{m}^i - b^*\tilde{m}_i D\tilde{n}^i + b\tilde{m}_i^* D\tilde{n}^i \quad . \tag{1.2.25'}$$

The component Φ_0 of the background electromagnetic field tensor in the tetrad undergoing parallel transport along the given congruence of isotropic geodesics is equal to $2Ab\tilde{\Phi}_1 \exp(\mathrm{i}\theta)$. According to (1.2.25, 25′), we have

$$D(Ab\exp(\mathrm{i}\theta)) = A\exp(\mathrm{i}\theta)[\tilde{m}_i D\tilde{l}^i + b^2\tilde{m}_i^* D\tilde{n}^i] \quad . \tag{1.2.26}$$

From Maxwell's equations (1.1.28–31), written in terms of the canonical tetrad, we obtain

$$D\tilde{\bar{\Phi}}_1 = 2A\tilde{\bar{\Phi}}_1(\tilde{\varrho} + b^*\tilde{\tau} - |b^2|\tilde{\mu} - b\tilde{\pi}) \ . \tag{1.2.27}$$

It follows from (1.2.26, 27) that

$$D \ln \Phi_0 = A\left\{\frac{1}{b}[\tilde{k} + 3|b^2|\tilde{\tau} + b^*\tilde{\sigma} + 3b\tilde{\varrho}] - b[3\tilde{\pi} + |b^2|\tilde{\nu} + 3b^*\tilde{\mu} + b\tilde{\lambda}]\right\} \ . \tag{1.2.28}$$

To obtain a convenient form for this equation, it is sensible to rewrite it by expressing the parameters A and b of the transformations in terms of the components of the vector $l_i = u_{,i}$ in the tetrad $\tilde{\boldsymbol{l}}$, $\tilde{\boldsymbol{n}}$, $\tilde{\boldsymbol{m}}$, $\boldsymbol{m}^*$:

$$\begin{aligned} D \ln \Phi_0 = {} & 3\left(u_{,(\tilde{m})}\tilde{\pi} - u_{,(\tilde{m}^*)}\tilde{\tau} + u_{,(\tilde{n})}\tilde{\varrho} - u_{,(\tilde{l})}\tilde{\mu}\right) \\ & + \frac{u^2_{,(\tilde{m}^*)}}{u_{,(\tilde{l})}}\tilde{\sigma} - \frac{u_{,(\tilde{n})}u_{,(\tilde{m}^*)}}{u_{,(\tilde{l})}}\tilde{k} - \frac{u^2_{,(\tilde{m})}}{u_{,(\tilde{n})}}\tilde{\lambda} + \frac{u_{,(\tilde{l})}u_{0(\tilde{m})}}{u_{,(\tilde{n})}}\tilde{\nu} \ . \end{aligned} \tag{1.2.28'}$$

If the bivector of the background electromagnetic field is degenerate, i.e., both of its invariants are equal to zero, then in some tetrad it can be reduced to the form $\tilde{F}_{ij} = \tilde{\Phi}_0 U_{ij}$. The corresponding expressions for this case were given in [1.26].

If $\arg \Phi_0$ is to remain unchanged along an arbitrary isotropic geodesic, the following conditions must be imposed on the background gravitational field:

$$\varrho = \varrho^*, \quad \mu = \mu^*, \quad \pi + \tau^* = 0, \quad k = \sigma = \nu = \lambda = 0 \ . \tag{1.2.29}$$

These conditions lead to a Weyl tensor of Petrov type D, for which all solutions of Einstein's vacuum equations have now been found [1.46].

By virtue of (1.2.22, 23), the variation of the phase of Φ_0 gives rise to mutually consistent variations of the phases of the functions f and P.

Making the substitutions

$$\sqrt[4]{g}\sqrt{8G}f = c^2\mathcal{F}, \quad \mathcal{P} = \sqrt[4]{g}\dot{P}, \quad c^2\varphi = \Phi_0\sqrt{2G} \ , \tag{1.2.30}$$

in (1.2.22, 23), we obtain

$$\frac{d}{d\alpha}\mathcal{P} + \varphi^*\mathcal{F} = 0, \quad \frac{d}{d\alpha}\mathcal{F} - \varphi\mathcal{P} = 0, \quad \nabla_i l^i = \frac{d \ln \sqrt{g}}{d\alpha} \ . \tag{1.2.31}$$

We represent the complex function φ in trigonometric form and replace the affine parameter α by the variable x defined by the relation $dx = |\varphi| d\alpha$. Eliminating the unknown $\mathcal{P}$ from (1.2.31), we obtain

$$\frac{d^2\mathcal{F}}{dx^2} + \mathcal{F} + \mathrm{i}\frac{d \arg\varphi}{dx}\frac{d\mathcal{F}}{dx} = 0 \ . \tag{1.2.32}$$

This equation relates the variation of the amplitude $|\mathcal{F}|$ of the electromagnetic wave and the angle of rotation $\arg \mathcal{F}$ of its plane of polarization.

In the particular case in which the conditions (1.2.29) hold, the argument of the function φ does not depend on x, and the system of equations (1.2.31) admits the simple solutions

$$\mathcal{F} = A\cos\left(\int \varphi\, d\alpha + \gamma\right), \quad \mathcal{P} = A\sin\left(\int \varphi\, d\alpha + \gamma\right) , \tag{1.2.33}$$

where the complex number A and the phase shift γ are constant along a fixed ray. It follows from the form of the solutions (1.2.33) that the amplitudes of the electromagnetic and gravitational waves are sinusoidally modulated with a frequency determined by the equation

$$2\pi = \int \varphi\, d\alpha = \frac{\sqrt{G}}{c^2}\int \sqrt{T_{ij}\, dx^i dx^j} \ .$$

Before entering the region with a strong electromagnetic field $|\varphi| \sim 1$, the gravitational and electromagnetic waves from sources localized in the same region propagate independently, with identical surfaces of the wave fronts, but in general with different planes of polarization. In the region $|\varphi| \sim 1$, these waves experience only partial mutual modulation, and the plane of polarization of each of the waves experiences a rotation with respect to the tetrad which undergoes parallel transport along the rays. Formally, this happens because the phase γ in the expressions (1.2.33) is in general complex.

The case of a real phase γ corresponds to a situation in which either the planes of polarization of the gravitational and electromagnetic short waves were coincident before the waves entered the region $|\varphi| \sim 1$ or only one of these waves was initially incident. In this case, the solutions (1.2.33) describe total mutual conversion of the waves over a length equal to the period calculated from the equation $\int |\varphi| d\alpha = 2\pi$, since the process of propagation of the waves takes place as a periodic appearance and disappearance of the gravitational and electromagnetic waves with conservation of their total intensity. This effect of periodic transfer of "energy" from one mode of oscillation to the other is analogous to the spatial beats of electromagnetic waves in two wave guides connected by a narrow gap [1.47, 48] and to the beats in a system of two pendulums connected by a weak spring [1.49].

1.2.5 Algebraic Structure of Perturbations of the Weyl Tensor in the Case of High-Frequency Waves

It follows from (1.2.17) that the perturbation of the metric corresponding to short waves has the asymptotic form

$$h_{ij} = Pm_i m_j + P^* m_i^* m_j^*, \quad \tilde{g}_{ij} = g_{ij} + h_{ij}/\omega \ . \tag{1.2.34}$$

Then from the definition of the tetrad components ψ_0, ψ_1, ψ_2, ψ_3, ψ_4 of the Weyl tensor it follows that the values of these quantities for the perturbed metric (1.2.34) have the asymptotic structure

$$\begin{aligned}
&\tilde{\psi}_4 = \dot{\psi}_4 - \tfrac{1}{2}\omega\ddot{P} + \dots ,\\
&\tilde{\psi}_3 = \dot{\psi}_3 + \tfrac{1}{2}(\delta + 3\beta - \alpha^*)\dot{P} + \kappa\Phi_1^*/(4\pi) + \dots ,\\
&\tilde{\psi}_2 = \dot{\psi}_2 + \tfrac{1}{2}\sigma\dot{P} + \dots , \quad \tilde{\psi}_1 = \dot{\psi}_1 + \mathrm{O}(1/\omega), \quad \tilde{\psi}_0 = \dot{\psi}_0 + \mathrm{O}(1/\omega),\\
&\dot{P} \equiv \partial P/\partial\xi, \quad \xi \equiv \omega u \ .
\end{aligned} \tag{1.2.34$'$}$$

It can be seen from these expressions that it is by no means obligatory to derive the relation (1.2.22) from Einstein's equation $\Phi_{20} = \kappa\Phi_2\Phi_0^*/(4\pi)$ [see (1.1.63)]. This same equation could have been obtained more elegantly from the Bianchi identity (1.1.46). In addition, it follows from (1.2.34′) that the Weyl tensor in the case of short waves has the algebraic structure

$$\tilde{W}_{iklm} = \dot{W}_{iklm} + \omega N_{iklm} + II_{iklm} + (1/\omega)I_{iklm} \quad ,$$

where $\dot{W}_{iklm}$ is the Weyl tensor of the background space, and N_{iklm}, II_{iklm}, and I_{iklm} are Weyl tensors of Petrov types N, II, and I, respectively.

Thus, we can say that when the perturbations in the Weyl tensor are expanded in inverse powers of the frequency, the degree of algebraic degeneracy of the successive terms becomes smaller. This fact is completely analogous to the properties of first-order breaks in the gravitational field (see Sect. 1.2.2).

1.2.6 Behavior of Short-Wave Perturbations of the Gravitational Field Near Caustic Surfaces

Waves attain the largest intensity at the focal points, where $g = 0$; this means that a treatment using the approximation of geometrical optics is not possible, since this approximation leads to fictitious features in the amplitudes of the waves.

We have shown (see Sect. 1.2.1) that the appearance of focal points on light rays is not an exotic phenomenon in general relativity, since the area of any convergent element of a wave front decreases monotonically, becoming equal to zero at a focal point.

The set of focal points forms a caustic surface. Here we consider only the simplest stable type of caustic (from the Greek *kaustikos* — burning) — the fold. From the point of view of the classification of the singularities of mappings of Lagrangian manifolds (which occur in the representation of solutions in the form of rapidly oscillating integrals [1.50]), only the following types of singularities (apart from the fold) can be stable in a three-dimensional space: the (Whitney) cusp, the "swallowtail", and the hyperbolic and elliptic umbilics [1.49]. The maximum intensity rises as we go from one type of singularity to the next, with the following orders of magnitude as a function of frequency ω: $\omega^{1/6}$ for the fold, $\omega^{1/4}$ for the cusp, $\omega^{3/10}$ for the swallowtail, and $\omega^{1/3}$ for both umbilics (at normal points, we assume the order of magnitude $\omega^0 = 1$).

For a given congruence of light rays having zero rotation and "carrying" short-wave perturbations, we go over to the comoving isotropic-geodesic system of coordinates α, u, ξ^1, ξ^2 [see (1.2.3)]. Exploiting the arbitrariness in the choice of the origin for the affine parameter, we shall measure it from the caustic surface.

We shall consider the asymptotic behavior of the metric coefficients near the caustic surface of a fold in the isotropic-geodesic coordinate system.

We construct a field of null tetrads in this system as follows. Suppose that the isotropic vectors $\boldsymbol{l}$, $\boldsymbol{n}$, and $\boldsymbol{m}$ have the following components:

$$
\begin{aligned}
&l_\alpha = 0, \quad l_u = 1, \quad l_A = 0;\\
&n_\alpha = 1, \quad n_u = \tfrac{1}{2}(g_0 + g_A g^A), \quad n_B = 0;\\
&m_\alpha = 0, \quad m_u = (2 - g_{22})^{-1/2}[g_2 + \mathrm{i} g^1 \sqrt{g}];\\
&m_1 = -(2g_{22})^{-1/2}[g_{12} + \mathrm{i}\sqrt{g}], \quad m_2 = -\sqrt{g_{22}/2},\\
&m^{*1} = -\mathrm{i}\sqrt{g_{22}/(2g)}, \quad m^{*2} = (1 - \mathrm{i} g_{12}/\sqrt{g})/\sqrt{2g_{22}},\\
&g \equiv g_{11}g_{22} - g_{12}^2, \quad A, B = 1, 2 \ .
\end{aligned}
\tag{1.2.35}
$$

The indices $A, B = 1, 2$ of the coefficients g_A and g^B are raised or lowered by means of the two-dimensional metric g_{AB}.

For such a tetrad field, the rotation coefficients can be expressed in terms of the metric as follows:

$$
k = 0, \quad \varrho = \varrho^* = -\partial \ln \sqrt[4]{g}/\partial\alpha; \quad 2\sigma = -\partial \ln(g_{22}/\sqrt{g})\partial\alpha + 4\varepsilon \ ; \tag{1.2.36}
$$

$$
\begin{aligned}
&\varepsilon = -\varepsilon^* = -\mathrm{i} g_{22}(4\sqrt{g})^{-1}\partial(g_{12}/g_{22})/\partial\alpha \ ;\\
&\beta + \alpha^* = \tau = \pi^* = \frac{1}{2\sqrt{2g_{22}}}\left[(g_{12} + \mathrm{i}\sqrt{g})\frac{\partial g^1}{\partial\alpha} + g_{22}\frac{\partial g^2}{\partial\alpha}\right];\\
&\mu = \frac{1}{2}\varrho(g_0 + g_A g^A) + \frac{1}{4}\left[2\nabla_A g^A - \frac{\partial \ln g}{\partial u}\right],\\
&\lambda = \frac{1}{2}\sigma^*(g_0 + g_A g^A) + \frac{m^{*A}m^{*B}}{2}\left(\nabla_A g_B + \nabla_B g_A - \frac{\partial g_{AB}}{\partial u}\right) \ .
\end{aligned}
\tag{1.2.37}
$$

For this field of null tetrads, (1.2.52, 60) have the form

$$
\Psi_0 = (D - 4\varepsilon - 2\varrho)\sigma, \quad \Phi_{00} = (D - \varrho)\varrho - \sigma\sigma^* \ . \tag{1.2.38}
$$

We assume that the component Ψ_0 of the Weyl tensor and the component $\Phi_{00} = \kappa T_{ij} l^i l^j$ of the energy-momentum tensor have the following orders of magnitude near the caustic: $\Phi_{00} = \mathrm{O}(\sigma\sigma^*)$, $\Psi_0 = \mathrm{O}(\sigma\varrho)$. Then it follows from (1.2.38) that the leading terms in the expansions of ϱ and σ with respect to α have the form

$$
\varrho = -\frac{1}{2\alpha} + \mathrm{O}(1), \quad \sigma = \left(\frac{1}{2\alpha} + \mathrm{O}(1)\right)\exp\left(4\int \varepsilon \, d\alpha\right) \ . \tag{1.2.39}
$$

Expressing ϱ in terms of the metric according to (1.2.36) yields

$$
g = G_0\alpha^2 + \mathrm{O}(\alpha^3), \quad G_0 = G_0(u, x^1, x^2) \ .
$$

Using the fact that the imaginary part of σ is 2ε, from (1.2.39) we obtain

$$4i\alpha\varepsilon = \sin\left(i\int 4\varepsilon\, d\alpha\right) .$$

It follows from this that $\varepsilon = i\varepsilon_0$, $\varepsilon_0 = \varepsilon_0(u, x^1, x^2)$. By equating the real parts of σ determined by (1.2.36, 39), we arrive at

$$\frac{\partial \ln(\sqrt{g}/g_{22})}{\partial\alpha} = \frac{\cos(4i\varepsilon_0\alpha)}{\alpha} \approx \frac{1}{\alpha} . \tag{1.2.40}$$

Expressing ε in terms of the metric, we obtain

$$2\varepsilon\sqrt{g} = g_{12}\partial \ln g_{22}/\partial\alpha - \partial g_{12}/\partial\alpha . \tag{1.2.41}$$

It follows from (1.2.39–41) that the matrix g_{AB} has the following form for $\alpha \to 0$:

$$\begin{pmatrix} a(\varphi_{,x^1})^2(1+\beta_1\alpha)+O(\alpha^2) & a\varphi_{,x^1}\varphi_{,x^2} - 2\varepsilon_0\sqrt{G_0}\,\alpha^2 + O(\alpha^3) \\ a\varphi_{,x^1x^2} - 2\varepsilon_0\sqrt{G_0}\,\alpha^2 + O(\alpha^3) & a(\varphi_{,x^2})^2(1-\beta_1\alpha)+O(\alpha^2) \end{pmatrix} . \tag{1.2.42}$$

Here a, φ, β_1, and β_2 are arbitrary functions of u, x^1, and x^2 $(\partial\varphi/\partial x^2 \neq 0)$. An isotropic-geodesic coordinate system retains the same form under transformations of the type $x^{A\prime} = f^A(x^1, x^2, u)$. If we go over to a new system $y^1 = f$, $y^2 = \varphi$, where f is an arbitrary function of u, x^1, and x^2, we obtain from (1.2.42) the expressions

$$\begin{aligned} g_{11} &= G_{11}\alpha^2 + O(\alpha^3), \\ g_{12} &= -2\varepsilon_0\sqrt{G_{11}G_{22}}\,\alpha^2 + O(\alpha^3), \\ g_{22} &= G_{22} + O(\alpha) , \end{aligned} \tag{1.2.43}$$

where G_{11} and G_{22} are arbitrary real functions of u, x^1, and x^2.

To see how the metric coefficients g^A $(A = 1, 2)$ depend on α, we make use of the equation

$$\Psi_1 + \Phi_{10}^* = (D - 2\varrho - 2\varepsilon)\tau - 2\sigma\tau^* , \tag{1.2.44}$$

which, for the quoted field of tetrads, results from (1.1.51, 61). We assume that $\Psi_1 + \Phi_{10}^* = O(\varrho\pi)$ near a caustic. Then from (1.2.44) in a first approximation we obtain

$$\tau \approx (M + iN/\alpha^2)\exp(2i\varepsilon_0\alpha) ,$$

where M and N are arbitrary real functions of u, x^1, and x^2.

By expressing τ in terms of the metric and using (1.2.37, 44), we obtain

$$g^1 = -\frac{N}{\alpha^2}\sqrt{\frac{2}{G_{11}}} + O\left(\frac{1}{\alpha}\right), \quad g^2 = G^2 + 2\alpha M\sqrt{\frac{2}{G_{22}}},$$
$$G^2 = G^2(u, x^1, x^2) .$$

Dropping the indices on g^A, we find

$$g_1 = -N\sqrt{2G_{11}} + \mathrm{O}(\alpha), \quad g_2 = G^2 G_{22} + \mathrm{O}(\alpha) \ . \tag{1.2.45}$$

We now determine the asymptotic behavior of $g_0 + g_A g^A$. To do this, we consider the relation (1.1.55):

$$(D - \varrho)\mu - (\delta - \alpha^* + \beta + \pi^*)\pi - \sigma\lambda + R/12 = \psi_2 \ . \tag{1.2.46}$$

Let us assume that $\psi_2 - R/12 = \mathrm{O}(\pi\pi^*)$. Then according to the preceding calculations it follows from (1.2.46) in a first approximation that

$$\frac{1}{2}\varrho\frac{\partial}{\partial\alpha}(g_0 + g_A g^A) \approx \pi\pi^* \ .$$

Hence

$$g_0 + g_A g^A \approx \frac{2N^2}{\alpha^2} + \mathrm{O}\left(\frac{1}{\alpha}\right) \ . \tag{1.2.47}$$

In the approximation of geometrical optics, the d'Alembertian $\nabla_i \nabla^i$, defined in terms of the background metric and acting on a scalar function ψ of the fast variable ωu and the slow variables α, x^1, and x^2, is equal asymptotically to the operator

$$2(D - 2\varrho)\psi_{,u} \ .$$

Let us differentiate (1.2.23) with respect to the fast variable. Then the system of equations (1.2.22, 23) can be rewritten in the form

$$\nabla_i \nabla^i P + 4G\Phi_0^* F/c^4 = 0, \quad F \equiv \omega f \ , \tag{1.2.22'}$$

$$\nabla_i \nabla^i F - \tfrac{1}{2}\Phi_0 P_{,uu} = 0 \ . \tag{1.2.23'}$$

Unlike the system of equations (1.2.22, 23), the system (1.2.22′, 23′) is meaningful even near caustic surfaces, where the approximation of geometrical optics breaks down.

In a particular case, along a family of bicharacteristics we have $d \arg \Phi_0/d\alpha = 0$. Then by means of a rotation of the tetrad we can make the component Φ_0 of the external electromagnetic field a real function for all α.

We now differentiate (1.2.22′) with respect to the fast variable (the differential operator $\partial/\partial u$ and the d'Alembertian commute for high-frequency waves). Then the system (1.2.22′, 23′) can be rewritten as a pair of closed equations for χ_+ and χ_-:

$$\left(\nabla_i \nabla^i + \mathrm{i}\Phi_0\sqrt{\frac{2G}{c^4}}\frac{\partial}{\partial u}\right)\chi_\pm = 0, \quad \chi_\pm \equiv F \pm \sqrt{\frac{c^4}{8G}}\,\mathrm{i}\frac{\partial P}{\partial u} \ . \tag{1.2.48}$$

Near a caustic surface $\alpha = 0$, the second derivatives of the solutions with respect to the affine parameter (which are negligibly small outside the neighborhood of the caustic) come into play in the d'Alembertian. In the isotropic-geodesic coordinate system (1.2.3), the d'Alembertian (operating on a scalar) becomes asymptotically equal to the operator

$$\frac{2}{\sqrt[4]{g}}\frac{\partial}{\partial\alpha}\sqrt[4]{g}\frac{\partial}{\partial u} - \frac{1}{\sqrt{g}}\frac{\partial}{\partial\alpha}\left[(g_0 + g_A g^A)\sqrt{g}\frac{\partial}{\partial\alpha}\right] \quad . \tag{1.2.49}$$

At normal points of isotropic geodesics, the second term in (1.2.49) is small and can be omitted[5]. Near caustics, we can rewrite (1.2.22, 23) in a form which takes into account the dependence of the metric coefficients on α:

$$\begin{aligned} &\frac{2}{\sqrt{\alpha}}\frac{\partial}{\partial\alpha}\left(\sqrt{\alpha}\frac{\partial P}{\partial u}\right) + \frac{G_0}{\alpha}\frac{\partial}{\partial\alpha}\left(\frac{1}{\alpha}\frac{\partial}{\partial\alpha}P\right) = \frac{8G}{c^4}\Phi_0^* f, \\ &\frac{2}{\sqrt{\alpha}}\frac{\partial}{\partial\alpha}\left(\sqrt{\alpha}\frac{\partial f}{\partial u}\right) + \frac{G_0}{\alpha}\frac{\partial}{\partial\alpha}\left(\frac{1}{\alpha}\frac{\partial}{\partial\alpha}f\right) + \Phi_0\frac{\partial^2 P}{\partial u^2} = 0 \quad . \end{aligned} \tag{1.2.50}$$

Let us make the change of variables

$$y = \frac{\alpha^2}{\sqrt[3]{4G_0^2}}, \qquad u - \frac{\alpha^3}{3G_0} = x \quad .$$

Then from (1.2.50) we obtain

$$\begin{aligned} &\frac{\partial^2 P}{\partial y^2} - y\frac{\partial^2 P}{\partial x^2} = \frac{4G}{c^4}\sqrt[3]{2G_0}\,\Phi_0^* f, \\ &\frac{\partial^2 f}{\partial y^2} - y\frac{\partial^2 f}{\partial x^2} = -\frac{1}{2}\sqrt[4]{2G_0}\,\Phi_0\frac{\partial^2 P}{\partial x^2} \quad . \end{aligned} \tag{1.2.51}$$

In the absence of external electromagnetic fields, when $\Phi_0 = 0$, this system decomposes into two independent Tricomi equations.

For monochromatic waves, the system of equations (1.2.51) reduces to the system

$$\begin{aligned} &\frac{\partial^2 P}{\partial y^2} + \omega^2 y P - \frac{4G}{c^4}\sqrt[3]{2G_0}\,\Phi_0^* f = 0, \\ &\frac{\partial^2 f}{\partial y^2} + \omega^2 y f + \frac{1}{2}\omega^2\sqrt[3]{2G_0}\,\Phi_0 P = 0 \quad . \end{aligned}$$

The "irradiated" region, i.e., the region accessible by means of some light ray from the congruence, corresponds to $y > 0$. For the "shadow" region, i.e.,

---

[5] We note that many excellent studies (for example, [1.49–51]) have been devoted to the investigation of asymptotic solutions of *linear* hyperbolic systems, particularly in the approximation of geometrical optics.

the region consisting of points inaccessible on the congruence for real values of the affine parameter α, we have $y < 0$.

In comparison with normal points on the rays, at focal points the luminosity of the waves is in general increased by a factor $\omega^{1/6}$.

1.3 Interaction of Gravitational and Electromagnetic Waves

1.3.1 Curvature of Space-Time in a Plane Electromagnetic Wave

As is well known, plane waves which propagate in one direction without change of form are distinguished by the property that the light rays corresponding to them have zero expansion, $\mathrm{Re}\{\varrho\} = 0$, rotation, $\mathrm{Im}\{\varrho\} = 0$, and shear, $\sigma = 0$. Space-time must admit a group of motions along the light rays, which are the orbits of this group. Therefore, for exact plane-wave solutions of the Einstein-Maxwell equations in empty space, there must exist an absolutely parallel isotropic vector field $\boldsymbol{l}$. This means that in the isotropic-geodesic coordinate system (1.2.3) co-moving with the light rays the metric coefficients must not depend on the affine parameter α.

The α-independent form of the metric (1.2.3) is invariant under coordinate transformations of the form

$$u' = u, \quad \alpha' = \alpha + \Omega(u, \xi^1, \xi^2), \quad \xi^{\alpha'} = f^a(u, \xi^1, \xi^2), \quad a = 1, 2 \ ,$$

where $\Omega(u, \xi^1, \xi^2)$ and $f^a(u, \xi^1, \xi^2)$ are arbitrary functions. By means of an appropriate choice of these functions, it is possible to make the coefficients g_{12} and g_2 of the metric (1.2.3) equal to zero, and the coefficient g_{11} equal to unity [1.52]. Then the metric (1.2.3) takes the form

$$ds^2 = A\,du^2 + 2\,du\,d\alpha + 2S\,du\,dx - E\,dx^2 - dy^2 \ , \tag{1.3.1}$$

where $A \equiv g_0(u, x, y)$, $S \equiv g_1(u, x, y)$, $E \equiv g_{11}(u, x, y,)$, $\xi_1 = x$, $\xi_2 = y$.

Below, following [1.53], we shall show that in a certain coordinate system the general solution of the system of Maxwell's and Einstein's equations for plane waves takes the form

$$ds^2 = A(u, x, y)\,du^2 + 2\,du\,d\alpha - dx^2 - dy^2 \ , \tag{1.3.2}$$

in which the function $A(u, x, y)$ satisfies the equation

$$\frac{\partial^2 A}{\partial x^2} + \frac{\partial^2 A}{\partial y^2} = \frac{4G}{c^4}\Phi_2\Phi_2^*, \quad \Phi_2 = f(x + \mathrm{i}y, u) \ , \tag{1.3.3}$$

where f is an arbitrary analytic function of $x + \mathrm{i}y$, and the dependence of f on u is also arbitrary.

In the literature, one can find several particular solutions of the Einstein-Maxwell equations describing plane waves, for example, the solutions of Takeno,

Petrov, Kaĭgorodov and Pestov, Rosen, and Peres (see the references in the book of *Zakharov* [1.12]).

In order to show that all these solutions take the form (1.3.2, 3) in some coordinate system, we make use of the formalism of null tetrads.

Let us choose a field of null tetrads in the coordinate system (1.3.1) with the following components:

$$l_i(1,0,0,0), \quad n_i(A/2,1,S,0), \quad m_i(0,0,\sqrt{E/2},\mathrm{i}/\sqrt{2}) \ . \tag{1.3.4}$$

Then the nonzero rotation coefficients of a tetrad can be expressed in terms of the functions A, E, and S as follows:

$$\begin{aligned}
&\lambda = \frac{1}{4}\frac{\partial}{\partial u}\ln E, \quad \mu = \frac{1}{4}\frac{\partial}{\partial u}\ln E - \frac{\mathrm{i}}{2\sqrt{E}}\frac{\partial S}{\partial y},\\
&\alpha = \beta = -\frac{\mathrm{i}}{4\sqrt{2}}\frac{\partial \ln E}{\partial y}, \quad \nu = \frac{1}{\sqrt{2E}}\frac{\partial S}{\partial u} - \frac{1}{2\sqrt{2E}}\frac{\partial A}{\partial x} - \frac{\mathrm{i}}{2\sqrt{2E}}\frac{\partial A}{\partial y},\\
&\gamma = -\frac{\mathrm{i}}{4\sqrt{E}}\frac{\partial S}{\partial y} \ .
\end{aligned} \tag{1.3.5}$$

We shall assume that only the component Φ_2 of the electromagnetic field tensor F_{ij} is nonzero. Einstein's equations (1.1.65–68) take the form

$$\begin{aligned}
&(\delta - \delta^*)\alpha + 4\alpha^2 = 0, \quad \delta\gamma - \Delta\alpha - 2\lambda\alpha = 0,\\
&(\delta - \delta^*)\lambda + 4\alpha\lambda - 2\delta^*\gamma = 0 \ ,
\end{aligned} \tag{1.3.6}$$

$$\delta\nu - \Delta\mu + 2\alpha\nu - \mu^2 - \lambda\lambda^* = \frac{4G}{c^4}\Phi_2\Phi_2^* \ . \tag{1.3.7}$$

Using (1.3.4, 5) we can write (1.3.6) in the form

$$\begin{aligned}
&\frac{\partial^2 \ln E}{\partial y^2} + \frac{1}{2}\left(\frac{\partial \ln E}{\partial y}\right)^2 = 0, \quad \frac{\partial}{\partial y}\left(\frac{1}{\sqrt{E}}\frac{\partial S}{\partial y}\right) = 0,\\
&\frac{\partial}{\partial x}\left(\frac{1}{\sqrt{E}}\frac{\partial S}{\partial y}\right) + 2\frac{\partial^2\sqrt{E}}{\partial u\,\partial y} = 0 \ ,
\end{aligned} \tag{1.3.8}$$

which, after integration, gives

$$\begin{aligned}
&\sqrt{E} = a(x,u)y + b(x,u),\\
&S = c(u,x)\left[\frac{a(u,x)y^2}{2} + b(u,x)y\right] + d(u,x) \ ,
\end{aligned} \tag{1.3.9}$$

$$\frac{\partial a}{\partial u} + \frac{\partial c}{\partial x} = 0 \ , \tag{1.3.10}$$

where a, b, c, and d are functions of the indicated variables. Making use of these last equations, we shall list the sequence of transformations which reduces the metric (1.3.1) to the form (1.3.2). We consider first the case $a \neq 0$, for which we give each transformation and then, below it, the form of the interval after the indicated transformation (the coefficient A is different in certain transformations, but we always denote it by the same symbol):

$$\begin{aligned}
&1)\quad x' = \int a\,dx, \quad y' = y, \quad \alpha' = \alpha,\\
&\qquad ds^2 = A\,du^2 + 2\,du\,d\alpha' + [c(y'^2 + 2by') + d]\,du\,dx'\\
&\qquad\qquad - (y' + b)^2\,dx'^2 - dy'^2.\\
&2)\quad \alpha' = \alpha + \int (d - cb^2)\,dx, \quad y' = y, \quad x' = x,\\
&\qquad ds^2 = A\,du^2 + 2\,du\,d\alpha' + c(y' + b)^2\,du\,dx'\\
&\qquad\qquad - (y' + b)^2\,dx'^2 - dy'^2.\\
&3)\quad x' = y\cos x - \int b\sin x\,dx, \quad y' = -y\sin x - \int b\cos x\,dx,\\
&\qquad \alpha' = \alpha,\\
&\qquad ds^2 = A\,du^2 + 2\,du\,d\alpha' + f(u, x', y')\,du\,dx'\\
&\qquad\qquad + g(u, x', y')\,du\,dy' - dx'^2 - dy'^2.\\
&4)\quad \alpha' = \alpha + \int g(u, x, y)\,dy, \quad y' = y, \quad x' = x,\\
&\qquad ds^2 = A\,du^2 + 2\,du\,d\alpha' + 2\tilde{S}(u, x', y')\,du\,dx' - dx'^2 - dy'^2 .
\end{aligned} \tag{1.3.11}$$

The form of the metric (1.3.11) is identical to (1.3.1) if for E and S in (1.3.9) we take $a = 0$ and $b = 1$, but in this case we find, using (1.3.10), that the function $\tilde{S}$ in (1.3.11) has the form

$$S = c(u)y.$$

$$\begin{aligned}
&5)\quad \alpha' = \alpha + c(u)xy, \quad x = r\cos\varphi, \quad y = r\sin\varphi,\\
&\qquad ds^2 = A\,du^2 + 2\,du\,d\alpha' + 2c(u)r^2\,du\,d\varphi - dr^2 - r^2 d\varphi^2.\\
&6)\quad \varphi' = \varphi - \int c(u)\,du, \quad \alpha' = \alpha, \quad r' = r,\\
&\qquad ds^2 = A\,du^2 + 2\,du\,d\alpha' - dr'^2 - r'^2 d\varphi'^2 .
\end{aligned}$$

This form of the metric is identical to (1.3.2). If, however, $a = 0$, then to reduce the metric (1.3.1) to the final form (1.3.2) it is necessary to begin with the transformation $x' = \int b\,dx$ and then carry out the transformations 4–6.

1.3.2 Nonlinear Interaction of Plane Waves

Above, we considered the case of a plane wave traveling in one direction without change of form.

An exact formulation of the problem of the nonlinear interaction of plane gravitational and electromagnetic waves has been given by several authors [1.54–56]. In order to study the interaction of plane waves, it is necessary to consider classes of metrics having a more general form than (1.3.1). We shall consider pseudo-Riemannian spaces which admit two-parameter commutative groups of motions with two-dimensional space-like transitivity surfaces. We take the interval in the form

$$\begin{aligned} ds^2 = \mathrm{e}^{-M} du\, dv - \mathrm{e}^{-U}[&\mathrm{e}^{V} \cosh W\, dx^2 \\ &+ \mathrm{e}^{-V} \cosh W\, dy^2 - 2 \sinh W\, dx\, dy] \ . \end{aligned} \tag{1.3.12}$$

Here the unknown real functions M, U, V, and W depend on u and v alone. The Einstein-Maxwell system of equations remains self-consistent if we set the component Φ_1 of the electromagnetic field tensor equal to zero.

The isotropic surfaces u = const and v = const are characteristic of the hyperbolic system of Einstein-Maxwell equations.

In our case, the metric for the mutually incident waves before their interaction can be conveniently chosen in the form

$$ds^2 = 2\, du\, dv - g_{11}(u)\, dx^2 - 2g_{12}(u)\, dy\, dx - g_{22}(u)\, dy^2 \tag{1.3.13}$$

for one plane wave and

$$ds^2 = 2\, du\, dv - g_{11}(v)\, dx^2 - 2g_{12}(v)\, dy\, dx - g_{22}(v)\, dy^2 \tag{1.3.14}$$

for the other. Then each of the incident waves will have the particular form (1.3.12), in which the dependence on either u or v is absent. In the region of interaction of the waves, the coefficients of the metric (1.3.12) will have an essential dependence on both of the variables u and v.

The vectors of the null tetrad associated with the metric (1.3.12) have, with respect to the coordinates u, v, x, y, the components

$$\begin{aligned} &l^i(0, \mathrm{e}^{M/2}, 0, 0), \quad n^i(\mathrm{e}^{M/2}, 0, 0, 0), \\ &m^i(0, 0, \mathrm{e}^{(U-V)/2} \cosh(W + \mathrm{i}\pi/4), \mathrm{e}^{(U+V)/2} \sinh(W + \mathrm{i}\pi/4)) \ . \end{aligned} \tag{1.3.15}$$

The nonzero rotation coefficients of this tetrad field can be expressed in terms of the coefficients of the metric as follows:

$$\begin{aligned} &\varrho = \tfrac{1}{2}\mathrm{e}^{M/2} U_{,v}, \quad \sigma = -\tfrac{1}{2}\mathrm{e}^{M/2}[V_{,v} \cosh W - \mathrm{i}W_{,v}], \\ &\mu = -\tfrac{1}{2}\mathrm{e}^{M/2} U_{,u}, \quad \lambda = \tfrac{1}{2}\mathrm{e}^{M/2}[V_{,u} \cosh W + \mathrm{i}W_{,u}], \\ &2\varepsilon = -\tfrac{1}{2}\mathrm{e}^{M/2}[\mathrm{i}V_{,v} \sinh W + M_{,v}], \\ &2\gamma = \tfrac{1}{2}\mathrm{e}^{M/2}[-\mathrm{i}V_{,u} \sinh W + M_{,u}] \ . \end{aligned}$$

Instead of the tetrad components Φ_0 and Φ_2 of the electromagnetic field tensor, we introduce the functions $\tilde{\Phi}_0$ and $\tilde{\Phi}_2$ defined by the equations $\tilde{\Phi}_0 = e^{M/2}\Phi_0$, $\tilde{\Phi}_2 = e^{M/2}\Phi_2$. Then (1.1.28, 29) take the form

$$\tilde{\Phi}_{2,v} = -\tfrac{1}{2}(V_{,u}\cosh W + iW_{,u})\tilde{\Phi}_0 + \tfrac{1}{2}(U_{,v} + iV_{,v}\sinh W)\tilde{\Phi}_2 \quad , \tag{1.3.16}$$

$$\tilde{\Phi}_{0,u} = -\tfrac{1}{2}(V_{,v}\cosh W - iW_{,v})\tilde{\Phi}_2 + \tfrac{1}{2}(U_{,u} + iV_{,u}\sinh W)\tilde{\Phi}_0 \quad , \tag{1.3.17}$$

Equations (1.1.30, 31) are satisfied identically, as are Einstein's equations (1.1.61, 62, 66, 67). Equations (1.1.63, 64) are equivalent. Equations (1.1.60, 68, 65, 63) take respectively the forms

$$2U_{,vv} - U_{,v}^2 + 2U_{,v}M_{,v} - W_{,v}^2 - V_{,v}^2\cosh^2 W = \frac{4G}{c^4}\tilde{\Phi}_0\tilde{\Phi}_0^* \quad , \tag{1.3.18}$$

$$2U_{,uu} - U_{,u}^2 + 2U_{,u}M_{,u} - W_{,u}^2 - V_{,u}^2\cosh^2 W = \frac{4G}{c^4}\tilde{\Phi}_2\tilde{\Phi}_2^* \quad , \tag{1.3.19}$$

$$2M_{,uv} + U_{,u}U_{,v} - W_{,u}W_{,v} - V_{,u}V_{,v}\cosh W = 0 \quad , \tag{1.3.20}$$

$$\begin{aligned}&(2V_{,uv} - U_{,u}V_{,v} - U_{,v}V_{,u})\cosh W + 2(V_{,u}W_{,v} + V_{,v}W_{,u})\sinh W\\ &\quad + i[2W_{,uv} - W_{,u}U_{,v} - W_{,v}U_{,u} - 2\sinh W\cosh W V_{,u}V_{,v}]\\ &\quad = \frac{4G}{c^4}\tilde{\Phi}_2\tilde{\Phi}_0^* \quad .\end{aligned} \tag{1.3.21}$$

Taking the component of Einstein's equations $R_{uv} = 0$, we obtain the equation

$$U_{,uv} = U_{,u}U_{,v} \quad , \tag{1.3.22}$$

which has the general solution

$$U = -\ln[f(u) + g(v)] \quad , \tag{1.3.23}$$

where $f(u)$ and $g(v)$ are arbitrary functions of their arguments.

The relations (1.3.16, 17, 21) form a closed system of equations for the functions V, W, $\tilde{\Phi}_0$, and $\tilde{\Phi}_2$, in which U must be replaced by the expression (1.3.23)[6]. Using the resulting solution of this system, the function M can be found by means of (1.3.20).

In what follows, we confine ourselves to the particular case in which $W = 0$. According to (1.3.21), this situation can occur when $\mathrm{Im}\{\Phi_2\Phi_0^*\} = 0$. We shall assume that the forward front of one of the incident waves before the interaction is described by the equation $u = 0$ with $v \leq 0$, while the forward front of the other wave corresponds to $v = 0$ with $u \leq 0$. The instant at which they meet corresponds to the intersection of the characteristics $u = 0$ and $v = 0$. Since (1.3.16–21) have characteristics $u = \text{const}$ and $v = \text{const}$, breaks run along the characteristics $u = 0$, $v \geq 0$ and $v = 0$, $u \geq 0$; these correspond to decay of the

[6] The efforts of many authors have now led to the creation of effective analytic methods of "multiplication" of solutions of this system, i.e., methods of obtaining a countable set of solutions, starting from one known solution. The general solution has been found by the present author (see Sect. 3.3). Recently the interesting results were obtained by Chandrasekhar and Xanthopoulos, Ferrari and Ibanez Halil and Nutku, Garcia, see for references monograph of Griffiths.

break which occurs at the instant when the forward fronts of the waves meet. Thus, we have here a concrete realization of the situation described in a general form in Sect. 1.2.3, where we studied the Cauchy problem with initial data having breaks.

On the characteristic $u = 0$, $v \geq 0$, the breaks in $\tilde{\Phi}_2$ and $V_{,u}$ satisfy the conditions

$$\begin{aligned} [\tilde{\Phi}_2]_{,v} - \tfrac{1}{2}U_{,v}[\tilde{\Phi}_2] &= -\tfrac{1}{2}\tilde{\Phi}_0[V_{,u}], \\ [V_{,u}]_{,v} - \tfrac{1}{2}U_{,v}[V_{,u}] &= \frac{4G}{c^4}\tilde{\Phi}_0^*[\tilde{\Phi}_2] \ , \end{aligned} \tag{1.3.24}$$

where for the coefficients $U_{,v}$ and $\tilde{\Phi}_0$ we must take the values corresponding to the single-wave solution (1.3.13, 14).

Similarly, the breaks in Φ_0 and $V_{,v}$ along the surface $v = 0$, $u \geq 0$ must obey the conditions

$$\begin{aligned} [\tilde{\Phi}_0]_{,u} - \tfrac{1}{2}U_{,u}[\tilde{\Phi}_0] &= -\tfrac{1}{2}\tilde{\Phi}_2[V_{,v}], \\ [V_{,v}]_{,u} - \tfrac{1}{2}U_{,u}[V_{,v}] &= \frac{4G}{c^4}\tilde{\Phi}_2^*[\Phi_0] \ , \end{aligned} \tag{1.3.24'}$$

Now, using the explicit solution (1.3.23) for U, we shall reduce the system (1.3.16, 17, 21) to a simpler form. We shall assume that the derivatives of $f(u)$ and $g(v)$ do not vanish, so that in the region $u > 0$, $v > 0$ it is possible to go over to the new variables f and g, and to replace the functions $\tilde{\Phi}_0$ and $\tilde{\Phi}_2$ by the functions

$$\varphi_0 g'(v) = \tilde{\Phi}_0 e^{-U/2}, \quad \varphi_2 f'(u) = \tilde{\Phi}_2 e^{-U/2} \ .$$

Then the system of equations (1.3.16, 17, 21) takes the form

$$\{V_{,g}(f+g)\}_{,f} + \{V_{,f}(f+g)\}_{,g} = \frac{4G}{c^4}\varphi_2\varphi_0 \ ; \tag{1.3.25}$$

$$\varphi_{2,g} = -\tfrac{1}{2}V_{,f}\varphi_0; \quad \varphi_{0,f} = -\tfrac{1}{2}V_{,g}\varphi_2 \ . \tag{1.3.26}$$

From these last equations it follows that there exist potentials E and B such that

$$\varphi_2 = -e^{-V/2}E_{,f} = e^{V/2}B_{,f}; \quad \varphi_0 = -e^{-V/2}E_{,g} = e^{V/2}B_{,g} \ .$$

Going over now to the variables $f + g = q$, $f - g = t$, the system (1.3.25, 26) is reduced to the form

$$\begin{aligned} (qV_{,q})_{,q} - qV_{,tt} &= 2Gc^{-4}e^V(B_{,q}^2 - B_{,t}^2) = 2Gc^{-4}e^{-V}(E_{,t}^2 - E_{,q}^2), \\ (e^V B_{,q})_{,q} - (e^V B_{,t})_{,t} &= 0, \quad (e^{-V}E_{,q})_{,q} - (e^{-V}E_{,t})_{,t} = 0 \ . \end{aligned} \tag{1.3.27}$$

The system of equations for E and V can be simplified substantially by putting

$$e^V = (1 - \xi^2 - \mathrm{E}^2)q(1+\xi)^{-2}, \quad \sqrt{2G}\,c^{-2}E = \mathrm{E}(\xi+1)^{-1}, \quad \Phi = \xi + i\mathrm{E} \ .$$

In this case, we obtain for Φ the equation

$$\left(\frac{q\Phi_{,q}}{\Phi\Phi^* - 1}\right)_{,q} - \left(\frac{q\Phi_{,t}}{\Phi\Phi^* - 1}\right)_{,t} = 0 \ . \tag{1.3.28}$$

The system (1.3.27) has the t-independent particular solutions

$$qe^V = \frac{A^2}{\cosh^2 a}, \quad E = A \tanh a, \quad a = c_0 \ln(q/q_0) \ ,$$

where A, c_0, and q_0 are arbitrary constants.

Another important particular class of exact solutions of the system (1.3.27) consists of self-similar solutions, which can be sought in the form

$$V = \tilde{V}(\lambda) + \gamma \ln q, \quad E = q^{(1+\gamma)/2}\tilde{E}(\lambda), \quad B = q^{(1-\gamma)/2}\tilde{B}(\lambda) \ .$$

Here γ is an arbitrary constant, and $\lambda \equiv t/q$.

These solutions are analyzed later (see Sect. 1.3.4). Here we simply write the exact solutions for $\gamma = \pm 1$:

$$\begin{aligned} &qe^V = A^2 q^2 \cosh^2(c_0\theta/2), \quad E = Ac_0 q(\sin\theta)/2, \\ &B = A^{-1} \tanh(c_0\theta/2)(\gamma = 1), \\ &qe^V = \frac{A^2}{\cosh^2(c_0\theta/2)}, \quad E = \frac{c_0 A}{2} \tanh\frac{c_0\theta}{2}, \quad B = \frac{c_0}{2A} q \sin\theta(\gamma = -1) \ , \end{aligned}$$

where A and c_0 are arbitrary constants, and $-\lambda = \cos\theta$.

In the solution corresponding to the interaction of two electromagnetic waves having the form of a Heaviside step function, the components Φ_0 and Φ_2 of the electromagnetic field remain constant [1.57]. Before the waves meet, their metric has the form

$$\begin{aligned} &ds^2 = 2\,du\,dv - \cos^2 au(dx^2 + dy^2), \quad u < 0, \quad a^2 = \frac{G|\Phi_2|^2}{c^4} = \text{const}, \\ &ds^2 = 2\,du\,dv - \cos^2 bv(dx^2 + dy^2), \quad v < 0, \quad b^2 = \frac{G|\Phi_0|^2}{c^4} = \text{const} \ . \end{aligned}$$

In the interaction region, the metric has the form

$$ds^2 = 2\,du\,dv - \cos^2(au + bv)\,dx^2 - \cos^2(au - bv)\,dy^2 \ .$$

Other well-known exact solutions describe the interaction of a δ-like gravitational wave with a step-like electromagnetic or neutrino wave, or particular forms of interaction of neutrino waves [1.58, 59]. We note that when plane waves interact there is a mutual focusing of the light rays. In the case in which two gravitational waves interact, the effect of focusing of the rays described in Sect. 1.1 leads to the appearance of a singularity [1.54–56, 60].

1.3.3 Propagation of Weak Electromagnetic and Gravitational Waves in the Field of a Strong Electromagnetic Wave

For short-wave perturbations of the background of a strong electromagnetic plane wave, the situation is quite clear. In this case, we have all those effects which manifest themselves when short waves propagate in a vacuum in the presence of external electromagnetic fields (see Sect. 1.2.4), namely, partial or total mutual conversion of gravitational and electromagnetic waves, rotation of the plane of polarization of a wave with respect to a tetrad undergoing parallel transport, conservation of the total flux of energy in electromagnetic and gravitational waves, and distortion of the background through which short waves propagate.

In the case of plane waves, there is a particularly clear manifestation of the tendency, inherent in general relativity (see Sect. 1.1), for focusing of a pencil of isotropic geodesics without rotation: parallel isotropic geodesics passing through a nonlinear plane wave intersect on some surface, and rays from a momentarily active source which have passed through a plane wave meet at some point or on a space-like curve [1.27]. One might suppose that it would be possible to investigate perturbations near caustics on the basis of linear equations by means of an improved expansion with respect to the inverse frequency, as was done, for example, in Sect. 1.2. However, an exact treatment of the nonlinear interaction of two plane gravitational waves by a number of authors [1.54–56, 60] indicates the appearance of a singularity in the region of interaction of the waves, while the investigation made in [1.57] applies to a particular case and does not allow us to say that the appearance of a singularity is atypical for the case of the nonlinear interaction of electromagnetic waves.

Following [1.53], we shall show that the Einstein-Maxwell equations, linearized with respect to the background of a plane electromagnetic wave with circular polarization, reduce to closed second-order equations for certain combinations of the perturbations of the electromagnetic and gravitational fields, and we shall find an exact solution of these equations for the cases in which weak electromagnetic or gravitational waves with a backward front pass through a strong electromagnetic wave.

The concept of the cosmic background radiation as a combination of plane electromagnetic waves with a Planck distribution of the intensity with respect to the frequency becomes invalid at a high temperature of the radiation. The gravitational and electromagnetic radiation begin to interact with each other, and, as we shall show below, one electromagnetic wave passing through another is partially converted into a gravitational wave. In addition, owing to the weakness of the gravitational interaction, at a high temperature of the radiation a dominant role is played by quantum processes of pair production due to the nonlinear interactions of photons [1.61]. However, the classical gravitational effects described below should significantly modify the usual concepts of a radiation-dominant plasma at superhigh temperatures $T \gtrsim 10^8$ K.

Let us write the equations for small perturbations of the gravitational and electromagnetic fields on the background of a plane electromagnetic wave. In a coordinate system with the metric (1.3.2), the electromagnetic field in a monochromatic electromagnetic wave with frequency ω can be specified by the 4-potential $A_i(\varphi, 0, 0, 0)$, where $\varphi = B(u)(x \cos \omega u + y \sin \omega u)$. The tetrad component Φ_2 of the electromagnetic field tensor then has the form

$$\Phi_2 = -F_{ij} n^i m^{*j} = -2^{-1/2} B(u) \exp(-\mathrm{i}\omega u) \ , \tag{1.3.29}$$

where $B(u)$ is a real function. According to (1.3.3), the coefficient A in the metric (1.3.2) has the value

$$A(u, x, y) = \frac{G}{2c^4} B^2(u)(x^2 + y^2) \ . \tag{1.3.30}$$

In the tetrad field (1.3.4), space-time with the metric (1.3.2) will have only the following nonzero characteristics:

$$\nu = 2^{-3/2}(A_{,x} - \mathrm{i}A_{,y}) = \frac{GB^2(u)}{2\sqrt{2}\, c^4}(x - \mathrm{i}y) \ , \tag{1.3.31}$$

$$\psi_4 = \delta^* \nu = 2^{-2}(A_{,xx} - A_{,yy} - 2\mathrm{i}A_{,xy}) = 0 \ , \tag{1.3.32}$$

$$\Phi_{22} = 2^{-2}(A_{,xx} + A_{,yy}) = \frac{G}{2c^4} B^2(u) \ . \tag{1.3.33}$$

We shall retain the same symbols, but with primes, to denote perturbations of the characteristics of the field. Assuming that the perturbations of all the quantities are small and linearizing Maxwell's equations (1.1.30, 31), the Bianchi identities (1.1.39, 40), and the definition of ψ_0 given by (1.1.52), we obtain

$$D\Phi_1' - \delta^* \Phi_0' = -k' \Phi_2, \quad \delta\Phi_1' - \Delta\Phi_0' = -\sigma' \Phi_2 \ , \tag{1.3.34}$$

$$-D\psi_1' + \delta^* \psi_0' = -D\Phi_{01}' + \delta\Phi_{00}', \quad \Delta\psi_0' - \delta\psi_1' = -D\Phi_{02}' + \delta\Phi_{01}' \ , \tag{1.3.35}$$

$$\psi_0' = D\sigma' - \delta k' \ , \tag{1.3.36}$$

where

$$\Phi_{00}' = -\kappa T_{11}, \quad 2\Phi_{01}' = -\kappa T_{13}, \quad \Phi_{02}' = \kappa\Phi_0' \Phi_2^*/(8\pi) - \kappa T_{33}/2 \ ,$$

and T_{ij} is the energy-momentum tensor of the matter (excluding the energy-momentum tensor of the electromagnetic field).

After eliminating the perturbation Φ_1' from (1.3.34) and the perturbation ψ_1' from (1.3.35) and making use of (1.3.36), we obtain the closed system of equations

$$\Box\Phi_0' = -\sqrt{2}\, B(u) \exp(-\mathrm{i}u\omega)\psi_0', \quad \Box = 2(D\Delta - \delta\delta^*) \ , \tag{1.3.37}$$

$$\Box\psi_0' = -\frac{B(u)\sqrt{2}\,G}{c^4}e^{iu\omega}D^2\Phi_0' + \kappa L,$$
$$L \equiv D^2T_{33} + \delta^2T_{11} - 2D\delta T_{13} \ , \tag{1.3.38}$$

where $\Box$ is the d'Alembertian operator constructed from the metric (1.3.2) with the coefficient A determined by (1.3.30).

If we make a Fourier transformation with respect to the variable α [see (1.3.2)], on which the coefficients in (1.3.37, 38) do not depend, the d'Alembertian operator for the Fourier transforms ($\tilde{f}(k) = \int_{-\infty}^{+\infty}\exp(-ik\alpha)f(\alpha)\,d\alpha$) takes the form

$$-\Box = 2ik\frac{\partial}{\partial u} - b^2(u)r^2k^2 + \frac{\partial^2}{\partial x^2} + \frac{\partial^2}{\partial y^2},$$
$$b(u) \equiv \sqrt{G/(2c^4)}\,B(u) \ . \tag{1.3.39}$$

In the region in which the function $B(u)$ is nonzero and equal to a constant, it is convenient to go over to new variables D_a ($a = 1,2$) according to the relations

$$D_a = \sqrt{\frac{4G}{c^4}}\,e^{iu\omega}\tilde{\Phi}_0' + \frac{\eta_a}{k}\tilde{\psi}_0' \ , \tag{1.3.40}$$

$$b\eta_{1,2} = \omega \pm \sqrt{\omega^2 + 4b^2} \ . \tag{1.3.41}$$

After this, the system of equations (1.3.37, 38), written for the Fourier transforms $\tilde{\Phi}_0'$ and $\tilde{\psi}_0'$, decomposes into two independent equations:

$$\left(2ik\frac{\partial}{\partial u} - b^2r^2k^2 - \frac{4kb}{\eta_a}\frac{\partial^2}{\partial x^2} + \frac{\partial^2}{\partial y^2}\right)D_a = \kappa\eta_a\tilde{L} \ . \tag{1.3.42}$$

We note that an analogous situation occurs in the study of wave-like fields in the Nordström-Reissner field, where, as was shown in [1.62], the linearized system of Einstein's and Maxwell's equations also decomposes into independent second-order equations for certain combinations of the small perturbations of the metric and the components of the electromagnetic field tensor (see Chap. 4).

In regions of space in which there is no electromagnetic wave, the function $b(u)$ is equal to zero. In what follows, we shall consider the case in which the function $b(u)$ is piecewise constant, i.e., $b(u) = b =$ const for $0 < u < l$ and $b(u) = 0$ for $u < 0$ and $u > l$.

Interaction of a Nonlinear Electromagnetic Wave with Weak Gravitational and Electromagnetic Waves. Using (1.3.42), we can study the incidence of weak waves with arbitrary geometry on the nonlinear wave (1.3.3). In connection with the difficulties in solving the characteristic Goursat problem for the nonlinear hyperbolic system of equations written earlier in this section, we note that in the linearized formulation using (1.3.42) it is possible to study a larger class of problems for plane waves.

We shall consider two cases.

Case A. Suppose that a strong electromagnetic wave with the space-time metric (1.3.2, 30) is exposed to an incident weak gravitational wave in which the metric has the form

$$ds^2 = 2\,du\,dv + \varepsilon[\theta(v+a) - \theta(v-a)]xy\,dv^2 - dx^2 - dy^2,$$
$$\varepsilon = \text{const}, \quad a = \text{const},$$
$$\theta(v) = -1 \quad \text{for} \quad v < 0, \quad \theta(v) = +1 \quad \text{for} \quad v > 0 \ ,$$

where $\theta(v)$ is the Heaviside function. Simple calculations give for $u < 0$ the following expression for the only nonzero tetrad components of the Weyl tensor:

$$\psi_0 = \tfrac{1}{2}\varepsilon\mathrm{i}[\theta(v+a) - \theta(v-a)] \ .$$

For $0 < u < l$, the weak gravitational wave propagates in the field of the strong electromagnetic wave. When for $u > l$ the weak wave passes through the region of the nonlinear electromagnetic wave, it splits up into a weak gravitational wave and a weak electromagnetic wave. For $u > l$, these two waves propagate independently in the linear formulation.

For $u > l$, the structure of these waves is described by the following exact solution of (1.3.42):

$$\Phi_0' = \sqrt{\frac{c^4}{16G}}\,\varepsilon\eta_2\eta_1\left[\exp\left(-\frac{2\mathrm{i}bl}{\eta_2}\right) - \exp\left(-\frac{2\mathrm{i}bl}{\eta_1}\right)\right]\frac{1}{\Delta}\int\limits_{-\infty}^{v}\psi\,dv \ , \qquad (1.3.43)$$

where $\psi = 0$ for $|v - r^2C(u)| \geq a$, $\psi = 1$ for $|v - r^2C(u)| \leq a$, $C(u) = [2(u-l) - b^{-1}\cot(bl/2)]^{-1}$, $\Delta = (\eta_1 - \eta_2)[\cos(bl/2) - 2b(u-1)\sin(bl/2)]$, and η_1 and η_2 are related to the amplitude and frequency of the strong electromagnetic wave by (1.3.41).

The perturbation of the tetrad component ψ_0' of the Weyl tensor for $u > l$ is determined by the equation

$$\psi_0' = \frac{\mathrm{i}\varepsilon}{2\Delta}\left[\eta_2\exp\left(-\frac{2\mathrm{i}bl}{\eta_2}\right) - \eta_1\exp\left(-\frac{2\mathrm{i}bl}{\eta_1}\right)\right] \ . \qquad (1.3.44)$$

Case B. We assume now that a strong electromagnetic wave is exposed to a weak electromagnetic wave, which for $u < 0$ has only a single nonzero tetrad component

$$\Phi_0' = \frac{A}{\sqrt{2}}\mathrm{e}^{\mathrm{i}(v\omega_0+\varphi_0)}[\theta(v-a) - \theta(v+a)],$$
$$\omega_0 = \text{const}, \quad \varphi_0 = \text{const}, \quad a = \text{const}, \quad A = \text{const} \ .$$

In [1.53] it was shown, using the technique of integral transformations, that a gravitational wave also appears for $u > l$. With Ψ defined by (1.3.43) an exact solution of the linearized equations (1.3.42) is given in this case by the expressions

$$\Phi_0' = \frac{A}{\Delta} \exp\{i[\omega_0(v - r^2 C(u) - l) + \varphi_0]\}$$
$$\times \left[\eta_2 \exp\left(-\frac{2ibl}{\eta_2}\right) - \eta_1 \exp\left(-\frac{2ibl}{\eta_1}\right)\right] , \tag{1.3.45}$$

$$\psi_0' = \sqrt{\frac{2G}{c^4}} \frac{A}{\Delta} \exp[i\omega_0 v + i\varphi_0] \left[\exp\left(-\frac{2ibl}{\eta_1}\right) - \exp\left(-\frac{2ibl}{\eta_2}\right)\right]$$
$$\times \{ie^{i\omega_0 a}\delta(v - r^2 C - a) - ie^{-i\omega_0 a}\delta(v - r^2 C + a)$$
$$- \omega_0 e^{(-i\omega_0 r^2 C)}\Psi\} . \tag{1.3.46}$$

For the solutions (1.3.43–46), the surfaces $S_\pm : 0 = v - r^2 C(u) \pm a$ are characteristic surfaces, since $g^{ij}\partial_i S \partial_j S = 0$ on them. The gravitational and electromagnetic fields have breaks on these surfaces.

In the solution (1.3.43, 44) (Case A), the gravitational field has a second-order break on the indicated surfaces, while the electromagnetic wave which appears is continuous. In accordance with the general theory of breaks in the gravitational field (see Sect. 1.2.2), the intensity of the break $[\psi_0]$ satisfies the continuity equation $\nabla_i ([\psi_0]^2 g^{ij}\partial_j S) = 0$.

When a weak electromagnetic wave is incident on a strong electromagnetic wave (Case B), the gravitational field has first-order breaks on the surfaces $S_\pm = 0$ (the components of the electromagnetic field tensor have discontinuities, while the nonzero component of the Weyl tensor has a δ-function singularity). The structure of such solutions can be determined in accordance with the theory developed in Sect. 1.2. It is worth mentioning the review by Griffiths (in preparation) on the colliding plane-wave problem and new related results like the Nutku-Halil solution, the set of solutions by Ibanez, various families of solutions by Chandrasekhar and Xanthopoulos, by Ernst, Garcia and Hauser.

1.3.4 Oscillatory Character of Solutions Near a Singularity

For a self-similar interaction of electromagnetic waves in general relativity (due to the mutual conversion of electromagnetic and gravitational waves) the oscillatory behavior of the solutions in the vicinity of a singularity is discussed.

In Sect. 1.3.2 we pointed out that the system (1.3.2) has the self-similar solution

$$V = \gamma \ln q + V(\lambda), \quad \Phi = q^{(1-\gamma)/2} 2\sqrt{G}\, c^{-2}\varphi(\lambda), \quad \lambda = t/q .$$

If these expressions are substituted into (1.3.27), we obtain

$$\begin{aligned} &V''(\lambda^2 - 1) + \lambda V' \\ &\quad = 2e^V[(1-\gamma)^2\varphi^2/4 - (1-\gamma)\lambda\varphi\varphi' + (\lambda^2 - 1)\varphi'^2], \\ &\varphi''(\lambda^2 - 1) + \lambda\varphi' \\ &\quad = \varphi(\gamma - 1)^2/4 + V'((1-\gamma)\lambda\varphi/2 - (\lambda^2 - 1)\varphi') , \end{aligned} \tag{1.3.47}$$

where the primes denote derivatives with respect to λ. We note that this system of equations remains unchanged if V is replaced by $V+B$, and φ by $\varphi\exp(-B/2)$. Therefore an arbitrary constant in the expression for V is not essential.

We now go over to the new variable θ defined by the equation $-\cos\theta = \lambda$ $(0 < \theta < \pi)$. Then the system (1.3.47) can be rewritten in the form

$$\begin{aligned} &\ddot{V} + 2e^{V}[\varphi^2(1-\gamma)^2/4 + (1-\gamma)\cot\theta\varphi\dot{\varphi} - \dot{\varphi}^2] = 0, \\ &\ddot{\varphi} + (\gamma-1)^2\varphi/4 + (\gamma-1)\varphi\dot{V}(\cot\theta)/2 + \dot{V}\dot{\varphi} = 0 \ , \end{aligned} \tag{1.3.48}$$

where the dots indicate derivatives with respect to θ.

It is a remarkable fact that this system has a first integral, which describes the conservation of the intensity of the electromagnetic and gravitational waves during the process of their mutual conversion

$$\dot{V}^2 + e^{V}(\varphi^2(\gamma-1)^2 + 4\dot{\varphi}^2) = c_0^2 = \text{const} \ . \tag{1.3.49}$$

By means of this integral, it is possible to reduce the order of the system (1.3.48).

After making the substitutions

$$2(\ln\varphi)^{\cdot} = (1-\gamma)\cot\mu, \quad \dot{V} = c_0\cos\chi, \quad e^{V/2}\varphi(1-\gamma) = c_0\sin\chi\sin\mu$$

the system (1.3.48) can be rewritten in the form

$$\begin{aligned} 2\sin\theta\dot{\chi} &= c_0\sin\chi\sin(2\mu-\theta), \\ \sin\theta(2\dot{\mu}-1+\gamma) &= c_0\cos\chi[\cos(2\mu-\theta)-\cos\theta] \ . \end{aligned} \tag{1.3.50}$$

If we introduce the new dependent and independent variables $\Omega \equiv 2\mu - \theta$ and $x \equiv \ln\cot(\theta/2)$, for $\gamma = 0$ we obtain from (1.3.50)

$$-2\,d\chi/dx = c_0\sin\chi\sin\Omega \ , \tag{1.3.51}$$

$$-d\Omega/dx = c_0\cos\chi[\cos\Omega - (e^{2x}-1)/(e^{2x}+1)] \ . \tag{1.3.52}$$

In the limit $x \to \infty$, (1.3.52) takes the form

$$-d\Omega/dx = c_0\cos\chi(\cos\Omega - 1) + 2c_0e^{-2x}\cos\chi\, O(e^{-4x}) \ ,$$

in which, in the zeroth order approximation, we can neglect the terms of order $\exp(-2x)$. Then the solution of the asymptotic system (1.3.51,52) for $\lambda \to -1$ (or $x \to +\infty$) has the form

$$\begin{aligned} &\cos\chi = \sqrt{1-c_1^2}\cos(c_0c_1x+\beta), \quad \sin\chi\sin(\Omega/2) = c_1, \\ &\cos\Omega = \frac{(1-c_1^2)\sin^2(c_0c_1x+\beta) - c_1^2}{1-(1-c_1^2)\cos^2(c_0c_1x+\beta)} \ , \end{aligned} \tag{1.3.53}$$

where c_1 and β are arbitrary constants of integration, with $|c_1| < 1$.

This solution corresponds to an infinitely recurring nonlinear mutual conversion of the gravitational and electromagnetic waves as the singular point $\lambda = -1$ is approached. The function $\dot{V}$ oscillates around zero with amplitude $c_0\sqrt{1-c_1^2}$.

The corrections to the leading term in (1.3.53) for $\lambda \to -1$ are of order $\exp(-2x)$.

Similarly, the nonlinear mutual conversion of gravitational and electromagnetic waves is described asymptotically for $\lambda \to 1$ by the equation

$$\cos\chi = \sqrt{1-c_1^2}\cos[c_0\tilde{c}_1 x + \tilde{\beta}], \quad \tilde{c}_1 = \mathrm{const}, \quad \tilde{\beta} = \mathrm{const} \ .$$

Besides the general solution in which oscillations occur for both $\lambda \to +1$ and $\lambda \to -1$, the system (1.3.51, 52) has special one-parameter solutions which oscillate only for $\lambda \to +1$ and have a nonoscillating asymptotic behavior for $\lambda \to -1$, and vice versa.

The system (1.3.47) also has exact solutions corresponding to a purely electromagnetic wave:

$$\varphi = \mathrm{const}\sqrt{1 \pm \lambda}, \quad V = \mathrm{const} \ .$$

However, these solutions cannot be interpreted as interacting waves, since in them either Φ_0 or Φ_2 is zero.

We shall establish the form which the plane electromagnetic waves described in the interaction region by the self-similar solution must have before the interaction.

The tetrad components of the bivector of the electromagnetic field corresponding to the self-similar solution have the form

$$\Phi_0 = \mathrm{e}^{V/2}\frac{g'(v)}{f(u)+g(v)}\left(\frac{\varphi}{2} - (1+\lambda)\varphi'\right),$$
$$\Phi_2 = \mathrm{e}^{V/2}\frac{f'(u)}{f(u)+g(v)}\left(\frac{\varphi}{2} + (1-\lambda)\varphi'\right) \ .$$

In the region before the interaction, for the wave incident from the left we have $g(v) = 1/2$ and hence $\lambda = (2f(u)-1)/(2f(u)+1)$. For this wave, $\Phi_0 = M = 0$ [see (1.3.12)]. Of Einstein's equations, only (1.3.19) is not satisfied identically:

$$2U_{,uu} - U_{,u}^2 = V_{,u}^2 + 4G\Phi_2^2/c^4 \ .$$

In the self-similar case, it follows from this equation that

$$2u''(1-\lambda) = u'[3 + V'^2(1-\lambda)^2 + 2\mathrm{e}^V(\varphi/2 + (1-\lambda)\varphi')^2] \ . \qquad (1.3.54)$$

In this equation, $V(\lambda)$ and $\varphi(\lambda)$ must be replaced by the solutions of the system (1.3.47); this follows from the continuity of the functions $V(\lambda)$ and $\varphi(\lambda)$ on the characteristic $v = 0$. Equation (1.3.54) can be used to determine u as a function of λ by quadratures. After inversion of the function $u = u(\lambda)$, we find the required function $f(u) = (1+\lambda)/[2(1-\lambda)]$. Similar calculations can be made for the wave incident from the right.

Extensive numerical calculations carried out by L.Yu. Blazhennova-Mikulich have shown that the functions $f(u)$ and $g(v)$ decrease monotonically with increasing u and v, becoming equal to zero at finite values of u and v.

Finally, we note that the function $M(u, v)$ is nonzero in the region of interaction of the waves. Knowing V, the function M must be found from (1.3.20), which can be written in the form

$$2(M_{,qq} - M_{,tt}) + 1/q^2 = V_{,q}^2 - V_{,t}^2,$$
$$q = f(u) + g(v),$$
$$t = f(u) - g(v) \quad .$$

In the self-similar case, this equation reduces to

$$2\ddot{M} = \dot{V} + 1 = c_0^2 \cos^2 \chi + 1 \quad .$$

Hence, taking into account the boundary conditions $M = 0$ at $\theta = 0$ and $\theta = \pi$, we find that

$$M = \int_0^\theta (\theta - \theta_1)[c_0^2 \cos^2 \chi(\theta_1) + 1]\, d\theta_1 - \frac{\theta}{\pi} \int_0^\pi (\pi - \theta_1)[c_0^2 \cos^2 \chi(\theta_1) + 1]\, d\theta_1$$

1.4 Conditions on Surfaces with Strong Breaks in Theories of the Gravitational Field

At the present time, in both theory and practice, complicated models are introduced to describe continuous media, for which the parameters specifying the states of elementary volumes include, for example, the characteristics of deformations, the electromagnetic field, the composition of mixtures, the structure of individual molecules and molecular aggregates, etc. The introduction of such auxiliary parameters for a medium entails the introduction of new equations, which have the character of kinetic equations or relations similar to equations of state. These equations must be added to the system of universal conservation laws. In order to construct models and formulate problems in terms of these models, it is necessary to find the conditions which hold at strong breaks inside a region occupied by a medium and boundary conditions at the boundaries of this region.

Sedov[7] and representatives of his school [1.15, 63–67] have developed unified general regular methods of deriving closed systems of equations for specific models and conditions on the discontinuities, using the basis variational equation

---

[7] In writing this section, we relied heavily on the paper by *Sedov* [1.15].

$$\delta \int_{V_4} \Lambda \, dV + \delta W^* + \delta W = 0 \ , \qquad (1.4.1)$$

where Λ is a given Lagrange function, whose arguments are the parameters describing the state of a small macroscopic element of the medium and their derivatives with respect to the time and the coordinates. The relation of the basis equation (1.4.1) to Lagrange's classical variational principle and to the complete energy equation facilitates the construction of the Lagrange function and the functional δW^*. In particular, facts like this enable us to be guided by the methods and results of the thermodynamics of irreversible processes.

The presence of δW^* in (1.4.1) is due to the occurrence of irreversible effects in the interaction of a given element of the medium with the surrounding elements, as is taken into account by the nonholonomic character of the volume integral in δW^*. A given functional δW^* which is linear in the variations of the parameters may contain a volume integral over V_4, an integral over the surface Σ which bounds the four-dimensional volume V_4, and surface integrals over both sides of the hypersurface S_3 of the break, contained in V_4. The variations of the parameters on Σ and S_3 in (1.4.1) are nonzero.

The additional term δW appears in (1.4.1) in order to compensate the corresponding surface integrals in δW^* and in the integrals which occur after the variation of the Lagrange function is integrated by parts.

The invariant formulation of the basis equation (1.4.1) makes it possible to apply it to various relativistic theories of the gravitational field (see, for example, [1.52, 68], which rely on the concept of a space with torsion[8]). In spite of the self-consistency of the general theory of relativity, the absence of reliable experimental data make it impossible to ignore other theories of gravity which lead to the Newtonian law of universal gravitation in the nonrelativistic limit. From the mathematical standpoint, general relativity is the most highly developed theory of the gravitational field in which many exact solutions describing the properties of this field have been studied theoretically and interpreted. However, in the weak-field approximation, which alone is accessible to physical experiment, many effects of general relativity are also predicted by other theories of the gravitational field, in particular, by the above-mentioned.

In Riemannian spaces, the components of the metric tensor can be regarded as unknown parameters characterizing the intrinsic degrees of freedom of space-time. This raises the question of the actual selection of certain coordinate systems by means of various conditions and constructions, in particular, by means of special admissible assumptions about the functional form of the components of the metric tensor.

The definition of a comoving coordinate system involves particularization of the points of some material or, in general, mentally defined medium. The indi-

[8] Logunov and his school (see, for example, [1.69]) are developing a theory of the gravitational field by postulating a pseudo-Euclidean character of the physical space-time in order to define an energy-momentum tensor of the gravitational field.

vidual points of a material medium can be fixed by means of three Lagrangian coordinates ξ^1, ξ^2, ξ^3. The time-like coordinate line ξ^4 coincides with the world line of a fluid element. The coordinate system x^i of an observer in a nonsymmetric Riemannian space can be introduced by means of purely geometrical constructions; for example, a unique coordinate system can be defined on some chart by taking the coordinates to be the four invariants of the Weyl tensor.

A coordinate transformation $x^i = f^i(\xi^1, \xi^2, \xi^3, \xi^4)$ represents nothing other than the law of motion of a given medium with respect to the coordinate system of the observer. If the functions $g_{ij}(x^k)$ and $\hat{g}_{pq}(\xi^k)$ are known, the four functions f^i can be found from the tensor rule of transformation of the metric coefficients:

$$g_{ij}(x^k)\frac{\partial x^i}{\partial \xi^p}\frac{\partial x^j}{\partial \xi^q} = \hat{g}_{pq}(\xi^t) \quad . \tag{1.4.2}$$

Consequently, if equations for determining the metric of the comoving coordinate system can be formulated independently of the system of the observer, the law of motion $x^i = f^i(\xi^k)$ can be found from (1.4.2). These arguments enable us to understand why the law of motion follows in this case from the field equations, in the same way that the equations for the momentum and energy are conditions for compatibility of Einstein's equations. If the geometrical properties of space are known and simple, the equations which determine the law of motion of a medium may not be so simple, since they are not simple consequences of the metric properties of space, although these properties leave a clear imprint on the nature and form of these equations.

Bearing in mind what we have said above, we now consider models in which Λ and δW^* have the form

$$\Lambda = \Lambda(R, g_{ij}, \mu^A, \nabla_k \mu^A, x^i_k, K_B), \quad x^i_k \equiv \partial x^i/\partial \xi^k,$$

$$\delta W^* = -\int\limits_{V_4} M_A \delta\mu^A dV \quad ,$$

in which $g_{ij}(x^k)$, $\mu^A(x^k)$, $x^i(\xi^k)$ are unknown functions, μ^A being thermodynamical parameters characterizing the state of the medium, and $K_B(\xi^k)$ are known functions such as generalized physical constants.

We shall use the notation δA for the variation of the function A at fixed Lagrangian coordinates, and ∂A for the variation of A at fixed coordinates of the observer. It can be verified that these variations are related by the rule

$$\delta\mu^A = \partial\mu^A + \delta x^k \nabla_k \mu^A \quad .$$

For the components of the metric tensor, the two variations are identical. The variation of the element of volume is

$$\delta\, dV = (\partial(\ln\sqrt{-g}) + \nabla_l\, \delta x^l)\, dV \quad .$$

For the variation of the scalar curvature, we have the equations

$$\partial R = -R^{ij}\partial g_{ij} + \nabla_l W^l, \quad W^l = (g^{ik}g^{lj} - g^{ij}g^{lk})\nabla_k \partial g_{ij} \ .$$

Taking the variation in (1.4.1) on the basis of the foregoing equations, by considering the volume integral and equating the coefficients of the independent variations we obtain the following Euler equations: for ∂g_{ij},

$$\begin{aligned} &- R^{ij}\partial\Lambda/\partial R + g^{ij}\Lambda/2 + \partial\Lambda/\partial g_{ij} - \nabla_q B^{(ij)q} \\ &- (g^{ij}g^{qk} - g^{iq}g^{jk})\nabla_q \nabla_k \partial\Lambda/\partial R = 0 \ , \end{aligned} \tag{1.4.3}$$

with

$$B^{ijq} = \left(\frac{\partial\Lambda}{\partial\nabla_i \mu^A}F^{A[q}_{Bs}g^{j]s} + \frac{1}{2}\frac{\partial\Lambda}{\partial\nabla_q \mu^A}F^{Ai}_{Bs}g^{js}\right)\mu^B \ ; \tag{1.4.4}$$

for δx^i,

$$\nabla_s\left(\frac{\partial\Lambda}{\partial x^i_j}x^s_j\right) + \frac{\partial\Lambda}{\partial x^s_j}\nabla_i x^s_j + \frac{\partial\Lambda}{\partial K_B}\nabla_i K_B + M_A\nabla_i\mu^A = 0 \ ; \tag{1.4.5}$$

for $\delta\mu^A$,

$$\frac{\partial\Lambda}{\partial\mu^A} - \nabla_l\frac{\partial\Lambda}{\partial\nabla_l \mu^A} = M_A \ . \tag{1.4.6}$$

In (1.4.4) we have introduced the notation F^{Aj}_{Bs} for the corresponding products of Kronecker delta symbols δ^p_q, depending on the structure of the indices A in the tensor components of μ^A:

$$\nabla_i \mu^A \equiv \frac{\partial\mu^A}{\partial x^i} + F^{Aj}_{Bs}\Gamma^s_{ij}\mu^B \ .$$

The relations (1.4.3) are generalized equations for the gravitational field. If the function Λ depends nonlinearly on R, these equations contain fourth-order derivatives of the components of the metric tensor.

After taking the variation of Λ and integrating by parts, we obtain for δW an integral which extends over the three-dimensional hypersurface Σ which bounds the four-dimensional volume V_4, and also over each of the two sides of the hypersurface S_3 of the strong break. The required relations on the hypersurface of the strong break can be obtained by taking into account the arbitrariness of the volume V_4 and of the continuous (on S_3) functions δx^i and variations ∂g_{ij}, $\partial(\partial g_{ij}/\partial n)$, $\delta\mu^A$, and δx^i. In particular, δW is given by the expression

$$\begin{aligned} \delta W = &- \int\limits_{\Sigma + S_3^\pm} \Bigg\{\left[B^{(ij)k} + q^{ikjl}\nabla_l\left(\frac{\partial\Lambda}{\partial R}\right)\right]\partial g_{ij} - \frac{\partial\Lambda}{\partial R}q^{ikjl}\nabla_l(\partial g_{ij}) \\ &+ \frac{\partial\Lambda}{\partial(\nabla_k \mu^A)}\delta\mu^A + \left(\Lambda\delta^k_i + \frac{\partial\Lambda}{\partial x^i_s}x^k_s - \frac{\partial\Lambda}{\partial\nabla_k \mu^A}\nabla_i\mu^A\right)\delta x^i\Bigg\} n_k\, d\sigma; \\ & q_{ikjl} = g_{ij}g_{kl} - g_{ik}g_{jl} \ . \end{aligned} \tag{1.4.7}$$

Since $\delta\mu^A$ and δx^i in (1.4.7) are arbitrary, we have

$$[N_A^k n_k \sqrt{\mathcal{G}}] = 0, \quad [P_i^k n_k \sqrt{\mathcal{G}}] = 0 \quad ,$$

where N_A^k denotes the coefficient of $\delta\mu^A$ and P_k^i denotes the coefficient of δx^i in the surface integral for δW. The tensor P_i^k is the energy-momentum tensor of the continuous medium. The square brackets indicate the difference between the corresponding quantities on the two sides of the surface of the break. The continuity condition for $N_A^k n_k \sqrt{\mathcal{G}}$ at the break is a new condition which supplements the usual continuity conditions for the flux of energy and momentum at the break. The coefficients of $\delta\mu^A$ and δx^i are also continuous at a break in the absence of a gravitational field in the special theory of relativity.

An exposition of the main results obtained by applying the basis variational equation to various models of continuous media with no gravitational field, including both new models and those that are already used in theory and practice, can be found in the papers and textbook by *Sedov* [1.15, 65–67, 70, 71]. Here, following [1.15], we shall confine ourselves to a discussion of the new conditions on the discontinuities in the gravitational field which follow from the requirement that the variations ∂g_{ij} and their derivatives $\partial(\partial g_{ij})/\partial n$ along the normal to the surface of the break must be continuous.

For this purpose, we separate explicitly the normal derivative of ∂g_{ij} in the second term of (1.4.7) and make use of a special coordinate system, assuming that the normal to the surface of the break is space-like. We take the coordinate line x^1 along the normal to the surface S_3, and choose the remaining coordinate lines on the surface S_3. In such a coordinate system, the quadratic form at points of S_3 reduces to

$$d\sigma^2 = g_{11}(dx^1)^2 + g_{\alpha\beta}\, dx^\alpha dx^\beta \quad (\alpha, \beta = 2, 3, 4) \quad , \tag{1.4.8}$$

where the coordinates x^2, x^3, x^4 are parameters on the surface S_3. The derivative along the normal to the surface is given by

$$\frac{\partial f}{\partial n} \equiv \frac{1}{\sqrt{-g_{11}}} \frac{\partial f}{\partial x^1} \quad .$$

From the condition that the coefficient of ∂g_{ij} is continuous it follows that

$$[T^{ijk} n_k \sqrt{\mathcal{G}}] = 0 \quad , \tag{1.4.9}$$

where the components of the tensor T^{ijk} in the coordinate system defined by (1.4.8) have the form

$$\begin{aligned}
-T_{11}^1 &= B_{11}^1 + \frac{1}{2} g^{\alpha\beta} \frac{\partial g_{\alpha\beta}}{\partial x^1} \frac{\partial \Lambda}{\partial R} \quad , \\
-T^{1\gamma 1} &= -T^{\gamma 11} = B^{1\gamma 1} - g^{\gamma\beta} \frac{\partial}{\partial x^\beta} \left(\frac{\partial \Lambda}{\partial R} g^{11} \right) \quad , \\
-T^{\alpha\beta}{}_1 &= B^{\alpha\beta}{}_1 + g^{\alpha\beta} \frac{\partial}{\partial x^1} \frac{\partial \Lambda}{\partial R} - \frac{1}{2} \frac{\partial \Lambda}{\partial R} \frac{\partial g^{\alpha\beta}}{\partial x^1} \qquad (\alpha, \beta = 2, 3, 4) \quad .
\end{aligned} \tag{1.4.10}$$

In these expressions, the tensor B^{ijk} is assumed to be symmetrized with respect to the first and second indices.

From the condition that the normal derivative of the variation ∂g_{ij} is continuous it follows that

$$[\partial \Lambda / \partial R g^{\alpha\beta} \sqrt{\mathcal{G}}] = 0 \quad . \tag{1.4.11}$$

Equations (1.4.9, 11) show that at the discontinuities certain conditions must be imposed on the components of the metric tensor and their successive derivatives along the normal. On the surface S_3 the derivatives $\partial x^i / \partial \xi^k$ have breaks in the general case, so that the conditions (1.4.9, 11) depend on the coordinate systems of the observer on both sides of the surface S_3.

Recently, *Sedov* [1.67] gave relations between the algebraic properties of the Weyl tensor and the dynamical properties of the gravitational field.

2. The Classical Theory of Black Holes

Observable Manifestations of the X-ray Source Cygnus X−1. Sources of x-ray emission in our galaxy were discovered in June 1962 by means of a high altitude rocket containing special equipment for detecting the flux of x rays at a wavelength of 3 Å [2.1]. A systematic study of x-ray sources was begun after the launch on 12 December 1970 of the artificial satellite Uhuru, which carried an x-ray telescope. The third Uhuru catalog lists 161 discrete x-ray sources, starting with the brightest source Sco X−1 in the constellation Scorpius, with intensity 3×10^{-7} erg/cm^2, and ending with sources 10^4 times less bright. The luminosities of the x-ray sources lie in the range 10^{35}–10^{38} erg/s [2.2]. After careful analysis of the data, it was found that only the radiation from the x-ray source Cyg X−1 in the constellation Cygnus had characteristics consistent with theoretical ideas about "black holes". The radiation from Cygnus X−1 exhibits no periodicity at any time scale between 0.1 s and several days. It was established that fluctuations in the radiation occur both at relatively low energies (2–5 keV) and at high energies (5–12 keV), but the typical pulsations are stronger at high energies. What was most intriguing was the existence of bursts on the scale of milliseconds [2.2, 3]. This observed variability of the radiation indicates that the region from which it emanates is compact (an upper limit on the characteristic size of this region is obtained as the product of a millisecond and the speed of light, i.e., it is less than 300 km).

Extensive data from both optical and radio observations of the object Cyg X−1 have now been collected. It is interesting to note that the emission from Cygnus X−1 in the range of radio waves was not detected until its discovery at the end of March 1971 [2.2], when it was established that the x-ray emission with its accompanying radio emission was subject to variation; to be specific, the flux of energy at relatively low frequencies became four times weaker, while in the frequency range 10–20 keV it rose by a factor of two. The disappearance of the low-frequency component of the radiation is evidently due to a decrease of the plasma density.

The source Cyg X−1 has been identified with the optical star HDE 226868, which, judging by its emission spectrum, is a hot star with a surface temperature of about 25 000 K. The period of this binary system is 5.6 days. As is well known, from spectral observations of a single component of a binary system it is possible to obtain only the so-called mass function $f(M_X)$, whose value for Cyg X−1 was found to be approximately $0.2M_\odot$. [We recall that $f(M_X) = M_X^3 \sin^3 \gamma/(M_O + M_X)^2$, where M_X is the mass of the x-ray source, M_O is the mass of the optical star, and the angle γ is the inclination of the plane of the orbital motion of the binary star with respect to the line of sight.] Judging by its surface temperature, the optical star HDE 226868 must be a blue supergiant and must have mass $(20–30)M_\odot$. Calculations of M_X at a fixed value of M_O give a decrease of M_X with increasing angle γ. However, even at moderate angles $\gamma \sim 60°$ with $M_O \sim 10M_\odot$ a value $M_X \sim 4M_\odot$ is obtained for the mass of the x-ray source; if $\gamma = 90°$ and $M_O = 10M_\odot$, a value somewhat greater than three solar masses is obtained for all M_X.

Thus, the proof that the source Cygnus X−1 has a mass greater than three solar masses is based on the assertion that the optical companion is a typical blue supergiant with a mass exceeding ten solar masses. If it is assumed that a supergiant with mass $10M_\odot$ is situated at a distance 0.5 kpc from the Earth, its luminosity will correspond to the observed luminosity of the star HDE 226868. However, the distance to this star, measured on the basis of the reddening of the emission in the dust near the galactic plane, is approximately 1.5–2.5 kpc; i.e., the star HDE 226868 must be a supergiant with even greater mass $M_O \sim 20M_\odot$.

Thus, the occurrence of the millisecond variations in the radiation from Cygnus X−1 leaves no doubt that it is a compact object. The mass of this star cannot be less than three solar masses, and its emission in the optical range is weak [2.4]. What can we conclude from this? We shall describe qualitatively the contemporary ideas about the evolution of stars with large mass [2.5–10].

Stages in the Evolution of Stars with Large Mass. Numerous observations indicate that all young clusters of stars in which new stars are still being formed are situated in extended clouds of gas and dust [2.6]. It is therefore highly probable that the stars are formed as a result of gravitational condensation of the gas and dust, which undergo intense cooling during this process. We note also that photographs of these regions clearly reveal a flaky, irregular structure. Owing to the release of gravitational energy during the rapid contraction, a protostar flares up and becomes opaque for radiation. The energy of the contraction cannot be transmitted by radiation from the interior of opaque stars; inside a star, there are turbulent eddies which carry heat from the warmer interior regions to the cooler exterior regions. As a star contracts and its temperature rises, the gas becomes completely ionized, and the convective transport of energy is replaced by radiative transport. The larger the mass of the star, the earlier the radiative mechanism of energy transport begins to operate. In the first stage of their evolution, massive protostars fall in the same region of the mass-luminosity diagram as the red giants. However, the number of protostars in this region of the diagram is small in comparison with the number of red giants, since the characteristic time during which the protostars remain in this region is comparatively short.

As a protostar continues to contract and its temperature rises to eight million degrees, thermonuclear reactions which burn hydrogen into helium flare up within the body of the star (proton-proton reactions in stars with small mass, and reactions of the carbon-nitrogen cycle in massive stars). Stars in this period of evolution lie on the main sequence in the Hertzsprung-Russell diagram. The hydrogen burns more quickly in massive stars; we can say that the ratio of the times for burning of the hydrogen within two stars is equal to the inverse ratio of the squares of their masses. For stars whose mass is not very large, the characteristic burning time can even exceed the lifetime of the universe [$(1\text{–}2) \times 10^{10}$ yr]. Therefore, even in the "oldest" stars with mass less than that of the Sun the hydrogen has not yet had time to burn out, while in stars with mass equal to 15 solar masses the hydrogen burns out after "only" ten million years.

As the hydrogen in a star burns out, an isothermal helium core is formed, this being surrounded by a thin layer of hot hydrogen. The star expands, the temperature of its outer layers drops, the matter of the star becomes more opaque, and convective eddies appear in its outer layers. The star leaves the main sequence of the Hertzsprung-Russell diagram and becomes a red giant.

As the mass of the helium in the central core increases, the core contracts more and more, and its temperature rises. When the temperature crosses the threshold value at which the triple alpha process sets in, burning of helium and production of carbon begin. The concentration of mass then begins to rise more sharply near the center of the star, its radius expands even more, and it becomes a yellow or red supergiant. As the helium burns out in the central core of the star, the abundance of carbon, nitrogen, and oxygen rises. The neutrino luminosity begins to carry away an increasingly significant fraction of the thermonuclear energy, and the processes in the star become explosive.

Further reactions involving production of increasingly heavy elements up to iron nuclei can take place only in stars with sufficiently large mass. In the case of the Sun, the reactions taking place at the end of its evolution will involve helium, and its core will then consist primarily of carbon and nitrogen. Having been deprived of thermonuclear sources of energy, the Sun will contract, after passing through the stage of a red giant, since the strong radiation losses lead to an effective adiabatic index less than 4/3 in the equation of state of the matter[1].

---

[1] Stable equilibrium of a star is impossible with an adiabatic index $\gamma < 4/3$. In fact, the total energy of a star is equal to the sum of the internal and gravitational energies, $E = \int_V \varrho(u - Gm(r)/r)\,dV$, where u is the internal energy density, $m(r)$ is the mass inside a sphere of radius r, and dV is the element of volume. For a state to be stable, the total energy at equilibrium must be a minimum. If we imagine that the star is displaced from equilibrium and expand its energy up to the terms of second order in the increments of the parameters, we obtain for the energy increment E' the expression

$$E' = \int_V \varrho\,dV \left\{ \left(-\frac{p}{\varrho}\frac{dV'}{dV} + \frac{Gm}{3rV}V' \right) + \frac{\gamma}{2}\frac{p}{\varrho}\left(\frac{dV'}{dV}\right)^2 - 2\frac{Gm}{r}\left(\frac{V'}{3V}\right)^2 \right\} ,$$

where V' is the Lagrangian increment of the volume V. The linear terms in the expression for E' drop out, since the ground state is the equilibrium state. For homogeneous perturbations of the form $V' = \nu V$, where ν = const, the second variation of E, equal to E', takes the form $E' = (\nu^2/2) \int (\gamma - 4/3) p\,dV$ if we make use of the virial theorem. The stability condition implies that E' must be positive, i.e., the adiabatic index γ must be greater than 4/3 over most of the volume in which the pressure is not small.

In stars with mass equal to the mass of the Sun, the collapse is stopped by the pressure of the degenerate electron gas, and such a star becomes frozen in the state of a white dwarf. In stars with mass greater than that of the Sun, it is necessary to take into account the neutrino losses more accurately than at present. Nevertheless, on the basis of several calculations of the compression and explosion of stars, the following picture of the subsequent evolution of very massive stars emerges.

Inside a massive star, an iron core is formed, and this is surrounded by layers containing more light elements and fewer heavy elements as the periphery of the star is approached. Such a star therefore contains a rather large quantity of carbon and oxygen, as well as ionized hydrogen in the outer layers. After the formation of the iron core, the effective adiabatic index in the equation of state of the matter, averaged over the star, becomes less than 4/3. This leads to the onset of a catastrophic contraction of the star, during which a strong shock wave is formed. The temperature beyond the front of the wave rises sharply, and there is a nuclear detonation of the unburned elements, especially the oxygen. The star flares up as a supernova and discards its shell, in particular, as a result of absorption of the neutrino radiation in the shell. As it expands, the shell of the supernova becomes involved in the motion of the interstellar gas. At the initial stage of the expansion, the shell moves with speed $(3\text{–}6) \times 10^3$ km/s, which is much greater than the speed of sound. Calculations show that the radiative energy losses of the gas are large when its temperature is about 10^5 degrees. At higher and lower temperatures, the radiation losses can be neglected. In these cases, the well-known strong-explosion solution of *Sedov* [2.11] can be used to describe the discharge of the interstellar gas. By means of a rational choice of the initial energy of the explosion and the density of the interstellar medium, it is possible to bring this solution into agreement with the observed parameters of the remnants of supernova explosions, such as those of the Crab Nebula [2.12].

When a star contracts, its angular speed increases sharply. In this connection, neutron stars rotate with a large angular speed.

After discarding its shell, a star with mass up to $4M_\odot$ becomes either a neutron star or a black hole. In the case of a star with mass between $4M_\odot$ and $15M_\odot$, the explosion caused by the detonation of the oxygen is so strong that an appreciable fraction of the mass of the star is dispersed. Stars with even larger masses (from 15 to 40 solar masses) collapse so rapidly that the detonation does not have time to take effect. Such stars do not manage to discard their shells and "lose weight" in time to avoid the fate of being buried in a black hole.

The small dimensions of the source Cyg X-1 and its large mass for a small optical luminosity can be explained most naturally by the assumption that the x-ray source Cygnus X-1 is a collapsed star which has become frozen for an observer at the pseudo-Euclidean infinity at its Schwarzschild radius.

Some authors (see the references in the reviews of [2.8, 13]) assume that radiation can be generated by a strong magnetic field between two normal stars, which is frozen into the gas and heats it strongly as a result of magnetohydrodynamic turbulence. This picture is strengthened by the analogy with the solar wind, which "extracts" a magnetic field from the Sun [2.14, 15].

A model of disk accretion of gas onto a black hole in the nonrelativistic approximation was proposed by *Sunjaev* and *Shakura* [2.8] and by *Pringle* and *Rees* [2.16], and has been studied by many other authors [2.17–19][2].

The reason for the possible appearance of a disk in the case of Cygnus X–1 is the same as for ordinary close binary systems [2.20]. The blue supergiant HDE 226868, the companion of Cyg X–1, is a highly inflated star, which completely fills its Roche lobe. It is easier for the stellar wind from this star to overcome the force of gravity in the region of the inner Lagrange point, where the gas stream is picked up by the gravitational field of the black hole. Owing to the orbital rotation of the binary system, the gas stream does not fall to the center of the black hole, but spirals around it. This leads to the formation of a disk, in which the particles move in practically circular orbits. Because of turbulent friction, the particles lose angular momentum and slip into orbits closer and closer to the center. The turbulent viscosity leads to strong heating of the gas, which, in fact, had a temperature of 25 000 K at the surface of the supergiant. It can be assumed that all the thermal energy from the turbulent friction and Compton scattering in the inner regions of the disk is radiated.

The plasma currents attain the greatest speeds in the inner regions of the disk, whose dimensions are of the order of 200 km and from which about 80 % of the total radiation is emitted.

Owing to the turbulent friction, the magnetized plasma in the inner regions is quite capable of being heated to a temperature of between 5 and 500 million degrees. The observed x-ray spectrum corresponds to just this range of temperatures.

The foregoing picture of the turbulent motion of hot gas inside a disk can in principle lead to the observed fluctuations in the emission on any time scale (from several milliseconds to several days).

Owing to the thermal instability, hot spots appear on the disk, which rotate in circular Kepler orbits. The directional character of the x-ray emission from the hot spots can be explained by the Compton emission mechanism and by the gravitational focusing effect of a black hole. Radiation from hot spots moving with relativistic speeds at sufficiently small radii is subject to a strong Doppler shift. Therefore the x-ray emission from the hot spots should occur in pulses, with a period between the pulses equal to the period of rotation of a spot on the disk around the black hole [2.21].

In the general case, a star which has exhausted its nuclear fuel and, having a supercritical mass, is doomed to end its evolutionary path in the state of a black hole may be charged. Mechanisms by which a star surrounded by a hot plasma can acquire a charge were proposed in [2.22–25]. Electrostatic fields can appear

[2] The semiempirical theory of disk accretion rests on observations and experiment, since the viscosity of a turbulent magnetized plasma has not yet been calculated by means of a consistent theory.

in stars as a result of polarization of a static plasma in the gravitational field [2.26].

It has been shown [2.27] that the production of electron-positron pairs in strong electrostatic fields has the consequence, according to Schwinger's formula, that the charge of a black hole cannot exceed $G^2M^2m_e^2/(e\hbar c)$. Nevertheless, the existence of even a small charge in a black hole can lead to the appearance of new effects which qualitatively alter the processes of collapse and of emission and propagation of waves in the immediate vicinity of the black hole.

The plausible physical arguments outlined above are aimed at giving an element of realism to the geometrical objects studied in this chapter, which serves as an introduction to the theory of weakly asymptotically simple manifolds. In Sect. 2.1 we introduce the basic definitions and give an account of some general theorems. In Sect. 2.2 we give the elementary facts from the theory of Lie groups and the theory of exterior forms on manifolds which are needed to read Sect. 2.3. In Sect. 2.3 we discuss theorems on the properties of stationary and static weakly asymptotically simple manifolds of general relativity (the theorems of Carter, Hawking, Israel, Lichnerowitz, Papapetrou, and Robinson). In Sect. 2.4 we examine a formal analogy between the properties of a black hole and a certain thermodynamic system. Detailed discussions of the physical properties of black holes can be found in the fundamental monographs by *Chandrasekhar* [1.72], *Gal'tsov* [1.73] and *Novikov* and *Frolov* [1.74].

2.1 Asymptotically Flat Gravitational Fields

In this chapter, we shall study space-time manifolds in the theory of relativity which have a pseudo-Euclidean structure far from the sources, in other words, manifolds which are *asymptotically flat.* With increasing distance from the sources, the scale of variation of the perturbations caused by nonstationarity of the sources becomes smaller than the scale of variation of the background, which tends to become flat quite rapidly. Therefore we can speak of gravitational waves emitted by a bounded material system and impose the condition that there are no convergent waves at infinity.

For asymptotically flat gravitational fields, we have the situation described in Sect. 1.2. Consequently, far from the sources of the perturbation, the Weyl tensor of the perturbed gravitational field has the algebraic structure of a tensor of Petrov type N. *Sachs* [2.28] discovered the property of increasing degree of algebraic degeneracy of the Weyl tensor for waves in asymptotically flat spaces with increasing distance from the sources. In particular, he established that with increasing distance from the sources along any isotropic geodesic $\mathcal{L}$ the Weyl tensor W has the asymptotic structure

$$W = \frac{N}{r} + \frac{\mathrm{III}}{r^2} + \frac{\mathrm{II}}{r^3} + \frac{\mathrm{I}}{r^4} + \frac{\mathrm{I}'}{r^5} \ ,$$

where r is the affine parameter along the geodesic $\mathcal{L}$, measured in the direction from the region of the perturbations, and N, III, II, and I are Weyl tensors of the corresponding Petrov types, which can be reduced to the canonical form in a null tetrad, one of whose real isotropic vectors is directed along the tangent to $\mathcal{L}$. For Weyl tensors of Petrov types II, III, and N, there is a degenerate principal direction (see Sect. 1.1) along the tangent to $\mathcal{L}$. The Weyl tensor I′ has principal directions not coincident with the tangent to $\mathcal{L}$. The mass and angular momentum of a material system in asymptotically flat fields can be found by investigating the asymptotic behavior of the gravitational field far from the system. The loss of mass or energy of the system can be determined from the flux of energy in gravitational waves which propagate along isotropic geodesics at the pseudo-Euclidean infinity.

Penrose proposed a geometric definition of the pseudo-Euclidean infinity as the smooth conformal boundary of the manifold M_4 formed by the "initial" and "final" points of each isotropic geodesic of the manifold M_4.

We shall explain the concept of a conformal boundary (which makes it possible to study infinity locally, as it were, without recourse to awkward asymptotic transitions) for the example of Minkowski space [2.29]. If we go over from the system of spherical coordinates r, θ, φ, t to the coordinates p, q, θ, φ, where $\tan p = ct + r$, $\tan q = ct - r$, the metric takes the form

$$ds^2 = c^2dt^2 - dr^2 - r^2(d\theta^2 + \sin^2\theta\, d\varphi^2) = (1 + \tan^2 p)(1 + \tan^2 q)\, d\tilde{s}^2,$$
$$d\tilde{s}^2 \equiv dp\, dq - \sin^2(p - q)(d\theta^2 + \sin^2\theta\, d\varphi^2)/4 \ .$$

The points of Minkowski space in the space of the coordinates p, q, θ, φ form a finite domain $-\pi/2 \le q \le p \le \pi/2$. The metric $d\tilde{s}^2$ is completely regular on the conformal boundary $p = \pi/2$, $q = -\pi/2$, which consists of isotropic hypersurfaces. Each isotropic geodesic in Minkowski space begins on the boundary $q = -\pi/2$ (of the isotropic infinity of the past, $\mathcal{B}^-$) and ends on the boundary $p = \pi/2$ (of the isotropic infinity of the future, $\mathcal{B}^+$).

The manifold M_4 with the metric ds^2 is said to be *asymptotically simple* if the following conditions hold:

1) M_4 can be embedded in $\tilde{M}_4$ as a manifold with the boundary ∂M_4;

2) there exists on $\tilde{M}_4$ a smooth pseudo-Riemannian metric $d\tilde{s}^2$ such that the metric ds^2 is conformal to the metric $d\tilde{s}^2$, i.e., $d\tilde{s}^2 = \Omega^2 ds^2$;

3) on the boundary ∂M_4 we have $\Omega = 0$, $d\Omega \neq 0$;

4) each isotropic geodesic begins and ends on the boundary ∂M_4;

5) near ∂M_4, Einstein's vacuum equations $R_{ij} = 0$ hold.

Under these conditions, the boundary ∂M_4 is an isotropic hypersurface in the manifold $\tilde{M}_4$, since the scalar curvatures $\tilde{R}$ and R constructed from the metrics $\tilde{g}_{ij}$ and g_{ij} are related by the equation

$$\Omega^2\tilde{R} = R - 6\Omega(\tilde{g}^{ik}\tilde{\nabla}_i\, \tilde{\nabla}_k\, \Omega) + 3\tilde{g}^{ik}\Omega_{,i}\Omega_{,k} \ .$$

Near ∂M_4, we have $R = 0$ according to the last condition. Therefore, in passing to the limit with $\Omega \to 0$, we obtain for Ω an eikonal equation in the metric $\tilde{g}_{ij}$ on the boundary ∂M_4. Asymptotically simple spaces are suitable for formulating the Cauchy problem, since they admit a global Cauchy hypersurface and are topologically equivalent to a Euclidean space. The boundary ∂M_4 consists of two isotropic surfaces $\mathcal{B}^+$ and $\mathcal{B}^-$, each of which has the topology of the product of a sphere and a line [2.29].

In order to study manifolds with a nontrivial topology, we must weaken the requirements (1–5). A manifold M_4 is said to be *asymptotically simple in the weak sense* if there exists an asymptotically simple manifold M_4' whose neighborhoods $\mathcal{B}^+$ and $\mathcal{B}^-$ are isometric with the analogous neighborhoods of M_4.

We recall that for a given point p we have adopted the notation $\mathcal{I}^+(p)$ and $\mathcal{I}^-(p)$ (see the end of Sect. 1.2.1) for the sets of points of the manifold M_4 which can be connected to the point p by time-like or isotropic curves directed from p into the future or into the past, respectively. Then $\mathcal{I}^-(\mathcal{B}^+)$ is the set of points about which an observer at the pseudo-Euclidean infinity $\mathcal{B}^+$ can in principle acquire any information, and $\mathcal{I}^+(\mathcal{B}^-)$ is the set of points which can be reached by a signal from a distant observer.

The boundary of the set $\mathcal{I}^-(\mathcal{B}^+)$, if it is not empty, is called the *event horizon* $H^+ \equiv \partial[\mathcal{I}^-(\mathcal{B}^+)]$.

I) Let $\mathcal{L}$ be a space-like surface, with no boundary, in a manifold which is asymptotically simple in the weak sense. The Cauchy region $D^+(\mathcal{L})$ for the surface $\mathcal{L}$ is defined as the set of points for which each non-space-like curve directed into the past necessarily intersects the surface $\mathcal{L}$ [2.29, 30].

For manifolds which are asymptotically simple in the weak sense, we also assume the validity of the principle of *strong asymptotic predictability*:

1) One of the boundaries of the Cauchy region $D^+(\mathcal{L})$ coincides with the pseudo-Euclidean boundary $\mathcal{B}^+$.

2) The evolution of the events in the neighborhood of the horizon can be predicted by means of Cauchy data on $\mathcal{L}$, for which it is necessary that

$$\mathcal{I}^+(\mathcal{L}) \cap \mathcal{I}^-(\mathcal{B}^+) \subset D^+(\mathcal{L}) \ .$$

It can be shown [2.30] that for an arbitrary space-like surface $\mathcal{L}$ with no boundary it is possible to construct a family of space-like surfaces $\mathcal{L}(\tau)$ homeomorphic to $\mathcal{L}$ and covering the region $D^+(\mathcal{L})$, with the following properties: a) the union of all $\mathcal{L}(\tau)$ ($\mathcal{L}(0) = \mathcal{L}, 0 \le \tau < \infty$) is identical to the region $D^+(\mathcal{L})$; b) the surface $\mathcal{L}(\tau_2)$ lies in the future of the surface $\mathcal{L}(\tau_1)$, i.e., $\mathcal{L}(\tau_2) \subset \mathcal{I}^+(\mathcal{L}(\tau_1))$ for $\tau_2 > \tau_1$; c) for each τ the intersection of $\mathcal{L}(\tau)$ with the pseudo-Euclidean boundary $\mathcal{B}^+$ has the topology of a space-like sphere, and the sphere $\mathcal{L}(\tau_2) \cap \mathcal{B}^+$ lies strictly in the future of the sphere $\mathcal{L}(\tau_1) \cap \mathcal{B}^+$ for $\tau_2 > \tau_1$.

If there exists a nonempty horizon H^+, the surfaces $\mathcal{L}(\tau)$ will intersect H^+ on compact two-dimensional space-like surfaces $L(\tau)$, which represent the boundaries of black holes. The area of a connected component of $L(\tau)$ is called the *surface area of a black hole.*

The most important property of a black hole is that its surface area cannot decrease with time. To prove this, it is sufficient to show that the expansion parameter ϱ of the isotropic geodesics of the horizon H^+ is less than zero for all H^+. If in some neighborhood $U \subset H^+$ the parameter ϱ were positive, then as the affine parameter of the isotropic geodesics of the horizon increases, the geodesics would go from the horizon into the interior of the black hole (see Sect. 1.2.1: property C of the isotropic geodesics). On the other hand, for $\varrho > 0$ in $U \subset H^+$ in the neighborhood of the horizon it would be possible to find a two-dimensional surface Σ such that the parameter ϱ is positive in both families of isotropic geodesics passing through Σ. In that case, however, the isotropic geodesics orthogonal to Σ could not remain on $\partial\mathcal{I}^+(\Sigma)$ (the boundary of the future set for Σ) for all values of the affine parameter (property C of the isotropic geodesics).

An event horizon is formed during the process of collapse of a sufficiently compact mass, but not for any topology of space-time. In spatially compact models (for example, models of closed Friedman universes), the abundance of matter is sufficient to focus any congruence of isotropic geodesics without rotation [2.30].

II) We consider now the topology of a two-dimensional compact region with zero expansion of the isotropic geodesics (of an apparent event horizon).

For a certain time after the sources flare up, on a space-like, connected, compact, orientable, two-dimensional surface Σ there exists an outward light wave front as well as an inward wave front.

The corresponding two characteristic surfaces are envelopes of the light cones constructed at each point of Σ. Let u_1 and u_2 be the "outer" and "inner" characteristic surfaces, respectively. Following [2.30], we shall now establish the topological structure of the surface Σ on which the expansion parameter ϱ of the "outer" characteristic u_1 vanishes. Such surfaces Σ (more precisely, their envelopes) are called *apparent event horizons*. They lie inside the "true" horizons H^+ and coincide with the latter only in stationary fields.

We shall assume that for a continuous fibration of some neighborhood of u_1 by means of a family of isotropic surfaces $\{u\}$ for surfaces lying outside u_1 ($u - u_1 > 0$) the optical parameter has a value $\varrho < 0$ (i.e., the surfaces expand), while for surfaces inside u_1 ($u - u_1 < 0$) it has a value $\varrho > 0$. Then $d\varrho/d\omega < 0$, where ω is any space-like direction orthogonal to Σ (the exterior normal).

The set of space-like directions orthogonal to Σ at some point $Q \in \Sigma$ forms a two-dimensional vector tangent subspace spanned by $\boldsymbol{l}$ and $\boldsymbol{n}$, where $\boldsymbol{l}$ is an isotropic vector at the point Q tangential to u_1 ($l_i = f\nabla_i u_1$), and $\boldsymbol{n}$ is an isotropic vector orthogonal to Σ such that $\boldsymbol{l} \cdot \boldsymbol{n} = 1$.

Suppose that the vectors $\boldsymbol{l}$ undergo parallel transport along themselves: $D\boldsymbol{l} = 0$. Different assignments of the function f on the compact manifold Σ correspond to different space-like fields of normals $\boldsymbol{\omega} = (\boldsymbol{l} - \boldsymbol{n})/\sqrt{2}$ when the two-dimensional surfaces Σ are embedded in three-dimensional space-like hypersurfaces.

We shall assume that a null tetrad can undergo parallel transport along isotropic geodesics on the characteristic u_1, i.e., $Dm_i = Dn_i = 0$ for $i = 1,2,3,4$. By means of the Newman-Penrose equations, we shall find an expression for $d\varrho/d\omega = (D\varrho - \Delta\varrho)/\sqrt{2}$. To do this, we write down the equations (1.1.55, 60) from the formalism of null tetrads, taking into account the fact that $\pi = \varepsilon = k = 0$ and $\varrho = 0$ at the points of Σ:

$$\begin{aligned} &\Phi_{00} + \sigma\sigma^* = D\varrho, \\ &\psi_2 - R/12 + \sigma\lambda + \tau\tau^* - (\delta^* + \beta^* - \alpha)\tau = -\Delta\varrho \ . \end{aligned} \tag{2.1.1}$$

We note that the operator $\delta^* + \beta^* - \alpha$ has the form of a divergence,

$$(\delta^* + \beta^* - \alpha)\psi = \nabla_A(m^{*A}\psi), \quad A = 1,2 \ ,$$

where the covariant differentiation is taken in the sense of the interior geometry on the two-dimensional surface Σ.

From (2.1.1) we have

$$\sqrt{2}\frac{d\varrho}{d\omega} = \Phi_{00} + \sigma\sigma^* + \tau\tau^* - (\delta^* + \beta^* - \alpha)\tau + \psi_2 - \frac{R}{12} + \sigma\lambda \ . \tag{2.1.2}$$

Let us consider the equation which follows from the Newman-Penrose equations (1.1.56, 65),

$$\begin{aligned} 2\psi_2 - 2\Phi_{11} + R/12 &= 2((\delta^* + \beta^* - \alpha)\beta - (\delta + \beta - \alpha^*)\alpha - \sigma\lambda) \\ &= \nabla_A[m^{*A}(\tau + \delta \ln f) - m^A(\tau^* + \delta^* \ln f)] \\ &\quad - K - 2\sigma\lambda \ , \end{aligned} \tag{2.1.3}$$

where the term in the square brackets is purely imaginary and is a divergence term, and K is the Gaussian curvature of the two-dimensional surface,

$$\begin{aligned} -K &\equiv (\delta + \beta - \alpha^*)(\beta^* - \alpha) + (\delta^* + \beta^* - \alpha)(\beta - \alpha^*) \\ &= \nabla_A[m^A\nabla_B m^{*B} + m^{*A}\nabla_B m^B] \ . \end{aligned}$$

Equation (2.1.3) provides an embedding of the two-dimensional surface Σ in a pseudo-Riemannian manifold M_4, i.e., a connection between the interior geometry on Σ and the "exterior" geometry on M_4.

Substituting the value of $\psi_2 + \sigma\lambda$ from (2.1.3) into (2.1.2), we obtain

$$\sqrt{2}\, d\varrho/d\omega = \Phi_{00} + \sigma\sigma^* + \tau\tau^* + \nabla_A(B^A/2) + \Phi_{11} - R/8 - K/2 \ . \tag{2.1.4}$$

Here the continuous vector field B_A on the manifold Σ is defined, according to (2.1.2), as $B_A = n^i\nabla_A l_i$.

Let us integrate (2.1.4) over the surface Σ. Then the left-hand side of (2.1.4) will be negative, the first three terms on the right-hand side will be positive, and the fourth term will vanish after integration over the surface. The sum $\Phi_{11} - R/8$ in an arbitrary orthonormalized tetrad e_1, e_2, e_3, e_4 will be equal to

$$2\pi G[T_1^1 + T_4^4]/c^4 \quad ,$$

where the coordinate vectors e_2 and e_3 are tangential to Σ. Under Hawking's so-called *energy dominance condition* $T_4^4 > |T_1^1|$, the sum $\Phi_{11} - R/8$ will always be positive [2.31].

We now make use of the well-known theorem [2.32] which asserts that every compact, orientable, connected, two-dimensional manifold Σ with no boundary is homeomorphic to a sphere with n handles.

The Euler characteristic of a manifold Σ homeomorphic to a sphere with n handles is proportional to the integral of the Gaussian curvature K over Σ:

$$\int_{\Sigma} K \, d\Sigma = 4\pi(1-n) \quad . \tag{2.1.5}$$

Therefore the integral of the last term in (2.1.4) over Σ is negative only for a sphere ($n = 0$); for a torus ($n = 1$) and for spheres with larger numbers of handles, this integral is non-negative.

Thus, for $n \geq 1$ we arrive at a contradiction, since the integral of the left-hand side of (2.1.4) is negative, while the integral of the right-hand side is positive. The only possible topological form for a compact orientable two-dimensional surface Σ (inside which $\varrho > 0$, while $\varrho = 0$ on the surface Σ itself) is a sphere.

III) We consider now a two-dimensional compact space-like surface Σ on which both of the families of isotropic geodesics lying on isotropic hypersurfaces passing through Σ have a negative rate of expansion, i.e., a parameter value $\varrho > 0$. The existence of such *trapping* surfaces Σ is, as was discovered by *Penrose* [2.29], an indication that "something unpleasant" will happen in the subsequent history of this region.

In particular, we shall show that the following requirements on space-time are incompatible.

1) There exists a noncompact global Cauchy hypersurface $\mathcal{L}$ which any time-like geodesic intersects once and only once.

2) There exists a compact trapping surface Σ in the Cauchy region $D^+(\mathcal{L})$: $\Sigma \subset D^+(\mathcal{L})$.

3) The property of focusing of isotropic geodesics with zero rotation holds.

4) Each isotropic geodesic in M_4 can be extended into the future up to an arbitrarily large value of the affine parameter.

By virtue of assumption 4, the future set $\mathcal{I}^+(\Sigma)$ has a boundary $\partial\mathcal{I}^+(\Sigma)$ with the following properties: a) no two points of $\partial\mathcal{I}^+(\Sigma)$ can be connected by a time-like curve; b) the boundary $\partial\mathcal{I}^+(\Sigma)$ is a three-dimensional closed edge-free manifold embedded in M_4 [2.29]. Such a manifold is said to be *semi-space-like*.

Both isotropic hypersurfaces passing through Σ contain focused normal families of isotropic geodesics. Therefore on each light ray emitted orthogonally to Σ a focal point is formed for a bounded variation of the affine parameter. Owing to the compactness of Σ, the semi-space-like set $\partial\mathcal{I}^+(\Sigma)$ must possess a boundary, as is most readily seen by mapping it onto the Cauchy surface $\mathcal{L}$ by means of an arbitrary family of time-like curves intersecting $\partial\mathcal{I}^+(\Sigma)$ only once. Since $\mathcal{L}$ is noncompact (assumption 1), the image of the compact manifold $\partial\mathcal{I}^+(\Sigma)$ will possess a boundary in $\mathcal{L}$. This contradicts property (b) of the boundary $\partial\mathcal{I}^+(\Sigma)$ of the future set.

Thus, when a trapping surface is formed in space-time, the subsequent development leads to either a Cauchy horizon[3] or infinite limits for certain invariants of the curvature tensor.

2.2 Basic Elements of the Theory of Lie Groups and Exterior Forms

2.2.1 The Concept of Lie Groups

Let M_4 be a sufficiently smooth differentiable 4-dimensional manifold. We say that an N-parameter Lie group G_N acts in the manifold M_4 if the manifold admits motions in itself (automorphisms) joining points lying on transitivity surfaces: $y^i = f^i(x^k, a_1, \ldots, a_N)$, where $a_1, a_2 \ldots, a_N$ are parameters of the group. A set of continuous transformations $y = f(x, a)$ forms a group if and only if they satisfy the following conditions:

a) there exists an identity transformation $a = e$: $x^i = f^i(x^k, e)$;

b) the transformation $x = \varphi(y, a)$ which is the inverse of a given transformation $y = f(x, a)$ belongs to the group [in other words, there must exist a set of parameters $b_1, \ldots, b_N$ in terms of which the transformation $x = \varphi(y, a)$ can be written in the form $x = f(y, b)$];

c) the laws of composition and associativity hold, which means that there must exist a set of parameters $c_1, \ldots, c_N$, $c = (a, b)$, such that

$$z = f(f(x,a),b) = f(x,(a,b)), \quad (a,(b,c)) = ((a,b),c) \ . \tag{2.2.1}$$

The relations (2.2.1) can be regarded as functional constraints on the four functions $f^i(x^k, a)$ $(i, k = 1,2,3,4)$. Important consequences follow from (2.2.1). The derivative of the transformation $y = f(x, a)$ with respect to the group parameter a_ν admits the separation of variables

$$\frac{\partial y^i(x,a)}{\partial a^\nu} = \xi^i_\sigma(y) A^\sigma_\nu(a) \ . \tag{2.2.2}$$

[3] The *Cauchy horizon* in the future of the set $\mathcal{L}$ refers to the boundary of the Cauchy region $D^+(\mathcal{L})$ in the future. Cauchy horizons consist of segments of isotropic geodesics which can be extended up to the boundary of $\mathcal{L}$.

The Greek indices in (2.2.2) run through the values from 1 to N, where N is the number of parameters of the group. This separation of variables is unique, apart from the transformation

$$\xi'^{i}_{\sigma}(y) = \xi^{i}_{\gamma}(y)L^{\gamma}_{\sigma}, \quad A'^{\sigma}_{\nu}(a) = A^{\sigma}_{\gamma}\tilde{L}^{\gamma}_{\nu} \ ,$$

where the matrices L and $\tilde{L}$ are inverses of one another and are formed from constants.

We require now that the system of equations (2.2.2) is consistent, i.e.,

$$\frac{\partial^2 y^i}{\partial a^\nu \partial a^\sigma} = \frac{\partial^2 y^i}{\partial a^\sigma \partial a^\nu} \ ,$$

and we obtain

$$\begin{aligned} &\frac{\partial \xi^{i}_{\nu}(x)}{\partial x^k}\xi^{k}_{\sigma}(x) - \frac{\partial \xi^{i}_{\sigma}(x)}{\partial x^k}\xi^{k}_{\nu} \\ &= \xi^{i}_{\gamma}(x)\left(\frac{\partial A^{\gamma}_{\beta}(a)}{\partial a^\alpha} - \frac{\partial A^{\gamma}_{\alpha}(a)}{\partial a^\beta}\right)\tilde{A}^{\alpha}_{\nu}(a)\tilde{A}^{\beta}_{\sigma}(a) \ , \end{aligned} \tag{2.2.3}$$

where $\tilde{A}^{\alpha}_{\nu}(a)$ is the inverse of the matrix $A^{\alpha}_{\beta}(a)$.

If the equalities (2.2.3), which contain functions of the variables x and a, are to hold, it is necessary that

$$\frac{\partial \xi^{i}_{\nu}(x)}{\partial x^k}\xi^{k}_{\sigma}(x) - \frac{\partial \xi^{i}_{\sigma}(x)}{\partial x^k}\xi^{k}_{\nu}(x) = \xi^{i}_{\gamma}(x)C^{\gamma}_{\nu\sigma} \ , \tag{2.2.4}$$

$$\frac{\partial A^{\gamma}_{\beta}(a)}{\partial a^\alpha} - \frac{\partial A^{\gamma}_{\alpha}(a)}{\partial a^\beta} = C^{\gamma}_{\nu\sigma}A^{\nu}_{\alpha}(a)A^{\sigma}_{\beta}(a) \ . \tag{2.2.5}$$

The constants $C^{\gamma}_{\nu\sigma}$ (the structure constants of the group) completely determine the individuality of the group.

If the set of numbers $C^{\gamma}_{\nu\sigma}(\gamma, \nu, \sigma = 1, \dots, N)$ is to define a group, equalities which follow from the definition of $C^{\gamma}_{\nu\sigma}$ must hold. According to (2.2.4, 5) and the associativity property, we have

$$C^{\gamma}_{\nu\sigma} = -C^{\gamma}_{\sigma\nu}; \quad C^{\nu}_{\mu\alpha}C^{\mu}_{\beta\gamma} + C^{\nu}_{\mu\beta}C^{\mu}_{\gamma\alpha} + C^{\nu}_{\mu\gamma}C^{\mu}_{\alpha\beta} = 0 \ .$$

According to the definition of the action of the group on a Riemannian manifold, we have the equality

$$g_{ij}(x) = \frac{\partial y^m}{\partial x^i}\frac{\partial y^n}{\partial x^j}g_{mn}(y) \ , \tag{2.2.6}$$

where $y = f(x, a)$ is a transformation belonging to the group. For a transformation $y = f(x, a)$ which differs infinitesimally from the identity transformation, we have

$$y^i \approx f^i(x^k, e) + \frac{\partial f^i}{\partial a^\nu}\delta a^\nu = x^i + \xi^{i}_{\nu}\delta a^\nu \ . \tag{2.2.7}$$

The vectors $\xi^i_\sigma(\sigma = 1, \ldots, N)$ are called the *Killing vectors* of the group G_N on the manifold M_4.

The integral curves of the vector fields $\xi^i_\sigma(x)C^\sigma$, where $C^\sigma(\sigma = 1, \ldots, N)$ is a set of N arbitrary constants, will be called the *trajectories* of the group G_N on the manifold M_4.

Apart from infinitesimal quantities of higher order, (2.2.6) can be rewritten in the form of the so-called Killing equations

$$\frac{\partial g_{ij}}{\partial x^k}\xi^k_\nu + g_{ik}\frac{\partial \xi^k_\nu}{\partial x^j} + g_{jk}\frac{\partial \xi^k_\nu}{\partial x^i} = \nabla_i \xi_{j\nu} + \nabla_j \xi_{i\nu} = 0 \ . \tag{2.2.8}$$

In principle, the procedure of finding the metric corresponding to a given group of motions is as follows: 1) from the structure constants of the group, we find the Killing vectors $\xi^i_\nu(x)$ by integrating (2.2.4); 2) the set of equations (2.2.8) is regarded as a system of partial differential equations for the components of the matrix g_{ij}, and from the known functions $\xi^i_\nu(x)$ we find solutions for the metric depending on arbitrary functions of the appropriate number of variables.

2.2.2 The Concept of Skew and Differential Forms

Let $\boldsymbol{\xi}_1, \ldots, \boldsymbol{\xi}_N (N \le 4)$ be independent variable vectors and $\Omega_N(\boldsymbol{\xi}_1, \ldots, \boldsymbol{\xi}_N)$ be a scalar function of them, linear in each of its vector arguments. The semilinear function Ω_N is said to be a skew N-form if it changes sign when any two arguments are interchanged.

Any N-form can be represented uniquely as a sum

$$N!\Omega_N = a_{i_1 i_2 \ldots i_N}\xi_1^{i_1}\xi_2^{i_2}\ldots\xi_N^{i_N} \ ,$$

where the coefficients $a_{i_1 \ldots i_N}$ are antisymmetric in each pair of indices.

Suppose that we are given two skew forms: a k-form $\Omega_k(\xi_1, \ldots, \xi_k)$ and an l-form $\Omega_l(\eta_1, \ldots, \eta_l)$. The *exterior product* $\Omega_k \wedge \Omega_l$ of the forms Ω_k and Ω_l is defined as the result of antisymmetrization of their product with respect to all $k + l$ arguments:

$$\Omega_k \wedge \Omega_l = \frac{1}{(k+l)!}\sum \pm \Omega_k(\boldsymbol{\xi}_{i_1}, \ldots, \boldsymbol{\xi}_{i_k})\Omega_l(\boldsymbol{\xi}_{i_{k+1}}, \ldots, \boldsymbol{\xi}_{i_{k+l}}) \ .$$

Here $\boldsymbol{\xi}_1 \equiv \xi_1, \ldots, \boldsymbol{\xi}_k \equiv \xi_k$; $\boldsymbol{\xi}_{k+1} \equiv \eta_1, \ldots, \boldsymbol{\xi}_{k+l} \equiv \eta_l$; the summation is taken over all permutations $i_1, i_2, \ldots, i_{k+l}$ of the integers $1, 2, \ldots, k+l$. If a permutation is even, we must take the plus sign; if it is odd, the minus sign.

A k-form Ω_k is said to be *simple* if it can be represented as an exterior product of k linear 1-forms $\overset{1}{\omega}(\boldsymbol{\xi}_1), \ldots, \overset{k}{\omega}(\boldsymbol{\xi}_k)$,

$$\Omega_k(\boldsymbol{\xi}_1, \ldots, \boldsymbol{\xi}_k) = \frac{1}{k!}\sum \pm \overset{1}{\omega}(\boldsymbol{\xi}_{i_1})\ldots\overset{k}{\omega}(\boldsymbol{\xi}_{i_k}) = \overset{1}{\omega} \wedge \overset{2}{\omega} \wedge \ldots \wedge \overset{k}{\omega} \ ,$$

where the summation goes over all permutations $i_1, \ldots, i_k$ of the integers $1, \ldots, k$, with the plus sign for an even permutation and the minus sign for an odd permutation.

The criterion for linear dependence of linear forms can be expressed as the vanishing of their exterior product.

A *differential k-form* at a given point is defined as a skew form of arguments $\boldsymbol{\xi}_1, \ldots, \boldsymbol{\xi}_k$ which are the vectors of k infinitesimal displacements from the given point to neighboring points (so that the components of the vector $\boldsymbol{\xi}_i$ are equal to the differentials of the coordinates of the point for the i–th infinitesimal displacement):

$$k!\,\Omega(\boldsymbol{d}_1, \ldots, \boldsymbol{d}_k) = a_{i_1 \ldots i_k}(x)\, d_1 x^{i_1} \ldots d_k x^{i_k} \quad . \tag{2.2.9}$$

Two coordinate systems in a region $\mathcal{D}$ have the same orientation if the Jacobian of the transformation from one to the other is positive in $\mathcal{D}$. A 4-form can be integrated in an invariant manner over a given oriented region, and $\boldsymbol{d}_1, \ldots, \boldsymbol{d}_4$ can be taken to be the infinitesimal displacements along the coordinate lines $x^1, \ldots, x^4$.

For invariant integration over oriented N-dimensional surfaces, it is necessary to use differential N-forms. An N-dimensional surface can be parameterized by means of parameters $u^1, \ldots, u^N$:

$$x^k = f^k(u^1, \ldots, u^N), \quad k = 1,2,3,4 \quad .$$

Then for the differentials $d_1, \ldots, d_N$ we can take the infinitesimal displacements along the lines $u_1, \ldots, u_N$.

The *exterior differential* of a differential k-form $\Omega_k(\boldsymbol{d}_1, \ldots, \boldsymbol{d}_k)$ is defined as the $(k+1)$-form $\tilde{\Omega}_{k+1}(\boldsymbol{d}, \boldsymbol{d}_1, \ldots, \boldsymbol{d}_k)$ constructed from the form Ω_k as follows [2.33].

In the form Ω_k, each coefficient $a_{i_1 \ldots i_k}$ is replaced by its total differential $(\partial a/\partial x^j)\, dx^j$, and the resulting expression is antisymmetrized with respect to all the symbols $\boldsymbol{d}, \boldsymbol{d}_1, \ldots, \boldsymbol{d}_k$. In expanded form, the exterior differential $d\Omega_k$ can be written as

$$\begin{aligned} d\Omega_k = {} & \frac{1}{(k+1)!}(a_{i_1 \ldots i_k, i} - a_{i i_2 \ldots i_k, i_1} - \ldots - a_{i_1 \ldots i_{k-1} i, i_{k+1}}) \\ & \times dx^1 d_1 x^{i_1} \ldots d_k x^{i_k} \quad . \end{aligned} \tag{2.2.10}$$

It follows from the definition (2.2.10) that the exterior differential of a form which is in turn an exterior differential is equal to zero.

The most important application of the concept of an exterior differential is the multidimensional Stokes formula

$$\int\limits_{\mathcal{D}} d\Omega_k = \int\limits_{\partial\mathcal{D}} \Omega_k \quad , \tag{2.2.11}$$

which states that the integral of the exterior differential of a form Ω_k over a $(k+1)$-dimensional oriented, simply connected region $\mathcal{D}$ is equal to the integral of the k-form Ω_k over the boundary of $\mathcal{D}$. Under the conditions of the theorem, in the region $\mathcal{D}$ and on its boundary the coefficients of the form Ω_k are single-valued functions of the class C^1.

Differential forms can be used to study the topology of a manifold. Here it is important that the result of two successive evaluations of the exterior differential of a skew-symmetric form is identically equal to zero. Thus, a local criterion for a given $(k+1)$-form to be the exterior differential of some k-form is the vanishing of the $(k+2)$-form equal to the exterior differential of the given $(k+1)$-form. This lemma does not in any way imply the global existence of a form on the entire manifold.

We say that a differential form Ω_k is *closed* if its exterior differential is equal to zero. Closed forms generate a linear space. Two closed k-forms are said to be equivalent if they differ by the exterior differential of some form Ω_{k-1}.

The equivalence classes of closed k-forms generate the so-called k-dimensional cohomology group of the manifold M.

2.2.3 Frobenius's Theorem

Suppose that on some chart of the manifold M_4 we are given a field of linearly independent vectors $\xi_1, \ldots, \xi_N$ $(N < 4)$. Under what differential conditions on $\xi_1, \ldots, \xi_N$ does there exist a $(4-N)$-dimensional surface which intersects these vectors orthogonally?

Let dx^i be the infinitesimal increments of the coordinates between two neighboring points on the required surface. Then it follows from the condition of orthogonality that the differential forms $\overset{1}{\omega} = \xi_{1i}dx^i, \ldots, \overset{N}{\omega} = \xi_{Ni}dx^i$ generated by $\xi_1, \ldots, \xi_N$ vanish on this surface.

Frobenius's theorem asserts that there exists a surface which intersects a field of linearly independent vectors orthogonally if and only if the exterior product of the exterior differential of each of the N linear forms with all the other forms is equal to zero. In coordinate form, this reduces to the requirement that

$$\xi_{k[i,j}\xi_{1i_1} \ldots \xi_{Ni_N]} = 0, \quad k = 1, \ldots, N \quad \text{or} \quad d\overset{k}{\omega} \wedge \overset{1}{\omega} \wedge \ldots \wedge \overset{N}{\omega} = 0 . \tag{2.2.12}$$

The square brackets in (2.2.12) signify antisymmetrization.

In the case $N = 1$, (2.2.12) takes the form

$$\xi_{[i}\xi_{k,j]} = 0 \quad \text{or} \quad d\omega \wedge \omega = 0, \quad \omega \equiv \xi_i dx^i \ .$$

This is the condition for the existence of a three-dimensional hypersurface which intersects a field of linearly independent vectors orthogonally. In the case $N = 2$, the condition for the existence of a two-dimensional surface which intersects the vector fields ξ and η orthogonally takes the form

$$\xi_{[i}\eta_j\xi_{k,l]} = \xi_{[i}\eta_j\eta_{k,l]} = 0 \quad \text{or} \quad d\overset{1}{\omega} \wedge \overset{1}{\omega} \wedge \overset{2}{\omega} = 0, \quad d\overset{2}{\omega} \wedge \overset{1}{\omega} \wedge \overset{2}{\omega} = 0 \ .$$

Definition. The *Lie derivative* $\mathcal{L}_\xi T$ of a tensor field T with respect to a field ξ is the tensor quantity defined as

$$\mathcal{L}_\xi T^{i\ldots}_{j\ldots} = \xi^k \partial_k T^{i\ldots}_{j\ldots} - T^{k\ldots}_{j\ldots} \partial_k \xi^i + \ldots + T^{i\ldots}_{k\ldots} \partial_j \xi^k \quad .$$

Here the right-hand side contains one term for each upper and each lower index of the tensor $T^{i\ldots}_{j\ldots}$, and the partial derivatives ∂_k can be replaced by covariant derivatives ∇_k. The Lie derivative characterizes the rate of variation of the tensor $T^{i\ldots}_{j\ldots}$ when the region in which the vector field ξ is given is mapped into itself by means of a mapping $F(t)$. The mapping $F(t)$ is generated by the integral curves of the vector field ξ.

Problems [2.34]. 1) Prove that a Killing vector k_i satisfies the relations

$$-\nabla_i \nabla_j k_l = R^m_{ijl} k_m ,$$
$$\nabla^l \{(\nabla_{[i} k_j) k_{l]}\} = \tfrac{2}{3} k^l R_{l[i} k_{j]} \quad .$$

2) Suppose that k_i and t_i are Killing vectors of some group $\mathcal{G}$. Prove that

$$\nabla^m \{t_{[m} k_i \nabla_j k_{l]}\} = \tfrac{1}{2} k^m R_{m[i} k_j t_{l]} \quad .$$

Hint: Use the fact that the Lie derivative of the tensor $k_{[i} \nabla_j k_{l]}$ is equal to zero[4].

3) Suppose that F_{ij} is an antisymmetric second-rank tensor (with zero Lie derivative along the Killing vector field $\boldsymbol{k}$). Prove that

$$\nabla^l \{F_{[ij} k_{l]}\} = \tfrac{2}{3} (\nabla^l F_{l[i}) k_{j]} \quad .$$

4) Suppose that the group $\mathcal{G}_2$ is commutative, i.e., $\nabla^l (k_{[l} t_{j]}) = 0$. Prove that a tensor F_{ij} satisfying the conditions of Problem 3 obeys the relations

$$\nabla^l F_{[ij} k_l t_{m]} = \tfrac{1}{2} (\nabla^l F_{l[i}) k_j t_{m]} \quad .$$

5) Prove that a Killing vector $\boldsymbol{k}$ obeys the relation

$$\nabla_i (\omega^i / V^2) = 0, \quad 2\omega^i \equiv \varepsilon^{ijlm} k_j \nabla_l k_m , \quad V = k_l k^l \quad .$$

Hint: $-(k^n \nabla_n k_i) \varepsilon^{ijlm} k_j \nabla_l k_m = -\frac{1}{4} (k_n k^n) \varepsilon^{ijlm} \cdot \nabla_i k_j \nabla_l k_m$.

6) Prove the equality

$$k_{[i} t_j \nabla_l k_{m]} = \tfrac{1}{2} [k_{[i} t_j \nabla_{l]} k_m - k_m t_{[i} \nabla_j k_{l]} + t_m k_{[i} \nabla_j k_{l]}] \quad ,$$

where k_i and t_j are two Killing vectors.

---

[4] Lie derivatives along Killing vectors of vector and tensor fields on manifolds admitting groups of motions are equal to zero.

Solution:

$$\begin{aligned} k_{[i}t_j\nabla_l\, k_{m]} &= \tfrac{1}{6}\{k_{[i}t_{j]}\nabla_l\, k_m + k_{[j}t_{l]}\nabla_i\, k_m + k_{[l}t_{i]}\nabla_j\, k_m + k_{[m}t_{i]}\nabla_l\, k_j \\ &\quad + k_{[l}t_{m]}\nabla_i\, k_j + k_{[j}t_{m]}\nabla_l\, k_i\} \\ &= \tfrac{1}{2}k_{[i}t_j\nabla_{l]}\, k_m + \tfrac{1}{12}k_m\{t_i\nabla_l\, k_j + t_l\nabla_j\, k_i + t_j\nabla_i\, k_l\} \\ &\quad + \tfrac{1}{12}t_m\{k_i\nabla_j\, k_l + k_l\nabla_i\, k_j + k_j\nabla_l\, k_i\} \ . \end{aligned}$$

7) Suppose that the Killing vectors $\boldsymbol{k}$ and $\boldsymbol{t}$ of a commutative group $\mathcal{G}_2$ satisfy the Frobenius conditions

$$k_{[i}t_j\nabla_l\, k_{m]} = k_{[i}t_j\nabla_l\, t_{m]} = 0 \ .$$

Prove that under these conditions

$$\begin{aligned} &A_{[i,j]} = B_{[i,j]} = 0, \quad A_i \equiv \sigma^{-1}(-Xk_i + Wt_i), \quad B_i \equiv \sigma^{-1}(Wk_i - Vt_i), \\ &X \equiv t^l t_l, \quad V \equiv k^l k_l, \quad W \equiv t^l k_l, \quad \sigma \equiv W^2 - XV \ . \end{aligned}$$

2.3 Stationary Gravitational Fields

A manifold M_4 which is asymptotically simple in the weak sense is said to be stationary if there acts in M_4 a one-parameter group R whose trajectories are time-like, at least in the neighborhood of the conformal boundaries $\mathcal{B}^+$ and $\mathcal{B}^-$.

Definition. The *domain of exterior stationarity* **U** is the maximal connected region having a time-like Killing vector and containing the neighborhoods of the conformal pseudo-Euclidean boundaries $\mathcal{B}^+$ and $\mathcal{B}^-$.

The *ergosphere* $\mathcal{E}$ is the boundary of the region **U**, and, by definition, the Killing vector on the ergosphere is isotropic: $k_l k^l = 0$.

Certain stationary fields are static fields, in which, by definition, there exist hypersurfaces intersecting orthogonally the trajectories of the group R on M_4. In this case, Frobenius's theorem implies the condition

$$k_{[i}\nabla_j\, k_{l]} = 0 \ . \qquad (2.3.1)$$

We now consider *static gravitational fields.*

1) Theorem A (by Carter) [2.34]. For static fields, the ergosphere $\mathcal{E}$ coincides with the event horizon H if the latter exists.

Indeed, from the orthogonality condition (2.3.1) it is easy to show that

$$V_{,[i}k_{j]} = V\nabla_{[j}\, k_{i]}, \quad V \equiv k_l k^l \ . \qquad (2.3.2)$$

Therefore the ergosphere, on which $V = 0$, is an isotropic surface, since it follows from (2.3.2) that the normal to the surface $V = 0$ with the components $V_{,i}$ is parallel to the isotropic vector k_i.

2) Theorem B (due to Lichnerowitz and Carter) [2.34, 35]. In static gravitational fields, the energy-momentum tensor T_{ij} of the matter satisfies the condition

$$k^l T_{l[i} k_{j]} = 0 \quad . \tag{2.3.3}$$

It follows from the result of Problem 1 that if the condition (2.3.1) holds, then $k^l R_{l[i} k_{j]} = 0$. Making use of Einstein's equations the Ricci tensor is replaced by the matter tensor T_{ij} to obtain the assertion of the theorem.

3) Theorem C (by Lichnerowitz) [2.35]. The condition (2.3.3) is not only necessary but also sufficient for the static character of space, provided that the following additional restrictions hold: a) the manifold M_4 is asymptotically simple (see Sect. 2.1); b) the vector k_i is time-like throughout the region **U**.

Hawking [2.36] generalized this result of Lichnerowitz to the case of weakly asymptotic spaces in which the ergosphere coincides with the horizon: if an event horizon exists, the fact that it coincides with the ergosphere is a feature which distinguishes static spaces from other stationary spaces.

To prove the theorem, we introduce, following [2.34], the rotation vector

$$\omega^i = \tfrac{1}{2}\varepsilon^{ijlm} k_j \nabla_l k_m \quad . \tag{2.3.4}$$

It follows from the result of Problem 1 that

$$\nabla_{[i}\,\omega_{j]} = -\tfrac{1}{2}\varepsilon_{ijlm} k^l R^{mn} k_n \quad . \tag{2.3.5}$$

Therefore, if the conditions (2.3.3, 5) are satisfied, we obtain $\nabla_{[i}\,\omega_{j]} = 0$.

It follows from this that there exists a function U such that $\omega_i = U_{,i}$, and on the horizon we can put $U = 0$, since $\omega_i|_H = 0$ by virtue of the assumption that the surface $V = 0$ is isotropic. Consider the identity

$$\nabla_i (U\omega^i/V^2) = \omega_i\omega^i/V^2 + U\nabla_i(\omega^i/V^2) \quad , \tag{2.3.6}$$

where ω^i is a space-like vector according to the definition (2.3.4) and is orthogonal to the vector k_i, which, by the conditions of Theorem C, is time-like off the event horizon.

It follows from the result of Problem 5 that the second term in (2.3.6) is equal to zero.

We multiply (2.3.6) by k^j and make use of the equality $k^j\nabla_i A^i = 2\nabla_i (A^{[i}k^{j]})$, which holds because the Lie derivative of the vector A_i is equal to zero. Then (2.3.6) can be written in the form

$$\nabla_i (U\omega^{[i}k^{j]}/V^2) = \omega^i\omega_i k^j/(2V^2) \quad . \tag{2.3.7}$$

We define an *intersecting surface* Σ in M_4 as a submanifold Σ which intersects all the trajectories of the group R with a Killing vector $\boldsymbol{k}$ in the region U (off the horizon). Let us choose an intersecting surface Σ in the form of a space-like hypersurface. Then, as a consequence of Stokes's theorem, the integral of the divergence of the bivector on the left-hand side reduces to an integral over the boundary of Σ, consisting of two parts: the intersection of Σ with the horizon H and an infinitely remote two-dimensional surface S_∞,

$$\int_\Sigma \nabla_i \left(\frac{U\omega^{[i}k^{j]}}{V^2} \right) n_j d\Sigma$$

$$= \int_{\partial\Sigma} \frac{U\omega_{[i}k_{j]}\sqrt{-g}}{V^2} \left(\frac{\partial x^i}{\partial u}\frac{\partial x^j}{\partial v} - \frac{\partial x^j}{\partial u}\frac{\partial x^i}{\partial v} \right) du\, dv \quad , \tag{2.3.8}$$

where $\partial\Sigma = H + S_\infty$, and u and v are parameters on the two-dimensional surfaces H and S_∞. We shall assume that the pseudo-Euclidean asymptotic limit is approached sufficiently rapidly to ensure that the integral (2.3.8) over S_∞ vanishes.

On the horizon, the function U and its derivatives $U_{,i} = \omega_i$ are equal to zero. As the horizon is approached, the function $U\omega_i$ tends to zero more rapidly than V^2 (by the condition on the horizon, the vector k_i is parallel to the normal $V_{,i}$ to the isotropic surface $V = 0$). Thus, it follows from (2.3.7) that

$$\int_\Sigma [\omega_i\omega^i k^j n_j \, d\Sigma/V^2] = 0 \quad . \tag{2.3.9}$$

From the fact that the integrand in (2.3.9) has a definite sign and from the condition of Theorem C ($\omega_i\omega^i \le 0$, $k_j n^j > 0$) it follows that $\omega_i = 0$ everywhere on Σ and hence also everywhere off the horizon H.

Thus, from the condition (2.3.3) for T_{ij} and from the stationary character of the Killing vector in the strict sense, $k_i k^i > 0$ everywhere off the horizon, we deduce the static character of the gravitational field itself.

4) Theorem D of Israel [2.37]. The Schwarzschild solution (the Nordström-Reissner solution in the presence of an electrovacuum field) is the only solution of Einstein's equations in the class of static, weakly asymptotic simple manifolds possessing a nondegenerate, nonsingular, compact, simply connected event horizon.

Under static conditions, there exists a function $t(x^i)$ such that $k_i = \alpha\nabla_i t$.

According to Theorem A, the coordinate lines of t coincide with the orbits of the action of the group R and are time-like everywhere up to the horizon H. Let x^1, x^2, x^3 be coordinates on the hypersurface $t = \text{const}$, which intersects the coordinate lines of t orthogonally. Then it follows from Killing's equations that the coefficients of the metric

$$ds^2 = p^2 dt^2 - g_{\alpha\beta}\, dx^\alpha dx^\beta, \quad p^2 = k_i k^i, \quad \alpha, \beta = 1,2,3 \quad , \tag{2.3.10}$$

will not depend on the time t.

We assume that the horizon is a connected isotropic surface with a compact, orientable, space-like, two-dimensional cross section. The component R_{tt} of the

Ricci tensor for the metric (2.3.10) is $-p\,\Delta p$, where the Laplacian operator Δ is formed from the spatial part of the metric (2.3.10). Therefore in empty space the function p is harmonic, and $p \to 1$ as $r \to \infty$; on the horizon, $p = 0$ according to Theorem A. In the region **U**, the harmonic function p cannot have singularities and cannot attain its maximum or minimum value (equal to 1 and 0, respectively). Consequently, the function p can be chosen for one of the spatial coordinates, whose variation from 0 to 1 describes motion from the horizon to the pseudo-Euclidean infinity. We choose the other two spatial coordinates θ^a $(a = 1, 2)$ on the surfaces $p = \text{const}$ in such a way that the coordinate lines of p intersect the surfaces $p = \text{const}$ orthogonally; then we have

$$ds^2 = p^2 dt^2 - q^2 dp^2 - g_{ab}\, d\theta^a\, d\theta^b, \quad a, b = 1, 2 \ , \tag{2.3.11}$$

We introduce a field of null tetrads as follows. Suppose that complex vectors $\boldsymbol{m}$ and $\boldsymbol{m}^*$ are constructed from real vectors lying on the surfaces $p = \text{const}$, and let the vectors $\boldsymbol{l}$ and $\boldsymbol{n}$ have components

$$l_i(p/\sqrt{2}, -q/\sqrt{2}, 0, 0), \quad n_i(p/\sqrt{2}, q/\sqrt{2}, 0, 0) \ .$$

Then the rotation coefficients of the null tetrad will have the values

$$\varrho = \mu = -\frac{1}{2\sqrt{2}\,q^2}\frac{\partial q}{\partial p}, \quad k = -\tau = \pi^* = -\nu^* = \frac{\delta q}{2q},$$

$$\varepsilon = \gamma^* = \frac{1}{2\sqrt{2}\,qp} + \frac{1}{2\sqrt{2}\,q} m^{*a}\left(\frac{\partial m_a}{\partial p} - \frac{1}{2} m^b \frac{\partial g_{ab}}{\partial p}\right),$$

$$\sigma = \lambda^*, \quad \alpha = -\beta^* = -\frac{1}{2}\nabla_a\, m^{*a} \ .$$

Here and in the remainder of this subsection, covariant differentiation with Latin indices $a, b, \ldots$ is taken in the sense of the interior geometry on the two-dimensional surface $p = \text{const}$, and $\Delta^{(2)}$ is the Laplacian operator on the surface $p = \text{const}$.

We shall derive two equations [see (2.3.13, 16) below] as follows. Equation (1.1.60) for the component Φ_{00} of the Ricci tensor in the static case takes the form

$$\Phi_{00} = \frac{1}{\sqrt{2}\,q}\frac{\partial \varrho}{\partial p} - \varrho^2 - \frac{1}{\sqrt{2}\,pq}\varrho - \nabla_a\,(m^{*a}k) - \sigma\sigma^* - 2kk^* \ . \tag{2.3.12}$$

This equation can be rewritten as

$$-\frac{1}{2q}\frac{\partial}{\partial p}\left(p^{-1}\frac{\partial q^{-1/2}}{\partial p}\right) + \frac{1}{p}\left\{\frac{1}{2}\Delta^{(2)}\sqrt{q} + (\sigma\sigma^* + kk^* + \Phi_{00})\sqrt{q}\right\} = 0 \ . \tag{2.3.13}$$

In (2.3.13) we have used the identity

$$\sqrt{q}\,\nabla_a (m^{*a}k) \equiv \tfrac{1}{2}\Delta^{(2)}\sqrt{q} - kk^*\sqrt{q} \ .$$

Just as in the proof of Hawking's theorem concerning the topology of an apparent horizon (Sect. 2.1), we subtract (1.1.65) from (1.1.56):

$$\frac{R}{24} + \psi_2 - \Phi_{11} = \varrho^2 - \frac{1}{2}\overset{(2)}{R} - \sigma\sigma^* \ . \tag{2.3.14}$$

Here $\overset{(2)}{R}$ is the Gaussian curvature of the two-dimensional surface $p = \text{const}$:

$$\overset{(2)}{R} = -\nabla_a [m^{*a}\nabla_b m^b + m^a \nabla_b m^{*b}] \ .$$

Substituting the expression (1.1.55) for ψ_2 into (2.3.14), we obtain

$$\begin{aligned} &-\nabla_a (m^{*a}k) - 2kk^* + \frac{1}{\sqrt{2}\,q}\frac{\partial}{\partial p}\varrho - 2\varrho^2 \\ &+ \frac{1}{\sqrt{2}\,pq}\varrho + \frac{1}{2}\overset{(2)}{R} - \left(\Phi_{11} - \frac{R}{8}\right) = 0 \ . \end{aligned} \tag{2.3.15}$$

Multiplying (2.3.12) by two and subtracting (2.3.15) from it, we obtain

$$\begin{aligned} &\frac{1}{4q}\frac{\partial}{\partial p}\left(-p\frac{\partial q^{-1}}{\partial p} + \frac{4}{q}\right) \\ &+ p\left[2\sigma\sigma^* + \frac{1}{2}\overset{(2)}{R} + 2kk^* + \frac{1}{4}\Delta \ln q + 2\Phi_{00} - \Phi_{11} + \frac{R}{8}\right] = 0 \ . \end{aligned} \tag{2.3.16}$$

We now consider the boundary conditions at the horizon and at the pseudo-Euclidean infinity.

a) Using the fact that according to Einstein's equation the Laplacian operator gives zero when acting on the function p ($R_{00} = -p\,\Delta p$) in the coordinate system (2.3.11), we obtain

$$\frac{\partial}{\partial p}\left(\frac{\sqrt{g}}{q}\right) = 0, \quad \text{where} \quad g = g_{11}g_{22} - g_{12}^2 \ . \tag{2.3.17}$$

Since the horizon is nondegenerate, the area of the two-dimensional cross section of the horizon H is not identically zero, so that $\sqrt{g} \neq 0$ at the horizon.

At the pseudo-Euclidean infinity, a static gravitational field is characterized in a first approximation by the Newtonian potential $\varphi = GM/r$:

$$p = \sqrt{g_{44}} = \sqrt{1 - 2\varphi/c^2} \approx 1 - m/r, \quad m \equiv GM/c^2 \ .$$

Using the fact that $q^2dp^2 \approx dr^2$ as $r \to \infty$, we obtain $q \approx r^2/m$ as $r \to \infty$. Let us integrate (2.3.17) with respect to p from 0 to 1:

$$\left.\frac{\sqrt{g}}{q}\right|_H = \sin\theta m \ . \tag{2.3.18}$$

It follows from (2.3.18) that at the horizon the function q is nonzero, $q \to q_0$ as $p \to 0$.

b) Let us now calculate the invariant $I_1 = R_{iklm}R^{iklm}$ of the curvature tensor. By definition, the invariant I_1 has a finite value at the event horizon. Calculating this invariant by means of the field equations in empty space, we obtain

$$I_1 = \frac{12}{V^2}\nabla_\alpha \nabla_\beta V \nabla_\gamma \nabla_\delta V g^{\alpha\gamma} g^{\beta\delta} \quad .$$

Using the coordinate system (2.3.11), we finally obtain for I_1 the expression

$$I_1 = \frac{24}{(pq)^2}[\sigma\sigma^* + kk^* + 4\varrho^2] \quad . \tag{2.3.19}$$

Since (2.3.19) remains finite as $p \to 0$ and $q \to q_0 < \infty$ as $p \to 0$, we obtain from (2.3.19)

$$\sigma \to 0, \quad k \to 0, \quad \varrho \to 0 \quad \text{as} \quad p \to 0 \ .$$

From the fact that $k = 0$ it follows that $\delta q_0 = 0$, and q_0 is a constant at the horizon.

Integrating (2.3.18) over the surface of the horizon, we obtain

$$A = 4\pi m q_0 \quad , \tag{2.3.20}$$

where A is the surface area of the horizon or black hole.

c) Let us eliminate $\partial\varrho/\partial p$ from (2.3.12, 15). This gives

$$-\varrho^2 + \sigma\sigma^* + \Phi_{00} - \left(\Phi_{11} - \frac{R}{8}\right) + \frac{\sqrt{2}}{pq}\varrho + \frac{\overset{(2)}{R}}{2} = 0 \quad . \tag{2.3.21}$$

In empty space, $\Phi_{00} = \Phi_{11} = R = 0$, so that in the limit $p \to 0$ it follows from (2.3.21) and the results of step (b) above that

$$\lim_{p \to 0} \frac{1}{p}\frac{\partial q}{\partial p} q^{-3} = \overset{(2)}{R} \quad .$$

d) By repeating Hawking's arguments (see Part II of Sect. 2.1) concerning the topology of the event horizon (the apparent horizon for stationary spaces coincides with the true horizon), it can be seen from (2.3.15) that the surface of the horizon is homeomorphic to a sphere. Therefore, according to the well-known theorem of Gauss and Bonnet, we have

$$\int_H \overset{(2)}{R} \sqrt{g}\, d\theta^1 d\theta^2 = 4\pi \quad .$$

e) We now multiply (2.3.13, 16) by $\sqrt{g}$, integrate over the surface $p = \text{const}$, and then integrate with respect to p from 0 to 1 using the results of steps (c) and (d).

Then from (2.3.13, 16), respectively, we obtain

$$q_0 \geq 4m, \quad \pi q_0^2 \leq A \ . \tag{2.3.22}$$

The equality is possible only if $\sigma \equiv 0$ and $k \equiv 0$ in the region U. However, taking into account (2.3.20), the strict inequalities (2.3.22) are incompatible. Therefore the system (2.3.22) admits only solutions with the equalities, from which it follows that $\sigma = 0$ and $k = 0$ throughout the region U. The condition $k = 0$ means that the function q does not depend on the coordinates θ^1 and θ^2, i.e., $q = q(p)$. The conditions $\sigma = k = \lambda = \nu = 0$ imply zero values of the components of the Weyl tensor $\psi_0 = \psi_1 = \psi_3 = \psi_4$. Thus, a static field with a nondegenerate horizon must have a Weyl tensor of Petrov type D.

It follows from (2.3.21) that the Gaussian curvature $\overset{(2)}{R}$ of the two-dimensional surface $p = \text{const}$ is a quantity which is independent of the coordinates θ_1 and θ_2, i.e., the surface $p = \text{const}$ is a sphere.

The foregoing criteria correspond to the spherically symmetric Schwarzschild solution, and only to this solution.

We now consider *stationary gravitational fields*. Here we dispense with the assumption of static conditions described by (2.3.1), assuming nevertheless that there is a group R with time-like trajectories in the domain of exterior stationarity U.

5) Theorem E by Hawking [2.30, 36]. A stationary predictable manifold which is asymptotically simple in the weak sense and has a nondegenerate event horizon necessarily possesses axial symmetry.

In other words, no rotating black hole can be stationary until all the fields become axially symmetric.

The property of predictability implies the existence of a Cauchy surface $\mathcal{L}$ such that each world line emanating from points of $\mathcal{B}^+$ and directed into the past returns to $\mathcal{L}$ without running into any singularity along the way.

By virtue of Theorem C, in a nonstatic but stationary manifold with a nondegenerate horizon there exists a region off the horizon (inside the ergosphere) in which the Killing vector is space-like. On the horizon, which is an isotropic surface, the isotropic generators do not coincide with the trajectories k^i of the group R.

Under translations in time, the horizon goes into itself. Therefore the corresponding trajectories of the group R wind onto the horizon. In moving along an isotropic geodesic of the horizon, an observer moves (rotates) with respect to the stationary coordinate system determined by the trajectories of the group R.

It follows from Einstein's equations and the Bianchi identities in the Newman-Penrose formalism that $\psi_0 = \psi_1 = \varrho = \sigma = 0$ on the horizon, the scalar ψ_2 is constant along a null generator of the horizon, and the rotation coefficient ε is constant over the entire horizon.

Making use of the axial symmetry of all the field functions on the horizon, it can be shown by means of a local analytic continuation that the solution is also axially symmetric in some region off the horizon. For a proof of Theorem E, the reader is referred to the monograph of *Hawking* and *Ellis* [2.30].

6) Thus, let us consider stationary manifolds with a nondegenerate horizon. According to Theorem E, in such gravitational fields there acts a subgroup $\mathbf{1} \times S^1$ with closed trajectories and a Killing vector t_i.

We shall assume that the group R $\times S^1$ is Abelian, i.e., that the vectors $\boldsymbol{k}$ and $\boldsymbol{t}$ commute:

$$k^i t^j_{,i} - t^i k^j_{,i} = 0 \quad \text{or} \quad \nabla^i (k_{[i} t_{j]}) = 0 \ . \tag{2.3.23}$$

In what follows (in this chapter), we shall confine ourselves to a study of fields in which there exist two-dimensional surfaces F which intersect the integral curves of both Killing vectors k^i and t^i orthogonally. According to Frobenius's theorem, this is possible if

$$(\nabla_{[i} k_j) \varrho_{kl]} = 0, \quad (\nabla_{[i} t_j) \varrho_{kl]} = 0, \quad \varrho_{ij} \equiv 2k_{[i} t_{j]} \ . \tag{2.3.24}$$

Theorem F of Carter [2.34]. The boundary of the region $\mathcal{R}$ in which the two-dimensional surface F intersecting the orbits of $\boldsymbol{k}$ and $\boldsymbol{t}$ orthogonally is space-like coincides with the event horizon (provided that the latter is nondegenerate).

To prove the theorem, we introduce the notation

$$V = k_i k^i, \quad X = t_i t^i, \quad W = k_i t^i, \quad \sigma = -\tfrac{1}{2} \varrho_{ij} \varrho^{ij} = W^2 - VX \ .$$

Let $\boldsymbol{a}$ be a vector in the subspace $\mathcal{A}$ spanned by the vectors $\boldsymbol{k}$ and $\boldsymbol{t}$:

$$a_i = \alpha k_i + \beta t_i \ .$$

In the linear subspace $\mathcal{A}$, there exist two isotropic vectors when $\sigma \equiv W^2 - VX > 0$ and one isotropic vector when $\sigma = 0$. This follows from the equation

$$a^2 = \alpha^2 V + 2\alpha\beta W + \beta^2 X = 0 \ .$$

It is obvious that the region $\mathcal{R}$ in which the surface F is space-like is characterized by the inequality $\sigma > 0$, and σ vanishes on the boundary $\partial\mathcal{R}$.

We make use of the result of Problem 6 and rewrite Frobenius's conditions (2.3.24) for the existence of the two-dimensional surface F in the form

$$k_{[i} t_j \nabla_{k]} k_l = k_l t_{[i} \nabla_j k_{k]} - t_l k_{[i} \nabla_j k_{k]} \ , \tag{2.3.25}$$

$$k_{[i} t_j \nabla_{k]} t_s = -t_s k_{[i} \nabla_j t_{k]} + k_s t_{[i} \nabla_j t_{k]} \ . \tag{2.3.26}$$

We multiply (2.3.25) by t_s and (2.3.26) by k_l, and add the results. In the equation which is obtained, we then replace the index l by s and, conversely, s by l, and subtract the result from the preceding equation.

We then obtain

$$k_{[i}t_j\nabla_{k]}(k_lt_s - t_lk_s) = (k_lt_s - t_lk_s)\nabla_{[i}\,k_jt_{k]} \quad , \tag{2.3.27}$$

or

$$\varrho_{[ij}\nabla_{k]}\,\varrho_{ls} = \varrho_{ls}\nabla_{[i}\,\varrho_{jk]} \quad . \tag{2.3.27$'$}$$

Multiplying both sides of (2.3.27′) by ϱ^{ls}, we obtain

$$\varrho_{[ij}\nabla_{k]}\,\sigma = 2\sigma\nabla_{[i}\,\varrho_{jk]} \quad . \tag{2.3.28}$$

We have $\sigma = 0$ on the boundary $\partial\mathcal{R}$, and from (2.3.28) it follows that

$$\varrho_{[ij}\nabla_{k]}\,\sigma = 0 \quad . \tag{2.3.29}$$

It follows from (2.3.29) that the vectors k_i, t_i, and $\nabla_i\,\sigma$ are linearly dependent:

$$\nabla_i\,\sigma = \alpha k_i + \beta t_i \quad . \tag{2.3.30}$$

On the other hand,

$$k^i\frac{\partial\sigma}{\partial x^i} = 2Wk^i\frac{\partial W}{\partial x^i} - Xk^i\frac{\partial V}{\partial x^i} - Vk^i\frac{\partial X}{\partial x^i} = 0 \quad , \tag{2.3.31}$$

since $k^i\partial W/\partial x^i = k^i\partial V/\partial x^i = k^i\partial X/\partial x^i = 0$. Similarly, it follows from the commutation condition for the vectors $\boldsymbol{k}$ and $\boldsymbol{t}$ and from the conditions $\nabla_{(i}\,k_{j)} = 0$ that $t^i\partial\sigma/\partial x^i = 0$.

Multiplying (2.3.30) by αk^i and βt^i in turn and adding the results, using (2.3.31), we have

$$\alpha^2V + 2\alpha\beta W + \beta^2X = 0 \quad . \tag{2.3.32}$$

On the other hand,

$$(\nabla_i\,\sigma)(\nabla_j\,\sigma)g^{ij} = \alpha^2V + 2\alpha\beta W + \beta^2X \quad .$$

Therefore it follows from (2.3.32) that the surface $\sigma = 0$ is isotropic, with the expansion parameter ϱ equal to zero. Consequently, the surface $\sigma = 0$ is the exterior event horizon.

7) Definitions. A vector field n_i is said to be circular if it satisfies the conditions

$$n_{[i}k_jt_{k]} = 0 \quad .$$

A bivector field F_{ij} is said to be circular if it satisfies the conditions $F_{[ij}k_lt_{m]} = 0$, $F_{ij}k^it^j = 0$.

A symmetric tensor field T_{ij} is said to be circular if it satisfies the conditions

$$k^i T_{i[j} k_m t_{n]} = t^i T_{i[j} k_m t_{n]} = 0 \ . \qquad (2.3.33)$$

Lemma 1. From Frobenius's conditions for the existence of a surface F intersecting the integral curves of the vectors $\boldsymbol{k}$ and $\boldsymbol{t}$ orthogonally it follows that the Ricci tensor is circular.

To prove this, it is sufficient to make use of the result of Problem 2.

Lemma 2. From the condition that the electromagnetic field tensor F_{ij} is circular it follows that the current vector is circular (Problem 3).

Indeed, from the commutation property of $\boldsymbol{k}$ and $\boldsymbol{t}$ and from the conditions (2.3.24) it follows that

$$\nabla^l (F_{[ij} k_m t_{l]}) = \tfrac{1}{2} \nabla_i F^l{}_{[m} k_i t_{j]} = 2\pi j_{[m} k_i t_{j]}/c \ ,$$

since $\nabla^l F_{lk} = 4\pi j_k/c$ by virtue of Maxwell's equations.

Lemma 3 [2.34]. There exists a constant Ω such that the linear combination of Killing vectors $\boldsymbol{k} + \Omega \boldsymbol{t}$ is an isotropic vector on the horizon whenever the conditions (2.3.24) hold.

The vector $\boldsymbol{k} - V\boldsymbol{t}/W$ is isotropic on the horizon, which coincides with the surface $\sigma = 0$ by virtue of Theorem F. We shall show that the function V/W is constant on the horizon. Then the required constant Ω will obviously be $-V/W$, where the function V/W is evaluated on the horizon.

In fact,

$$\begin{aligned} \nabla_k (-V/W) &= W^{-2}(V \nabla_k W - W \nabla_k V) \\ &= W^{-2}(V \nabla_k (k_l t^l) - 2W k^l \nabla_k k_l) \ . \end{aligned} \qquad (2.3.34)$$

Using (2.3.23), we have

$$\nabla_k (k_l t^l) = t^l \nabla_k k_l + k^l \nabla_k t_i = (\nabla_k k_l) t^l - (\nabla_l t_k) k^l = 2 t^l \nabla_k k_l \ .$$

Therefore, multiplying (2.3.34) by $k_i t_j$ and antisymmetrizing with respect to the indices i, j, k, we obtain

$$k_{[i} t_j \nabla_{k]} (-V/W) = 2W^{-2}(k_{[i} t_j \nabla_{k]} k_l)(t^l V - k^l W) \ . \qquad (2.3.35)$$

Using (2.3.25), we rewrite (2.3.35) in the form

$$k_{[i} t_j \nabla_{k]} (-V/W) = (W^2 - XV) k_{[i} \nabla_j k_{k]} \ .$$

Therefore on the horizon, where $W^2 - VX = 0$, the vector $\nabla_k (-V/W)$ is a linear combination of the vectors $\boldsymbol{k}$ and $\boldsymbol{t}$:

$$\nabla_i (-V/W) = \alpha k_i + \beta t_i \ .$$

On the other hand, the derivatives of $-V/W$ along the Killing vectors $\boldsymbol{k}$ and $\boldsymbol{t}$ are equal to zero. This completes the proof.

8) Theorem G [2.38]. Lemma 1 admits a converse. Papapetrou proved that if the energy-momentum tensor of the matter satisfies the condition (2.3.33) for a circular tensor, then Frobenius's conditions (2.3.24) for the existence of a surface F with an orthogonal intersection are satisfied.

Indeed, if the energy-momentum tensor of the matter satisfies the condition for a circular tensor, then, by virtue of Einstein's equations, the Ricci tensor satisfies this same condition.

It follows from the result of Problem 2 that in this case

$$\nabla^i(t_{[i}k_j\nabla_k k_{l]}) = \nabla^i(k_{[i}t_j\nabla_k t_{l]}) = 0 \ . \tag{2.3.36}$$

We introduce vectors ω^i and ψ^i characterizing the rotation of the vectors $\boldsymbol{k}$ and $\boldsymbol{t}$:

$$2\omega_i = \varepsilon_{ijkl}k^j\nabla^k k^l, \quad 2\psi_i = \varepsilon_{ijkl}t^j\nabla^k t^l \ .$$

Any fourth-rank tensor which is completely antisymmetric is proportional to ε_{ijkl}. Therefore

$$t_{[i}k_j\nabla_k k_{l]} = \frac{1}{4!}\omega^m t_m\varepsilon_{ijkl}; \quad k_{[i}t_j\nabla_k t_{l]} = \frac{1}{4!}\psi^m k_m\varepsilon_{ijkl} \ . \tag{2.3.37}$$

Then the conditions (2.3.36) can be written as

$$\nabla_i(t_l\omega^l) = 0, \quad \nabla_i(k_l\psi^l) = 0 \ .$$

Therefore the pseudoscalars $t_l\omega^l$ and $k_l\psi^l$ are constants. However, on the axis of rotation the vectors t_k and ψ_k vanish, so that $t_l\omega^l = 0$, $k_l\psi^l = 0$, and, as a consequence of (2.3.37), Frobenius's conditions (2.3.24) are satisfied.

9) Canonical Form of the Metric Tensor. It follows from Frobenius's conditions (2.3.24) that there exists a two-dimensional surface F such that any direction orthogonal to it is a linear function of the Killing vectors $\boldsymbol{k}$ and $\boldsymbol{t}$.

It follows from the results of Problem 7 that there exist potentials for the vectors A_i and B_i, where $A_i \equiv \sigma^{-1}(-Xk_i + Wt_i)$, $B_i \equiv \sigma^{-1}(Wk_i - Vt_i)$; we denote these potentials by t and φ, respectively:

$$\nabla_i t = \sigma^{-1}(-Xk_i + Wt_i), \quad \nabla_i\varphi = \sigma^{-1}(Wk_i - Vt_i) \ .$$

Let θ^a $(a = 1,2)$ be parameters on the two-dimensional surfaces $F = \text{const.}$ We go over to a system of coordinates t, φ, θ^1, θ^2. Since the coordinate lines of t and φ are orthogonal to the coordinate lines of θ^1 and θ^2, the metric coefficients g_{ta} and $g_{\varphi a}$ are equal to zero. It follows from Killing's equations that the remaining coefficients do not depend on the coordinates t and φ. Thus, the metric form has the structure

$$ds^2 = V\,dt^2 + 2W\,dt\,d\varphi + X\,d\varphi^2 - g_{ab}d\theta^a d\theta^b \ , \tag{2.3.38}$$

where the functions V, W, X, and g_{ab} do not depend on t and φ.

Lemma 4. In the presence of a stationary electromagnetic field, the function $p = \sqrt{\sigma}$ is harmonic on the surface F with the interior metric g_{ab}.

To prove this, we calculate the expression $Q = g^{tt}R_{tt} + 2g^{t\varphi}R_{t\varphi} + g^{\varphi\varphi}R_{\varphi\varphi}$. From Einstein's equations $R_{ij} = \kappa T_{ij}$, where T_{ij} is the energy-momentum tensor of the electromagnetic field [calculated according to the canonical form (2.3.46) given below], it follows that the function Q is equal to zero. On the other hand, from a direct calculation of Q by means of the expression for the components of the Ricci tensor in terms of the metric form (2.3.38) it follows that $Q = p^{-1}\nabla_a \nabla^a p$. Consequently, p is a harmonic function which is equal to zero on the horizon according to Theorem F.

We note that at the pseudo-Euclidean infinity p tends to the radius in cylindrical coordinates. As a harmonic function, p cannot have a maximum or minimum in any region off the horizon which contains no infinitely remote point, and it attains a minimum value $p = 0$ on the horizon.

It follows from the condition that p is harmonic that there exists a function q which is harmonically conjugate to the function p. The lines p = const are orthogonal to the lines q = const on the surfaces F = const. Therefore, if p and q are chosen as parameters on the surface, the metric form finally becomes

$$ds^2 = V\,dt^2 + 2W\,dt\,d\varphi + X\,d\varphi^2 - \Sigma(p,q)(dp^2 + dq^2) \quad , \qquad (2.3.38')$$

where $p^2 = W^2 - VX$.

10) Canonical Form of the Electromagnetic Field Tensor. We introduce the vectors of the electric and magnetic field strengths, E_i and B_i:

$$E_i = F_{ij}k^j, \quad B_i = \tfrac{1}{2}\varepsilon_{ijkl}k^j F^{kl}, \quad V F_{ij} = 2k_{[i}E_{j]} + \varepsilon_{ijkl}k^k B^l \quad .$$

Contracting Maxwell's equations $\nabla_{[i} F_{jl]} = 0$ with the vector k^l, we obtain

$$\nabla_i E_j - \nabla_j E_i + (k^l \nabla_l F_{ij} + F_{lj}\nabla_i k^l + F_{il}\nabla_j k^l) = 0 \quad . \qquad (2.3.39)$$

The bracketed expression in (2.3.39) is the Lie derivative of the bivector F_{ij}, which is equal to zero.

Contracting (2.3.39) with the vector k^l and using the fact that the Lie derivative of the vector E_i along the Killing vector t^i is equal to zero, we obtain

$$\nabla_i (t^j E_j) - (t^j \nabla_j E_i + E_j \nabla_i t^j) = \nabla_i (t^j E_j) = 0 \quad .$$

From the fact that the scalar $t^i E_i$ vanishes on the symmetry axis it follows that $t^j E_j$ is equal to zero everywhere.

It follows from the result of Problem 4 that

$$\nabla_i (B_k t^k) = 4\pi \varepsilon_{ijkl} k^j t^k j^l \quad .$$

Using the fact that the current vector is circular and that the expression $B_k t^k$ vanishes on the symmetry axis, we obtain $B_k t^k = 0$. Thus, we have proved the following lemma.

Lemma 5 [2.34]. From the fact that the current vector is circular it follows that the electromagnetic field tensor satisfies the conditions for a circular tensor:

$$F_{ik}k^i t^k = 0, \quad F_{[ij}t_k k_{l]} = 0 \ . \tag{2.3.40}$$

It follows from the conditions (2.3.40) that the bivector F_{ij} can be represented in the form

$$F_{ij} = 2l_{[i}k_{j]} + 2f_{[i}t_{j]} \ , \tag{2.3.41}$$

where $l_i k^i = l_i t^i = f_i k^i = f_i t^i = 0$.

It follows from (2.3.39) that there exists a potential E for the vector E_i:

$$E_i = \nabla_i E \ . \tag{2.3.42}$$

Contracting the equations $\nabla_{[i} F_{jk]} = 0$ with t^k, we obtain, by analogy with the condition (2.3.42),

$$F_{jl}t^l = \nabla_j B \ . \tag{2.3.43}$$

Finally, we shall show that the vector

$$A_i = Et_{,i} + B\varphi_{,i} \tag{2.3.44}$$

is a vector potential for the electromagnetic field.

From (2.3.44) we have

$$A_{[i,j]} = t_{,[i}E_{,j]} + \varphi_{,[i}B_{,j]} \ . \tag{2.3.45}$$

Using (2.3.41–43), it is easy to show that, according to (2.3.45), in fact $2A_{[i,j]} = F_{ij}$.

It follows from the expression (2.3.44) for the 4-potential of the electromagnetic field that the 1-form $A_i dx^i$ can be represented as

$$A_i dx^i = E\, dt + B\, d\varphi \ .$$

Then the 2-form $F_{ij}dx^i \wedge dx^j$ of the electromagnetic field has the structure

$$F_{ij} = 2E_{,a}dx^a \wedge dt + 2B_{,a}dx^a \wedge d\varphi \ . \tag{2.3.46}$$

11) Ernst Equations for a Stationary Gravitational Field with Axial Symmetry [2.39]. The metric form (2.3.38′) can also be written as

$$ds^2 = f(dt - \omega\, d\varphi)^2 - f^{-1}[e^{2\gamma}(dp^2 + dq^2) + p^2 d\varphi^2] \ , \tag{2.3.47}$$

where we have introduced the notation

$$f \equiv V, \quad X \equiv f\omega^2 - p^2/f, \quad W = -f\omega, \quad f^{-1}e^{2\gamma} \equiv \Sigma \ .$$

To obtain Maxwell's equations (in the absence of currents) and the equations of gravitation, it is convenient to begin with a variational principle.

The Lagrangian density function for the metric in the form (2.3.47) is

$$\begin{aligned}\mathcal{L}_{\mathrm{gr}} &= \frac{\sqrt{-g}\, g^{ij}}{2\kappa}(\Gamma^k_{ij}\Gamma^l_{kl} - \Gamma^k_{il}\Gamma^l_{jk}) \\ &= \frac{1}{4\kappa}[pf^{-2}(\nabla f)^2 - p^{-1}f^2(\nabla\omega)^2] \ ,\end{aligned} \tag{2.3.48}$$

where we have introduced the notation $(\nabla f)^2 \equiv (f_{,p})^2 + (f_{,q})^2$.

The Lagrangian density function for the electromagnetic field is $\mathcal{L}_{\mathrm{em}} = \sqrt{-g}\, F_{ij}F^{ij}/(8\pi)$. Calculating it for a field having the form (2.3.46), we obtain

$$\mathcal{L}_{\mathrm{em}} = [(p^2/f - f\omega^2)(\nabla E)^2 - 2f\omega\nabla E\nabla B - f(\nabla B)^2]/(8\pi p) \ . \tag{2.3.49}$$

Equating to zero the first variation of the total action, we obtain the pair of Maxwell's equations and the pair of Einstein's equations [the function γ in (2.3.47) is calculated after integration of these equations by means of any field equation $R_{ij} = \kappa T_{ij}$ which contains this function]

$$\nabla(p^{-1}f^2\nabla\omega) - \kappa f(\omega\nabla E + \nabla B)\cdot\nabla E/(2\pi p) = 0 \ , \tag{2.3.50}$$

$$\begin{aligned}&\nabla(p\nabla f) - pf^{-1}(\nabla f)^2 + p^{-1}f^3(\nabla\omega)^2 \\ &\quad + \kappa[(f\omega\nabla E + f\nabla B)^2 + p^2(\nabla E)^2]/(4\pi p) = 0 \ ,\end{aligned} \tag{2.3.51}$$

$$\nabla[p^{-1}f(\omega\nabla E + \nabla B)] = 0 \ , \tag{2.3.52}$$

$$\nabla(pf^{-1}\nabla E) - p^{-1}f\nabla\omega(\omega\nabla E + \nabla B) = 0 \ . \tag{2.3.53}$$

It follows from (2.3.52) that there exists a function D such that

$$\omega E_{,p} + B_{,p} = pf^{-1}D_{,q}, \quad \omega E_{,q} + B_{,q} = -pf^{-1}D_{,p} \ . \tag{2.3.54}$$

Eliminating the function B from (2.3.54), we obtain

$$\nabla(pf^{-1}\nabla D) = \omega_{,q}E_{,p} - \omega_{,p}E_{,q} \ . \tag{2.3.55}$$

Equation (2.3.53) can be written in the form

$$\nabla(pf^{-1}\nabla E) = -(\omega_{,q}D_{,p} - \omega_{,p}D_{,q}) \ . \tag{2.3.56}$$

We introduce the complex potential $\Phi \equiv E + \mathrm{i}D$.

Multiplying (2.3.55) by the imaginary unit and adding the result to (2.3.56), we obtain for the potential Φ the equation

$$f\nabla(p\nabla\Phi) = (pf_{,p} + \mathrm{i}f^2\omega_{,q})\Phi_{,p} + (pf_{,q} - \mathrm{i}f^2\omega_{,p})\Phi_{,q} \ . \tag{2.3.57}$$

We introduce the new independent variables z and z^* defined by

$$2z = p + \mathrm{i}q, \quad 2z^* = p - \mathrm{i}q \ .$$

Then (2.3.57) takes the form

$$p^{-1} f \left[\frac{\partial}{\partial z^*} \left(p \frac{\partial}{\partial z} \Phi \right) + \frac{\partial}{\partial z} \left(p \frac{\partial}{\partial z^*} \Phi \right) \right] = G_+ \frac{\partial \Phi}{\partial z} + G_- \frac{\partial \Phi}{\partial z^*} \ , \tag{2.3.58}$$

$$G_+ \equiv \frac{\partial f}{\partial z^*} + \frac{f^2}{p} \frac{\partial \omega}{\partial z^*}, \quad G_- \equiv \frac{\partial f}{\partial z} - \frac{f^2}{p} \frac{\partial \omega}{\partial z} \ . \tag{2.3.59}$$

We now transform (2.3.50, 51). Equation (2.3.50) can be written in the form

$$\frac{\partial}{\partial z} \left(G_+ + \frac{\kappa}{4\pi} \Phi^* \frac{\partial \Phi}{\partial z^*} \right) = \frac{\partial}{\partial z^*} \left(G_- + \frac{\kappa}{4\pi} \Phi^* \frac{\partial \Phi}{\partial z} \right) \ . \tag{2.3.50'}$$

It follows from (2.3.50′) that there exists a function ε such that

$$G_+ + \frac{\kappa}{4\pi} \Phi^* \frac{\partial \Phi}{\partial z^*} = \frac{\partial \varepsilon}{\partial z^*}, \quad G_- + \frac{\kappa}{4\pi} \Phi^* \frac{\partial \Phi}{\partial z} = \frac{\partial \varepsilon}{\partial z} \ . \tag{2.3.60}$$

In the new variables, (2.3.51) can be written in the form

$$\frac{f}{p} \left[\frac{\partial}{\partial z} (pG_+) + \frac{\partial}{\partial z^*} (pG_-) \right] - 2G_+G_- + \frac{\kappa f}{4\pi} \left[\frac{\partial \Phi}{\partial z^*} \frac{\partial \Phi^*}{\partial z} + \frac{\partial \Phi}{\partial z} \frac{\partial \Phi^*}{\partial z^*} \right] = 0 \ . \tag{2.3.61}$$

Substituting the expressions (2.3.60) for G_+ and G_- in terms of ε into (2.3.61), we finally obtain

$$\frac{f}{p} \left[\frac{\partial}{\partial z^*} \left(p \frac{\partial}{\partial z} \varepsilon \right) + \frac{\partial}{\partial z} \left(p \frac{\partial}{\partial z^*} \varepsilon \right) \right] = G_+ \frac{\partial \varepsilon}{\partial z} + G_- \frac{\partial \varepsilon}{\partial z^*} \ . \tag{2.3.62}$$

It follows from the definitions (2.3.59, 60) that

$$\frac{\partial}{\partial z} (\varepsilon + \varepsilon^*) = \frac{\kappa}{4\pi} \frac{\partial}{\partial z} (\Phi \Phi^*) + G_- + G_+^* = \frac{\kappa}{4\pi} \frac{\partial}{\partial z} (\Phi \Phi^*) + 2 \frac{\partial f}{\partial z} \ .$$

Hence $f = \operatorname{Re} \varepsilon - \kappa \Phi \Phi^* / (8\pi)$. Thus, (2.3.58, 60, 62) form a closed system of complex equations in the Ernst form:

$$f \Delta \Phi = \nabla \Phi \cdot \boldsymbol{M}, \quad f \Delta \varepsilon = \nabla \varepsilon \cdot \boldsymbol{M}, \quad \boldsymbol{M} \equiv \nabla \varepsilon - \kappa \Phi^* \nabla \Phi / (4\pi) \ . \tag{2.3.63}$$

After going over to the required functions in the projective space of the variables u, v, w ($\varepsilon = (w-u)/(w+u)$, $\sqrt{G}\,\Phi = vc^2/(u+w)$), the system (2.3.63) takes the form

$$uDw - wDu = 0, \quad (u - w)Dv - v(Du - Dw) = 0 \ . \tag{2.3.63'}$$

Here we have introduced the D-operator notation

$$D \equiv (u^*u + v^*v - w^*w)\Delta - 2(u^* \nabla u + v^* \nabla v - w^* \nabla w) \nabla \ .$$

According to (2.3.63′), any solution of the system

$$DV_i = 0(V_1 = u,\ V_2 = v,\ V_3 = w) \tag{2.3.64}$$

will determine a solution for ε and Φ. It is interesting that the system (2.3.64) retains its form under arbitrary unitary rotations in the complex space of the variables u, v, w.

Thus, if the vector V_0 is a solution of the system (2.3.64), then the vector V with components $V_i = A_{ij}V_{0j}$, where A_{ij} is a constant matrix with the property $A_{ij}A^*_{ik} = \eta_{ik}$, $\eta_{11} = \eta_{22} = -\eta_{33} = 1$, $\eta_{ik} = 0$, $i \neq k$, is also a solution of the system. Consequently, starting from a known particular solution of the system (2.3.64), it is possible to make use of unitary rotations in the projective space of the variables u, v, w to obtain a new solution depending on several parameters [2.40].

If we transform from the variables p and q to the variables x and y ($p = p_0\sqrt{(x^2-1)(1-y^2)}$, $q = p_0xy$), the system (2.3.64) takes the form

$$\begin{aligned}\eta^{ik}\{V_iV_k^*[((x^2-1)V_{j,x})_{,x} + ((1-y^2)V_{j,y})_{,y}]\\ - 2V_k^*[(x^2-1)V_{i,x}V_{j,x} + (1-y^2)V_{i,y}V_{j,y}]\} = 0\ .\end{aligned}$$

For vacuum solutions, $v = 0$. In this case, we can put $w = i$, and the equation for u takes the form [2.41]

$$\begin{aligned}(uu^*-1)[((x^2-1)u_{,x})_{,x} + ((1-y^2)u_{,y})_{,y}]\\ = 2u^*[(x^2-1)u^2_{,x} + (1-y^2)u^2_{,y}]\ .\end{aligned} \tag{2.3.65}$$

Those solutions of the electrovacuum system (2.3.61) in which the functions v and w are proportional, $v = \text{const}\cdot w$, reduce to solutions of Einstein's vacuum equations.

The solution of Kerr and of Kerr and Newman for a rotating charged black hole corresponds to the linear solution of (2.3.65) given by

$$u = x\cos\lambda - iy\sin\lambda, \quad \lambda = \text{const}\ . \tag{2.3.66}$$

12) We now consider *charged black holes.*

a) In the general case, stationary black holes are completely characterized by three parameters: mass, charge, and angular momentum. They are described by the exact Kerr-Newman solution of the system of Einstein's and Maxwell's equations. The corresponding metric tensor can be reconstructed from the solution (2.3.66).

In the coordinates of *Boyer* and *Lindquist* [2.42], the Kerr-Newman solution has the form[5]

[5] Particular cases of the solution (2.3.67) are: a) the Schwarzschild solution ($I = Q = 0$, $M \neq 0$); b) the Kerr solution ($Q = 0$, $I \neq 0$, $M \neq 0$); c) the Nordström-Reissner solution ($I = 0$, $Q \neq 0$, $M \neq 0$).

$$
\begin{aligned}
ds^2 &= \Sigma^{-1}\{(b - a^2\sin^2\theta)c^2dt^2 - 2(r^2 + a^2 - b)a\sin^2\theta c\,dt\,d\varphi \\
&\quad - \Sigma^2 dr^2/b - \Sigma^2 d\theta^2 - [(r^2+a^2)^2 - ba^2\sin^2\theta]\sin^2\theta\,d\varphi^2\} \ ; \\
\Sigma &\equiv r^2 + a^2\cos^2\theta, \quad b \equiv r^2 - 2mr + a^2 + q^2, \quad \Delta = -g^{rr}, \\
m &\equiv GM/c^2, \quad a \equiv I/(Mc), \quad q^2 = GQ^2/c^4 \ ,
\end{aligned}
\tag{2.3.67}
$$

where M is the mass, I is the angular momentum, and Q is the charge of the black hole. The corresponding 2-form for the electromagnetic field has the form

$$
\begin{aligned}
\Sigma^2 \boldsymbol{F} &= 2Q\{-(r^2 - a^2\cos^2\theta)\,dr \wedge (dt - a\sin^2\theta\,d\varphi) \\
&\quad + 2ar\sin\theta\cos\theta\,d\theta \wedge [a\,dt - (r^2+a^2)\,d\varphi]\} \ .
\end{aligned}
\tag{2.3.68}
$$

For the solution (2.3.67, 68), the basis vectors of the complex tetrad field can be chosen in the form

$$
\begin{aligned}
l^i &= \left\{\frac{r^2+a^2}{\Delta}, 1, 0, \frac{a}{\Delta}\right\}, \quad n^i = \frac{1}{2\Sigma}\{r^2+a^2, -\Delta, 0, a\}, \\
m^i &= \frac{1}{\sqrt{2}\,(r + \mathrm{i}a\cos\theta)}\left\{\mathrm{i}a\sin\theta, 0, 1, \frac{\mathrm{i}}{\sin\theta}\right\} \ .
\end{aligned}
\tag{2.3.69}
$$

In the Newman-Penrose formalism, the nonzero tetrad components of the Weyl, Ricci, and electromagnetic-field tensors and the nonzero rotation coefficients have the values

$$
\begin{aligned}
&\psi_2 = m\varrho^3 + q^2\varrho^3\varrho^*, \quad \Phi_1 = -\frac{1}{\sqrt{2}}q\varrho^2, \quad \Phi_{11} = \frac{1}{2}q^2\varrho^2\varrho^{*2}, \\
&\varrho = -(r - \mathrm{i}a\cos\theta)^{-1}, \quad \gamma = \mu + (r-m)\varrho\varrho^*/2, \quad \mu = \varrho^2\varrho^*\Delta/2, \\
&2\sqrt{2}\,\beta = -\cot\theta\varrho^*, \quad \pi\sqrt{2} = \mathrm{i}a\sin\theta\varrho^2, \quad \tau\sqrt{2} = -\mathrm{i}a\sin\theta\varrho\varrho^* \ .
\end{aligned}
\tag{2.3.70}
$$

In the Boyer-Lindquist coordinates, the surface of the ergosphere is determined by the equation

$$
r^2 - 2mr + a^2\cos^2\theta + q^2 = 0 \ . \tag{2.3.71}
$$

The surfaces of the event horizons H satisfy the equation

$$
r^2 - 2mr + a^2 + q^2 = 0 \ . \tag{2.3.72}
$$

The condition for the existence of a nondegenerate event horizon has the form

$$
m^2 \geq a^2 + q^2 \ .
$$

Equation (2.3.72) has two roots, corresponding to the exterior and interior event horizons. From the point of view of an external observer, the surface of a collapsing body possessing charge and angular momentum "freezes" on the exterior event horizon.

b) The solution for a nonrotating charged black hole was obtained by Reissner and Nordström. It has the form

$$ds^2 = c^2 dt^2 A - A^{-1} dr^2 - r^2(d\theta^2 + \sin^2\theta\, d\varphi^2),$$
$$A \equiv 1 - \frac{r_g}{r} + \frac{GQ^2}{(c^4 r^2)} \ .$$

Here r_g is the Schwarzschild radius, given by $r_g \equiv 2GM/c^2$, and Q is the total charge of the black hole. The solution is written in a coordinate system in which the coordinate lines of the time are directed along the trajectories of the Killing vector $\boldsymbol{k}$. Therefore in this coordinate system the solution becomes meaningless inside the horizon, where the Killing vector becomes space-like.

In order to obtain an analytic continuation of the solution into the interior of the horizon, we must go over to a coordinate system in which the metric has no singularities at $r = r_\pm$, where r_+ and r_- are the larger and smaller roots of the equation (2.3.72) with $a = 0$. For this, following [2.43], we go over to the coordinates u and v defined by

$$u = u(r^* + ct), \quad v = v(r^* - ct), \quad dr^*/dr \equiv A^{-1} \ .$$

Then the metric takes the form

$$ds^2 = -f^2 du\, dv - r^2(d\theta^2 + \sin^2\theta\, d\varphi^2), \quad f^2 = A/(u'v') \ . \tag{2.3.73}$$

If the metric is to have no singularities at $r = r_+$ or $r = r_-$, the zeros of the function A must coincide with the simple zeros of the function $u'(r^* + ct)v'(r^* - ct)$. This requires that the function $u'v'$ can be represented in the form $\varphi(r^*)\psi(t)$, from which we obtain

$$\begin{aligned} &u = \alpha \exp[\gamma(ct + r^*)], \quad v = \beta \exp[\gamma(r^* - ct)], \\ &f^2 = e^{-2\gamma r} r^* (r - r_+)^{n_+} (r - r_-)^{n_-} / (r^2 \gamma^2 \alpha\beta), \\ &n_+ \equiv 1 - 2\gamma r_+^2/(r_+ - r_-), \quad n_- \equiv 1 + 2\gamma r_-^2/(r_+ - r_-) \ . \end{aligned} \tag{2.3.73'}$$

It is clear from the expression obtained for f^2 that it is impossible to choose a coordinate system with no singularities at both $r = r_+$ and $r = r_-$. To avoid singularities in the metric at $r = r_+$, we must put $n_+ = 0$. Therefore a metric which is appropriate for analytic continuation beyond the exterior horizon $r = r_+$ has the form (2.3.73′), where $n_+ = 0$, $n_- = 1 + r_-^2/r_+^2$, $\gamma_+ = (r_+ - r_-)/(2r_+^2)$. The metric (2.3.73) is written in one chart from the atlas of the manifold, where u and v vary from $-\infty$ to $+\infty$. The interior horizon $r = r_-$ corresponds to the Cauchy horizon. To obtain the metric on the other chart of the manifold, we must put

$$n_- = 0, \quad n_+ = 1 + r_+^2/r_-^2, \quad \gamma_- = (r_- - r_+)/(2r_-^2) \ .$$

This new chart with the coordinate grid $\tilde{u}$, $\tilde{v}$ has an intersection with the preceding chart on the open set $u > 0$, $v > 0$. The metric in the new coordinates $\tilde{u}$ and $\tilde{v}$ has no singularities at $r = r_-$ and covers part of the manifold near the singularity $r = 0$. It is interesting to note that, in contrast to the case of uncharged nonrotating black holes, the singularity at $r = 0$ for charged black holes has a time-like character, and the time-like geodesics intersecting $r = r_-$ first approach the singularity and then recede from it.

By systematically matching the charts in the regions in which they have a nonzero intersection, we obtain the maximal analytic continuation of the Nordström-Reissner solution.

We note that the continuation of the solution from one chart to another is not determined by the field equations in general relativity; the particular continuation described above was distinguished by the requirement of analyticity and nonextensibility. However, even analytic continuation does not, in general, ensure that the solution is uniquely determined globally: in the general case, it is difficult to establish which events reached on the various non-space-like curves can be identified.

A collapsing cloud can pass through the two horizons r_+ and r_- and find itself in a new universe. Moreover, there is no singularity inside the cloud, though a time-like singularity occurs outside the cloud! After the cloud crosses the Cauchy horizon, we can say nothing about its subsequent fate, since it becomes possible to influence the entire evolutionary process by means of signals from the time-like singularity.

It follows from the linearized formulation that small perturbations of the initial data at the Cauchy horizon grow without limit, but do they necessarily lead to the formation of singularities at Cauchy horizons [2.30, 44]? Because of analytic difficulties, this question has not been fully investigated.

13) We consider now the *uniqueness theorem* for stationary electrovacuum solutions with axial symmetry. The Lagrangian of the electrovacuum equations for the metric coefficients X and W [see (2.3.38)] and the potentials E and B of the electromagnetic field [see (2.3.46)] has the form

$$
\begin{aligned}
L = p[(\nabla \ln X)^2 - (\nabla B)^2/X] \\
- p^{-1}[X^2(\nabla(W/X))^2 - X(\nabla E + W\nabla B/X)^2] \ ,
\end{aligned}
\tag{2.3.74}
$$

which, apart from divergence terms, follows from the expressions for the Lagrangians (2.3.48, 49). Equating to zero the coefficients of the independent variations $\delta(W/X)$ and δE, we obtain the field equations

$$
\begin{aligned}
&p\nabla[p^{-1}X^2\nabla(W/X)] + X\nabla B(\nabla E + W\nabla B/X) = 0, \\
&\nabla[p^{-1}X(\nabla E + W\nabla B/X)] = 0 \ .
\end{aligned}
$$

These equations imply the existence of Ernst-type potentials Y and Φ, i.e.,

$$
\begin{aligned}
&\varepsilon_{ij}X^2(W/X)_{,j} = p[Y_{,j} + 2(\Phi B_{,i} - B\Phi_{,j})], \\
&\varepsilon_{ij}X(E_{,i} + WB_{,i}/X) = 2\Phi_{,j}p \ ;
\end{aligned}
\tag{2.3.75}
$$

here i and j indicate the coordinates p and q, and ε_{ij} is the Levi-Civita symbol, with $\varepsilon_{11} = \varepsilon_{22} = 0$, $\varepsilon_{12} = -\varepsilon_{21} = 1$. For the functions Y and Φ introduced in (2.3.75), we obtain, after eliminating W/X and E, the system

$$
\begin{aligned}
&\nabla[pX^{-2}(\nabla Y + 2\Phi\nabla B - 2B\nabla\Phi)] = 0, \\
&\nabla[pX^{-1}\nabla\Phi] + pX^{-2}\nabla B \cdot M = 0, \\
&M \equiv \nabla Y + 2\Phi\nabla B - 2B\nabla\Phi \ .
\end{aligned}
\tag{2.3.76}
$$

From (2.3.74), by considering a variation with respect to X and B (for fixed W/X and E) and making use of (2.3.75), we obtain

$$
\begin{aligned}
&\nabla(pX^{-1}\nabla X) + p\left[\frac{M^2}{X^2} - \frac{2(\nabla\Phi)^2}{X} - \frac{2(\nabla B)^2}{X}\right] = 0, \\
&\nabla(pX^{-1}\nabla B) - X^{-2}p\nabla\Phi \cdot M = 0 \ .
\end{aligned}
\tag{2.3.76$'$}
$$

As is readily verified, the system (2.3.76, 76′) can be derived from a variational principle with a Lagrangian L:

$$
L = p\left\{(\nabla \ln X)^2 + \frac{M^2}{X^2} - \frac{4(\nabla\Phi)^2}{X} - \frac{4(\nabla B)^2}{X}\right\} \ . \tag{2.3.77}
$$

The function X must be less than zero, and equal to zero on the symmetry axis. This follows from the requirement that there are no closed time-like curves, whose existence would contradict the principle of causality.

We shall show that the functional L is convex, i.e., the second variation $\delta^2 L$ is positive definite, apart from divergence terms. We denote by $\mathcal{P}$ the set of parameters X, Y, Φ, and B, and write $\partial\mathcal{P}$ for a perturbation of the state $\mathcal{P}$. Then, apart from cubic terms in $\delta\mathcal{P}$, we have

$$
L(\mathcal{P} + \delta\mathcal{P}) = L(\mathcal{P}) + \delta L + \delta^2 L + \mathrm{O}((\delta\mathcal{P})^3) \ .
$$

We introduce the notation

$$
\delta X/X \equiv q_1, \quad (\delta Y + 2\Phi\delta B - 2B\delta\Phi)/X \equiv q_2, \quad \delta\Phi \equiv q_3, \quad \delta B \equiv q_4 \ .
$$

Apart from divergence terms, the linear part δL is equal to zero by virtue of the field equations for the unperturbed state $\mathcal{P}$.

We write the quadratic corrections $\delta^2 L$ for the individual terms in (2.3.77):

$$
p\delta^2(\nabla \ln X)^2 = p(\nabla q_1)^2 + q_1^2\nabla(p\nabla \ln X) - \nabla[pq_1^2\nabla \ln X] \ ,
$$

$$
\begin{aligned}
p\delta^2[((\nabla\Phi)^2 + (\nabla B)^2)/X] = pX^{-1}\{&(\nabla q_3)^2 + (\nabla q_4)^2 - \nabla\Phi\nabla q_3 q_1 \\
&+ \nabla\Phi q_3\nabla q_1 - \nabla B \cdot M q_3 q_1 X^{-1} - \nabla B\nabla q_4 q_1 \\
&+ \nabla B q_4 \nabla q_1 + \nabla E \cdot M q_4 q_1 X^{-1} + q_1^2[(\nabla\Phi)^2 \\
&+ (\nabla B)^2]\} - \nabla\{pX^{-1}[(\nabla\Phi)q_3 q_1 \\
&+ (\nabla B)q_4 q_1]\} \ ,
\end{aligned}
$$

$$\begin{aligned}\delta^2(M^2/X^2) = &(\nabla q_2)^2 + \mu^2 + 4M(q_3\nabla q_4 - q_4\nabla q_3)X^{-2}\\ &+ \nabla q_2(4q_3\nabla B - 4q_4\nabla\Phi - 2q_1 M)X^{-1}\\ &- q_2(4\nabla q_3\nabla B - 4\nabla q_4\nabla\Phi - 2\nabla q_1 M)X^{-1}\\ &+ 4q_2 M(q_3\nabla\Phi + q_4\nabla B)X^{-2} - 2q_1 q_2\nabla X M X^{-2}\\ &- 2Mq_1(q_2\nabla X + 4q_3\nabla B - 4q_4\nabla\Phi)\\ &- q_2^2 p^{-1}\nabla(p\nabla\ln X) + 2M^2 q_1^2 X^{-2}\\ &+ p^{-1}\nabla\{pX^{-1}\nabla q_2(\mu - q_1 M/X)\} \ ,\end{aligned}$$

$$\mu \equiv q_2\nabla X + 4(q_3\nabla B - q_4\nabla\Phi) - Mq_1 \ .$$

Combining the terms in these expressions, it can be shown that

$$\delta^2 L = p\mathcal{G} + \operatorname{div} pA \ ,$$

$$\begin{aligned}\mathcal{G} \equiv &[\nabla q_1 + (q_2 M - 2q_3\nabla\Phi - 2q_4\nabla B)X^{-1}]^2\\ &+ [\nabla q_2 - (q_1 M - 2q_3\nabla B + 2q_4\nabla\Phi)X^{-1}]^2\\ &- 2X^{-1}(\nabla q_3 + q_4 M X^{-1} - q_1\nabla\Phi)^2\\ &- 2X^{-1}(\nabla q_3 - q_3\nabla\ln X + q_2\nabla B)^2\\ &- 2X^{-1}(\nabla q_4 - q_3 M X^{-1} - q_1\nabla B)^2\\ &- 2X^{-1}(\nabla q_4 - q_4\nabla\ln X - q_2\nabla\Phi)^2\\ &+ [q_1 M X^{-1} - q_2\nabla\ln X - 2(q_3\nabla B - q_4\nabla\Phi)]^2\\ &+ 12(q_3\nabla B - q_4\nabla\Phi)^2 X^{-2} \ ,\end{aligned}$$

$$A \equiv -q_2^2\nabla\ln X + X^{-1}q_1[\nabla\Phi q_3 + \nabla B q_4] + X^{-1}\nabla q_2(\mu - q_1 M/X) \ .$$

Thus, the equations for small perturbations q_A ($A = 1,2,3,4$) can be obtained from the variational principle $\delta\int p\mathcal{G}\,dp\,dq = 0$. From this, we have

$$-p\delta\mathcal{G}/\delta q_A \equiv \nabla(p\partial\mathcal{G}/\partial\nabla q_A) - p\partial\mathcal{G}/\partial q_A = 0, \quad A = 1,2,3,4 \ .$$

Multiplying $\delta\mathcal{G}/\delta q_A$ by q_A, summing over all A, and making use of the fact that $\mathcal{G}$ is a homogeneous quadratic function of q_A and ∇q_A, we obtain

$$-p\sum_{A=1}^{4} q_A\delta\mathcal{G}/\delta q_A = \nabla(pq_A\partial\mathcal{G}/\partial\nabla q_A) - 2p\mathcal{G} = 0 \ . \tag{2.3.78}$$

Theorem I by Carter-Robinson [2.45]. If two neighboring solutions of the electrovacuum system (2.3.76, 76′) are equal on the boundary of a simply connected region $\mathcal{D}$, then it follows from (2.3.78) that they are equal throughout the interior of the region $\mathcal{D}$.

Indeed, let us integrate (2.3.78) over the region $\mathcal{D}$. The integral of the divergence part of (2.3.78) reduces to an integral over the boundary of the region

$\mathcal{D}$ and therefore vanishes. The positive definite function $\mathcal{G}$ can be equal to zero only if each of the eight squares of which $\mathcal{G}$ consists is equal to zero. However, in that case all the q_A must be equal to zero identically.

We shall apply Theorem I to a region whose boundary is the pseudo-Euclidean infinity and the horizon of a black hole. For this purpose, it is convenient to transform to the variables μ and λ defined by the relations

$$p^2 = (\lambda^2 - c_1^2)(1 - \mu^2), \quad q = \lambda\mu, \quad c_1 \le \lambda < +\infty, \quad -1 \le \mu \le 1 \ .$$

As was shown by *Carter* [2.34], the pseudo-Euclidean condition is ensured by the following asymptotic behavior of the functions X, Y, Φ, and B as $\lambda \to +\infty$:

$$\begin{aligned} X &= (\mu^2 - 1)\lambda^2 + O(\lambda), \quad Y = 2I\mu(3 - \mu^2) + O(\lambda^{-1}), \\ \Phi &= -Q\mu + O(\lambda^{-1}), \qquad B = O(\lambda^{-1}) \ , \end{aligned} \tag{2.3.79}$$

where I and Q are, respectively, the angular momentum and charge of the black hole (in a system of units with $c = G = 1$), and the magnetic monopole is assumed to be equal to zero. A solution of the system (2.3.76, 76′) has an event horizon if it is regular at $\lambda = c_1$. It follows from Theorem I that there cannot exist two neighboring solutions which are regular at $\lambda = c_1$ and have the same asymptotic behavior (2.3.79). Therefore all stationary, axially symmetric, asymptotically flat electrovacuum fields which are regular at $\lambda = c_1$ form discrete families of solutions depending, at the most, on three parameters (mass, angular momentum, and charge).

In addition to this theorem, *Robinson* [2.46] managed to prove a global uniqueness theorem for stationary axially symmetric solutions of Einstein's vacuum equations which differ finitely inside a region $\mathcal{D}$ and are identical on the boundary of $\mathcal{D}$. This theorem implies that the Kerr solution with a nondegenerate event horizon is unique among all solutions of Einstein's vacuum equations which are regular at $\lambda = c_1$ and asymptotically flat as $\lambda \to \infty$. The corresponding result for the electrovacuum has not yet been proved but is apparently valid.

2.4 Energetics of Black Holes

2.4.1 Temperature of a Black Hole

According to Lemma 3, there exists a constant Ω such that the Killing vector $\boldsymbol{l} = \boldsymbol{k} + \Omega \boldsymbol{t}$ is time-like off the horizon and isotropic on the horizon H.

We go over from the group coordinate φ in the canonical form of the stationary metric (2.3.38) to the variable $\varphi^+ = \varphi - \Omega t$; then the metric (2.3.38) takes the form

$$ds^2 = V^+ dt^2 + 2W^+ dt\, d\varphi^+ + X\, d\varphi^{+2} - g_{ab} d\theta^{ab} d\theta^b \ , \tag{2.4.1}$$

$$W^+ = W + X\Omega, \quad V^+ = V + 2W\Omega + X\Omega^2, \quad a, b = 1, 2 \ . \tag{2.4.2}$$

The coefficient V^+ is equal to the scalar square of the vector $\boldsymbol{l}$ and therefore vanishes on the horizon. Thus, the surface $V^+ = 0$ is isotropic, and the normal to it must be directed along the vector $\boldsymbol{l}$:

$$\nabla_i V^+ = -2\Lambda l_i \quad . \tag{2.4.3}$$

Theorem K. The scalar Λ is constant on each connected component of the horizon.

This property of Λ was discovered [2.42] when the Kerr solution was transformed from the original coordinate system (in which it was first found [2.47]) to the canonical form (2.3.38). The fact that the scalar Λ is constant on the horizon has also been proved [2.34, 48] without using the theorem on the uniqueness of the Kerr solution for the description of rotating black holes.

The scalar Λ can be interpreted [2.49] as a quantity proportional to the temperature of a black hole: a black hole can emit particles as a quantum system with temperature $T_{\mathrm{H}} = \hbar\Lambda/(2\pi kc)$, where $\hbar$, k, and c are Planck's constant, Boltzmann's constant, and the speed of light in empty space, respectively [2.50–55].

2.4.2 Electrostatic Potential of a Black Hole

If φ is replaced by $\varphi^+ \equiv \varphi - \Omega t$, the 1-form of the potential of the electromagnetic field can be written as

$$E\,dt + B\,d\varphi = \Phi^+ dt + B\,d\varphi^+, \quad \Phi^+ \equiv E + \Omega B \quad . \tag{2.4.4}$$

We shall show that the function Φ^+ must be constant on the horizon.

Indeed, $\nabla_i \Phi^+ = F_{ij} l^j$. The component of the energy-momentum tensor of the electromagnetic field $T_{ij} l^i l^j$ must vanish on the horizon, since the area of the horizon of a stationary black hole with $T_{ij} l^i l^j \neq 0$ increases with time, which contradicts the assumption of stationarity. On the other hand, $T_{ij} l^i l^j = (4\pi)^{-1} F_{ij} l^i F^j_k l^k$. Therefore on the horizon the vector $F_{ij} l^j$ is isotropic; moreover, this vector is orthogonal to the vector l^i (which is time-like off the horizon and isotropic on the horizon). Hence we find that the vector $F_{ij} l^j$ is parallel to the vector l_i:

$$F_{ij} l^j = \nabla_i \Phi_+ = \alpha l_i \quad . \tag{2.4.5}$$

It follows from the condition (2.4.5) that the derivative of Φ_+ along any direction on the horizon must vanish. This proves the following lemma.

Lemma 7. The potential Φ_+ of the electric intensity $F_{ij} l^j$ maintains a constant value on each connected component of the horizon [2.34].

Definition. The value of Φ_+ on the horizon is called the electrostatic potential of the black hole.

2.4.3 Formula for the Mass of a Black Hole

In the general case of a stationary axially symmetric system, the total mass, total angular momentum, and total charge of the system can be determined from the requirement that the asymptotic behavior at large distances must correspond to the Newton analog:

$$\kappa M c^2 = 2 \oint_{S_\infty} \nabla^i k^j ds_{ij}, \quad Q = -(4\pi)^{-1} \oint_{S_\infty} F^{ij} ds_{ij} \ , \tag{2.4.6}$$

$$\kappa I = - \oint_{S_\infty} \nabla^i t^j ds_{ij} \ ; \tag{2.4.7}$$

here the integration extends over an asymptotically remote two-dimensional space-like surface S_∞ at the pseudo-Euclidean infinity.

When a large mass collapses, there is a loss of information about this star for an external observer, for whom the resulting field of the black hole (as the final state of the collapse) is characterized by only three parameters — the mass, the angular momentum, and the charge. From the point of view of thermodynamics, the final state of a collapsing system must be characterized by the maximum value of the entropy. It has been shown by numerous authors [2.56–64] that in the process of weakly nonspherical collapse the details of the internal structure of the star have an influence on the external field which vanishes according to a power law as the surface of the star approaches the horizon (see Chap. 4), and that no fields other than the gravitational and electrostatic fields can exist outside the black hole.

Following [2.48], we shall derive a relation between the scalar Λ (the temperature of the black hole), the area of the black hole, the angular speed of its rotation, the electrostatic potential, and the parameters associated with the conservation laws (2.4.6, 7), without making use of the proof given by *Robinson* [2.45] that the Kerr-Newman solution is locally unique among the stationary solutions of the electrovacuum equations possessing a nondegenerate horizon.

For this, we construct a three-dimensional space-like surface Σ. According to the generalized Stokes's theorem [see (2.2.11)], the expression for the energy (2.4.6) can be transformed to

$$\kappa M c^2 = 2 \oint_H \nabla^i k^j ds_{ij} - 2 \int_\Sigma \nabla_j \nabla^j k^i d\Sigma_i \ . \tag{2.4.8}$$

Using the result of Problem 1 in Sect. 2.2.3, this expression can be rewritten

$$\kappa M c^2 = 2 \oint_H \nabla^i k^j ds_{ij} + 2 \int_\Sigma R^i_j k^j d\Sigma_i \ . \tag{2.4.9}$$

The integral over the horizon H on the right-hand side of (2.4.9) can be transformed as follows:

$$\oint_H \nabla^i k^j ds_{ij} = \int_H \nabla^i l^j ds_{ij} - \Omega \oint_H \nabla^i t^j ds_{ij} . \tag{2.4.10}$$

The differential 2-form ds_{ij} is related to the element of area dA by the equation

$$ds_{ij} = l_{[i} n_{j]}\, dA ,$$

where l_i and n_i are isotropic vectors orthogonal to the surface H, with $l_i n^i = 1$. Using (2.4.3), we obtain from (2.4.9) the following result for the coefficient Λ:

$$\kappa M c^2 = 2 \oint_H \Lambda\, dA + 2\Omega I_{\mathrm{H}} \kappa + 2 \int_\Sigma R^i_j k^j d\Sigma_i , \tag{2.4.11}$$

$$\kappa I_{\mathrm{H}} = - \oint_H \nabla_i t_j\, ds^{ij} . \tag{2.4.12}$$

The constant I_{H} is called the *angular momentum of the black hole.* The scalar I_{H} defined by (2.4.12) is identical to the total angular momentum I defined by (2.4.7) if, off the horizon, there are no nongravitational fields or matter which give additional contributions to the angular momentum. In what follows, we shall assume that outside a black hole there exists only an electric field with sources equal to zero, and $I_{\mathrm{H}} = I$ in view of the absence of magnetic charges.

Using Theorem K, we can remove the scalar Λ from the integral sign and write (2.4.11) in the form

$$\frac{\kappa M c^2}{2} = \int_\Sigma R^i_j k^j d\Sigma_i + \kappa \Omega I + \Lambda A . \tag{2.4.13}$$

We choose the surface Σ such that the normal to it is orthogonal to the vector t. Then

$$\int_\Sigma R^i_j k^j d\Sigma_i = \int_\Sigma R^i_j l^j d\Sigma_i . \tag{2.4.14}$$

Suppose that

$$l^j R^i_j = \frac{\kappa}{8\pi} F^{il} \nabla_l \Phi^+ . \tag{2.4.15}$$

Using (2.4.15), the expression (2.4.14) can be rewritten in the form

$$\int_\Sigma l^j R^i_j d\Sigma_i = \frac{\kappa}{8\pi} \int_\Sigma \nabla_l (\Phi^+ F^{il})\, d\Sigma_i = -\frac{\kappa}{8\pi} \oint_H \Phi^+ F^{il} ds_{il} , \tag{2.4.16}$$

since the integral over S_∞ tends to zero because of the properties of the asymptotic behavior of the electric field.

According to Lemma 7, the potential Φ^+ is a constant on the horizon, and therefore Φ^+ can be removed from the integral sign. Thus, (2.4.16) takes the form

$$\int l^j R^i_j d\Sigma_i = \frac{\Phi^+ Q\kappa}{2} ,$$

since in the absence of sources off the horizon we have

$$\oint_{S_\infty} F^{il} ds_{il} = \oint_{H} F^{il} ds_{il} = -4\pi Q .$$

Therefore the formula for the mass of a black hole finally takes the elegant form

$$Mc^2 = 2\Omega I + \Phi^+ Q + 2\Lambda A/\kappa . \qquad (2.4.17)$$

In the Kerr-Newman solution, all the parameters of a black hole can be expressed explicitly in terms of quantities satisfying conservation laws: the mass, the angular momentum I, and the charge Q. If, however, the independent parameters are taken to be the area A of the horizon and the quantities I and Q, then for Ω and Φ^+ we have the expressions

$$\Omega = \frac{4\pi I}{MA}, \quad \Phi^+ = \frac{4\pi Q}{AMc^2}\left(\frac{A}{\kappa} + \frac{Q^2}{2}\right) . \qquad (2.4.18)$$

The area of a black hole is related to its parameters by an equation discovered by *Christodoulou* and *Ruffini* [2.65, 66]:

$$(Mc^2)^2 = 4\pi \left\{\frac{A}{\kappa^2} + \frac{Q^2}{\kappa} + \frac{Q^4}{4A} + \frac{I^2c^2}{A}\right\} . \qquad (2.4.19)$$

Using (2.4.17–19), we find for the scalar Λ the expression

$$\Lambda = \frac{4\pi}{Mc^2 A}(A/\kappa + Q^2/2 - GM^2) .$$

2.4.4 "Thermodynamics" of Black Holes

If the collapse of a sufficiently large mass takes place in a restricted region, the final state of the collapsing mass can evidently be only a black hole in which the singularity is hidden from a distant observer by an event horizon. This hypothesis is highly plausible; in any case, it is true for a body with small deviations from spherical symmetry (see Sect. 4.3). In the general theory of relativity, many solutions belonging to the class of so-called "bare singularities" are known: the Weyl solutions and the Kerr-Newman solution (in particular cases, the Kerr and Nordström-Reissner solutions) with $GM^2 < Q^2 + I^2c^2/(GM^2)$. It is physically obvious that collapse cannot occur if a mass rotates too rapidly (when the centrifugal forces prevent compression into a compact object) or if the matter is charged (when collapse is prevented by the electrostatic repulsion).

Theorem L due to Hawking [2.30, 49]. For all classical interactions of black holes, the total area of their event horizons cannot decrease.

As we have already pointed out in Sect. 2.1, in a space-time manifold which is asymptotically simple in the weak sense there exists a family of space-like surfaces $\{\mathcal{L}(\tau)\}$ which intersect the horizon $H^+(H^+ \equiv \partial \mathcal{I}^-(\mathcal{B}^+))$ on connected compact two-dimensional surfaces $L_i(\tau)$, $i = 1, 2, \ldots, N$, and the conformal pseudo-Euclidean boundary $\mathcal{B}^+$ on a two-dimensional sphere. The number of black holes N can change with time, since black holes can "fuse", with the formation of a common event horizon. Black holes cannot break up, since in the Cauchy region for the surface $\mathcal{L}$: $D^+(\mathcal{L})$ each time-like curve directed into the past must intersect the surface $\mathcal{L}$ outside the black holes, in the set $\mathcal{L} \cap \mathcal{I}^-(\mathcal{B}^+)$.

Suppose that at $\tau = \tau_0$ the boundary H^+ consisted of two components L_1 and L_2 (i.e., at $\tau = \tau_0$ there were two black holes) and that at $\tau = \tau_1$ the boundary H^+ consisted of one connected component L_3. As we noted in Sect. 2.1, the isotropic geodesics which form the horizon H^+ can only expand. The closed set L_3 consists of the closed set of points belonging to the isotropic geodesics which at $\tau = \tau_0$ passed through L_1 and L_2, as well as the open set of points belonging to the isotropic geodesics having initial points between the surfaces $\mathcal{L}(\tau_0)$ and $\mathcal{L}(\tau_1)$. Therefore the area of the two-dimensional surface L_3 is strictly greater than the sum of the areas of L_1 and L_2 of the black holes before their fusion.

Imagine that for a certain time a black hole was subjected to external influences, as a result of which its initial parameters M, I, and Q have changed to values $M + dM$, $I + dI$, and $Q + dQ$ in a final stationary state, and that before this state was reached all the perturbations not associated with the three basic parameters have died away. It follows from the Christodoulou-Ruffini formula (2.4.19) that the differentials dM, dI, dQ, and dA are related by the equation

$$d(Mc^2) = \frac{4\pi}{Mc^2 A}\left[\frac{A}{\kappa} + \frac{Q^2}{2} - GM^2\right] d\frac{A}{\kappa} + \frac{4\pi Q}{AMc^2}\left[\frac{A}{\kappa} + \frac{Q^2}{2}\right] dQ + \frac{4\pi I}{AM}\, dI \ . \tag{2.4.20}$$

Making use of (2.4.18), from (2.4.20) we obtain

$$d(Mc^2) = \frac{\Lambda\, dA}{\kappa} + \Omega\, dI + \Phi^+ dQ \ . \tag{2.4.21}$$

We compare (2.4.21) with the first law of thermodynamics for a charged, ideal, axially symmetric conductor rotating with angular speed Ω:

$$dE = T\, dS + \Omega\, dI + \Phi\, dQ \ . \tag{2.4.21'}$$

Here Φ is the potential of the conductor, E is its energy, T is the temperature, Q is the charge, and I is the angular momentum. According to the second law, the entropy S in an adiabatically isolated body can only increase.

The analogy between the corresponding terms in (2.4.21) and (2.4.21′) is quite transparent. Comparing (2.4.21) with (2.4.21′), Theorem L offers grounds for supposing that the role of the entropy in the "thermodynamics" of a black hole is played by a quantity proportional to the area of its horizon (or a monotonic function of A). By means of traditional thermodynamic arguments (relating to the work of a heat engine in a Carnot cycle), it can be shown [2.67, 68] that the entropy of a black hole must be proportional to its area. The exact value of the coefficient of proportionality can be found only on the basis of arguments involving quantum field theory [2.31, 49]. Using the relation between the temperature of a black hole and the scalar Λ mentioned in Sect. 2.4.1, we find from the equation $T_{\rm H} dS_{\rm H} = \Lambda\, dA/\kappa$ an expression for the entropy of the black hole in the form $S_{\rm H} = 2\pi kcA/(\kappa\hbar)$, where k and $\hbar$ are Boltzmann's and Planck's constants, respectively.

Equation (2.4.21) can be obtained differently by relating the change in the area of the black hole to the flux of energy and angular momentum of the matter through the event horizon in the linear approximation.

Let us consider the perturbed equation (1.1.60) on the horizon, where in the unperturbed stationary solution the rotation coefficients ϱ, σ, and k are equal to zero on the horizon. In the case of a perturbation of the horizon, we assume that the vector $\boldsymbol{l}$ remains isotropic and tangential to the light rays from which the horizon is formed. Then in the perturbed state we have a vanishing coefficient $k = 0$, and this equation takes the form

$$\frac{d\varrho}{dv} = \Lambda\varrho + \frac{\kappa}{2} T_{ij} l^i l^j , \quad \Lambda = \varepsilon + \varepsilon^* \ . \tag{2.4.22}$$

The parameter ϱ has the meaning of an index giving the rate of convergence of the light rays (see Sect. 1.1):

$$2\varrho = -\partial \ln \sqrt{g}/\partial v \ ,$$

where g is the determinant of the metric on a two-dimensional space-like cross section of the horizon. Therefore, if A is the area of this cross section, we have

$$dA/dv = -2\varrho A \ . \tag{2.4.23}$$

Solving the linear equation (2.4.22) for ϱ, we obtain

$$\varrho = -\frac{\kappa}{2} \int_v^\infty T_{ij} l^i l^j \exp[\Lambda(v - v')]\, dv' \ . \tag{2.4.24}$$

In (2.4.24) it is also assumed that when $v \to +\infty$ the black hole reaches a stationary state, so that $\varrho \to 0$ for $v \to \infty$. Substituting (2.4.24) into (2.4.23) and integrating with respect to v from $-\infty$ to $+\infty$, we readily obtain

$$\delta A = \frac{\kappa}{\Lambda} \int T^{ij} l_i\, d\Sigma_j , \quad d\Sigma_j = l_j\, dA\, dv , \quad l_j = k_j + \Omega t_j \ . \tag{2.4.25}$$

We now make use of the fact that the vector $T_{ij}k^i$ is the energy flux density, while the vector $-T_{ij}t^i$ is the angular-momentum flux density. Therefore the integrated flux of energy and angular momentum of uncharged matter through the horizon must lead to changes in the mass and angular momentum of the black hole by the amounts

$$\delta(Mc^2) = \int T_{ij}k^i d\Sigma^j, \quad \delta I = -\int T_{ij}t^i d\Sigma^j \ .$$

Then for the change in area of the black hole we obtain from (2.4.25) the final expression

$$\Lambda\delta(A/\kappa) + \Omega\delta I = \delta(Mc^2) \ . \tag{2.4.25'}$$

When $\delta Q = 0$, (2.4.25') is identical to (2.4.21), which we obtained formally using Christodoulou's formula.

We now consider the *extraction of energy from black holes*.

a) Consider the following thought experiment [2.69]. Suppose that a test particle falls freely in the field of a stationary black hole described by the Kerr solution, and that inside the ergosphere (as a result of some internal mechanism) it breaks up into two particles, one of which is captured by the black hole, while the other escapes to the pseudo-Euclidean infinity. Let us calculate what energy and angular momentum are carried away by the escaping particle. Since the field of the black hole is assumed to be stationary, the quantity $p_i k^i = E_0$ is constant along the trajectory of the particle (the time-like vector p_i is the 4-momentum of the particle). Outside the ergosphere the vector k^i is time-like, so that the constant E_0 can only be greater than zero. However, inside the ergosphere the expression $p_i k^i$ can be negative, since the vector k^i is space-like inside the ergosphere.

If the particle breaks up into two uncharged particles, the energy and angular momentum must be conserved:

$$E_0 = E_1 + E_2, \quad I_0 = I_1 + I_2; \quad I_0 \equiv p_i t^i \ . \tag{2.4.26}$$

It is possible to choose the parameters of the breakup in such a way that one of the produced particles has negative energy, i.e., $E_1 < 0$, and moves in the direction opposite to that of the rotation of the black hole, i.e., $I_1 < 0$. In that case, if the original particle was moving in the direction of rotation of the black hole, the second produced particle (which escapes outward) will have values of the energy and angular momentum exceeding those of the initial particle. Therefore the second particle extracts energy and angular momentum from the black hole, carrying them away to the pseudo-Euclidean infinity.

The black hole "swallows" the first particle and decreases its mass and angular momentum:

$$\delta(Mc^2) = E_1 < 0, \quad \delta I = I_1 < 0 \ .$$

The process takes place with the maximum gain of energy in the case of breakup of the incident particle near the horizon, when the first particle has 4-momentum which is asymptotically parallel to the vector $l_i = k_i + \Omega t_i$. However, in this case E_1 and I_1 are related by the equation $E_1 = \Omega I_1$ (where Ω is the angular speed of rotation of the black hole). It is only in this limiting case that the area of the black hole will not increase, according to (2.4.25′). In the general case, when a particle breaks up in the ergosphere with fulfillment of the conservation laws (2.4.26), the surface area A of the black hole can only increase. In the first (limiting) case it is natural to say that the process is *reversible*, whereas in the general case the process is *irreversible* in view of the previously mentioned analogy between the area A and the entropy S.

Using (2.4.19), we can readily estimate the maximum energy which can be extracted from an uncharged black hole. When a black hole loses its rotation, it becomes an uncharged black hole described by the Schwarzschild solution. The maximum amount of energy can be extracted from a black hole by means of reversible processes, when the final (irreducible) mass of the black hole is related to the original area A by the equation $\kappa c^2 M_0 = \sqrt{4\pi A}$. Therefore the upper limit on the energy which can in principle be extracted from a black hole by means of reversible processes is

$$\kappa M c^2 - \kappa M_0 c^2 = \sqrt{4\pi A + 4\pi\kappa^2 c^2 I^2/A} - \sqrt{4\pi A} \ . \tag{2.4.27}$$

b) Another classical process in which energy of rotational motion can be extracted from a black hole was proposed by *Press* [2.70]. Suppose that a black hole is situated in an external stationary field. Inside the ergosphere of the black hole, this field cannot change in a stationary manner from the point of view of a distant observer, since inside the ergosphere the Killing vector k^i becomes space-like.

Therefore, if a rotating black hole is situated in an external stationary force field, a local observer will see a flux of energy through the horizon. Because of this, the black hole evolves into another configuration (with mass not less than the irreducible mass). The direction of rotation of the black hole is aligned by the external force field in such a way that in the final configuration the total flux of energy through the horizon is equal to zero.

c) The Penrose method of extracting rotational energy from a black hole can be modified somewhat by considering, instead of incident particles, incident waves, such as scalar, electromagnetic, or gravitational waves [2.62, 71–74]. Since the main background is stationary and axially symmetric, solutions of the equations for the perturbations can be sought in terms of a superposition of solutions of the form $\exp[i(\omega t + m\varphi)]$. For a monochromatic wave, the ratio of ω to m gives the ratio of the flux of energy to the flux of angular momentum in the incident wave, and this same ratio will be preserved in the reflected and trapped waves.

Therefore the ratio of the increment in the energy of the black hole to the increment in its angular momentum will be ω/m:

$$\frac{\delta M c^2}{\delta I} = \frac{\omega}{m} \ . \tag{2.4.28}$$

On the other hand, it follows from (2.4.21) that

$$\delta(Mc^2) \geq \Omega \delta I \quad , \tag{2.4.29}$$

since $\delta A \geq 0$ according to Theorem L. It follows from (2.4.28, 29) that the reflected wave will carry away energy of the black hole, i.e., δM will be less than zero, only if $\omega < \Omega m$ [2.72]. We emphasize that this criterion of Starobinsky for enhancement of a wave after reflection from a rotating black hole does not depend on the specific form of the wave — scalar, electromagnetic, etc. This phenomenon of enhancement of waves after reflection from a rotating black hole has been called "superradiation" [2.74].

d) *Hawking* [2.75] has presented elegant arguments concerning an upper limit on the emitted energy when black holes collide.

When two black holes approach each other, it can happen that there is an isotropic surface S having a compact space-like cross section and containing the horizons of both black holes in its interior, when the light rays of the surface S are unable to overcome the forces of gravity and escape to the pseudo-Euclidean infinity. This process can be treated as the fusion of two black holes with areas A_1 and A_2 into a third black hole with area A_3.

It is obvious that a portion of the energy $(M_1 + M_2 - M_3)c^2$ must be emitted in the process of fusion. Using (2.4.19), we can readily estimate an upper limit on the emitted energy. If the original black holes with masses M_1 and M_2 and the resulting one with mass M_3 are not charged and do not rotate, each of them obeys the relation

$$A = 16\pi G^2 M^2 / c^4 \quad . \tag{2.4.30}$$

According to Theorem L,

$$A_1 + A_2 \leq A_3 \quad .$$

Using (2.4.30), this inequality can be rewritten

$$M_1^2 + M_2^2 \leq M_3^2 \quad .$$

Therefore we have the following inequality for the fraction of the emitted energy:

$$\frac{M_1 + M_2 - M_3}{M_1 + M_2} \leq 1 - \frac{\sqrt{M_1^2 + M_2^2}}{M_1 + M_2} \leq 1 - \frac{1}{\sqrt{2}} \quad .$$

Thus, when uncharged nonrotating black holes collide, up to $(1 - 1/\sqrt{2})100\,\% \approx 29\,\%$ of their initial energy can be emitted.

In the general case of the fusion of charged and rotating black holes with masses M_1 and M_2, up to $(1 - 1/\sqrt{8})100\,\% \approx 65\,\%$ of the initial energy $(M_1 + M_2)c^2$ of the colliding black holes can in principle be emitted.

We note that the *actually calculated* fraction of the energy emitted in a collision of two black holes may be much lower than the upper limits given here [2.76].

3. Stationary Axially Symmetric Fields in General Relativity

In addition to gravitational and electromagnetic fields, neutrino fields are objects of fundamental research in theoretical physics. The subject of this chapter is the nature of the interaction of these three types of material fields in fairly general situations.

In Sect. 1.2 we showed that in a region of neutrino emission there is a rotation of the polarization vector of linearly polarized gravitational waves. If, besides the neutrino emission, an external electromagnetic field is present, then gravitational and electromagnetic waves propagate as if there were no neutrino emission but there existed a certain effective electromagnetic field different from the given field. As was first shown in [3.1] (see Sect. 1.2), the process of propagation of waves in arbitrary external electromagnetic fields takes place as a gradual mutual conversion of gravitational and electromagnetic waves, with a simultaneous rotation of their polarization vectors.

In Sect. 3.1 we derive the canonical equations of the neutrino-electromagnetic vacuum with an Abelian group of motions G_2 on V_2. For the canonical equations obtained in Sect. 3.1, in Sect. 3.2 we study a Lie algebra with a countable number of parameters and carry out its exponentiation. We also give the general solution of the electrovacuum equations for Ernst data on the axis of symmetry, using an integral equation obtained by the present author for a unique function, and we present new classes of exact solutions of the equations for the neutrino-gravitational vacuum (Sect. 3.3). Use is made here of a model of the free neutrino field described by Weyl's equations. Like the Dirac model of the electron, this model has a well-known defect (an indefinite sign of the energy density), and this creates a number of difficulties, which can be eliminated by second quantization. Experiments to detect a possible neutrino mass and to study solar neutrinos may necessitate a radical revision of the existing concepts of neutrino physics. The classical approach developed here for the neutrino field would then be merely asymptotic.

Finally, in Sect. 3.4 we study a relationship among the integrable systems of equations of mathematical physics, which consists in the possibility of a linear representation of their internal-symmetry algebras (Lie-Bäcklund algebras) by infinite-dimensional matrices.

3.1 Canonical Equations of Massless Fields Admitting Abelian Two-Parameter Groups of Motions

We shall consider a space filled with free electromagnetic and neutrino fields which interact with free gravitational fields — the so-called neutrino electrovacuum. The complete system of equations describing this vacuum consists of Einstein's equations with the energy-momentum tensor of the neutrino and electromagnetic fields, Weyl's equations, and Maxwell's vacuum equations.

We shall consider the form of the equations of this system for solutions possessing an Abelian group of motions G_2. The situation in which one of the Killing vectors is time-like, while the second one is space-like and has closed orbits, corresponds to the case of stationary fields with axial symmetry. The situation in which both commuting Killing vectors are space-like corresponds to the case of plane or cylindrical waves.

In what follows, we shall refer to these cases as cases (a) and (b), respectively.

a) We write the square of the interval in the form

$$ds^2 = 2(l_i n_j - m_i m_j^*)\, dx^i dx^j \quad , \tag{3.1.1}$$

where the asterisk indicates complex conjugation, and in this case the vectors $\boldsymbol{l}$, $\boldsymbol{n}$, $\boldsymbol{m}$, $\boldsymbol{m}^*$ forming the complex tetrad do not depend on the time t or the coordinate angle φ. In what follows, we shall use capital Latin letters $A, B, \ldots$ $(A, B, \ldots = 1, 2)$ for the indices corresponding to the coordinates t and φ, and Greek letters $\mu, \nu, \ldots$ $(\mu, \nu, \ldots = 3, 4)$ for the remaining coordinates.

We shall write the complete system of equations in terms of projections onto the indicated tetrad field. This will enable us to make use of the well-known calculations of Newman and Penrose (see Sect. 1.1).

Let D, Δ, δ, and δ^* be the operators of differentiation along the vectors $\boldsymbol{l}$, $\boldsymbol{n}$, $\boldsymbol{m}$, and $\boldsymbol{m}^*$, respectively.

Weyl's equations for the neutrino field have the form

$$\begin{aligned} D\Phi + \delta^*\Psi &= (\varrho - \varepsilon)\Phi + (\alpha - \pi)\Psi \quad , \\ \delta\Phi + \Delta\Psi &= (\tau - \beta)\Phi + (\gamma - \mu)\Psi \quad , \end{aligned} \tag{3.1.2}$$

where Φ and Ψ are complex functions describing the neutrino field.

In order to calculate the rotation coefficients α, β, γ, ε, ... of the complex tetrad, we reduce the arbitrariness in the choice of the coordinate subsystem x^μ by requiring a conformal Euclidean character of the metric coefficients $g^{\mu\nu}$, after which, instead of the real coordinates x^3 and x^4, we introduce the complex coordinates $\sqrt{2}\,\xi = x^3 + \mathrm{i}x^4$, $\sqrt{2}\,\xi^* = x^3 - \mathrm{i}x^4$. The square of the interval then takes the form

$$\begin{aligned} ds^2 = {} & g_{AB}(dx^A + \theta g^A d\xi + \theta g^{A*} d\xi^*)(dx^B + \theta g^B d\xi + \theta g^{B*} d\xi^*) \\ & - 2\theta^2 d\xi\, d\xi^* \quad , \end{aligned} \tag{3.1.3}$$

where g_{AB} and θ are real functions of x^μ, and g^A are complex functions of x^μ. We note that in the presence of a neutrino field there do not exist two-dimensional surfaces which intersect both Killing vectors orthogonally.

We introduce l_A and n_B by means of the definition

$$g_{AB} = l_A n_B + l_B n_A \quad .$$

We shall raise and lower capital Latin indices by means of the metric g_{AB} and introduce the notation $r \equiv \sqrt{|\det g_{AB}|}$.

From the definition of the rotation coefficients of the tetrad field (see Sect. 1.1), we have

$$\begin{aligned}
&\tau = -\pi^* = -\frac{1}{2\theta r}\frac{\partial r}{\partial \xi} \ , \quad k = -\frac{1}{\theta} l^A \frac{\partial}{\partial \xi} l_A \ , \quad \nu = \frac{1}{\theta} n^A \frac{\partial}{\partial \xi} n_A \ , \\
&\varrho = -\varrho^* = 2\varepsilon = \frac{1}{2\theta^2} l_B \left[\frac{\partial}{\partial \xi^*}(\theta g^B) - \frac{\partial}{\partial \xi}(\theta g^{B*})\right] \ , \\
&\mu = -\mu^* = 2\gamma = \frac{1}{2\theta^2} n_B \left[\frac{\partial}{\partial \xi^*}(\theta g^B) - \frac{\partial}{\partial \xi}(\theta g^{B*})\right] \ , \\
&\alpha = \frac{1}{4\theta}\left[n^A \frac{\partial}{\partial \xi^*} l_A - l^A \frac{\partial}{\partial \xi^*} n_A\right] - \frac{1}{2\theta^2}\frac{\partial \theta}{\partial \xi^*} \ , \\
&\beta = \frac{1}{4\theta}\left[n^A \frac{\partial}{\partial \xi} l_A - l^A \frac{\partial}{\partial \xi} n_A\right] + \frac{1}{2\theta^2}\frac{\partial \theta}{\partial \xi} \ , \qquad \sigma = \lambda = 0 \ .
\end{aligned} \tag{3.1.4}$$

The tetrad components of the energy-momentum tensor of the neutrino field, $4\pi T^{(\nu)}_{ij} l^i m^j = \Phi^{(\nu)}_{01}$ and $4\pi T^{(\nu)}_{ij} n^i m^j = \Phi^{(\nu)}_{12}$, can be expressed in terms of the components of the spinor field, Φ and Ψ, as follows[1]:

$$\begin{aligned}
\Phi^{(\nu)}_{01} &= 4\pi \mathrm{i}[\Psi\delta\Psi^* - \Psi^*\delta\Psi - k\Phi\Phi^* + \varrho\Phi^*\Psi + (\beta - \alpha^* - \pi^*)\Psi\Psi^*] \ , \\
\Phi^{(\nu)}_{12} &= 4\pi \mathrm{i}[\Phi\delta\Phi^* - \Phi^*\delta\Phi - \nu^*\Psi\Psi^* - \mu\Psi^*\Phi + (\alpha^* - \beta - \tau)\Phi\Phi^*] \ .
\end{aligned}$$

(We are using a system of units in which the speed of light and the gravitational constant are equal to unity.)

The expressions for these same tetrad components of the Ricci tensor, according to its definition, have the form

$$\begin{aligned}
\Phi_{01} &= (\delta - \alpha^* - \beta - 3\tau)\varrho/2 + k\mu/2 \ , \\
\Phi_{12} &= (\delta + \alpha^* + \beta - 3\tau)\mu/2 + \nu^*\varrho/2 \ .
\end{aligned}$$

For what follows, it is important that the corresponding components of Einstein's equations [taking into account Weyl's equations (3.1.2)] can be written in the form of the system

[1] For the electromagnetic field, the components $\Phi^{(\mathrm{em})}_{01}$ and $\Phi^{(\mathrm{em})}_{12}$ of the energy-momentum tensor are equal to zero; here and in what follows, the superscripts (ν) and (em), respectively, label the neutrino and electromagnetic components of the energy-momentum tensor.

$$(\delta - \alpha^* - \beta - 3\tau)A - kB = 0 \ ,$$
$$(\delta + \alpha^* + \beta - 3\tau)B + \nu^* A = 0 \ , \tag{3.1.5}$$

where

$$A \equiv \varrho + 8\pi i \Psi\Psi^* \ , \quad B \equiv \mu - 8\pi i \Phi\Phi^* \ .$$

Here we use only the trivial solutions of the system (3.1.5):

$$A = B = 0 \ . \tag{3.1.6}$$

In this case, Weyl's equations (3.1.2) can be rewritten in the form

$$\delta^*\Psi = -4\pi i \Psi^*\Psi\Phi + (\alpha - \pi)\Psi \ ; \quad \delta\Phi = -4\pi i \Phi\Phi^*\Psi + (\tau - \beta)\Phi \ . \tag{3.1.2'}$$

Using the expressions (3.1.4) for α, β, τ, and π, we find that the system (3.1.2′) has the important "first" integral

$$\Phi\Psi^* = \frac{i}{8\pi\theta r}\frac{\partial w}{\partial \xi^*} \ , \tag{3.1.7}$$

where w is an arbitrary real harmonic function, and the factor $i/8\pi$ and the expression for an arbitrary analytic function of ξ^* in the form of the derivative of the function w with respect to ξ^* are introduced for convenience in the calculations.

Let us consider the tetrad components of the energy-momentum tensor of the neutrino field, $\Phi_{00}^{(\nu)} = 4\pi T_{ij}^{(\nu)} l^i l^j$ and $\Phi_{22}^{(\nu)} = 4\pi T_{ij}^{(\nu)} n^i n^j$:

$$\Phi_{00}^{(\nu)} = 8\pi i[k\Phi\Psi^* - k^*\Psi\Phi^* + \underline{(\varepsilon - \varepsilon^*)\Psi\Psi^*}] \ ,$$
$$\Phi_{22}^{(\nu)} = 8\pi i[\nu^*\Phi\Psi^* - \nu\Psi\Phi^* + \underline{(\gamma^* - \gamma)\Phi\Phi^*}] \ . \tag{3.1.8}$$

The underlined terms in (3.1.8) cancel with the terms $-\varrho^2$ and $-\mu^2$ of the Ricci tensor in Einstein's equations, by virtue of the solutions (3.1.6). The corresponding components of Einstein's equations (after multiplication by θ) will contain only the components of the energy-momentum tensor of the electromagnetic field, the metric coefficients g_{AB}, and the harmonic function w.

Before writing down these equations explicitly, let us consider the following tetrad component of Einstein's equations: $4\pi T_{ij} l^i n^j = \Phi_{11} = -\varrho\mu - (\delta - 2\tau + \beta - \alpha^*)\pi$. The electromagnetic constituent of this component of the energy-momentum tensor vanishes (in the stationary axially symmetric case, the electromagnetic field is described by the two complex Newman-Penrose scalars Φ_0 and Φ_2, since $\Phi_1 = 0$). The neutrino constituent of this component is $4\pi i[(\varepsilon^* - \varepsilon)\Phi\Phi^* + (\gamma - \gamma^*)\Psi\Psi^*] = -\varrho\mu$, and it follows from Einstein's equations that

$$(\delta - 2\tau - \alpha^* + \beta)\pi = 0 \ , \tag{3.1.9}$$

from which, using (3.1.4), we find at once that $\partial^2 r/\partial\xi\,\partial\xi^* = 0$. Therefore the real function $r \equiv \sqrt{|\det g_{AB}|}$ is a harmonic function.

The choice of the coordinates ξ and ξ^* is subject to a conformal transformation $\xi' = f(\xi)$, where $f(\xi)$ is an analytic function. In order to eliminate this arbitrariness, we can, for example, require that the harmonic function r be equal to the real part of $\xi\sqrt{2}$.

For what follows, it is convenient to go over to a tensor notation for the tetrad components of Einstein's equations which appear in (3.1.8,9). We shall substitute the expressions (3.1.4) for the coefficients into Einstein's equations

$$\begin{aligned}
\Phi_{00} &= 4\pi(T^{(\nu)}_{ij} + T^{(\mathrm{em})}_{ij})l^i l^j \quad , \\
\Phi_{22} &= 4\pi(T^{(\nu)}_{ij} + T^{(\mathrm{em})}_{ij})n^i n^j \quad , \\
\Phi_{11} &= 4\pi T^{(\nu)}_{ij} l^i n^j \quad .
\end{aligned}$$

In general, a symmetric second-rank tensor M_{AB} in a two-dimensional space is determined uniquely by the three projections $M_{AB}l^A l^B$, $M_{AB}n^A n^B$, and $M_{AB}l^A n^B$. However, the coordinates t and φ themselves for a fixed determinant $g_{AB} = l_A n_B + n_A l_B$ are fixed only with accuracy up to an arbitrary transformation from the matrix group SL(2, R). This group of transformations leaves invariant the Levi-Civita symbols ε_{AB} and $\varepsilon^{AB}(\varepsilon_{12} = -\varepsilon_{21} = \varepsilon^{12} = -\varepsilon^{21} = 1,\ \varepsilon_{AA} = \varepsilon^{AA} = 0,\ A = 1,2)$.

It is easy to derive the relations

$$\begin{aligned}
rn^A &= \varepsilon^{AC} n_C \quad , \quad rl^A = \varepsilon^{AC} l_C \quad , \\
n_A &= r\varepsilon_{AC} n^C \quad , \quad l_A = r\varepsilon_{AC} l^C
\end{aligned} \tag{3.1.10}$$

for a fixed orientation $l_1 n_2 - n_1 l_2 = r$.

It can be readily verified that we have a symmetric tensor with the projections

$$\begin{aligned}
\Phi_{00} + \varrho^2 &= -(\delta^* - 3\alpha - \beta^* + \pi)k + \tau k^* \quad , \\
\Phi_{11} + \varrho\mu &= -(\delta - 2\tau + \beta - \alpha^*)\pi \quad , \\
\Phi_{22} + \mu^2 &= (\delta + 3\beta + \alpha^* - \tau)\nu + \pi\nu^*
\end{aligned}$$

in the case of the tensor R_{AB}/θ^2 with

$$\begin{aligned}
R_{AB} \equiv \frac{1}{2}\Bigg[&\frac{\partial^2}{\partial\xi\,\partial\xi^*} g_{AB} + \frac{1}{2r}\frac{\partial r}{\partial\xi}\frac{\partial g_{AB}}{\partial\xi^*} + \frac{1}{2r}\frac{\partial r}{\partial\xi^*}\frac{\partial g_{AB}}{\partial\xi} \\
&- \frac{1}{2}\frac{\partial g_{AC}}{\partial\xi}\frac{\partial g_{BD}}{\partial\xi^*} g^{CD} - \frac{1}{2}\frac{\partial g_{AC}}{\partial\xi^*}\frac{\partial g_{BD}}{\partial\xi} g^{CD} \Bigg] \quad .
\end{aligned} \tag{3.1.11}$$

It follows from (3.1.7,8) that to reconstruct the energy-momentum tensor of the neutrino field from the three projections it is sufficient to find a symmetric tensor χ_{AB} from the projections

$$l^A l^B \chi_{AB} = \theta k = -l^A \frac{\partial l_A}{\partial \xi} \quad , \quad n^A n^B \chi_{AB} = \frac{1}{\theta} n^A \frac{\partial n_A}{\partial \xi} \quad .$$

Bearing in mind (3.1.10), it is easy to verify that such a tensor is

$$\chi_{AB} = \frac{1}{4r} \left[g_{AC} \frac{\partial g^{CD} r}{\partial \xi} \varepsilon_{BD} + g_{BC} \frac{\partial g^{CD} r}{\partial \xi} \varepsilon_{AD} \right] \quad .$$

According to Einstein's equations, we have

$$R_{AB} = -\left(\chi_{AB} \frac{\partial w}{\partial \xi^*} + \overset{*}{\chi}_{AB} \frac{\partial w}{\partial \xi} \right) + T^{(\mathrm{em})}_{AB} \theta^2 \quad ,$$

where $T^{(\mathrm{em})}_{AB}$ are the components of the energy-momentum tensor of the electromagnetic field. Contracting both sides of these equations with rg^{BC}, they can be reduced to the form

$$\frac{\partial}{\partial \xi} \left[rg^{BC} \frac{\partial (g_{AC} + w\varepsilon_{AC})}{\partial \xi^*} \right] + \frac{\partial}{\partial \xi^*} \left[rg^{BC} \frac{\partial (g_{AC} + w\varepsilon_{AC})}{\partial \xi} \right]$$
$$= 16\pi \theta^2 r T^{(\mathrm{em})}_{AC} g^{BC} \quad . \tag{3.1.12}$$

In the stationary axially symmetric case, only the components A_B $(B = 1,2)$ of the 4-potential A_i of the electromagnetic field are nonzero. Maxwell's vacuum equations in tensor form for the components F^{iB} of the electromagnetic field tensor have the form

$$0 = \nabla_i F^{iB} = \frac{1}{\sqrt{-\tilde{g}}} \frac{\partial}{\partial x^\mu} \left[\sqrt{-\tilde{g}} \, (\tilde{g}^{\mu\nu} \tilde{g}^{BC} - \tilde{g}^{\mu C} \tilde{g}^{B\nu}) \frac{\partial A_C}{\partial x^\nu} \right] \quad , \tag{3.1.13}$$

where the tildes over the contravariant components of the metric tensor indicate that these components of the metric are calculated as components of the matrix which is the inverse of g_{ij} $(i,j = 1,2,3,4)$. As is readily verified, in the coordinate system (3.1.3) it follows from the equations $\nabla_i F^{i\mu} = 0$ that

$$r\Phi_1 = r \left(g^{B*} \frac{\partial A_B}{\partial \xi^*} - g^B \frac{\partial A_B}{\partial \xi} \right) = \mathrm{const} = C_1 \quad . \tag{3.1.14}$$

In the particular case in which $C_1 = 0$, which is the only case considered here, the Newman-Penrose scalar Φ_1 vanishes.

In what follows, we shall agree to raise and lower the indices of f_{AB} and φ_A by means of the tensor ε_{AB} [3.2]:

$$f_{AB} \equiv g_{AB} \quad , \quad f_A^B = f_{AC} \varepsilon^{BC} \quad ,$$
$$f^{AB} = f_{DC} \varepsilon^{BC} \varepsilon^{AD} = -g^{AB} r^2 \quad , \quad f_A^B f_{BC} = r^2 \varepsilon_{AC} \quad .$$

Using (3.1.14), we can rewrite (3.1.13) in the form

$$\frac{\partial}{\partial \xi} \left[\frac{1}{r} f_B^C \frac{\partial}{\partial \xi^*} A_C \right] + \frac{\partial}{\partial \xi^*} \left[\frac{1}{r} f_B^C \frac{\partial}{\partial \xi} A_C \right] = 0 \quad . \tag{3.1.15}$$

We go over to the real variables x^3 and x^4, and introduce the operator $\nabla(\partial/\partial x^3, \partial/\partial x^4)$ to denote the gradient (when applied to a scalar) or the divergence (when applied to a vector). We now write (3.1.15) in the form

$$\nabla(r^{-1} f_A^C \nabla A_C) = 0 \ . \tag{3.1.15$'$}$$

It follows from this that there exist potentials B_A such that

$$r^{-1} f_A^C \nabla A_C = \tilde{\nabla} B_A \ , \tag{3.1.16}$$

where $\tilde{\nabla} \equiv (\partial/\partial x^4, \ -\partial/\partial x^3)$.

We introduce a potential φ_A [3.3] in the form $\varphi_A \equiv A_A + \mathrm{i} B_A$. Using (3.1.16), it is easy to show that the complex vector potential φ_A satisfies the equation

$$\mathrm{i} r \nabla \varphi_A = f_A^B \tilde{\nabla} \varphi_B \ . \tag{3.1.17}$$

We now find expressions for the components of the energy-momentum tensor of the electromagnetic field, $T_{AB}^{(\mathrm{em})} g^{BC}$, in the coordinate system (3.1.3), in order to obtain a closed system of equations for the neutrino-electromagnetic vacuum:

$$\begin{aligned} 4\pi T_{AB}^{(\mathrm{em})} &= -F_{\mu A} F_B^{\mu} + \frac{1}{2} g_{AB} F_{\mu C} F^{\mu C} \\ &= \frac{1}{\theta^2} \left[\frac{\partial A_E}{\partial \xi} \frac{\partial A_D}{\partial \xi^*} + \frac{\partial A_E}{\partial \xi^*} \frac{\partial A_D}{\partial \xi} \right] \left[\delta_A^E \delta_B^D - \frac{1}{2} g^{ED} g_{AB} \right] \ . \end{aligned}$$

Contracting these expressions with g^{BC}/θ^2 (not to be confused with $\tilde{g}^{BC}/\theta^2$!), we obtain

$$4\pi \theta^2 T_A^{C(\mathrm{em})} = r \left[\frac{\partial A_E}{\partial \xi} \frac{\partial A_D}{\partial \xi^*} + \frac{\partial A_E}{\partial \xi^*} \frac{\partial A_D}{\partial \xi} \right] \left[\delta_A^E g^{DC} - \frac{1}{2} g^{ED} \delta_A^C \right] \ .$$

Making use of the equations (3.1.15) for the electromagnetic field, we rewrite the expression on the right-hand side in divergence form:

$$\nabla \left\{ r \left[\delta_A^E g^{DC} - \tfrac{1}{2} g^{ED} \delta_A^C \right] A_E \nabla A_D \right\} \ .$$

Using the fact that $\delta_A^C f^{ED} A_E \nabla A_D \equiv A_A f^{CD} \nabla A_D - A^C f_A^D \nabla A_D$, we rewrite this expression in the form

$$\nabla \left[\frac{1}{2r} \left(A_A f^{DC} \nabla A_D + A^C f_A^D \nabla A_D \right) \right] \ .$$

Finally, Einstein's equations can be written as follows:

$$\nabla \left[\frac{1}{r} f_A^B \nabla \left(f_B^C + w \delta_B^C \right) - 2 A_A f^{DC} \nabla A_D - 2 f_A^D A^C \nabla A_D \right] = 0 \ . \tag{3.1.18}$$

It follows from this that there exists a matrix potential (ψ_A^C):

$$r\tilde{\nabla}\psi_A^C = f_A^B \nabla \left(f_B^C + w\delta_B^C\right) - 2A_A f^{DC} \nabla A_D - 2f_A^D A^C \nabla A_D \ .$$

We introduce a complex matrix potential (H_A^B):

$$\nabla H_A^B = \nabla \left(f_A^B + w\delta_A^B\right) - \varphi^{B*}\nabla\varphi_A - \frac{\mathrm{i}}{r} f_A^C \left[\tilde{\nabla}\left(f_C^B + w\delta_C^B\right) - \varphi^{B*}\nabla\varphi_C\right] \ . \tag{3.1.19}$$

Taking into account (3.1.16, 17), the condition for the existence of such a potential is provided by (3.1.18). It follows from (3.1.19) that the matrix potential (H_A^B) satisfies the equation

$$\mathrm{i}r\nabla H_A^B = f_A^C \tilde{\nabla} H_C^B \ . \tag{3.1.20}$$

Following [3.3], we introduce potentials L^B and K such that

$$\nabla L^B = 2\varphi^{C*}\nabla H_C^B \ , \quad \nabla K = 2\varphi^{C*}\nabla\varphi_C \ .$$

The condition for the existence of these potentials is provided by the equations

$$\tilde{\nabla}\varphi^{C*}\nabla H_C^B = 0 \ , \quad \tilde{\nabla}\varphi^{C*}\nabla\varphi_C = 0 \ .$$

Following [3.5] (the Kinnersley-Chitre equations in 3×3 matrix form have also been used by *Alekseev* [3.6]), we form a complex matrix (H_a^b) $(a, b = 1, 2, 3)$ as follows:

$$(H_a^b) = \begin{pmatrix} H_A^B & \varphi_A \\ L^B & K \end{pmatrix} \ .$$

From (3.1.17, 19, 20) we obtain the following matrix equations for (H_a^b):

$$2\mathrm{i}r\nabla^2 H_a^b = \nabla H_a^c \tilde{\nabla} H_c^b \ , \quad a, b, c, = 1, 2, 3 \ . \tag{3.1.21}$$

Thus, in the stationary axially symmetric case the Weyl-Einstein-Maxwell system leads to the closed system of equations (3.1.21), which, as we shall show in Sect. 3.2, possesses remarkable properties of group symmetry. A consequence of this matrix equation is the relation

$$\nabla^2 H_a^a = 0 \ .$$

It follows from the definition of (H_a^b) and from (3.1.19) that $\nabla H_a^a = 2(\nabla w - \mathrm{i}\tilde{\nabla} r)$. We introduce a harmonic function z conjugate to the harmonic function r ($\tilde{\nabla} r = -\nabla z$). The trace of the matrix (H_a^b) is

$$H_a^a = 2(w + \mathrm{i}z) \ . \tag{3.1.22}$$

Consider the components H_1^2 and H_1^3 of (3.1.21). Using (3.1.17, 20, 22), we obtain from (3.1.21) a system which is closed with respect to these components:

$$\begin{aligned}
&(\varphi\varphi^* + \mathrm{Re}\{H\})\left[\nabla^2 + \frac{\nabla(r+i\tilde{w})}{r}\nabla\right]\varphi = \nabla\varphi(\nabla H + 2\varphi^*\nabla\varphi) \ , \\
&(\varphi\varphi^* + \mathrm{Re}\{H\})\left[\nabla^2 + \frac{\nabla(r+i\tilde{w})}{r}\nabla\right]H = \nabla H(\nabla H + 2\varphi^*\nabla\varphi) \ , \\
&H \equiv -H_1^2 \ , \quad \varphi \equiv \varphi_1 \ , \quad \nabla\tilde{w} = \tilde{\nabla} w \ ,
\end{aligned} \tag{3.1.23}$$

where $\tilde{w}$ is the function which is harmonically conjugate to the function w.

b) We now consider the case in which both commuting Killing vectors are space-like (the case of plane or cylindrical waves). The square of the interval in such pseudo-Riemannian spaces can be written in the form

$$ds^2 = 2\theta^2 du\, dv - g_{AB}(dx^A + g_u^A du + g_v^A dv)(dx^B + g_u^B du + g_v^B dv) \ ,$$

where θ, g_u^B, g_v^B, g_{AB} are unknown functions of u and v. In this case, the two-dimensional metric g_{AB} is positive definite and can be represented in the form

$$g_{AB} = m_A m_B^* + m_B m_A^* \ , \quad m_A m^A = m_A^* m^{*A} = 0 \ , \quad m_A m^{*A} = 1$$

(the indices are lowered and raised here by means of the matrix g_{AB}). In this case, it is natural to define a field of isotropic tetrads as follows:

$$\begin{aligned}
&m_i(\theta m_A g_u^A, \theta m_A g_v^A, -m_1, -m_2) \ , \\
&l_i(\theta, 0, 0, 0,) \ , \quad n_i(0, \theta, 0, 0) \ , \quad m^i(0, 0, m^1, m^2) \ , \\
&l^i(0, \theta^{-1}, g_v^1, g_v^2) \ , \quad n^i(\theta^{-1}, 0, g_u^1, g_u^2) \ , \quad i = 1,2,3,4 \ .
\end{aligned}$$

The operators δ and δ^* give zero when applied to the various scalar characteristics of the fields under consideration. The corresponding Lie derivatives of the tensor characteristics are also equal to zero.

We write down the nonzero rotation coefficients of the tetrad for this case:

$$\begin{aligned}
&\tau = \pi^* = 2\beta = 2\alpha^* = \frac{m^A}{\theta}\left(\frac{\partial g_{Aj}}{\partial v}n^j - \frac{\partial g_{Aj}}{\partial u}l^j\right) \ , \\
&\varrho = -\frac{1}{2r\theta}\frac{\partial}{\partial v}r \ , \quad \sigma = -\frac{1}{\theta}m^A\frac{\partial m_A}{\partial v} \ , \quad \mu = \frac{1}{2r\theta}\frac{\partial}{\partial u}r \ , \\
&\lambda = \frac{1}{\theta}m^{A*}\frac{\partial m_A^*}{\partial u} \ , \quad \varepsilon = \frac{1}{2}\left[\frac{1}{\theta^2}\frac{\partial\theta}{\partial v} + \frac{1}{2\theta}m^{A*}\frac{\partial m_A}{\partial v} - \frac{1}{2\theta}m^A\frac{\partial m_A^*}{\partial v}\right] \ , \\
&\gamma = \frac{1}{2}\left[-\frac{1}{\theta^2}\frac{\partial\theta}{\partial u} + \frac{1}{2\theta}m^{A*}\frac{\partial m_A}{\partial u} - \frac{1}{2\theta}m^A\frac{\partial m_A^*}{\partial u}\right] \ , \quad k = \nu = 0 \ ,
\end{aligned}$$

where $r = \sqrt{\det g_{AB}}$.

It is interesting that the components Φ_{10} and Φ_{21} of Einstein's equations, to which the electromagnetic field does not contribute, can be represented in the form

$$(D - 3\varrho + \varepsilon - \varepsilon^*)(\alpha + 4\pi\mathrm{i}\Phi\Psi^*) - \sigma^*(\alpha^* - 4\pi\mathrm{i}\Phi^*\Psi) = 0$$

and

$$(\Delta + 3\mu + \gamma - \gamma^*)(\alpha + 4\pi\mathrm{i}\Phi\Psi^*) + \lambda(\alpha^* - 4\pi\mathrm{i}\Phi^*\Psi) = 0$$

respectively. This system of equations, rewritten for the quantity $\alpha + 4\pi\mathrm{i}\Phi\Psi^*$, has the trivial solution

$$\alpha + 4\pi\mathrm{i}\Phi\Psi^* = 0 \ . \tag{3.1.24}$$

Using this solution, Weyl's equations take the form

$$D\Phi = (\varrho - \varepsilon)\Phi + 4\pi\mathrm{i}\Phi\Psi^*\Psi \ ,$$
$$\Delta\Psi = -4\pi\Psi\Phi\Phi^* + (\gamma - \mu)\Psi \ .$$

These equations have the "first" integrals

$$\Phi\Phi^* = \frac{C_1(u)}{8\pi r\theta} \ , \quad \Psi\Psi^* = \frac{C_2(v)}{8\pi r\theta} \ ,$$

where C_1 and C_2 are arbitrary functions of their arguments. The component $\Phi_{20}^{(\nu)}$ of the energy-momentum tensor of the neutrino field has the form

$$\Phi_{20}^{(\nu)} = 8\pi\mathrm{i}[\sigma^*\Phi\Phi^* + 2\alpha\Phi\Psi^* + \lambda\Psi\Psi^*] \ . \tag{3.1.25}$$

The same tetrad component of the Einstein tensor, $R_{ij} - g_{ij}R/2$, is equal to the expression

$$\left(\frac{1}{\theta}\frac{\partial}{\partial v} - \frac{1}{2r\theta}\frac{\partial}{\partial v}r + \frac{1}{\theta^2}\frac{\partial\theta}{\partial v} + \frac{1}{2\theta}m^{A*}\frac{\partial m_A}{\partial v} - \frac{1}{2\theta}m^A\frac{\partial m_A^*}{\partial v}\right)\lambda$$
$$- 4\alpha^2 + \frac{1}{2r\theta}\left(\frac{\partial}{\partial u}r\right)m^{A*}\frac{\partial m_A^*}{\partial v} \ .$$

We note that the corresponding Einstein equation after multiplication by θ^2 contains only the components of the tensor g_{AB}, the functions $C_1(u)$ and $C_2(v)$ characterizing the neutrino field, and the components of the electromagnetic field tensor, since the terms containing the factor α cancel with each other on both sides of Einstein's equations by virtue of (3.1.24).

The component Φ_{11} of the Ricci tensor is

$$\frac{1}{r\theta^2}\frac{\partial^2 r}{\partial u\,\partial v} - 4\alpha\alpha^* \ . \tag{3.1.26}$$

On the other hand, the same tetrad component of the energy-momentum tensor of the neutrino field is

$$8\pi\mathrm{i}(\alpha^*\Phi\Psi^* - \alpha\Phi^*\Psi) \ .$$

Using (3.1.26), we obtain the wave equation for the function $r \equiv \sqrt{\det g_{AB}}$, from which it follows that

$$r = f(u) + g(v) \quad .$$

We represent the expression $\Phi_{20}+4\alpha^2$ in the form $m^{A*}m^{B*}R_{AB}$, where R_{AB} is a symmetric second-rank tensor. Then the tensor R_{AB} for fixed r obeying the wave equation can be reconstructed with accuracy up to a transformation from SL(2, R) ($x^{1'} = Ax^1 + Bx^2$; $x^{2'} = Cx^1 + Dx^2$, $AD - CB = 1$). We carry out the same procedure in the case of the expression $\Phi_{20}^{(\nu)} + 4\alpha^2$ for the neutrino field:

$$\begin{aligned}
&4r\theta^2(\Phi_{20}^{(\nu)} + 4\alpha^2)\\
&= m^{A*}m^{B*}\left[C_1(u)\left(g_{AD}\frac{\partial(g^{DC}r)}{\partial v}\varepsilon_{CB} + g_{BD}\frac{\partial(g^{DC}r)}{\partial v}\varepsilon_{CA}\right)\right.\\
&\left.- C_2(v)\left(g_{AD}\frac{\partial(g^{DC}r)}{\partial u}\varepsilon_{CB} + g_{BD}\frac{\partial g^{DC}r}{\partial u}\varepsilon_{CA}\right)\right] \quad .
\end{aligned}$$

Here we have used the fact that $rm^{B*}\varepsilon_{CB} = \mathrm{i}m_C^*$ and $m_1m_2^* - m_2m_1^* = \mathrm{i}r$.

After this, the calculations become analogous to those already carried out in case (a), and therefore we give only the final result for R_{AB} for the mixed components $A, B = 1, 2$ of Einstein's equations:

$$\begin{aligned}
&\left[rg^{AC}(g_{BC} + C\varepsilon_{BC})_{,u}\right]_{,v} + \left[rg^{AC}(g_{BC} + C\varepsilon_{BC})_{,v}\right]_{,u}\\
&= 16\pi g^{AC}T_{BC}^{(\mathrm{em})}\theta^2 r \ ; \quad C_{,u} = C_1(u) \ , \quad C_{,v} = C_2(v) \ , \qquad (3.1.27)
\end{aligned}$$

where $T_{AB}^{(\mathrm{em})}$ denotes the components of the energy-momentum tensor of the electromagnetic field.

Let A_B be the nonzero components of the 4-potential of the electromagnetic field. Maxwell's vacuum equations have the form

$$\left(rg^{AB}A_{B,u}\right)_{,v} + \left(rg^{AB}A_{B,v}\right)_{,u} = 0 \quad . \qquad (3.1.28)$$

The indices of f_{AB}, φ_A, and A_B will be shifted by means of the tensor ε_{AB}. Calculating the components of the energy-momentum tensor of the electromagnetic field and substituting them into (3.1.27), we obtain

$$\begin{aligned}
&\left\{r^{-1}\left[f_A^C\left(f_C^B + C\delta_C^B\right)_{,u} - 2f_A^C A_{C,u}A^B - 2f^{BC}A_{C,u}A_A\right]\right\}_{,v}\\
&+\left\{r^{-1}\left[f_A^C\left(f_C^B + C\delta_C^B\right)_{,v} - 2f_A^C A_{C,v}A^B - 2f^{BC}A_{C,v}A_A\right]\right\}_{,u} = 0 \ ,\\
&f_A^C f_C^B = -r^2\delta_A^B \ , \quad A_{C,u} \equiv \partial A_C/\partial u \quad . \qquad (3.1.29)
\end{aligned}$$

In order to obtain a complete system of equations, we rewrite (3.1.28) in the form

$$\left(r^{-1}f_A^C A_{C,u}\right)_{,v} + \left(r^{-1}f_A^C A_{C,v}\right)_{,u} = 0 \quad . \qquad (3.1.28')$$

We introduce variables x and t by means of the relations

$$u = \frac{x+t}{\sqrt{2}} \quad , \quad v = \frac{x-t}{\sqrt{2}} \quad .$$

Since (3.1.28′) has a divergence form, there exists a vector potential B_A:

$$r^{-1} f_A^C A_{C,x} = B_{A,t} \quad , \quad r^{-1} f_A^C A_{C,t} = B_{A,x} \quad . \tag{3.1.30}$$

Forming the combination $A_B + \mathrm{i} B_B$, we have from (3.1.30) an equation of self-duality for the gradients of φ_A:

$$r \nabla \varphi_A = \mathrm{i} f_A^C \tilde{\nabla} \varphi_C \quad , \tag{3.1.31}$$

where

$$\nabla = \left(\frac{\partial}{\partial t}, \frac{\partial}{\partial x} \right) \quad , \quad \tilde{\nabla} = \left(\frac{\partial}{\partial x}, \frac{\partial}{\partial t} \right) \quad .$$

It follows from (3.1.29) that there exists a complex matrix potential (H_A^B):

$$\nabla H_A^B = \nabla \left(f_A^B + C \delta_A^B \right) - \varphi^{B*} \nabla \varphi_A + \frac{\mathrm{i}}{r} \left[\tilde{\nabla} \left(f_C^B + C \delta_C^B \right) - \varphi^{B*} \tilde{\nabla} \varphi_C \right] f_A^C \quad . \tag{3.1.32}$$

From the definition of (H_A^B) it follows that

$$r \nabla H_A^B = \mathrm{i} f_A^C \tilde{\nabla} H_C^B \quad . \tag{3.1.33}$$

Taking the divergence of both sides of (3.1.31, 33), we obtain the equations

$$2\mathrm{i} r \eta^{ik} \frac{\partial^2 \varphi_A}{\partial x^i \partial x^k} = \varepsilon^{ik} \frac{\partial \varphi_C}{\partial x^i} \left(\frac{\partial}{\partial x^k} H_A^C + 2 \varphi^{C*} \frac{\partial}{\partial x^k} \varphi_A \right) \quad , \qquad (\eta^{ik}) = \begin{pmatrix} 1 & 0 \\ 0 & -1 \end{pmatrix} \quad ; \tag{3.1.34}$$

$$2\mathrm{i} r \eta^{ik} \frac{\partial^2 H_A^B}{\partial x^i \partial x^k} = \varepsilon^{ik} \frac{\partial H_C^B}{\partial x^i} \left(\frac{\partial}{\partial x^k} H_A^C + 2 \varphi^{C*} \frac{\partial}{\partial x^k} \varphi_A \right) \quad , \qquad (\varepsilon^{ik}) = \begin{pmatrix} 0 & 1 \\ -1 & 0 \end{pmatrix} \quad . \tag{3.1.35}$$

Introducing, as in [3.5], the 3×3 matrix (H_a^b) $(a, b = 1,2,3)$ such that $H_a^b = H_A^B$ $(a = A = 1,2;\ b = B = 1,2)$, $H_A^3 = \varphi_A$, $\nabla H_3^B = 2\varphi^{C*} \nabla H_C^B$, and $\nabla H_3^3 = 2\varphi^{C*} \nabla \varphi_C$, from (3.1.34, 35) we finally obtain the following equation for the matrix (H_a^b):

$$\eta^{ik} \frac{\partial^2 H_a^b}{\partial x^i \partial x^k} = \frac{\mathrm{i}}{2r} \varepsilon^{ik} \frac{\partial}{\partial x^i} H_a^c \frac{\partial}{\partial x^k} H_c^b \quad . \tag{3.1.36}$$

From the definition of (H^b_a) and (H^B_A) in (3.1.32) it follows that $H^a_a = 2(C + \mathrm{i}z)$, where z is defined by the equation $\tilde{\nabla} r = -\nabla z$.

For the components $H^2_1 = -H$ and $H^3_1 = \varphi$, we obtain from (3.1.31, 33–35) the complete system of equations

$$\eta^{ik}(\mathrm{Re}H + \varphi\varphi^*)\left[\frac{\partial^2 H}{\partial x^i\,\partial x^k} + \frac{1}{r}\frac{\partial(r+\mathrm{i}\tilde{C})}{\partial x^i}\frac{\partial}{\partial x^k}H\right]$$
$$-\frac{\partial}{\partial x^i}H\left(\frac{\partial}{\partial x^k}H + \varphi^*\frac{\partial}{\partial x^k}\varphi\right) = 0 \ ,$$

$$\eta^{ik}(\mathrm{Re}H + \varphi\varphi^*)\left[\frac{\partial^2 \varphi}{\partial x^i\,\partial x^k} + \frac{1}{r}\frac{\partial(r+\mathrm{i}\tilde{C})}{\partial x^i}\frac{\partial}{\partial x^k}\varphi\right]$$
$$-\frac{\partial}{\partial x^i}\varphi\left(\frac{\partial}{\partial x^k}H + \varphi^*\frac{\partial}{\partial x^k}\varphi\right) = 0 \ ,$$

where $\tilde{C}$ is defined by the equation $\tilde{\nabla} C = \nabla\tilde{C}$.

Our canonical equations for the neutrino electrovacuum, (3.1.17, 18, 20, 21, 23, 29, 31, 33, 36, 37), are generalizations of the corresponding electrovacuum equations [3.2, 3, 5] to the case in which a neutrino field is present.

3.2 Infinite-Dimensional Algebra and Lie Group of the Equations for the Neutrino Electrovacuum

The description of the interaction of free gravitational, electromagnetic, and neutrino fields in the general theory of relativity (the neutrino electrovacuum) possesses, for solutions with an Abelian group of motions G_2 on V_2, remarkable analytic properties, which make it possible to reduce the Weyl-Maxwell-Einstein system to a system of linear singular integral equations. This reduction is possible because of the existence of an infinite-dimensional Lie group which transforms one solution of the equations for the neutrino electrovacuum into another solution of these same equations. An arbitrary orbit of the group in its domain of analyticity can be approximated by rational functions. The corresponding exact solutions can be expressed in terms of the solutions of a system of linear algebraic equations (see Sect. 3.3). In contrast to the well-known studies of *Geroch* [3.2], *Kinnersley* and *Chitre* [3.3], and *Hauser* and *Ernst* [3.5], we develop here a new approach which permits a generalization to the case in which neutrino fields are present.

We begin with the system to which the Weyl-Maxwell-Einstein system is reduced in the stationary axially symmetric case (see Sect. 3.1):

$$\mathrm{i}r\nabla H^B_A = f^C_A\tilde{\nabla}H^B_C \ , \quad \mathrm{i}r\nabla\Phi_A = f^C_A\tilde{\nabla}\Phi_C \ ; \quad A,B,C = 1,2 \ . \qquad (3.2.1)$$

(To consider the case of plane waves, it is sufficient to make the substitution $\mathrm{i}z \to t$ in the final results.)

We introduce a complex matrix potential H_{AB} according to the definition

$$\nabla H_{AB} = \nabla(f_{AB} + w\varepsilon_{AB}) - 2\Phi_B^* \nabla \Phi_A - \frac{\mathrm{i}}{r} f_A^C \tilde{\nabla}(f_{CB} + w\varepsilon_{CB}) \ . \tag{3.2.2}$$

A 3×3 matrix $\boldsymbol{H}$ with components H_a^b is defined as follows:

$$\begin{aligned} &H_a^b = H_A^B \ , \quad a = A \ , \quad b = B \ , \quad H_A^3 = \Phi_A \ , \\ &\nabla H_3^B = 2\Phi^{C*} \nabla H_C^B \ , \quad \nabla H_3^3 = 2\Phi^{C*} \nabla \Phi_C \ . \end{aligned} \tag{3.2.3}$$

The existence of H_{AB} satisfying (3.2.2) follows from Einstein's equations, and the existence of H_3^B and H_3^3 follows from (3.2.1).

It is remarkable that the relations (3.2.1) and the definitions (3.2.3) can be obtained from the single 3×3 matrix equation

$$(\boldsymbol{\varepsilon D} + \boldsymbol{M}\nabla)\boldsymbol{H} = 0 \ ; \tag{3.2.4}$$

here

$$\boldsymbol{D} = 2\mathrm{i}(r\tilde{\nabla} - z\nabla) \ , \quad \boldsymbol{M} = \boldsymbol{M}^+ = \boldsymbol{\varepsilon H} - \boldsymbol{H}^+\boldsymbol{\varepsilon} - \tfrac{1}{2}\boldsymbol{\Pi} \ ,$$

$$\boldsymbol{\varepsilon} = \begin{pmatrix} 0 & 1 & 0 \\ -1 & 0 & 0 \\ 0 & 0 & 0 \end{pmatrix} \ , \quad \boldsymbol{\Pi} = \begin{pmatrix} 0 & 0 & 0 \\ 0 & 0 & 0 \\ 0 & 0 & 1 \end{pmatrix} \ ,$$

and the superscript + indicates Hermitian conjugation.

Using (3.2.1, 2), it can be shown that

$$2\mathrm{i}r\nabla^2 \boldsymbol{H} = \nabla \boldsymbol{H} \tilde{\nabla} \boldsymbol{H} \ . \tag{3.2.5}$$

It follows from (3.2.5) that there exists an infinite hierarchy of left and right matrix potentials $\boldsymbol{Q}_m$ and $\boldsymbol{H}_m$, respectively, which can be defined recursively:

$$\begin{aligned} \nabla \boldsymbol{H}_m &= \nabla \boldsymbol{H} \boldsymbol{H}_{m-1} - \boldsymbol{D}\boldsymbol{H}_{m-1} \ , \\ \nabla \boldsymbol{Q}_m &= \boldsymbol{Q}_{m-1} \nabla \boldsymbol{H} + \boldsymbol{D}\boldsymbol{Q}_{m-1} \ . \end{aligned} \tag{3.2.6}$$

Each of the matrices $\boldsymbol{H}_n$ ($n = 2, 3, \ldots$) is uniquely defined, apart from a constant matrix $\boldsymbol{C}_n$. Therefore in the hierarchy of potentials we can perform transformations of the form

$$\boldsymbol{H}'_m = \boldsymbol{H}_m + \boldsymbol{H}_{m-1}C_1 + \boldsymbol{H}_{m-2}C_2 + \ldots + \boldsymbol{H}_1 \boldsymbol{C}_{m-1} + \boldsymbol{C}_m \ . \tag{3.2.7}$$

The matrix $\boldsymbol{C}_1$ satisfies the condition

$$\boldsymbol{\varepsilon} C_1 = C_1^+ \boldsymbol{\varepsilon} \ , \tag{3.2.8}$$

which holds as a consequence of the invariance of (3.2.4). The matrices $\boldsymbol{H}_n$ which we have introduced, like $\boldsymbol{H}_1$, satisfy the equations

$$(\boldsymbol{\varepsilon D} + \boldsymbol{M}\nabla)\boldsymbol{H}_n = 0 \ . \tag{3.2.9}$$

In fact, owing to (3.2.4, 6), we have

$$\begin{aligned}(\varepsilon \boldsymbol{D} + M\nabla)\boldsymbol{H}_{n+1} &= \varepsilon \boldsymbol{D}\boldsymbol{H}_{n+1} + M(\nabla \boldsymbol{H}\boldsymbol{H}_n - \boldsymbol{D}\boldsymbol{H}_n)\\ &= \varepsilon \boldsymbol{D}\boldsymbol{H}_{n+1} - (\varepsilon \boldsymbol{D}\boldsymbol{H} + M\boldsymbol{D})\boldsymbol{H}_n\\ &= 2\mathrm{i}r[\varepsilon\tilde{\nabla}\boldsymbol{H}_{n+1} - (\varepsilon\tilde{\nabla}\boldsymbol{H} + M\tilde{\nabla})\boldsymbol{H}_n]\\ &\quad - 2\mathrm{i}z[\varepsilon\nabla\boldsymbol{H}_{n+1} - (\varepsilon\nabla\boldsymbol{H} + M\nabla)\boldsymbol{H}_n]\\ &= [-2\mathrm{i}r(\varepsilon\tilde{\boldsymbol{D}} + M\tilde{\nabla}) + 2\mathrm{i}z(\varepsilon\boldsymbol{D} + M\nabla)]\boldsymbol{H}_n \quad .\end{aligned}$$

This expression is equal to zero, by hypothesis.

We introduce a generating function for the matrix potentials $\boldsymbol{H}_n$:

$$\boldsymbol{F}(r,z,s) = \sum_{m=0}^{\infty} \boldsymbol{H}_m(-\mathrm{i}s)^m \quad , \qquad \boldsymbol{H}_0 = 1 \quad , \qquad \boldsymbol{H}_1 = \boldsymbol{H} \quad .$$

The expansion coefficients of the matrix which is the inverse of $\boldsymbol{F}$ satisfy the recurrence relations (3.2.6), and we therefore identify them in what follows with the coefficients in the relation

$$\boldsymbol{F}^{-1} = \sum_{m=0}^{\infty} \boldsymbol{Q}_m(\mathrm{i}s)^m \quad .$$

In the transformations (3.2.7), the matrix $\boldsymbol{F}$ is multiplied on the right by the matrix $C(t)$:

$$\boldsymbol{F}' = \boldsymbol{F}C(t) \quad , \qquad C(t) = \sum_{n=0}^{\infty}(-\mathrm{i}t)^n C_n \quad , \qquad C_0 = 1 \quad . \tag{3.2.10}$$

It follows from the relations (3.2.6) that the matrices $\boldsymbol{F}$ satisfy a redefined linear system, the consistency conditions for which are the second-order equations (3.2.5) [this situation is unusual in that the first-order equations (3.2.4) must also be satisfied, and the conditions (3.2.5) are differential consequences of (3.2.4)]:

$$\begin{aligned}\frac{\partial}{\partial\xi}\boldsymbol{F} &= -\frac{\mathrm{i}s}{1+2\mathrm{i}\xi s}\left(\frac{\partial}{\partial\xi}\boldsymbol{H}\right)\boldsymbol{F} \quad ,\\ \frac{\partial}{\partial\xi^*}\boldsymbol{F} &= -\frac{\mathrm{i}s}{1-2\mathrm{i}\xi^* s}\left(\frac{\partial}{\partial\xi^*}\boldsymbol{H}\right)\boldsymbol{F} \quad .\end{aligned} \tag{3.2.11}$$

It follows from (3.2.11) that the generating matrix $\boldsymbol{F}$ has branch points at $s = \mathrm{i}/2\xi$ and $s = -\mathrm{i}/2\xi^*$. In fact, from (3.2.11) we have

$$\int_{c_1}\left(\frac{\partial}{\partial\xi}\boldsymbol{F}\right)\boldsymbol{F}^{-1}ds = \frac{\pi}{2\xi^2}\frac{\partial\boldsymbol{H}}{\partial\xi} \quad , \qquad \int_{c_2}\left(\frac{\partial}{\partial\xi^*}\boldsymbol{F}\right)\boldsymbol{F}^{-1}ds = \frac{\pi}{2\xi^{*2}}\frac{\partial\boldsymbol{H}}{\partial\xi^*} \quad ,$$

where c_1 and c_2 are closed contours in the complex s plane around the points $s = \mathrm{i}/2\xi$ and $s = -\mathrm{i}/2\xi^*$, respectively.

We shall demonstrate the existence of a Lie algebra of solutions of (3.2.4), linearized around an arbitrary solution $\boldsymbol{H}$.

We find from (3.2.4) that the perturbations $\delta\boldsymbol{H}$ satisfy the equations

$$(\boldsymbol{\varepsilon D} + \boldsymbol{M}\nabla)\delta\boldsymbol{H} + (\boldsymbol{\varepsilon}\delta\boldsymbol{H} - \delta\boldsymbol{H}^{+}\boldsymbol{\varepsilon})\nabla\boldsymbol{H} = 0 \ . \tag{3.2.12}$$

The system (3.2.12) has solutions of the form ($\boldsymbol{\Gamma} = \text{const}$)

$$\delta\boldsymbol{H} = \chi_n(\boldsymbol{\Gamma}) = \sum_{k=0}^{n}(-1)^k\boldsymbol{H}_k\boldsymbol{\Gamma}\boldsymbol{G}_{n-k} \ , \quad \boldsymbol{G}_k = \boldsymbol{H}_k^{+}\boldsymbol{\varepsilon} + \boldsymbol{H}_{k-1}^{+}\boldsymbol{M} \tag{3.2.13}$$

under the additional condition $\boldsymbol{\Gamma} + (-1)^n\boldsymbol{\Gamma}^{+} = 0$ (it is easy to see that the $\boldsymbol{G}_k$ satisfy the same recurrence relations as the $\boldsymbol{Q}_k$).

In fact, we have

$$\delta\boldsymbol{M} = \sum_{k=0}^{n}(-1)^k\boldsymbol{\varepsilon}[\boldsymbol{H}_k\boldsymbol{\Gamma}\boldsymbol{H}_{n-k}^{+} + \boldsymbol{H}_{n-k}\boldsymbol{\Gamma}^{+}\boldsymbol{H}_k^{+}]\boldsymbol{\varepsilon}$$
$$+ \sum_{k=0}^{n-1}(-1)^k(\boldsymbol{\varepsilon}\boldsymbol{H}_k\boldsymbol{\Gamma}\boldsymbol{H}_{n-1-k}^{+}\boldsymbol{M} - \boldsymbol{M}\boldsymbol{H}_{n-1-k}\boldsymbol{\Gamma}^{+}\boldsymbol{H}_k\boldsymbol{\varepsilon}) \ .$$

Changing the order of summation in the second and fourth sums, and using the fact that $\boldsymbol{\Gamma} + (-1)^n\boldsymbol{\Gamma}^{+} = 0$, we find for $\delta\boldsymbol{M}$ the expression

$$\delta\boldsymbol{M} = \sum_{k=0}^{n-1}(-1)^k[\boldsymbol{\varepsilon}\boldsymbol{H}_k\boldsymbol{\Gamma}\boldsymbol{H}_{n-1-k}^{+}\boldsymbol{M} - \boldsymbol{M}\boldsymbol{H}_k\boldsymbol{\Gamma}\boldsymbol{H}_{n-1-k}\boldsymbol{\varepsilon}] \ .$$

Therefore, for $(\boldsymbol{\varepsilon D} + \boldsymbol{M}\nabla)\delta\boldsymbol{H} + \delta\boldsymbol{M}\nabla\boldsymbol{H}$ we obtain

$$\sum_{k=0}^{n}(-1)^k\boldsymbol{\varepsilon}\boldsymbol{H}_k\boldsymbol{\Gamma}(\boldsymbol{D}\boldsymbol{G}_{n-k} + \boldsymbol{H}_{n-k-1}^{+}\boldsymbol{M}\nabla\boldsymbol{H})$$
$$+ \sum_{k=0}^{n}(-1)^k\boldsymbol{M}\boldsymbol{H}_k\boldsymbol{\Gamma}(\nabla\boldsymbol{G}_{n-k} - \boldsymbol{H}_{n-1-k}^{+}\boldsymbol{\varepsilon}\nabla\boldsymbol{H}) \ .$$

This expression vanishes, since the expressions in parentheses are equal to zero.

The solutions (3.2.13) are related to each other by the recurrence relations

$$\chi_n\nabla\boldsymbol{H} - \nabla\boldsymbol{H}\chi_n + \boldsymbol{D}\chi_n = \nabla\chi_{n+1} \ . \tag{3.2.14}$$

We shall now show that the solutions of the type (3.2.13) form an infinite-dimensional Lie algebra. For this, we must find the perturbations $\delta\boldsymbol{H}_m$ and $\delta\boldsymbol{G}_m$ due to the perturbation $\delta\boldsymbol{H} = \chi_n$. On the basis of (3.2.6), it is easy to prove by induction that

$$\delta\boldsymbol{H}_m = \chi_n\boldsymbol{H}_{m-1} - \chi_{n+1}\boldsymbol{H}_{m-2} + \ldots + (-1)^{m-1}\chi_{n+m-1} \ , \tag{3.2.15}$$

$$\delta G_m = G_{m-1}\chi_n + G_{m-2}\chi_{n+1} + \ldots + \chi_{n+m-1} \quad . \tag{3.2.16}$$

In order to simplify the resulting expressions, we shall establish the existence of an integral of the system (3.2.11):

$$F^+(\varepsilon + \mathrm{i}s\boldsymbol{M})F = \gamma(s) \quad , \tag{3.2.17}$$

where $\gamma(s)$ is an anti-Hermitian matrix which is independent of r and z. Indeed,

$$\frac{\partial}{\partial\xi}[F^+(\varepsilon + \mathrm{i}s\boldsymbol{M})F] = F^+\mu F \quad ,$$

$$\mu \equiv \frac{\mathrm{i}s}{1+2\mathrm{i}\xi s}\left[\frac{\partial H^+}{\partial\xi}(\varepsilon + \mathrm{i}s\boldsymbol{M}) - (\varepsilon + \mathrm{i}s\boldsymbol{M})\frac{\partial H}{\partial\xi} + \frac{\partial \boldsymbol{M}}{\partial\xi}(1+2\mathrm{i}\xi s)\right] \quad .$$

The matrix μ is identically equal to zero, since, as a consequence of (3.2.4),

$$2\xi\frac{\partial\boldsymbol{M}}{\partial\xi} + \frac{\partial H^+}{\partial\xi}\boldsymbol{M} - \boldsymbol{M}\frac{\partial H}{\partial\xi} = 0 \quad , \quad \boldsymbol{M} = \varepsilon H - H^+\varepsilon - \frac{\Pi}{2} \quad .$$

Under the transformations (3.2.4), the matrix $\gamma(s)$ transforms as follows:

$$\gamma'(s) = C^+(s)\gamma(s)C(s) \quad .$$

We take advantage of the arbitrariness in the choice of $C(s)$ to reduce the matrix $\gamma(s)$ to its simplest form. For this, it is sufficient to note that the expansion of $\gamma(s)$ in powers of s has the form $\gamma(s) = \Omega + \mathrm{o}(s)$, where $\Omega \equiv \varepsilon - \mathrm{i}s\Pi/2$. The arbitrariness in choosing a function $C(s)$ which is analytic at the origin $[C(0) = 1]$ cannot affect the first two terms of the expansion. The remaining terms in $\gamma(s)$, denoted by $\mathrm{o}(s)$, can be made to vanish by choosing an appropriate matrix $C(s)$ in (3.2.10).

We rewrite (3.2.17) in the form

$$F^+(\varepsilon + \mathrm{i}s\boldsymbol{M}) = \Omega F^{-1} \tag{3.2.17$'$}$$

and equate the terms here with identical powers of s. This gives

$$G_k = \varepsilon Q_k - \frac{1}{2}\Pi Q_{k-1} \quad , \quad \sum_{k=0}^{n}(-1)^k G_k H_{n-k} = 0 \quad ; \quad n \geq 2 \quad ,$$
$$G_0 H_0 = \varepsilon \quad , \quad G_1 H_0 - G_0 H_1 = -\Pi/2 \quad , \tag{3.2.18}$$

since

$$G_0 = \varepsilon \quad , \quad H_0 = 1 \quad , \quad H_1 = H \quad , \quad G_1 = \varepsilon H - \Pi/2 \quad .$$

Consider the commutator of two solutions $\chi_p(\Gamma)$ and $\chi_p(\tilde{\Gamma})$ of (3.2.12):

$$\tilde{\delta}\chi_p(\Gamma) - \delta\chi_q(\tilde{\Gamma}) \quad .$$

It is easy to show that it can be transformed to the form

$$\sum_{k=0}^{p+q-1} (\chi_k(\Gamma)\chi_{p+q-i-k}(\tilde{\Gamma}) - \chi_k(\tilde{\Gamma})\chi_{p+q-1-k}(\Gamma)) \ . \tag{3.2.19}$$

Making use of the relations (3.2.18) and the expressions for χ_k, from (3.2.19) we obtain

$$\tilde{\delta}\chi_p(\Gamma) - \delta\chi_q(\tilde{\Gamma}) = \chi_{p+q-1}(g) - \tfrac{1}{2}\chi_{p+q-2}(\tilde{g}) \ . \tag{3.2.20}$$

Thus, the commutator of two solutions of the form (3.2.13) gives another solution of the same form, i.e., the relations

$$g + (-1)^{p+q-1}g^+ = \tilde{g} + (-1)^{p+q-2}\tilde{g}^+ = 0 \ ,$$
$$g \equiv \Gamma\varepsilon\tilde{\Gamma} - \tilde{\Gamma}\varepsilon\Gamma \ , \quad \tilde{g} \equiv \Gamma\Pi\tilde{\Gamma} - \tilde{\Gamma}\Pi\Gamma$$

are simple consequences of the conditions

$$\Gamma + (-1)^p\Gamma^+ = \tilde{\Gamma} + (-1)^q\tilde{\Gamma}^+ = 0 \ .$$

Thus, the solutions of the form (3.2.13) do indeed form an infinite-dimensional Lie algebra.

We shall now find the corresponding Lie group.

For this, we multiply both sides of (3.2.15) by $(-it)^m$ and sum over m from 0 to ∞. Using (3.2.18), we find that

$$\delta F = -it\sum_{p=n}^{\infty}(it)^p\chi_{p+n}F \ ,$$

from which it follows that

$$\delta F F^{-1} = -it\sum_{p=n}^{\infty}(it)^p\chi_{p+n} \ . \tag{3.2.21}$$

Making use of the integral (3.2.17′), we rewrite the expression (3.2.13) for χ_p in the form

$$\delta\chi_p = \sum_{k=0}^{p}(-1)^k H_k\Gamma\left(\varepsilon Q_{p-k} - \frac{1}{2}\Pi Q_{p-k-1}\right) \tag{3.2.22}$$

(we recall that $F^{-1} = \sum_{k=0}^{\infty}(is)^k Q_k$).

Applying the residue theorem, the expression for χ_p can be readily represented in the form

$$\delta\chi_p = \frac{1}{2\pi}\oint_L F(s)\Gamma\Omega(s)F^{-1}(s)\frac{ds}{(is)^{p+1}} \ . \tag{3.2.23}$$

Here L is a smooth contour which bounds a simply connected region L_+ in the complex s plane, containing the origin $s = 0$. By hypothesis, the series

$\sum_{k=0}^{\infty} \boldsymbol{H}_k(-\mathrm{i}s)^k$ converges in the region L_+. Substituting (3.2.23) into (3.2.21), we obtain

$$\delta \boldsymbol{F}\boldsymbol{F}^{-1} + \frac{1}{2\pi}\oint_L \boldsymbol{F}(s)\boldsymbol{\Gamma}\boldsymbol{\Omega}(s)\sum_{p=0}^{\infty}\frac{t^{p+1}ds}{s^{p+1}(\mathrm{i}s)^n}$$
$$= \delta \boldsymbol{F}\boldsymbol{F}^{-1} + \frac{1}{2\pi}\oint_L \boldsymbol{F}(s)\boldsymbol{\Gamma}\boldsymbol{\Omega}(s)\frac{t\,ds}{(s-t)(\mathrm{i}s)^n} \ . \tag{3.2.24}$$

In the general case, the solution of (3.2.12) consists of a sum of solutions of the form (3.2.13). Assigning the index $n-1$ to the matrix $\boldsymbol{\Gamma}$ in the solution (3.2.13),

$$\delta \boldsymbol{H} = \sum_{n=1}^{\infty}\sum_{k=0}^{n}(-1)^k \boldsymbol{H}_k\boldsymbol{\Gamma}_{n-1}\boldsymbol{G}_{n-k} \ ,$$

we obtain from (3.2.24) the expression

$$\delta \boldsymbol{F}\boldsymbol{F}^{-1} + \frac{1}{2\pi \mathrm{i}}\oint \boldsymbol{F}(s)\boldsymbol{\Gamma}(s)\boldsymbol{\Omega}(s)\boldsymbol{F}^{-1}(s)\frac{t\,ds}{s(s-t)} \ , \tag{3.2.25}$$

where we have introduced the generating function $\boldsymbol{\Gamma}(s) \equiv \sum_{k=0}^{\infty}(\mathrm{i}s)^{-k}\boldsymbol{\Gamma}_k$.

It is assumed that this series converges throughout the region $L_- = C - L_+$. We emphasize that the matrix $\boldsymbol{\Gamma}(s)$ is Hermitian, since $\boldsymbol{\Gamma}_k \mathrm{i}^k = (\boldsymbol{\Gamma}_k \mathrm{i}^k)^+$.

The expression (3.2.25) is the variation (for an infinitesimal variation of the group parameters) of the expression

$$\oint_L \boldsymbol{F}(s)\boldsymbol{u}(s)\dot{\boldsymbol{F}}^{-1}(s)\frac{t\,ds}{s(s-t)} = 0 \ , \quad \boldsymbol{u}(s) \equiv \exp[\boldsymbol{\Gamma}(s)\boldsymbol{\Omega}(s)] \ , \tag{3.2.26}$$

where $\boldsymbol{F}(s)$ is the result of a displacement of the initial solution $\dot{\boldsymbol{F}}(s)$ along the orbit of the group. For $\boldsymbol{F}$ differing slightly from $\dot{\boldsymbol{F}}$ and for small $\boldsymbol{\Gamma}(s)$, (3.2.26) again leads to (3.2.25) (if we make use of the analyticity of $\boldsymbol{F}$ and $\delta\boldsymbol{F}$ in L_+).

As an exponential of the product of a Hermitian matrix $\boldsymbol{\Gamma}(s)$ and an anti-Hermitian matrix $\boldsymbol{\Omega} = \boldsymbol{\varepsilon} - \mathrm{i}s\boldsymbol{\Pi}/2$, the matrix $\boldsymbol{u}$ obeys the Hauser-Ernst condition

$$\boldsymbol{u}^+\boldsymbol{\Omega}\boldsymbol{u} = \boldsymbol{\Omega} \ . \tag{3.2.27}$$

The central result of this section, which follows from (3.2.26), can be interpreted as the condition of analyticity of the matrix $\chi(s) \equiv \boldsymbol{F}(s)\boldsymbol{u}(s)\dot{\boldsymbol{F}}^{-1}(s)$ in the region L_-.

Thus, despite the presence of branch points in the matrices $\boldsymbol{F}$ and $\dot{\boldsymbol{F}}$ at $s = \mathrm{i}/2\xi$ and $s = -\mathrm{i}/2\xi^*$, the matrix $\chi(s)$ is analytic in L_-. Explicit exact solutions of (3.2.26) can be found for rational $\boldsymbol{u}(s)$ having pole singularities in L_+ (see Sect. 3.3).

We shall now find a condition satisfied by the matrix $\chi(s)$. Taking the Hermitian conjugate of the matrix χ^{-1}, we obtain

$$(\chi^{-1})^{+} = (F^{-1})^{+}(u^{-1})^{+}\dot{F}^{+} \ . \tag{3.2.28}$$

Using (3.2.27), we have $(u^{-1})^{+} = \Omega u \Omega^{-1}$. From (3.2.17) we obtain

$$(F^{-1})^{+} = (\varepsilon + \mathrm{i}sM)F\Omega^{-1} \ , \quad \dot{F}^{+} = \Omega\dot{F}^{-1}(\dot{M}\mathrm{i}s + \varepsilon)^{-1} \ . \tag{3.2.28$'$}$$

We note that $\det(\varepsilon + \mathrm{i}sM) = \det(\varepsilon + \mathrm{i}s\dot{M}) = (1-2zs)^2 + 4r^2s^2$ vanishes only at the branch points $2s = (z \pm \mathrm{i}r)^{-1}$. We shall show that $|\det\chi| = 1$. Indeed, from (3.2.11) we have

$$\frac{\partial}{\partial\xi}FF^{-1} = -\frac{\mathrm{i}s}{1+2\mathrm{i}\xi s}\frac{\partial}{\partial\xi}H \ , \quad \frac{\partial}{\partial\xi^*}FF^{-1} = -\frac{\mathrm{i}s}{1-2\mathrm{i}\xi^* s}\frac{\partial}{\partial\xi^*}H \ . \tag{3.2.29}$$

Taking the trace of both sides of the matrix equation (3.2.29), we have

$$\begin{aligned} \frac{\partial}{\partial\xi}(\ln\det F) &= -\frac{\mathrm{i}s}{1+2\mathrm{i}\xi s}\frac{\partial}{\partial\xi}H^a_a \ , \\ \frac{\partial}{\partial\xi^*}(\ln\det F) &= -\frac{\mathrm{i}s}{1-2\mathrm{i}\xi^* s}\frac{\partial}{\partial\xi^*}H^a_a \ . \end{aligned} \tag{3.2.30}$$

It follows from the definition of H^b_a that $H^a_a = \mathrm{Tr}\{H\} = 2(w + \mathrm{i}z)$. Therefore from (3.2.30) we have

$$\begin{aligned} &\det F = 1/\lambda \ , \quad \lambda = \sqrt{(1-2zt)^2 + 4r^2t^2}\,\exp(\mathrm{i}\sigma) \ , \\ &\sigma = 2t\left(\int\frac{\partial w}{\partial\xi}\frac{d\xi}{1+2\mathrm{i}\xi t} + \int\frac{\partial w}{\partial\xi^*}\frac{d\xi^*}{1-2\mathrm{i}\xi^* t}\right) \end{aligned} \tag{3.2.31}$$

[we set the constant of integration in (3.2.31) equal to unity, since F can be multiplied by a matrix $C(t)$].

From the definition of χ, it follows that $\det\chi = \det F \det u \det \dot{F}^{-1} = \det u$. From (3.2.27), we have $|\det u| = 1$. Substituting the expressions (3.2.28$'$) into (3.2.28), we obtain

$$\begin{aligned} &(\chi^{-1})^{+} = (\varepsilon + \mathrm{i}sM)\chi(\varepsilon + \mathrm{i}s\dot{M})^{-1} \quad \text{or} \\ &\chi^{+}(\varepsilon + \mathrm{i}sM)\chi = \varepsilon + \mathrm{i}s\dot{M} \ . \end{aligned} \tag{3.2.32}$$

Another property of the matrix χ is its analyticity at the branch points of the matrix F. From the definition of χ, we find

$$\begin{aligned} \frac{\partial}{\partial\xi}\chi &= -\frac{\mathrm{i}s}{1+2\mathrm{i}\xi s}\left(\frac{\partial}{\partial\xi}H\chi - \chi\frac{\partial}{\partial\xi}\dot{H}\right) \ , \\ \frac{\partial}{\partial\xi^*}\chi &= -\frac{\mathrm{i}s}{1-2\mathrm{i}\xi^* s}\left(\frac{\partial}{\partial\xi^*}H\chi - \chi\frac{\partial}{\partial\xi^*}\dot{H}\right) \ , \end{aligned} \tag{3.2.33}$$

Equation (3.2.32) is a first integral of the system (3.2.33).

In view of the analyticity of χ at the points $s = (z \pm \mathrm{i}\varrho)^{-1}$, it follows from (3.2.33) that

$$\left(\frac{\partial}{\partial\xi}H\chi(s)-\chi(s)\frac{\partial}{\partial\xi}\dot{H}\right)_{s=\mathrm{i}/2\xi}$$
$$=\left(\frac{\partial}{\partial\xi^*}H\chi(s)-\chi(s)\frac{\partial}{\partial\xi^*}\dot{H}\right)_{s=-\mathrm{i}/2\xi^*}=0\ . \tag{3.2.34}$$

Thus, we could find the matrix $\chi(s)$ directly from the conditions (3.2.32–34), assuming that it is a rational function of s.

The zeros and poles of $\chi(s)$ lie only in the region L_+ and are independent of r and z, since they arise from the zeros and poles of the matrix $u(s)$, which is independent of r and z.

Proof of the Geroch Hypothesis. Having discovered an infinite group of transformations for stationary axially symmetric gravitational fields, *Geroch* [3.2] put forward the hypothesis that an arbitrary asymptotically flat space (in the absence of electromagnetic and neutrino fields) can be obtained from Minkowski space by means of an appropriate transformation from the group. This hypothesis was proved by Xanthopoulos, and also by *Hauser* and *Ernst* [3.5].

Here we present a proof that arbitrary free gravitational, electromagnetic, and neutrino fields in general relativity with Abelian groups of motions G_2 on V_2 can be locally obtained from "Minkowski space" (generalized to the case in which a neutrino field is present) by means of a displacement along the orbit of some infinite-dimensional extended group of transformations L_∞.

The proof consists of three parts: (a) the construction of an arbitrary 3×3 matrix H on the symmetry axis by means of data on ε, Φ, and $\partial w/\partial r$ at $r=0$; (b) the construction of a 3×3 generating matrix $F(0,z,t)$ on the symmetry axis; (c) the derivation of a relation between the components of H and the parameters of the orbit of the extended group which carries the "initial" solution H into the solution H.

a) The construction of the matrix H near $r=0$. The 2×2 matrix f_{AB} is singular on the symmetry axis $r=0$, since $\det(f_{AB})=-r^2$ by definition. We assume that the 3×3 matrix H is an analytic function of r near the axis $r=0$, and that $H_1^2=-H=-\mathcal{E}$ and $H_1^3=\varphi=\Phi$ at $r=0$ are locally holomorphic functions of z. If we assume that $w\neq 0$ at $r=0$, then it follows from the condition of boundedness of H_{AB} at $r=0$ and from (3.2.2) that $f_{12}(0,z)=w(0,z)$. Then the closed orbits of the Killing vector t near the symmetry axis become time-like, since $t\cdot t=f_{22}(0,z)=w^2>0$. Therefore, if we are to avoid the appearance of closed time-like curves violating the principle of causality, we must assume that $w=0$ at $r=0$. It follows from the fact that w is a harmonic function that w in the neighborhood of $r=0$ has the form $w=\sum(-1)^n w_1^{(2n)}(z)r^{2n+1}/(2n+1)!$, and therefore $w=w_1(z)r+\mathrm{O}(r)$ near $r=0$.

The expansions of the components H_a^b $(a,b=1,2,3)$ in powers of r according to the system (3.2.1) and the definitions (3.2.2, 3) are completely determined by the functions $H_1^2(0,z)$, $H_1^3(0,z)$, and $w_1(z)$. Calculations give the following principal terms in the asymptotic forms of the components H_a^b near $r=0$:

$$H_1^1 \approx 2iz + 2w_1 r \ , \quad H_1^2 \approx -\mathcal{E} + irw_1 \partial\mathcal{E}/\partial z \ , \\ H_1^3 \approx \Phi + irw_1 \partial\Phi/\partial z \ , \tag{3.2.35}$$

$$H_2^1 \approx f_{22} \ , \quad H_2^2 \approx if_{22}\partial\mathcal{E}/2\partial z \ , \quad H_2^3 \approx -if_{22}\partial\Phi/2\partial z \ , \\ H_3^1 \approx -2f_{22}\Phi^* \ , \quad H_3^2 \approx -if_{22}\Phi^*\partial\mathcal{E}/\partial z \ , \quad H_3^3 \approx if_{22}\Phi^*\partial\Phi/\partial z \ . \tag{3.2.36}$$

Here $f_{22} \approx (w_1^2 - 1)r^2/f(0,z)$, $f(0,z) = \mathrm{Re}\{\mathcal{E}\} + \Phi\Phi^*$.

b) The construction of the matrix $\boldsymbol{F}$ near $r = 0$. We calculate the generating matrix $\boldsymbol{F}$ by means of the system

$$(1 + 2i\xi s)\boldsymbol{F}_{,\xi} = -is\boldsymbol{H}_{,\xi}\boldsymbol{F} \ , \quad (1 - 2i\xi^* s)\boldsymbol{F}_{,\xi^*} = -is\boldsymbol{H}_{,\xi^*}\boldsymbol{F} \ , \tag{3.2.37}$$

substituting into it the asymptotic form of $\boldsymbol{H}$ according to (3.2.35, 36). From the system (3.2.37) we find for $\boldsymbol{F}$, apart from terms $O(r)$, the asymptotic behavior $\boldsymbol{F} \approx (1 - iw_1 r\partial/\partial z)\boldsymbol{F}(0,z)$, where

$$\boldsymbol{F}(0,z) = \begin{pmatrix} (1-2zs)^{-1} & is\mathcal{E}(1-2zs)^{-1} & -is\Phi(1-2zs)^{-1} \\ 0 & 1 & 0 \\ 0 & 0 & 1 \end{pmatrix} \ . \tag{3.2.38}$$

c) The relation between the transformation of the internal symmetry of (3.2.4) and the initial values $\dot{\mathcal{E}}$, $\dot{\Phi}$ and transformed values $\mathcal{E}$, Φ of the solutions at $r = 0$.

A linear representation of the group L_∞ of nonlinear transformations of (3.2.4) is provided by the condition (3.2.26) of analyticity of the matrix function $\chi = \boldsymbol{F}\boldsymbol{u}\dot{\boldsymbol{F}}^{-1}$ in the region L_-, which includes the point at infinity. In L_-, we also have analyticity of the matrix function $\boldsymbol{u}(s)$ giving a displacement along the orbit of the group L_∞, which carries $\dot{\boldsymbol{F}}$ into $\boldsymbol{F}$ and is independent of r and z. The matrix function $\boldsymbol{u}(s)$ must be represented in the form (3.2.26) and must therefore satisfy (3.2.27).

The matrix functions $\dot{\boldsymbol{F}}$ and $\boldsymbol{F}$ are analytic in the region L_+, which includes the origin $s = 0$, and $\boldsymbol{F}(s = 0) = 1$, $(\partial\boldsymbol{F}/\partial s)_{s=0} = -i\boldsymbol{H}$. The regions L_+ and L_- are separated by a smooth contour L. The branch points of the matrix $\boldsymbol{F}$, $2s = (z \pm ir)^{-1}$, belong to the region L_-.

The conditions (3.2.34) are necessary and sufficient for analyticity of the matrix χ at the branch points.

From the conditions (3.2.34) at $r = 0$, we have

$$(H_a^c)'\chi_c^b - \chi_a^c(\dot{H}_c^b)' = 0 \ ; \tag{3.2.39}$$

here and in what follows, a prime indicates a derivative with respect to z, evaluated at $r = 0$.

We put $a = 2, 3$ in (3.2.39), using the expressions (3.2.36) for H_a^b. This gives $\chi_a^1 = 0$ $(a = 2, 3)$.

We now put $a = 1$ in (3.2.39). For $b = 1$ the condition (3.2.39) is satisfied identically, while for $b = 2, 3$ we have

$$2\mathrm{i}\chi_1^b = \chi_1^1(\dot{H}_1^b)' - (H_1^2)'\chi_2^b - (H_1^2)'\chi_s^b\Big|_{s=1/2z} \ . \tag{3.2.40}$$

Using the form (3.2.38) of the matrices $\boldsymbol{F}(0,z)$, we have

$$\chi_a^2 = u_a^2 - \mathrm{i}s\dot{\mathcal{E}}u_a^1 \ , \quad \chi_a^3 = u_a^3 + \mathrm{i}s\dot{\Phi}u_a^1 \ , \quad a = 2,3 \ .$$

The components χ_1^b for $b = 2,3$ at $r = 0$ have the form

$$\begin{aligned}
&\chi_1^2 = \frac{A}{1-2zs} \ , \quad \chi_1^3 = \frac{B}{1-2zs} \ , \\
&A = -\mathrm{i}s\dot{\mathcal{E}}(z)(u_1^1(s) + \mathrm{i}s\mathcal{E}(z)u_2^1(s) - \mathrm{i}s\Phi u_3^1(s)) + u_1^2(s) \\
&\qquad + \mathrm{i}s\mathcal{E}(z)u_2^2(s) - \mathrm{i}s\Phi(z)u_3^2(s) \ , \\
&B = \mathrm{i}s\dot{\Phi}(u_1^1 + \mathrm{i}s\mathcal{E}u_2^1 - \mathrm{i}s\Phi u_3^1) + u_1^3 + \mathrm{i}s\mathcal{E}u_2^3 - \mathrm{i}s\Phi u_3^3 \ .
\end{aligned} \tag{3.2.41}$$

For analyticity of χ_1^2 and χ_1^3, and also fulfillment of the condition (3.2.40) at $2s = 1/z$, the numerators on the right-hand sides of (3.2.41) must vanish:

$$A = B = 0 \quad \text{or} \quad 2s = 1/z \ . \tag{3.2.42}$$

We now factorize the group of linear transformations L_∞ as follows. Let us set $u_2^1 = u_3^1 = 0$, $u_3^3 = 1$. Then it follows from (3.2.26, 27) that the general form of the matrix $\boldsymbol{u}$ is given by

$$\boldsymbol{u} = \begin{pmatrix} a & as(\gamma - \mathrm{i}\alpha^*\alpha) & \mathrm{i}s\alpha a \\ 0 & 1/a^* & 0 \\ 0 & -2\alpha^* & 1 \end{pmatrix} \ . \tag{3.2.43}$$

Here $a(s)$ and $\alpha(s)$ are arbitrary complex functions of s, and $\gamma(s)$ is an arbitrary real function of s. All these functions are analytic in L_-.

To verify that the matrix $\boldsymbol{u}$ can be represented in the form of an exponential of the product of some Hermitian matrix and the anti-Hermitian matrix $\boldsymbol{\Omega}$, it is sufficient to take the logarithm of the matrix $\boldsymbol{u}$ by means of the Lagrange-Sylvester formula.

Substituting (3.2.43) into (3.2.42), we obtain

$$\begin{aligned}
&\mathcal{E} = aa^*(\dot{\mathcal{E}} + \mathrm{i}\gamma - 2\alpha^*\dot{\Phi} - \alpha\alpha^*) \\
&\Phi = a(\dot{\Phi} + \alpha) \ .
\end{aligned} \tag{3.2.44}$$

We note that $\mathcal{E}$ and Φ can be continued analytically and uniquely into some neighborhood of the axis $r = 0$ by means of the equations (see Sect. 3.1)

$$\mathcal{L}\mathcal{E} = 0 \ , \quad \mathcal{L}\Phi = 0 \ , \tag{3.2.45}$$

where the operator $\mathcal{L}$ is defined as follows:

$$\mathcal{L} \equiv (\mathrm{Re}\{\mathcal{E}\} + \Phi\Phi^*)[r\nabla^2 + (\nabla r + \mathrm{i}\nabla\tilde{\omega})\nabla] - r(\nabla\mathcal{E} + 2\Phi^*\nabla\Phi)\nabla \ .$$

Thus, on the basis of the relation between the parameters $a(s)$, $\alpha(s)$, $\gamma(s)$ of the group transformation and the functions $\mathcal{E}$ and Φ, it can be seen that for any pair of solutions $\dot{\boldsymbol{H}}$ and $\boldsymbol{H}$ which are analytic near $r = 0$ there exists at least one transformation of the internal symmetry of (3.2.4) which carries these solutions into each other.

As an application of the results (3.2.44), we shall show how it is possible to obtain a solution corresponding to a charged black hole with mass m and charge q in a neutrino field.

If we suppose that $\dot{\mathcal{E}} = 1$, $\dot{\Phi} = 0$ [this is an exact solution of (3.2.45)], the final solution for $r = 0$ takes the values

$$\mathcal{E} = \frac{z + z_0 - m}{z + z_0 + m} \quad , \quad \Phi = \frac{q}{z + z_0 + m} \quad , \quad z > m \tag{3.2.46}$$

(the origin is displaced on the z axis in order to ensure that the solution is analytic at $z = 0$).

According to (3.2.14, 46), we have

$$\begin{aligned} &\gamma = 0 \quad , \quad \omega = \frac{2qs}{1 + 2s(z_0 + m)} \quad , \\ &a^2 - \omega^2 = \frac{1 + 2s(z_0 - m)}{1 + 2s(z_0 + m)} \quad , \quad \omega = \alpha a \quad . \end{aligned} \tag{3.2.47}$$

The generating matrix $\dot{\boldsymbol{F}}^{-1}$ corresponding to $\dot{\mathcal{E}} = 1$, $\dot{\Phi} = 0$ has the nonzero components

$$\begin{aligned} &(\dot{F}^{-1})_1^1 = (-z + \mathrm{i}w)t + (1 + \lambda)/2 \quad , \\ &(\dot{F}^{-1})_1^2 = -\mathrm{i}t \quad , \\ &(\dot{F}^{-1})_2^1 = -\mathrm{i}z - w + (\lambda - 1)/2\mathrm{i}t \quad , \\ &(\dot{F}^{-1})_2^2 = (\dot{F}^{-1})_3^3 = 1 \quad , \\ &\lambda = \sqrt{(1 - 2tz)^2 + 4t^2r^2}\, \exp(2\mathrm{i}t) \left[\int \frac{\partial w}{\partial \xi} \frac{d\xi}{1 + 2\mathrm{i}\xi t} + \mathrm{c.c.} \right] \quad . \end{aligned} \tag{3.2.48}$$

We find the matrix $\boldsymbol{F}$ corresponding to the solution (3.2.46) from the equation

$$\int_L \boldsymbol{F}(s)\boldsymbol{u}(s)\dot{\boldsymbol{F}}^{-1}(s) \frac{t\,ds}{s(s - t)} = 0 \quad , \tag{3.2.49}$$

into which we substitute $\boldsymbol{u}(s)$ in the form (3.2.43) with the specific functions (3.2.47) and the matrix $\dot{\boldsymbol{F}}^{-1}$ defined by (3.2.48).

In the final solution, it is necessary to make the inverse displacement $z + z_0 \to z$. Then for $\mathcal{E}$ and Φ in the coordinates r and z we obtain

$$\Phi = \frac{2q}{2m + \lambda_+ + \lambda_-} \quad , \quad \mathcal{E} = \frac{\lambda_+ + \lambda_- - 2m}{\lambda_+ + \lambda_- + 2m} \quad ;$$

here

$$\lambda_+ = \sqrt{\left(z+\sqrt{m^2-q^2}\right)^2+r^2}\,\exp\left[\int \frac{\partial w}{\partial \xi}\frac{d\xi}{r+\mathrm{i}\left(z+\sqrt{m^2-q^2}\right)} - \text{c.c.}\right],$$

$$\lambda_- = \sqrt{\left(z-\sqrt{m^2-q^2}\right)^2+r^2}\,\exp\left[\int \frac{\partial w}{\partial \xi}\frac{d\xi}{r+\mathrm{i}\left(z-\sqrt{m^2-q^2}\right)} - \text{c.c.}\right].$$

To go over to the coordinates in which the Nordström-Reissner solution is usually written, we must make the substitutions

$$2R = 2m + \sqrt{\left(z+\sqrt{m^2-q^2}\right)^2+r^2} + \sqrt{\left(z-\sqrt{m^2-q^2}\right)^2+r^2},$$

$$2\sqrt{m^2-q^2}\cos\theta = \sqrt{\left(z+\sqrt{m^2-q^2}\right)^2+r^2} - \sqrt{\left(z-\sqrt{m^2-q^2}\right)^2+r^2}.$$

3.3 General Solution of the Einstein-Maxwell Equations for Ernst Data Regular Locally on the Symmetry Axis

In this section, we present a method of constructing the general solution of the Einstein-Maxwell system in the stationary axially symmetric case, and we give new classes of exact solutions of the Einstein-Weyl system. We mention an interesting paper [3.7] in which a solution of Einstein's equations was constructed for the vacuum with an arbitrary distribution of sources of mass and angular momentum on the symmetry axis. Comparatively few exact solutions of the combined system of Einstein-Weyl equations are known. An exact solution with a group of motions G_3 on V_2 has been found by *Golubiatnikov* [3.8]. Solutions with an isotropic Killing vector (the case of pure radiation) and with a zero energy-momentum tensor are known [3.9]. Solutions describing the interaction of a step-like neutrino wave and a δ-like gravitational wave have been found by *Griffiths* [3.10]. A solution depending on a single variable has been found by *Repchenkov* [3.11]. A self-similar interaction of neutrino waves in general relativity has been studied by Blazhennova-Mikulich. Cosmological solutions with neutrino fields have been studied by *Henneaux* [3.12].

Here we present (a) classes of solutions of the Einstein-Weyl equations described by a first-order equation (in the absence of neutrino fields, they reduce to solutions found by *Bitsadze* [3.13]), (b) a method of constructing the local general solution of the electrovacuum equations regular locally on the symmetry axis, (c) methods of constructing exact solutions of the equations for the neutrino vacuum, (d) a method of obtaining topologically nontrivial vacuum solutions of Einstein's equations, and (e) gravitational fields of rotating magnetized stars.

a) A Class of Solutions Described by a First-Order Equation. We shall consider a class of exact solutions of an equation obtained, as we have already shown [see (3.1.23)], from the Einstein-Weyl system in the stationary axially symmetric case for $\varphi = 0$:

$$r\nabla^2 H + \nabla H(\nabla r + \mathrm{i}\nabla\tilde{\omega}) - 2r(\nabla H)^2/(H + H^*) = 0 \ ,$$

where $\tilde{\omega}$ is a real harmonic function. We rewrite this equation in terms of the variables $\xi = r + \mathrm{i}z$, $\xi^* = r - \mathrm{i}z$:

$$\begin{aligned}
&\frac{\partial^2 H}{\partial\xi\,\partial\xi^*} + \frac{1}{(\xi+\xi^*)}\left[\frac{\partial H}{\partial\xi}\left(\frac{1}{2} + \mathrm{i}\frac{\partial\tilde{\omega}}{\partial\xi^*}\right) + \frac{\partial H}{\partial\xi^*}\left(\frac{1}{2} + \mathrm{i}\frac{\partial\tilde{\omega}}{\partial\xi}\right)\right] \\
&\quad - \frac{2}{(H+H^*)}\frac{\partial H}{\partial\xi}\frac{\partial H}{\partial\xi^*} \\
&= \frac{\partial\mathcal{A}}{\partial\xi^*} + \frac{\mathcal{A}}{(\xi+\xi^*)}\left(\frac{1}{2} + \mathrm{i}\frac{\partial\tilde{\omega}}{\partial\xi^*}\right) + \frac{\mathcal{A}^*}{(\xi+\xi^*)}\left(\frac{1}{2} + \mathrm{i}\frac{\partial\tilde{\omega}}{\partial\xi}\right) \\
&\quad - \frac{2}{(H+H^*)}\frac{\partial H}{\partial\xi^*}\mathcal{A} \\
&= \frac{\partial\mathcal{B}}{\partial\xi} + \frac{\mathcal{B}}{(\xi+\xi^*)}\left(\frac{1}{2} + \mathrm{i}\frac{\partial\tilde{\omega}}{\partial\xi}\right) + \frac{\mathcal{B}^*}{(\xi+\xi^*)}\left(\frac{1}{2} + \mathrm{i}\frac{\partial\tilde{\omega}}{\partial\xi^*}\right) \\
&\quad - \frac{2}{(H+H^*)}\frac{\partial H}{\partial\xi}\mathcal{B} = 0 \ , \quad \text{where}
\end{aligned}$$

$$\mathcal{A} \equiv \frac{\partial H}{\partial\xi} - \frac{(H+H^*)}{(\xi+\xi^*)}\left(\frac{1}{2} + \mathrm{i}\frac{\partial\tilde{\omega}}{\partial\xi}\right) , \quad \mathcal{B} \equiv \frac{\partial H}{\partial\xi^*} - \frac{(H+H^*)}{(\xi+\xi^*)}\left(\frac{1}{2} + \mathrm{i}\frac{\partial\tilde{\omega}}{\partial\xi^*}\right) .$$

Therefore solutions of the first-order equations $\mathcal{A} = 0$ or $\mathcal{B} = 0$ are also solutions of the original equation.

b) Construction of the General Solution of the Einstein-Maxwell Equations for the Stationary Axially Symmetric Case. In Sect. 3.2 we proved that the function $\chi \equiv \boldsymbol{F}\boldsymbol{u}\dot{\boldsymbol{F}}^{-1}$ is holomorphic outside any closed contour L which encloses all the singularities of the matrix $\boldsymbol{u}$ and excludes the branch points of the matrix $\boldsymbol{F}$. We assume that the branch points $s = (2z \pm 2\mathrm{i}r)^{-1}$ are the only singularities of the matrix $\boldsymbol{F}$ on the Riemann sphere and that the matrix $\boldsymbol{F}$ is holomorphic off the cut $\mathcal{L}$ joining these two points. For $\dot{\boldsymbol{F}}$ we take the generating matrix (3.2.48) of Minkowski space (the function w in this case is equal to zero). Suppose that the matrix $\boldsymbol{u}$ is factorized in accordance with (3.2.43). Then the components of $\boldsymbol{u}$ are related to the values of the Ernst potentials $\mathcal{E}$ and Φ on the symmetry axis by (3.2.44), in which $\dot{\mathcal{E}} = 1$, $\dot{\Phi} = 0$. Let $\mathcal{E}(0, z) \equiv e(z)$ and $\Phi(0, z) \equiv f(z)$ be arbitrary locally holomorphic functions (not only meromorphic ones on the Riemann sphere). We shall find locally the general solution $\mathcal{E}(r, z)$, $\Phi(r, z)$ of the Einstein-Maxwell system with these values on the axis.

The matrix function χ is analytic on the cut $\mathcal{L}$, and therefore $[\chi] \equiv (\chi_+ - \chi_-)_{\mathcal{L}} = 0$, where $\chi_\pm$ are the respective boundary values of χ on the left and right sides of the cut. Calculating the components of χ and making use of their continuity, we have

$$[F_a^2] = \mathrm{i}se[F_a^1] \ , \quad [F_a^3] = -\mathrm{i}sf[F_a^1] \ , \quad a = 1,2,3 \ , \tag{3.3.1}$$

$$\mathrm{i}s\{F_a^1\}\tilde{e}\,\{F_a^2\} - 2\tilde{f}\{F_a^3\} = 0 \ , \quad \{A\} \equiv A_+ + A_- \ . \tag{3.3.2}$$

In these relations, e and f are functions of the argument $\xi \equiv 1/2s$, and $\tilde{e}(\xi) \equiv (e(\xi^*))^*$, $\tilde{f}(\xi) \equiv (f(\xi^*))^*$. The matrix $\boldsymbol{F}$ is holomorphic off the cut $\mathcal{L}$. Therefore its components can be expressed in the form of the Cauchy integrals

$$\begin{aligned} F_a^1 &= \frac{1}{2\pi\mathrm{i}} \int\limits_{\mathcal{L}} \frac{ds}{(s-t)} [F_a^1] \ , \\ F_a^b &= \delta_a^b + \frac{t}{2\pi\mathrm{i}} \int\limits_{\mathcal{L}} \frac{ds}{s(s-t)} [F_a^b] \ , \quad b = 2,3 \ . \end{aligned} \tag{3.3.3}$$

In the expression for F_a^1, we have not yet made use of the condition $F_a^1(0) = \delta_a^1$. Therefore the discontinuity $[F_a^1]$ must satisfy the additional constraint

$$2\pi\mathrm{i}\delta_a^1 = \int\limits_{\mathcal{L}} \frac{ds}{s} [F_a^1] \ . \tag{3.3.3'}$$

The discontinuities $[F_a^1]$ become infinite at the ends of the cut $\mathcal{L}$. We introduce the unknown functions $\mu_a \equiv [F_a^1]\sqrt{(1-2zs)^2 + 4r^2s^2}$, which are continuous and bounded on the cut $\mathcal{L}$. We use for $[F_a^b]$ the conditions which follow from their representations (3.3.3), namely,

$$\begin{aligned} \{F_a^1\} &= \frac{1}{\pi\mathrm{i}} \fint\limits_{\mathcal{L}} \frac{ds}{(s-t)} [F_a^1] \ , \\ \{F_a^b\} - \delta_a^b &= \frac{t}{\pi\mathrm{i}} \fint\limits_{\mathcal{L}} \frac{ds}{s(s-t)} [F_a^b] \ , \quad b = 2,3 \ , \end{aligned} \tag{3.3.4}$$

where the principal values of the integrals are taken. In order to obtain singular integral equations for μ_a, we substitute the expressions (3.3.4) into (3.3.2) and use (3.3.1). In the integrals, it is very convenient to transform to the variable of integration σ defined by the relation $1/2s \equiv z + \mathrm{i}r\sigma$. This variable σ runs from -1 to $+1$, and the required integral equations take the elegant form[2]

$$\fint_{-1}^{1} \frac{d\sigma\, \mu_a(\xi)}{\sqrt{1-\sigma^2}\,(\sigma-\tau)} (e(\xi) + \tilde{e}(\eta) + 2\tilde{f}(\eta)f(\xi)) = 2\pi r(\delta_a^2 - 2\tilde{f}(\eta)\delta_a^3) \ , \tag{3.3.5}$$

where $\xi \equiv z + \mathrm{i}r\sigma$, $\eta \equiv z + \mathrm{i}r\tau$, $\tau \in [-1,1]$.

[2] For the case of cylindrical (or plane) waves, we must put $\xi \equiv t + r\sigma$, $\eta \equiv t + r\tau$ in (3.3.5–7).

The components of the required matrix H can be expressed in terms of the functions μ_a as follows:

$$H_a^1 = \frac{2}{\pi} \int_{-1}^{1} \frac{\xi \mu_a(\xi)\, d\sigma}{\sqrt{1-\sigma^2}} \ ,$$

$$H_a^2 = -\frac{1}{\pi} \int_{-1}^{1} \frac{e(\xi)\mu_a(\xi)\, d\sigma}{\sqrt{1-\sigma^2}} \ , \tag{3.3.6}$$

$$H_a^3 = \frac{1}{\pi} \int_{-1}^{1} \frac{f(\xi)\mu_a(\xi)\, d\sigma}{\sqrt{1-\sigma^2}} \ .$$

For a single-valued definition of μ_a, the additional condition (3.3.3′) takes the form

$$\int_{-1}^{1} \frac{\mu_a(\xi)\, d\sigma}{\sqrt{1-\sigma^2}} = \pi \delta_a^1 \ . \tag{3.3.7}$$

We shall give the general solution of the homogeneous equation (3.3.5) for $a = 1$, dropping the subscript on μ_1 and using the well-known formula

$$\frac{1}{\pi} \fint_{0}^{\pi} \frac{d\theta \cos n\theta}{(\cos\theta - \cos\varphi)} = \frac{\sin n\varphi}{sin\varphi} \ , \qquad n = 0, 1, 2 \ldots \ .$$

We expand the functions $\mu(\xi)$, $e(\xi)$, and $f(\xi)$ in Fourier series, making the substitution $\sigma = \cos\theta$:

$$\mu(\xi) = \sum_{k=0}^{\infty} \mu_k(r,z) \cos k\theta \ , \qquad e(\xi) = \sum_{k=0}^{\infty} e_k(r,z) \cos k\theta \ ,$$
$$f(\xi) = \sum_{k=0}^{\infty} f_k(r,z) \cos k\theta \ , \qquad \xi \equiv z + \mathrm{i} r \cos\theta \ . \tag{3.3.8}$$

From the condition (3.3.7) we obtain $\mu_0 = 1$. Substituting (3.3.8) into (3.3.6), for the Ernst potentials $\mathcal{E} = -H_2^1$, $\Phi = H_1^3$ we obtain the expressions

$$\mathcal{E} = e(z) + \frac{1}{2} \sum_{i=1}^{\infty} e_i \mu_i \ , \qquad \Phi = f(z) + \frac{1}{2} \sum_{i=1}^{\infty} f_i \mu_i \ . \tag{3.3.9}$$

Substituting (3.3.8) into (3.3.5), we obtain the following system of equations for $\mu_k(r,z)$:

$$\sum_{s=1}^{\infty} \mu_s T_{sn} = T_n \quad , \quad n = 1,2,3,\dots \quad ,$$

$$T_n \equiv -2 \left[e_n + \sum_{k=0}^{\infty} (f_{k+n} \tilde{f}_n - f_n \tilde{f}_{k+n}) + \sum_{k=0}^{n} f_k \tilde{f}_{n-k} \right] \quad ,$$

$$T_{ns} \equiv \sum_{k=0}^{n} f_{k+s} \tilde{f}_{n-k} + \sum_{k=0}^{\infty} (f_{k+n+s} \tilde{f}_k - f_k \tilde{f}_{k+n+s} - f_{k+s} \tilde{f}_{k+n})$$

$$- \sum_{k=0}^{s} f_{s-k} \tilde{f}_{k+n} + e_{s+n} - \tilde{e}_{s+n} + \tau_{sn} \quad .$$

For $s \leq n$ we have

$$\tau_{sn} = e_{n-s} + \tilde{e}_{n-s} + \sum_{k=0}^{n-s} f_k \tilde{f}_{n-s-k} + \sum_{k=0}^{s} f_{s-k} \tilde{f}_{n-k} + \sum_{k=0}^{\infty} f_{k+n-s} \tilde{f}_k \quad ,$$

while for $s > n$ we have

$$\tau_{sn} = e_{s-n} + \tilde{e}_{s-n} + \sum_{k=0}^{n} f_{s-k} \tilde{f}_{n-k} + \sum_{k=0}^{s-n} f_{s-n-k} \tilde{f}_k + \sum_{k=s-n}^{\infty} f_{n-s+k} \tilde{f}_k \quad .$$

The solutions (3.3.9) can be expressed in a more compact form. Introducing the notation $D \equiv \det T_{ik}$, $D_e \equiv \det(T_{ik} + e_i T_k)$, $D_f \equiv \det(T_{ik} + f_i T_k)$, these solutions can be written as ratios of determinants with infinitely many rows and columns:

$$\mathcal{E} = e(z) + (D_e - D)/(2D) \quad , \quad \Phi = f(z) + (D_f - D)/(2D) \quad . \tag{3.3.9'}$$

Thus, we have solved the problem of finding the general solution of the system of Einstein-Maxwell equations[3]! From (3.3.9′) we can calculate an arbitrary solution with any degree of accuracy. We note that solutions in the Prandtl theory of thin wings and hydroplanes [3.14] are frequently written in a similar form.

We shall now show how to find exact solutions of (3.3.5) for arbitrary *rational* functions $e(\xi)$ and $f(\xi)$. We must first find the roots $\xi_1, \xi_2, \dots, \xi_N$ of the equation $e(\xi) + \tilde{e}(\xi) + 2f(\xi)\tilde{f}(\xi) = 0$. We denote their multiplicities by $m_1, m_2, \dots, m_N$, respectively. There can be only real roots and pairs of complex-conjugate roots. The solution for $\mu(\xi)$ must be sought in the form

$$\mu(\xi) = A + \sum_{k=1}^{N} (A_k^1 (\xi - \xi_k)^{m_k - 1} + A_k^2 (\xi - \xi_k)^{m_k - 2} + \dots + A_k^{m_k})(\xi - \xi_k)^{-m_k} ,$$

where A_k^n and A are functions of r and z.

[3] For the general solution of the Cauchy problem in the case of cylindrical waves, the functions $e(\xi)$ and $f(\xi)$ can be related to the Cauchy data by means of (3.3.9′) by putting $t = 0$.

From (3.3.5, 7), using the formulas

$$\int_{-1}^{1} \frac{d\sigma}{(a+\mathrm{i}\sigma b)\sqrt{1-\sigma^2}} = \frac{\pi}{\sqrt{a^2+b^2}} ,$$

$$⨍ \frac{d\sigma}{(\sigma-\tau)(\sigma-\gamma)\sqrt{1-\sigma^2}} = \frac{\pi}{(\tau-\gamma)\sqrt{\gamma^2-1}} ,$$

where a, b, and γ are arbitrary constants, we obtain a closed linear system for the determination of $A_k^n (k = 1, \dots, N;\ n = 1, \dots, m_k)$ after making a partial-fraction decomposition of the rational functions and equating the coefficients of the independent partial fractions to zero. Suppose, for example, that $e(\xi) = (\xi - m - \mathrm{i}\nu)/(\xi + m - \mathrm{i}\nu)$ and $f(\xi) = q/(\xi + m - \mathrm{i}\nu)$, where m, ν, and q are constants. The equation $e + \tilde{e} + 2f\tilde{f} = 0$ in this case has roots $\xi = \pm\alpha$, $\alpha = \sqrt{m^2 - q^2 - \nu^2}$. We seek a solution for μ in the form

$$\mu = A + \frac{A_+}{(\xi+\alpha)} + \frac{A_-}{(\xi-\alpha)} . \tag{3.3.8'}$$

It follows from (3.3.7) that

$$1 = A + \frac{A_+}{r_+} + \frac{A_-}{r_-} , \qquad r_\pm = \sqrt{(z\pm\alpha)^2 + r^2} .$$

Substituting the solution in the form (3.3.8′) into (3.3.5), we obtain

$$A = \frac{A_+}{m-\alpha-\mathrm{i}\nu} + \frac{A_-}{m+\alpha-\mathrm{i}\nu} ,$$

$$\frac{A_+(m^2 - m\alpha - q^2 - \mathrm{i}\nu m)}{r_+[(m-\alpha)^2+\nu^2]} + \frac{A_-(m^2 + m\alpha - q^2 - \mathrm{i}\nu m)}{r_-[(m+\alpha)^2+\nu^2]} = 0 .$$

After finding A and $A_\pm$ and substituting them into the expressions for $\mathcal{E}$ and Φ calculated according to (3.3.6), we obtain

$$\mathcal{E} = A - \frac{A_+(m+\alpha+\mathrm{i}\nu)}{r_+(m-\alpha-\mathrm{i}\nu)} - \frac{A_-(m-\alpha+\mathrm{i}\nu)}{r_-(m+\alpha-\mathrm{i}\nu)}$$

$$= \frac{r_+ \mathrm{e}^{\mathrm{i}\theta} + r_- \mathrm{e}^{-\mathrm{i}\theta} - 2m\cos\theta}{r_+ \mathrm{e}^{\mathrm{i}\theta} + r_- \mathrm{e}^{-\mathrm{i}\theta} + 2m\cos\theta} ,$$

$$\Phi = q\left[\frac{A_+}{r_+(m-\alpha-\mathrm{i}\nu)} + \frac{A_-}{r_-(m+\alpha-\mathrm{i}\nu)}\right]$$

$$= \frac{2q\cos\theta}{r_+ \mathrm{e}^{\mathrm{i}\theta} + r_- \mathrm{e}^{-\mathrm{i}\theta} + 2m\cos\theta} ,$$

$$\sin\theta \equiv \frac{\nu}{\sqrt{m^2-q^2}} .$$

This is, in fact, the well-known Kerr-Newman solution. In the general case of rational $e(\xi)$ and $f(\xi)$, the expressions for $\mathcal{E}$ and Φ can be written as rational functions of $\sqrt{(z-\xi_k)^2+r^2}$ and $z-\xi_k$ with $k=1,2,\ldots,N$ (the solution of *Sato* and *Tomimatzu* [3.15] is obtained as a particular case).

c) Construction of New Solutions by Means of a Displacement According to Subgroups of an Infinite-Dimensional Group. The integral matrix equation (3.2.26) with $t \in L_+$ reduces to an integral equation for one unknown function in two significant cases: (1) when the matrix $\boldsymbol{\tau}(t) \equiv \boldsymbol{u} - \mathbb{1}$ has only one nonzero off-diagonal component; (2) when the matrix $\boldsymbol{\tau}(t)$ is diagonal. Although an arbitrary matrix $\boldsymbol{\tau}(t)$ can be represented in the form of the product of four matrices of the type 1, the study of transformations of the type 2 enables us to obtain physically interesting solutions by solving an integral equation only once.

In this subsection, we derive integral equations and give their solutions for matrices $\boldsymbol{\tau}(t)$ which are arbitrary rational functions that are analytic in the neighborhood of infinity, for the cases 1 and 2 defined above. With a particular choice of the neutrino field w in the form $w=\pm r$, it turns out to be possible to obtain a solution for an arbitrary displacement according to a one-parameter subgroup of an infinite-dimensional group.

1) We shall consider the integral matrix equation (3.2.26) with $t \in L_+$ for the component $F_1^2(t)$ under the assumption that the matrix $\boldsymbol{\tau}(t)$ has only one nonzero component $\tau_2^1(t)$. We adopt the notation $\tau_2^1(t) = T(t)$, $F_1^2(t) = -Y(t)t$. We obtain a Fredholm integral equation of the second kind:

$$Y(t) - \dot{Y}(t) = \frac{1}{2\pi\mathrm{i}} \int\limits_L Y(s)T(s)M(s,t)\,ds \ , \tag{3.3.10}$$

$$\begin{aligned} &M(s,t) \equiv \lambda(s)\mathrm{e}^{\mathrm{i}\sigma}[\dot{F}_2^2(s)\dot{F}_1^2(t) - \dot{F}_1^2(s)\dot{F}_2^2(t)]/(s-t) \ , \\ &tY(t) \equiv -F_1^2(t) \ , \quad t\dot{Y}(t) \equiv -\dot{F}_1^2(t) \ , \end{aligned} \tag{3.3.11}$$

where $\dot{F}_1^2(t)$, $\dot{F}_2^2(t)$ are the components of the generating matrix $\boldsymbol{F}(t)$ for the initial solution from which we intend to generate new solutions. It follows from the condition that $T(s)$ is bounded in the limit $s \to \infty$ and from its rational character that this function can be represented in the form

$$T(s) = \sum_{m=1}^{N} \sum_{n=0}^{\alpha_m} \frac{a_n^m}{(s-u_m)^{n+1}} + \text{const} \ , \tag{3.3.12}$$

where $\alpha_m + 1$ is the order of the pole at the point $s = u_m$, and N is the number of distinct poles $u_1, \ldots, u_N$. The restriction to the complex constants a_n^m and u_m is the condition that, after reduction to a common denominator, $T(s)$ can be represented in the form of a ratio of two polynomials with real coefficients, in which the degree of the polynomial in the numerator does not exceed the degree of the polynomial in the denominator.

Integrating by parts and calculating the residues at the points $s = u_m$, from (3.3.10, 12) we obtain

$$Y(t) - \dot{Y}(t) = \sum_{m=1}^{N} \sum_{n=0}^{\alpha_m} \frac{a_n^m}{n!} \left(\frac{\partial^n}{\partial s^n} Y(s) M(s,t) \right)_{s=u_m} . \tag{3.3.13}$$

We denote $\partial^k Y(s)/\partial s^k$ at $s = u_m$ by Y_{mk}. It follows from (3.3.13) that

$$\begin{aligned} Y(t) - \dot{Y}(t) &= \sum_{m=1}^{N} \sum_{k=0}^{\alpha_m} Y_{mk} \varphi^{mk}(t) , \\ \varphi^{mk}(t) &= \sum_{n=0}^{\alpha_m - k} a_{n+k}^m \left(\frac{\partial^n}{\partial s^n} \frac{M(s,t)}{k!n!} \right)_{s=u_m} . \end{aligned} \tag{3.3.14}$$

We differentiate (3.3.14) q times with respect to t and set $t = u_l$ (q takes the values from 0 to α_l, where $l = 1, 2, \ldots, N$). From (3.3.14) we obtain a closed linear algebraic system for Y_{lq}:

$$\begin{aligned} &Y_{lq} - \dot{Y}_{lq} = \sum_{m=1}^{N} \sum_{k=0}^{\alpha_m} Y_{mk} Q_{lq}^{mk} , \\ &q = 0, \ldots, \alpha_l \ ; \quad l = 1, \ldots, N \ , \quad Q_{lq}^{mk} = \left. \frac{\partial^q \varphi^{mk}(t)}{\partial t^q} \right|_{t=u_l} . \end{aligned} \tag{3.3.15}$$

The required solution $H = H(r, z)$ of (3.3.1) can be expressed in terms of the solution Y_{mk} as follows:

$$H = \mathrm{i}Y(0) = \mathrm{i} \sum_{m=1}^{N} \sum_{k=0}^{\alpha_m} Y_{mk} \sum_{n=0}^{\alpha_m - k} \frac{a_{n+k}^m}{k!n!} \left(\frac{\partial^n}{\partial s^n} [\dot{Y}(s) \lambda(s) \mathrm{e}^{\mathrm{i}\sigma}] \right)_{s=u_m} . \tag{3.3.16}$$

Using (3.2.48) for the initial solution, we obtain

$$H = \sum_{m=1}^{N} \sum_{k=0}^{\alpha_m} \frac{Y_{mk}}{k!} a_k^m . \tag{3.3.17}$$

In the absence of a neutrino field, the solutions (3.3.15, 16) include, as special cases, the solutions found previously by *Sato* and *Tomimatzu* [3.15], *Kinnersley* and *Chitre* [3.3], and *Hauser* and *Ernst* [3.5].

2) We note that in the general case the matrix $\boldsymbol{u} = \mathbb{1} + \boldsymbol{\tau}(t)$, as the exponential of a matrix whose trace is equal to zero, must have determinant equal to unity. Therefore in the general case the diagonal matrix $\mathbb{1} + \boldsymbol{\tau}(t)$ has the form

$$\mathbb{1} + \boldsymbol{\tau}(t) = \begin{pmatrix} \Gamma(t) & 0 \\ 0 & (\Gamma(t))^{-1} \end{pmatrix} ,$$

where $\Gamma(t)$ is an arbitrary analytic function of t, regular at infinity and real on the real axis.

In order to derive an integral equation containing a single unknown function, we make use of the result of Sect. 3.2 according to which there exists an analytic continuation $\chi(t)$ of the matrix function $\boldsymbol{F}(s)(\mathbb{1}+\boldsymbol{\tau}(s))(\dot{\boldsymbol{F}}(s))^{-1}$ given on the contour L, into the region L_-, by means of a Cauchy integral.

We write this equality on the contour L in terms of components:

$$\Gamma^{-1}\chi_1^1 = (F_1^1\dot{F}_2^2 - F_1^2\dot{F}_2^1\Gamma^{-2})\tilde{\lambda} \ , \quad \tilde{\lambda} = \lambda \exp i\sigma \ , \tag{3.3.13'}$$

$$\Gamma^{-1}\chi_1^2 = (-F_1^1\dot{F}_1^2 + F_1^2\dot{F}_1^1\Gamma^{-2})\tilde{\lambda} \ , \tag{3.3.14'}$$

$$\Gamma\chi_2^1 = (\Gamma^2 F_2^1\dot{F}_2^2 - F_2^2\dot{F}_2^1)\tilde{\lambda} \ , \tag{3.3.15'}$$

$$\Gamma\chi_2^2 = (-\Gamma^2 F_2^1\dot{F}_1^2 + F_2^2\dot{F}_1^1)\tilde{\lambda} \ . \tag{3.3.16'}$$

We emphasize that the functions $\Gamma^{-1}\chi_1^A$ and $\Gamma\chi_2^A$ $(A=1,2)$ are analytic in the region L_-.

We now make use of the well-known Privalov-Gakhov solution [3.16] of the nonhomogeneous Riemann Problem of finding functions $X_\pm$ which are analytic in the corresponding regions $L_\pm$ and related on the contour L by the linear equation

$$X_- = aX_+ + b \ ,$$

where $a = a(s)$ and $b = b(s)$ are functions specified arbitrarily on the contour L. In the case in which the index of the function $\ln a(s)$ is equal to zero, the Privalov-Gakhov solution has the form

$$X_- = X(t) \quad \text{for} \quad t \in L_- \ ,$$
$$X_+ = X(t) \quad \text{for} \quad t \in L_+ \ ,$$
$$X(t) = \exp(-\Phi(t)) \left[\text{const} - \frac{1}{2\pi i}\int\limits_L b(s)(\exp\Phi(s))\frac{ds}{(s-t)}\right] \ , \tag{3.3.17'}$$

where

$$\Phi(t) = \frac{1}{2\pi i}\int \frac{\ln a(s)}{s-t}\,ds \ .$$

Using (3.3.17'), from (3.3.13', 14') and the condition $F_A^B(0) = \delta_A^B$ we obtain

$$F_1^1(t)\dot{F}_2^2(t)\tilde{\lambda}(t) = 1 + \frac{1}{2\pi i}\int\limits_L \tilde{\lambda}(s)F_1^2(s)\dot{F}_2^1(s)\Gamma^{-2}(s)\frac{t\,ds}{s(s-t)} \ , \tag{3.3.18}$$

$$F_1^1(t)\dot{F}_1^2(t)\tilde{\lambda}(t) = \frac{1}{2\pi i}\int\limits_L \tilde{\lambda}(s)F_1^2(s)\dot{F}_1^1(s)\Gamma^{-2}(s)\frac{t\,ds}{s(s-t)} \ . \tag{3.3.19}$$

Eliminating the function $F_1^1(t)$ from (3.3.18, 19), we obtain the singular integral equation

$$\dot{Y}(t) = \frac{1}{2\pi i} \int_L Y(s) \Gamma^{-2}(s) W(s,t)\, ds \ , \tag{3.3.20}$$

where

$$\begin{aligned} &sY(s) = -F_1^2(s) \ , \quad s\dot{Y}(s) = -\dot{F}_1^2(s) \ , \\ &W(s,t) = \frac{\tilde{\lambda}(s)}{(s-t)} [\dot{F}_1^1(s)\dot{F}_2^2(t) - \dot{F}_2^1(s)\dot{F}_1^2(t)] \ . \end{aligned} \tag{3.3.21}$$

A similar singular integral equation can be obtained from the system (3.3.15′, 16′) by eliminating F_2^2:

$$\begin{aligned} &\dot{Z}(t) = \frac{1}{2\pi i} \int_L Z(s) \Gamma^2(s) W(s,t)\, ds \ , \\ &sZ(s) = F_2^1(s) \ , \quad s\dot{Z}(s) = F_2^1(s) \ , \end{aligned} \tag{3.3.22}$$

where $W(s,t)$ is given by (3.3.21), as before.

In the case $w = r$ (or $w = -r$), it is easy to obtain the solution of (3.3.20, 22) for an arbitrary analytic function $\Gamma^{-2}(s)$. In fact, in this case, according to (3.2.48), the components $\dot{F}_A^B(t)$ are given by the expressions

$$\dot{F}_1^1 = (1 + 2i\xi t)^{-1} \ , \quad \dot{F}_1^2 = \frac{it}{(1+2i\xi t)} \ , \quad \dot{F}_2^1 = 0 \ , \quad \dot{F}_2^2 = 1 \ , \quad \xi = r + iz \ .$$

Therefore $W(s,t) = 1/(s-t)$, and it follows from (3.3.20, 22) that

$$F_1^2(t) = \frac{it\Gamma^2(i/2\xi)}{1+2i\xi t} \ , \quad F_1^1(t) = \dot{F}_1^1 \ , \quad F_2^1(t) = 0 \ , \quad F_2^2(t) = 1 \ ,$$

Thus, in the case $w = r$ the function $H(r,z) = \Gamma^2(i/2\xi)$ is the required solution. Since $\Gamma = \Gamma(s)$ is arbitrary, H is an arbitrary analytic function of the complex variable $\xi = r + iz$, regular at the origin.

In the case of an arbitrary harmonic function w, we assume that $\Gamma^2(s)$ is an arbitrary rational function which is bounded but nonzero at infinity. We represent $\Gamma^{-2}(s)$ in the form

$$\Gamma^{-2}(s) = \frac{1}{\alpha} \frac{\prod_{k=1}^{N_1} (s-u_k)^{\alpha_k+1}}{\prod_{n=1}^{N_2} (s-v_n)^{\beta_n+1}} \ , \tag{3.3.23}$$

where $\alpha = \text{const}$, the exponent $\alpha_k + 1$ is the order of the zero at the point $s = u_k$, the exponent $\beta_n + 1$ is the order of the pole at the point $s = v_n$ ($u_k = \text{const}$, $v_n = \text{const}$), N_1 is the number of distinct zeros, N_2 is the number of distinct poles, and

$$\sum_{k=1}^{N_1}(\alpha_k+1)=\sum_{n=1}^{N_2}(\beta_n+1)=N \quad .$$

If the denominator in (3.3.23) has complex-conjugate roots, the corresponding gravitational fields have ring singularities. Integrating the right-hand side of (3.3.20) by parts and taking the residues at the points $s = v_n$, $n = 1, \ldots, N_2$, we obtain

$$\dot{Y}(t)=Y(t)\Gamma^{-2}(t)+\sum_{n=1}^{N_2}\frac{1}{\alpha\beta_n!}\frac{\partial^{\beta_n}}{\partial s^{\beta_n}}\left[Y(s)W(s,t)\frac{\prod_{k=1}^{N_1}(s-u_k)^{\alpha_k+1}}{\prod_{k\neq n}^{N_2}(s-v_k)^{\beta_k+1}}\right]_{s=v_n} \quad .$$

We simplify the form of this expression by introducing the notation

$$\frac{\partial}{\partial s^m}\left[Y(s)\frac{\prod_{k=1}^{N_1}(s-u_k)^{\alpha_k+1}}{\alpha\prod_{k\neq n}^{N_2}(s-v_k)^{\beta_k+1}}\right]_{s=v_n}=Y_{nm} \quad ,$$

$$m=0,1,\ldots,\beta_n \;; \quad n=1,2,\ldots,N_2 \quad ,$$

$$\frac{1}{m!(\beta_n-m)!}\left(\frac{\partial^{\beta_n-m}}{\partial s^{\beta_n-m}}W(s,t)\right)_{s=v_n}=Q_{nm}(t) \quad ,$$

from which we have

$$\dot{Y}(t)=Y(t)\Gamma^{-2}(t)+\sum_{n=1}^{N_2}\sum_{m=0}^{\beta_n}Y_{nm}Q^{nm}(t) \quad . \tag{3.3.24}$$

In order to obtain the required solution for H, we put $t = 0$ in (3.3.24), which gives

$$H=A\left[\dot{H}-\mathrm{i}\sum_{n=1}^{N_2}\sum_{m=1}^{\beta_n}Y_{nm}Q^{nm}(0)\right] \quad ,$$

$$A\equiv\alpha\frac{\prod_{n=1}^{N_2}v_n^{\beta_n+1}}{\prod_{k=1}^{N_1}u_k^{\alpha_k+1}} \quad . \tag{3.3.25}$$

Thus, the problem reduces to the determination of expressions for the quantities Y_{nm}. In order to obtain these expressions, we differentiate both sides of (3.3.24) l times with respect to t ($l = 0, 1, \ldots, \alpha_k$) and then put $t = u_k$. This gives

$$\frac{\partial^l \dot{Y}(u_k)}{\partial s^l} = \sum_{n=1}^{N_2} \sum_{m=0}^{\beta_n} Y_{nm} Q_{kl}^{nm} \ , \quad Q_{kl}^{nm} \equiv \left(\frac{\partial^l Q^{nm}(t)}{\partial t^l} \right)_{t=u_k} \ , \tag{3.3.26}$$

where $l = 0, 1, \ldots, \alpha_k$, and k varies from 1 to N_1.

Thus, the relations (3.3.26) form a linear algebraic system of order N

$$\left(N = \sum_{k=1}^{N_1} (\alpha_k + 1) = \sum_{n=1}^{N_2} (\beta_n + 1) \right)$$

for the N unknowns $Y_{nm} (m = 1, \ldots, \beta_n; \ n = 1, 2, \ldots, N_2)$.

We consider now the particular case of (3.3.23) in which $\alpha_k = \beta_n = 0$, i.e., the expression (3.3.23) has the form

$$\Gamma^{-2}(s) = \frac{1}{\alpha} \frac{\prod_{k=1}^{N} (s - u_k)}{\prod_{k=1}^{N} (s - v_k)} \ , \tag{3.3.27}$$

where u_k and v_k are distinct constants. We take the initial solution for $\dot{F}_A^B(t)$ (3.2.48) corresponding to the metric $\dot{g}_{11} = 1$, $\dot{g}_{12} = w$, $\dot{g}_{22} = w^2 - r^2$. We introduce the notation

$$X_l = \frac{\mathrm{i} F_1^2(v_l)}{\alpha v_l^2} \frac{\prod_{k=1}^{N} (v_l - u_k)}{\prod_{k \neq l}^{N} (v_l - v_k)} \ .$$

According to (3.3.26), the quantities X_l are determined by the linear system

$$1 + \sum_{l=1}^{N} \frac{X_l}{2} + u_m \sum_{l=1}^{N} \frac{\lambda_m^+ + \lambda_l^-}{v_l - u_m} X_l v_l = 0 \ , \quad m = 1, \ldots, N \ , \tag{3.3.28}$$

where

$$\begin{aligned} &2\lambda_m^+ = \tilde{\lambda}(u_m)/u_m \ , \quad 2\lambda_l^- = \tilde{\lambda}(v_l)/v_l \ , \\ &\tilde{\lambda}(t) = \sqrt{(1 - 2tz)^2 + (2rt)^2} \exp(\mathrm{i}\sigma(t)) \ , \\ &\sigma(t) = 2t \left[\int \frac{\partial w}{\partial \xi} \frac{d\xi}{(1 + 2\mathrm{i}\xi t)} + \int \frac{\partial w}{\partial \xi^*} \frac{d\xi^*}{(1 - 2\mathrm{i}\xi^* t)} \right] \ . \end{aligned} \tag{3.3.29}$$

When the constants u_m and v_l $(m, l = 1, \ldots, N)$ are real, it is convenient to replace them by the constants z_k and m_k given by the expressions

$$u_k = 1/2(z_k + m_k) \ , \quad v_k = 1/2(z_k - m_k) \ .$$

Then from (3.3.27) we obtain the linear algebraic system of equations

$$2+\sum_{l=1}^{N} X_l\left[1+\frac{(\lambda_k^+ + \lambda_l^-)}{(z_k - z_l + m_k + m_l)}\right] = 0 \ , \tag{3.3.30}$$

$$\begin{aligned}\lambda_k^+ &= \sqrt{(z_k + m_k - z)^2 + r^2}\exp(\mathrm{i}\sigma_k^+) \ , \\ \lambda_k^- &= \sqrt{(z_k - m_k - z)^2 + r^2}\exp(\mathrm{i}\sigma_k^-) \ ,\end{aligned} \tag{3.3.31}$$

$$\begin{aligned}\sigma_k^+ &\equiv \int \frac{\partial w}{\partial \xi}\frac{d\xi}{(z_k + m_k + \mathrm{i}\xi)} + \int \frac{\partial w}{\partial \xi^*}\frac{d\xi^*}{(z_k + m_k - \mathrm{i}\xi^*)} \ , \\ \sigma_k^- &\equiv \int \frac{\partial w}{\partial \xi}\frac{d\xi}{(z_k - m_k + \mathrm{i}\xi)} + \int \frac{\partial w}{\partial \xi^*}\frac{d\xi^*}{(z_k - m_k - \mathrm{i}\xi^*)} \ .\end{aligned} \tag{3.3.32}$$

The solution (3.3.30–32) corresponds to a solution for N black holes situated on the symmetry axis in an arbitrary neutrino field[4]. In the absence of a neutrino field, this solution reduces to the Belinskii-Zakharov N-soliton solution [3.17].

To find the unknown components H_A^B, we must make use of the formula

$$\left.\frac{\partial}{\partial t}F_A^B(t)\right|_{t=0} = -\mathrm{i}H_A^B \ .$$

As a result, we obtain

$$\begin{aligned}H_{11} &= A\left(1+\sum_{l=1}^{N} X_l\right) \ , \quad A = \alpha\prod_{n=1}^{N}(v_n/u_n) \ , \\ H_{12} &= 2(w+\mathrm{i}z) + \sum_{l-1}^{N} X_l[w + \mathrm{i}(z - z_l) + \mathrm{i}\lambda_l^- + \mathrm{i}m_l] \ .\end{aligned} \tag{3.3.33}$$

We note that in the case of a single black hole the explicit solution has the form

$$\begin{aligned}H_{11} &= \frac{\lambda^+ + \lambda^- - 2m}{\lambda^+ + \lambda^- + 2m} \ , \quad m = m_1 \ , \\ H_{12} &= 2(w+\mathrm{i}z) - 4m\frac{w + \mathrm{i}(z - z_1) + \mathrm{i}\lambda^- + \mathrm{i}m}{\lambda^+ + \lambda^- + 2m} \ .\end{aligned} \tag{3.3.34}$$

d) Method of Obtaining Topologically Nontrivial Solutions of Einstein's Equations with an Abelian Group of Motions G_2 on V_2 in the Absence of a Neutrino Field. As we found earlier [see (3.2.44)], the matrix function $\boldsymbol{u}(s)$ can be obtained by means of an analytic continuation, into the complex s plane, of the functions $\mathcal{E}(z)$ and $\Phi(z)$ specified on the symmetry axis. For this, we must replace z in the expressions for these functions by $\mathrm{i}/2s$. Then in the general

[4] In this solution for $m_k > 0$ ($k = 1, \ldots, N$) there exist δ-like singularities of the Ricci tensor on the symmetry axis between the black holes (so-called conical points) corresponding to rather mystical structures. The same singularities occur in the solution even if there is no neutrino field [3.17].

case we obtain multivalued functions. For example, we could specify $\mathcal{E}$ and Φ uniquely on the z axis by means of a selected branch of an algebraic curve whose branch points lie off the z axis. However, if an analytic continuation is made into the complex plane, the matrix function $\boldsymbol{u}(s)$ is found to be multivalued. We shall consider only the case in which the Riemann surface of the function $\boldsymbol{u}(s)$ has n sheets and is determined by some algebraic curve. The genus g of an algebraic curve is given by the formula $V = 2(n+g-1)$, where V is the number of branch points, including their multiplicities. Then, as in Sect. 3.2, the analysis of the group properties of the system of equations (3.2.1) makes it possible to deduce the analytic structure of the matrix function $\chi(s) = \boldsymbol{F}\boldsymbol{u}\dot{\boldsymbol{F}}^{-1}$ on the indicated Riemann surface. This surface is homeomorphic to a sphere with g handles. The function χ and the parameter s can be represented as single-valued functions of a uniformization parameter σ: $\chi = \chi(\sigma)$, $s = s(\sigma)$. For $g \geq 2$ the functions $\chi(\sigma)$ and $s(\sigma)$ are automorphic (Fuchsian, in Poincaré's terminology), for $g = 1$ they are elliptic, and for $g = 0$ they are rational. We recall that an automorphic function is one which is invariant with respect to some group of bilinear transformations. For $g \geq 2$, the functions $\chi(\sigma)$ and $s(\sigma)$ depend parametrically on $6g-6$ real moduli of an algebraic curve of the Riemann surface of $\boldsymbol{u}(s)$. According to the Riemann-Roch theorem (see below), the function χ cannot have less than $g+1$ distinct simple poles not situated at the Weierstrass points. Recently D.A. Korotkin (Theor. Math. Phys. 77, No.1, 25 (1988) [in Russian]) succeeded to find the exact solutions of the Ernst equations using the theta-function technique developed by Matveev, Its, Dubrovin, and Novikov.

e) Restrictions Imposed by the Topology of a Two-Dimensional Manifold on the Order of the Singularities of the Neutrino Field in the Stationary Axially Symmetric Case. We assume that a two-dimensional manifold S with local coordinates x^3 and x^4 for $t = \text{const}$, $\varphi = \text{const}$ is compact. The admissible changes of variables preserving the form of the metric (3.1.3) are analytic transformations $\xi' = f(\xi)$ ($\xi = x^3 + ix^4$) with a positive definite Jacobian. Consequently, the manifold S is orientable. From the topological point of view, such manifolds are homeomorphic to spheres with g handles (g is called the genus of the surface). From the point of view of conformal transformations, the surface is defined by the genus and by the $3g-3$ complex parameters which determine the conformally distinct compact Riemann surfaces with genus g [3.18].

Every compact orientable surface S can be realized as an n-sheeted Riemann surface of some algebraic function given by the irreducible polynomial

$$\varphi^n + r_1(\xi)\varphi^{n-1} + \ldots + r_n(\xi) = 0 \ , \tag{3.3.35}$$

where $r_i(\xi)$ ($i = 1, 2, \ldots, n$) are rational functions of the parameter ξ. Any other meromorphic function f on S can be written in the form

$$f = R_1(\xi)\varphi^{n-1} + \ldots + R_n(\xi) \ , \tag{3.3.36}$$

where $R_i(\xi)$ ($i = 1, 2, \ldots, n$) are rational functions of ξ.

We use the properties of compact Riemann surfaces in order to determine the restrictions on the order of the singularities of the neutrino field given by the singularities of the harmonic function w. We assume that the meromorphic function ξ, whose real part is equal to the harmonic function $\sqrt{|g_{AB}|}$, takes each value n times on S. Then this function establishes a one-to-one conformal mapping onto an n-sheeted Riemann surface.

If S is given by an algebraic curve (3.3.35), the analytic function $W = w + \mathrm{i}\tilde{w}$ must have the form (3.3.36). The function W is specified not in the entire complex ξ plane, but only in the half-plane $\mathrm{Re}\{\xi\} = r \geq 0$. In order to define this function in the whole plane, we require that $[W(\xi)]^* = -W(-\xi^*)$. This equality will be satisfied if the coefficients $r_i(\mu)$ and $R_i(\mu)$ ($\mu = \mathrm{i}\xi$) are rational functions of μ over the field of real numbers. We stress that meromorphic functions on S other than constants cannot be everywhere regular.

The arbitrariness in specifying the singularities of the meromorphic function $W(\xi)$ is restricted by the Riemann-Roch theorem. In order to formulate this theorem, let us recall several definitions. Suppose that near an arbitrary point Q on S the function W can be expanded in a series $W(\sigma) = a_n\sigma^n + a_{n+1}\sigma^{n+1} + \ldots$ with $a_n \neq 0$ (σ is the so-called uniformization parameter, and n is a positive or negative integer). The meromorphic function W on S can have only a finite number of zeros and poles $Q_1, Q_2, \ldots, Q_N$ with prescribed orders $\alpha_1, \alpha_2, \ldots, \alpha_N$ of the zeros or poles at all these points. We shall call the symbol $A = Q_1^{\alpha_1} Q_2^{\alpha_2} \ldots Q_N^{\alpha_N}$ the divider and adopt the notation (W) for the divider of the meromorphic function W. The degree of a divider $A = Q_1^{\alpha_1} \ldots Q_N^{\alpha_N}$, denoted by $d[A]$, is defined as the sum of the orders, $\alpha_1 + \alpha_2 + \ldots + \alpha_N$. For meromorphic functions, $d[(W)] = 0$.

The dimension of the complex vector space $L[A]$ of meromorphic functions f' having singularities at prescribed points $Q_1, Q_2, \ldots Q_N$ with orders $\alpha_i' \geq \alpha_i$ ($i = 1, 2, \ldots, N$) will be denoted by $r[A]$. The space $L[A]$ is empty for dividers with positive degree $d[A] > 0$, since the degree of the divider of an arbitrary meromorphic function is equal to zero.

For Abelian differentials with a local representation $\omega = (a_n\xi^n + a_{n+1}\xi^{n+1} + \ldots)\,d\xi$ at the points $P_1, P_2, \ldots, P_N$ ($a_0 \neq 0$ at all other points of S), we introduce the symbol (ω) to denote the divider A with prescribed orders $\alpha_1, \alpha_2, \ldots, \alpha_N$ of the zeros or poles at the points P_i ($i = 1, \ldots, N$): $A = P_1^{\alpha_1} P_2^{\alpha_2} \ldots P_N^{\alpha_N}$. The degree of an Abelian differential is related to the genus of the surface by the formula $d[(\omega)] = 2g - 2$. For a given divider A, the dimension of the complex vector space of Abelian differentials ω': $\{\omega'\} = \Omega[A]$ with singularities of orders $\alpha_i' \geq \alpha_i$ ($i = 1, \ldots, N$) at prescribed points $P_1, P_2, \ldots, P_N$ will be denoted by $i[A]$. We note that the space $\Omega[A]$ is empty if $d[A] \geq 2g - 1$.

The Riemann-Roch theorem asserts that a given divider A has the property

$$r[A^{-1}] = d[A] + i[A] - g + 1 \quad .$$

From this theorem it follows, in particular, that on a surface of genus g it is always possible to find g distinct points such that there does not exist a nonconstant meromorphic function whose singularities are poles of order not higher than the first at these points. This constitutes a restriction on the zeros and simple poles of the function $W = w + i\tilde{w}$.

As another corollary of the Riemann-Roch theorem, we give the assertion that for $g \geq 2$ it is possible to specify a pole of order not exceeding g as the only singularity of a meromorphic function $W(\xi)$ only at a finite number of points $P_1, \ldots, P_N$ (the so-called Weierstrass points).

As has been shown by Musin and the present author, in the general case a harmonic function w for the neutrino field must be multivalued if it satisfies the inequality $w^2 \leq r^2$ and thus makes it possible to avoid the appearance of closed time-like curves violating the fundamental physical principle of causality. Thus, in the presence of a neutrino field the two-dimensional space-time cross section t = const, φ = const has, in general, a nontrivial topology (the topology of a sphere with g handles in the compact case).

f) Gravitational Fields of Rotating Magnetized Stars. For the calculation of fields in the neighborhood of rotating magnetized stars a different form of our integral equation (3.3.5) (for $a = 1$) is more useful. We shall generate the functions $e(z)$ and $f(z)$ in that equation by integrals of the Cauchy type:

$$2\pi i e(z) = \int_a^b \varepsilon(\xi)\frac{d\xi}{(z-\xi)} \quad , \quad 2\pi i f(z) = \int_a^b \varphi(\xi)\frac{d\xi}{(z-\xi)} \quad .$$

These functions are holomorphic everywhere in the complex z plane except for the segment $[a, b]$ on the real axis, where due to Sohotzky's formulae they have the discontinuity $[e(\xi)] = \varepsilon(\xi)$, $[f(\xi)] = \varphi(\xi)$.

Let us show that these functions play the role of sources of gravitational and electromagnetic fields. For any given geometry of an axially symmetric star with given Ernst functions on its surface it is possible to obtain sources on the axis of symmetry which reproduce outside the star fields coinciding with those of the given star. It should be stressed that the Ernst equations are of elliptical type and that one cannot determine independently the Ernst functions and their transverse derivatives on the surface of the star. Thus for a solution of the interior problem one need not consider the exterior solution, because one can use the relationship between $\mathcal{E}$, Φ and their transverse derivatives on the boundary of the stars as boundary conditions for the solution of the interior problem. With these points in mind the integral equations (3.3.44, 45) as given below are very useful. We shall also show that the Hansen-Torne coefficients in the multipole expansions of Ernst potentials at infinity can be expressed in terms of the moments of the sources of the gravitational and electromagnetic fields on the symmetry axis.

Let us introduce functions $X_2(t)$ via the definition

$$\pi X_i(t) = -\sqrt{t^2-1}\int_{-1}^{1} \mu(z + irs)\chi_i(s)\frac{ds}{(s-t)}\sqrt{1-s^2} \quad ,$$

$$\chi_1 \equiv 1 \quad , \quad \chi_2 \equiv e(z + irs) \quad , \quad \chi_3 \equiv f(z + irs)$$

(for the notation see (3.3.1–6), $\mu(\xi) \equiv \mu_1(\xi)$). Then we arrive at

$$X_2^+ + X_2^- = e(\xi)(X_1^+ + X_1^-) \ , \quad X_3^+ + X_3^- = f(\xi)(X_1^+ + X_1^-) \ , \tag{3.3.37}$$

where X_i^+, X_i^- are the values of the functions X_i on different sides of the cut from -1 to $+1$ in the complex t plane and

$$X_1(\infty) = 0 \ , \quad X_2(\infty) = \mathcal{E} \ , \quad X_3(\infty) = \Phi \ .$$

Our integral equation (3.3.5) for $a = 1$ may be rewritten as

$$X_2^+ - X_2^- + \tilde{e}(\xi)(X_1^+ - X_1^-) + 2\tilde{f}(X_3^+ - X_3^-) = 0 \ . \tag{3.3.38}$$

Using $t = (\lambda + 1/\lambda)/2$ one maps the exterior of the cut from -1 to $+1$ of the t plane into the exterior or interior of the circle $|\lambda| = 1$. We define $X_2(\lambda)$ to be holomorphic outside the circle $|\lambda| = 1$, while $X_1(\lambda)$ and $X_2(\lambda)$ are holomorphic inside $|\lambda| = 1$ in the λ plane.

Then we have[5]

$$X_2(\lambda) - X_2\left(\frac{1}{\lambda}\right) = \tilde{e}(\xi)\left(X_1(\lambda) - X_1\left(\frac{1}{\lambda}\right)\right) + 2\tilde{f}\left(X_3(\lambda) - X_3\left(\frac{1}{\lambda}\right)\right) \ .$$

With the help of (3.3.37) equation (3.3.38) can be brought to

$$X_2(\lambda) = \left(\frac{e + \tilde{e}}{2} + f\tilde{f}\right) X_1(\lambda) + \left(\frac{e - \tilde{e}}{2} + f\tilde{f}\right) X_1\left(\frac{1}{\lambda}\right) - 2\tilde{f}X_3\left(\frac{1}{\lambda}\right) \ . \tag{3.3.39}$$

Let Y_+, Y_- be solutions of the Riemann problem:

$$Y_+ = \left(\frac{e + \tilde{e}}{2} + f f\right) Y_- \ , \quad Y_-(0) = 1 \ , \tag{3.3.40}$$

where Y_+ and Y_- are respectively holomorphic outside and inside $|\lambda| = 1$.

Equation (3.3.39) can be written as

$$\frac{X_2(\lambda)}{Y_+(\lambda)} = \frac{X_1(\lambda)}{Y_-(\lambda)} + \frac{X(1/\lambda)}{Y_+(\lambda)}\left(\frac{e - \tilde{e}}{2} + f\tilde{f}\right) - 2\tilde{f}\frac{X_3(1/\lambda)}{Y_+(\lambda)} \ . \tag{3.3.41}$$

We use the relationship $Y_+(1/\lambda) = Y_+(\infty)/Y_-(\lambda)$. Multiplying both sides of (3.3.41) by $\mu/\lambda(\lambda - \mu)$, where μ is inside $|\lambda| = 1$, and integrating the resulting expression along the curve $|\lambda| = 1$, we obtain

$$X_1(\mu) = Y_-(\mu)\Bigg[1 + \frac{1}{2\pi i}\int\limits_{|\lambda|=1} \frac{Y_-(\lambda)}{Y_+(\infty)} \times \left(X_1(\lambda)\left(\frac{e - \tilde{e}}{2} + f\tilde{f}\right) - 2\tilde{f}X_3(\lambda)\right)\frac{d\lambda}{(\lambda - 1/\mu)}\Bigg] \ . \tag{3.3.42}$$

---

[5] Another boundary problem is obtained by Neugebauer and Kramer see J. Phys. A **16**, 1927 (1983).

Similarly from equation (3.3.37) one arrives at

$$X_3(\mu) = \frac{(\mu - 1/\mu)}{2\pi i} \int\limits_{|\lambda|=1} f(\lambda) \frac{(\lambda X_1(\lambda) - \mu X_1(\mu))}{(\lambda - \mu)(\lambda - 1/\mu)} \, d\lambda \; . \tag{3.3.43}$$

Now we specify the assumption about e, f being represented by Cauchy type integrals and deform (contract) the curve $|\lambda| = 1$ to the interior image of the segment $[a, b]$ in the λ-plane using the holomorphic properties of $X_1(\lambda)$ and $X_3(\lambda)$ on the interior of $|\lambda| = 1$. Returning from the variable λ to the variable $\xi = z + ir(\lambda + 1/\lambda)/2$ and substituting $X_1(\xi) = A(\xi)w(\xi)$, $X_3(\xi) = B(\xi)w(\xi)$, where $w(\xi) \equiv \sqrt{(z - \xi)^2 + r^2}$, into equations (3.3.42, 43) we finally obtain

$$\begin{aligned} A(\xi) = \frac{Y_-(\xi)}{w(\xi)} \Bigg\{ 1 + \frac{1}{2\pi i} \int\limits_a^b \frac{Y_-(\eta)}{Y_+(\infty)} \Omega(\xi, \eta) \\ \times \left(A(\eta) \left[\frac{e - \tilde{e}}{2} + f\tilde{f} \right] - 2[\tilde{f}] B(\eta) \right) d\eta \Bigg\} \; , \end{aligned} \tag{3.3.44}$$

$$B(\xi) = \frac{1}{2\pi i} \int\limits_a^b [f] \frac{A(\xi) - A(\eta)}{\xi - \eta} \, d\eta \; , \tag{3.3.45}$$

where

$$Y_-(\xi) = \exp\left(\frac{1}{2\pi i}\right) \int\limits_a^b \left[\ln\left(\frac{e + \tilde{e}}{2} + f\tilde{f} \right) \right] \Omega(\xi, \eta) \frac{d\eta}{w(\eta)} \; ,$$

$$Y_+(\infty) = \exp \frac{1}{2\pi i} \int\limits_a^b \left[\ln\left(\frac{e + \tilde{e}}{2} + f\tilde{f} \right) \right] \frac{d\eta}{w(\eta)} \; ,$$

$$\Omega(\xi, \eta) \equiv \frac{1}{2} \left(1 + \frac{w(\xi) - w(\eta)}{\xi - \eta} \right) \; ,$$

while square brackets denote the discontinuity of the corresponding function on the segment $[a, b]$. After solving the system (3.3.44, 45), which can easily be written as a single integral equation, the Ernst potentials are be found to be

$$\begin{aligned} \mathcal{E} &= Y_+(\infty) + \frac{1}{2\pi i} \int\limits_a^b Y_-(\xi) \left\{ \left[\frac{e - \tilde{e}}{2} + f\tilde{f} \right] A(\xi) - 2[\tilde{f}] B(\xi) \right\} d\xi \; , \\ \Phi &= \frac{1}{2\pi i} \int\limits_a^b [f] A(\xi) \, d\xi \; . \end{aligned} \tag{3.3.46}$$

In the case of various stars to be modelled, one must introduce several segments $[a_i, b_i]$ $i = 1, 2, \dots N$ and give (3.3.44–46) in the form of a sum of integrals on the segments $[a_i, b_i]$ with sources $[e]_i$ and $[f]_i$.

Let us show how one can calculate the asymptotics of $\mathcal{E}$, Φ.

1) First we calculate $\mathcal{E}$, Φ in the neighborhood of the axis of symmetry outside the sources $z > \xi$, $\xi \in [a, b]$. The asymptotic behavior of A and B is as follows:

$$(z-\xi)A(\xi) \approx 1 - \frac{r^2}{2(z-\xi)}\left(\frac{e'(z) + 2\tilde{f}(z)f'(z)}{e(z) + \tilde{e}(z) + 2f(z)\tilde{f}(z)} + \frac{1}{z-\xi}\right) + \dots ,$$

$$(z-\xi)B(\xi) \approx f(z) - \frac{r^2}{4}f''(z) - \frac{r^2}{2(z-\xi)}\left(\frac{f(z)}{z-\xi} - f'(z)\right)$$

$$\times\left(\frac{e'(z) + 2\tilde{f}(z)f'(z)}{e(z) + \tilde{e}(z) + 2f(z)\tilde{f}(z)} + \frac{1}{z-\xi}\right) + \dots .$$

Then for the Ernst potentials from (3.3.46) we have the expansions:

$$\mathcal{E} \approx e(z) + \frac{r^2}{4}\left(-e'' + \frac{2e'(e' + 2\tilde{f}f')}{e + \tilde{e} + 2\tilde{f}f}\right) + \dots ,$$

$$\Phi \approx f(z) + \frac{r^2}{4}\left(-f'' + \frac{2f'(e' + 2\tilde{f}f')}{e + \tilde{e} + 2\tilde{f}f}\right) + \dots .$$

2) Next we consider asymptotic expansions of $A(\xi)$ and $B(\xi)$ for large $R = \sqrt{r^2 + z^2}$. We define $z = R\cos\varphi$, $r = R\sin\varphi$, $\cos\varphi \equiv t$ and introduce the notation for the moments of the sources

$$2\pi\mathrm{i}e_k = \int_a^b [e(\xi)]\xi^k d\xi , \quad 2\pi\mathrm{i}f_k = \int_a^b [f(\xi)]\xi^k d\xi .$$

The moments of the holomorphic function

$$2\pi\mathrm{i}\tilde{f}(z)f(z) = \int_a^b [f(\xi)\tilde{f}(\xi)]\frac{d\xi}{(z-\xi)}$$

are related to the moments f_k as follows:

$$\int_a^b [f(\xi)\tilde{f}(\xi)]\, d\xi = 0 ,$$

$$\int_a^b [f(\xi)\tilde{f}(\xi)]\xi^k d\xi = 2\pi\mathrm{i}(f_{k-1}\tilde{f}_0 + f_{k-2}\tilde{f}_1 + \dots + f_0\tilde{f}_{k-1}) .$$

The moments of the holomorphic function $\ln((e+\bar{e})/2+f\bar{f})$ can be expressed also in terms of the moments e_k, f_k. Then the solutions of the integral equations (3.3.44, 45) are

$$\begin{aligned}A(\xi) \approx{}& \frac{1}{R}+\frac{1}{R^2}\left(\frac{e_0}{2}(1-t)+\xi t\right)+\frac{1}{R^3}\left\{\frac{3t^2-1}{2}\xi^2+\frac{(1-t)(1+3t)}{4}e_0\xi\right.\\ &+\frac{(1-t)(1+3t)}{4}e_1-\frac{t(1-t)}{2}m_0^2-\frac{(1-t)^2}{2}a_0^2\\ &\left.+\frac{(1-t)(1-3t)}{4}\mathrm{i}a_0m_0+\frac{(1-t)(1+t)}{2}f_0f_0^*\right\}+\dots,\\ B(\xi) \approx{}& \frac{f_0t}{R^2}+\frac{3t^2-1}{2R^3}(\xi f_0+f_1)+\frac{(1-t)(1+3t)}{4R^3}e_0f_0+\dots .\end{aligned} \tag{3.3.47}$$

Substituting these expressions into (3.3.46) one obtains:

$$\begin{aligned}\mathcal{E} \approx{}& 1+\frac{e_0}{R}+\frac{1}{R^2}\left(te_1-\frac{(1-t)}{2}e_0^2\right)+\frac{3t^2-1}{2}\frac{e_2}{R^3}\\ &+\frac{(1-t)}{R^3}\left\{\frac{(1+3t)}{2}e_1e_0+\frac{1+t}{2}f_0f_0^*-\frac{m_0^3t}{2}\right.\\ &\left.+\mathrm{i}a_0\left(\frac{m_0^2(1-5t)}{4}-\frac{(1-t)}{4}a_0^2+\frac{(1-2t)}{2}\mathrm{i}a_0m_0\right)\right\}+\dots\\ \Phi ={}& \frac{f_0}{R}+\frac{1}{R^2}\left(tf_1+\frac{(1-t)}{2}e_0f_0\right)+\frac{(3t^2-1)}{2}\frac{f_2}{R^3}\\ &+\frac{(1-t)}{R^3}\left\{\left(\frac{1+3t}{4}\right)(e_1f_0+e_0f_1)\right.\\ &\left.+f_0\left(\frac{1-3t}{4}\mathrm{i}a_0m_0-\frac{(1-t)}{4}a_0^2-\frac{t}{2}m_0^2\right)+\frac{(1+t)}{2}f_0f_0^*\right\}+\dots .\end{aligned} \tag{3.3.48}$$

In (3.3.47, 48)

$$m_0 \equiv \mathrm{Re}\{e_0\}\ ,\quad a_0 \equiv \mathrm{Im}\{e_0\}\ ,\quad e_0 \equiv m_0+\mathrm{i}a_0\ .$$

3) Let us assume the Ernst function on the surface of the star to be given. Using the integral equations (3.3.44, 45) and taking into account (3.3.46), one can consider the problem of finding the sources $[e(\xi)]$, $[f(\xi)]$, which determine the fields for given functions $\mathcal{E}$, Φ on the surface of star. Thus one can express the transverse derivatives of $\mathcal{E}$, Φ as a functional of $\mathcal{E}$, Φ on the surface under consideration. Hence the Dirichlet problem for the Ernst equations of the electrovacuum is reduced to two linear problems:

a) determination of the sources $[e(\xi)]$, $[f(\xi)]$ (the discontinuities of the holomorphic generating functions $e(z)$, $f(z)$) according to the given values $\mathcal{E}$, Φ on the considered curve $r = r(z)$ with the help of (3.3.44–46);
b) determination of $\mathcal{E}$, Φ for any point (r, z) using (3.3.44–46) for sources$[e(\xi)]$, $[f(\xi)]$ derived in (a).

3.4 Lie-Bäcklund Groups of Integrable Systems of Mathematical Physics

Integrable systems of equations (or systems integrable in the kinematic sense) usually refer to equations which can be represented in the form of compatibility conditions for linear matrix equations [3.19]

$$\varphi_{,\xi} = U\varphi \ , \quad \varphi_{,\eta} = V\varphi \ , \quad U = U(\lambda, \xi, \eta) \ , \quad V = V(\lambda, \xi, \eta) \ , \tag{3.4.1}$$

$$U_{,\eta} - V_{,\xi} + UV - VU = 0 \ . \tag{3.4.1$'$}$$

Equation (3.4.1′) can be treated as the condition of zero curvature of a two-dimensional manifold with the connections U and V, admitting a covariantly constant vector field φ. It is important that the matrices U and V depend on an auxiliary analytic parameter λ. However, the matrix equation (3.4.1′) is satisfied if and only if there are a finite number of equations not depending on λ. These represent initial integrable systems.

In most applications, η and ξ have the meaning of the time and a space coordinate (on the entire real axis or on a circle in the case of a segment with identified ends). In relativistic models, it is convenient to interpret ξ and η as retarded and advanced times $\mp x + t$, respectively. For stationary two-dimensional integrable systems, it is convenient to take for ξ and η the complex coordinates $\xi = x + \mathrm{i}y$, $\eta = x - \mathrm{i}y$.

We now give some examples.

a) Suppose that the matrices U and V have poles in λ at one and the same point of the complex λ plane, of order n and m, respectively. Without loss of generality, we can choose this point to be the point at infinity. In nonrelativistic models of this type, ξ and η have the meaning of a coordinate and the time.

Example 1.

$$U = \mathrm{i}(\lambda\sigma_3 + R) \ , \quad V = \mathrm{i}(2\lambda^2\sigma_3 + 2\lambda R - \mathrm{i}\sigma_3 R_{,x} - \sigma_3 R^2) \ .$$

Here σ_a $(a = 1, 2, 3)$ are the standard 2×2 Pauli matrices

$$\sigma_1 = \begin{pmatrix} 0 & 1 \\ 1 & 0 \end{pmatrix} \ , \quad \sigma_2 = \begin{pmatrix} 0 & -\mathrm{i} \\ \mathrm{i} & 0 \end{pmatrix} \ , \quad \sigma_3 = \begin{pmatrix} 1 & 0 \\ 0 & -1 \end{pmatrix}$$

and R is an unknown 2×2 matrix $\begin{pmatrix} 0 & r \\ q & 0 \end{pmatrix}$.

The compatibility condition (3.4.1′) reduces to the equation

$$\mathrm{i}R_{,t} - \sigma_3 R_{,xx} - 2\sigma_3 R^3 = 0 \ . \tag{3.4.2}$$

Example 2.

$$U = \mathrm{i}(\lambda\sigma_3 + R) \ ,$$
$$V = \mathrm{i}(-4\lambda^3\sigma_3 - 4\lambda^2 R + 2\lambda\sigma_3(R^2 + \mathrm{i}R_{,x}) + \mathrm{i}RR_{,x} + R_{,xx} + 2R^3) \ .$$

Equation (3.4.1′) is equivalent to the equation

$$R_t + R_{,xxx} + 6R^2 R_{,x} = 0 \ . \tag{3.4.3}$$

Example 3.

$$U = \mathrm{i}(\lambda M + [H, M]) \ , \quad V = \mathrm{i}(\lambda N + [H, N]) \ ,$$

where $M = \mathrm{diag}(a_1, \ldots, a_n)$, $N = \mathrm{diag}(b_1, \ldots, b_n)$, the square brackets denote the commutator, and a_i and b_i are real constants. Here (3.4.1′) has the form

$$[M, H_{,t}] - [N, H_{,x}] - \mathrm{i}([MH], [NH]) = 0 \ . \tag{3.4.4}$$

Example 4.

$$U = \frac{\lambda}{2\mathrm{i}} S \ , \quad V = \frac{\mathrm{i}\lambda^2}{2} S + \frac{\lambda}{2} SS_{,x}$$

where we have the 2×2 matrix $S = \sum_{a=1}^{3} S_a \sigma_a$, in which σ_a $(a = 1, 2, 3)$ are the Pauli matrices, with $S_1^2 + S_2^2 + S_3^2 = 1$. Then (3.4.1′) is equivalent to the equation

$$2\mathrm{i}S_{,t} = SS_{,xx} - S_{,xx}S \ . \tag{3.4.5}$$

Example 5.

$$U = \mathrm{i}(\lambda J + R) \ , \quad V = \mathrm{i}(2\lambda^2 J + 2\lambda R - \mathrm{i}JR_{,x} - JR^2) \ ,$$

where U, V, J, and R are $n \times n$ matrices, with $J = \mathrm{diag}(1, \ldots, 1, -1)$, and in the matrix R only the last column and the last row with a zero diagonal element are nonzero[6]. Equation (3.4.1) has the form

$$\mathrm{i}R_t - JR_{xx} - 2JR^3 = 0 \ . \tag{3.4.6}$$

b) The matrices U and V have two simple poles. In relativistic models of this type, ξ and η have the meaning of either $t \mp x$ or $x \pm \mathrm{i}y$.

[6] In Example 5, we can take J to be an arbitrary matrix satisfying $J^2 = 1$ and $R = JM - MJ$, where M is an arbitrary matrix.

Example 1.

$$2iU = \alpha\sigma_3\varphi_{,\xi} + \lambda\sigma_1 \sin\left(\frac{\alpha\varphi}{2}\right) + \lambda\sigma_2 \cos\left(\frac{\alpha\varphi}{2}\right) ,$$
$$2iV = -\alpha\sigma_3\varphi_{,\eta} - \frac{1}{\lambda}\sigma_1 \sin\left(\frac{\alpha\varphi}{2}\right) + \frac{1}{\lambda}\sigma_2 \cos\left(\frac{\alpha\varphi}{2}\right) ,$$

for which the compatibility condition (3.4.1′) is equivalent to the equation

$$\alpha\varphi_{,\xi\eta} + \sin(\alpha\varphi) = 0 . \tag{3.4.7}$$

Example 2.

$$\begin{aligned} U &= \mathrm{diag}(\varphi_{1,\xi}, \varphi_{2,\xi},\ldots,\varphi_{n,\xi}) + \lambda e_a \frac{\exp(\varphi_{a+1} - \varphi_a)}{2} , \\ V &= -\mathrm{diag}(\varphi_{1,\eta}, \varphi_{2,\eta},\ldots,\varphi_{n,\eta}) - \frac{1}{\lambda} e_a \frac{\exp(\varphi_{a+1} - \varphi_a)}{2} . \end{aligned} \tag{3.4.8}$$

In the matrix e_a the component in row a and column $a+1$ is equal to unity, and the others are zero, while in the matrix e_{-a} the component in row $a+1$ and column a is equal to unity.

Here the compatibility condition (3.4.1′) has the form

$$\begin{aligned} &2\varphi_{a,\xi\eta} = \exp(\varphi_{a+1} - \varphi_a) - \exp(\varphi_a - \varphi_{a-1}) , \\ &a = 1, \ldots, n , \quad \varphi_{n+1} = \varphi_1 . \end{aligned} \tag{3.4.8′}$$

Example 3.

$$U = \frac{H_{,\xi}}{(\lambda - 1)} , \quad V = \frac{H_{,\eta}}{(\lambda + 1)}$$

with the following compatibility condition for the $n \times n$ matrix H:

$$2H_{,\xi\eta} + H_{,\xi}H_{,\eta} - H_{,\eta}H_{,\xi} = 0 . \tag{3.4.9}$$

Example 4.

$$U = \frac{H_{,\xi}}{(\lambda - \xi)} , \quad V = \frac{H_{,\eta}}{(\lambda - \eta)}$$

with the compatibility condition

$$(\xi - \eta)H_{,\xi\eta} + H_{,\xi}H_{,\eta} - H_{,\eta}H_{,\xi} = 0 . \tag{3.4.10}$$

The foregoing examples have an important practical significance in various reductions allowed by the integrable systems (3.4.2–10). Great importance attaches to the fact that the matrices U and V belong to some matrix algebra, since the operations of differentiation with respect to ξ and η and the commutator of two matrices remain within this matrix algebra.

For example, (3.4.2) with the reduction $R \in \mathrm{SU}(2)$ (the algebra of anti-Hermitian matrices) reduces to a nonlinear Schrödinger equation of the first kind. If, however, $R \in \mathrm{SU}(1,1)$, then (3.4.2) reduces to a Schrödinger equation of the second kind. Equation (3.4.3) with the additional reduction $R^* = \sigma_1 R \sigma_1$ reduces to a modified Korteweg-de Vries equation, which in turn reduces to the ordinary KdV equation through a Miura transformation.

In the cases (3.4.4, 6) we can impose the conditions H, $R \in \mathrm{SU}(p, n-p)$, where p can take one of the values $1, 2, \ldots, n$. In the case $n = 3$, the matrix equation (3.4.4) reduces to a system of equations for three wave interactions with various relations between the group velocities.

In the case (3.4.5, 7), U and V clearly belong to the SU(2) algebra (more precisely, to its algebra of currents; see below), and (3.4.5) describes the one-dimensional model of an isotropic Heisenberg ferromagnet, while (3.4.7) describes the model of self-induced transparency in laser physics. For (3.4.10) an additional reduction is given by the differential relation described by the first-order equation (3.2.4). If the trace of the matrix H is not fixed, then for $n = 3$ we obtain the general case of the neutrino electrovacuum, and for $\mathrm{Tr}\{H\} = \xi + \eta$ the neutrino field is "expelled" from the general case.

The expressions (3.4.8′) for U and V are obtained from the general matrices with simple poles at $\lambda = 0$ and $\lambda = \infty$ in the case of the so-called $\mathbb{Z}_n$ reduction:

$$U(\zeta\lambda) = Z^{-1}U(\lambda)Z \ , \quad V(\zeta\lambda) = Z^{-1}V(\lambda)Z \ ,$$

where $\zeta = \exp(2\pi i/n)$ is an n-th root of 1, and Z is a diagonal matrix of the powers of ζ:

$$Z_{ij} = \zeta^i \delta_{ij} \ , \quad i, j = 1, \ldots, n \ .$$

The divider of the matrix functions U and V (i.e., the pole structure of the rational functions U and V at definite points of the complex λ plane) is preserved by so-called gauge transformations. These are related to the arbitrariness in the choice of the redefined linear system with a given divider of the functions U and V associated with the substitution $\varphi' = \Omega\varphi$, where Ω does not depend on λ. The matrices U and V then change as the connections

$$U' = \Omega_{,\xi}\Omega^{-1} + \Omega U \Omega^{-1} \ , \quad V' = \Omega_{,\eta}\Omega^{-1} + \Omega V \Omega^{-1} \ . \tag{3.4.11}$$

For example, in the case of the reduction $R \in \mathrm{SU}(2)$ the equations (3.4.2, 5) with the same divider are gauge-equivalent. Similarly, the equation (3.4.9) of the principal chiral field is gauge-equivalent to the sine-Gordon equation (3.4.7) and the two-dimensional Toda model (3.4.8) with the reduction $\varphi_1 = -\varphi_2 = \varphi$ in the case $n = 2$.

In principle, in a given region D of the complex λ plane the divider of the functions U and V is also preserved by the transformations $\varphi' = \Omega\varphi$ when Ω depends on λ, provided that the matrix Ω as a function of λ is nondegenerate and holomorphic in the region D. Then the transformation (3.4.11), which we write in the form

$$U' = (\Omega\varphi)_{,\xi}\varphi^{-1}\Omega^{-1} \quad , \quad V' = (\Omega\varphi)_{,\eta}\varphi^{-1}\Omega^{-1} \quad , \tag{3.4.11'}$$

can be extended to the whole complex plane if there exists a function Ω_1, holomorphic outside D, such that

$$(\Omega\varphi)_{,\xi}\varphi^{-1}\Omega^{-1} = (\Omega_1\varphi)_{,\xi}\varphi^{-1}\Omega_1^{-1} \quad ; \quad (\Omega\varphi)_{,\eta}\varphi^{-1}\Omega^{-1} = (\Omega_1\varphi)_{,\eta}\varphi^{-1}\Omega_1^{-1}$$

on the boundary of D. It can be seen from this that on ∂D we have

$$\Omega\varphi = \Omega_1\varphi G(\lambda) \quad , \tag{3.4.12}$$

where $G(\lambda)$ is independent of ξ and η.

However, (3.4.12) can be interpreted as a classical Riemann boundary-value problem for the functions Ω and Ω_1, holomorphic in D and outside D, respectively, and related on the contour ∂D by the linear equation (3.4.12). By means of the indicated transformation (3.4.11′), which *Zakharov* and *Shabat* [3.20] called the procedure of "dressing the bare solution U and V", it is possible to use a given solution to obtain new solutions U' and V'.

The possibility of representing a given system of equations in the form of the condition of zero curvature is a major basic problem. After the fundamental studies of *Gardner* et al. [3.21] and *Lax* [3.22], the greatest successes here have been achieved by *Zakharov* et al. [3.23].

We now consider the general case of an integrable system with functions U and V which are rational in the parameter λ. It is remarkable that the system (3.4.1) admits exact solutions of the linearized equations. Suppose that φ_0, U_0, and V_0 are arbitrary exact solutions of the system (3.4.1). Then the following theorem holds.

Theorem 1. The following expressions for $\delta\varphi$, δU, and δV are solutions of the linearized equations (3.4.1, 1′):

$$2\pi i\delta\varphi(\mu) = \int_C \Phi(\lambda)(\lambda-\mu)^{-1}d\lambda\,\varphi_0(\mu) \quad ,$$
$$\Phi(\lambda) \equiv \varphi_0(\lambda)\Gamma(\lambda)\varphi_0^{-1}(\lambda) \quad , \tag{3.4.13}$$

$$2\pi i\delta U(\mu) = \int_C [U_0(\lambda) - U_0(\mu), \Phi(\lambda)](\lambda-\mu)^{-1}d\lambda \quad , \tag{3.4.13'}$$
$$2\pi i\delta V(\mu) = \int_C [V_0(\lambda) - V_0(\mu), \Phi(\lambda)](\lambda-\mu)^{-1}d\lambda \quad .$$

In these expressions, the arbitrary matrix $\Gamma(\lambda)$ is independent of ξ and η, and its form is determined solely by the additional reduction applied to the matrices

U and V. The contour $\mathcal{L}$ bounds a simply connected region in which, according to (3.4.13), the perturbations $\delta\varphi(\mu)$, $\delta U(\mu)$, $\delta V(\mu)$ have the same analyticity properties as the unperturbed solutions $\varphi_0(\mu)$, $U(\mu)$, $V(\mu)$ inside the contour[7] $\mathcal{L}$.

Theorem 2. The solutions (3.4.13) form an infinite-dimensional algebra of solutions if the matrices $\{\Gamma(\lambda)\}$ form a matrix algebra.

To prove this, we consider the second variation of the solution (3.4.13),

$$2\pi i \delta_1 \delta\varphi(\mu) = \int_{\mathcal{L}} (\delta_1\varphi(\lambda)\Gamma(\lambda)\varphi_0^{-1}(\lambda) - \Phi(\lambda)\delta_1\varphi(\lambda)\varphi_0^{-1}(\lambda))\frac{d\lambda}{\lambda-\mu}\varphi_0(\mu)$$
$$+ \int_{\mathcal{L}} \Phi(\lambda)\frac{d\lambda}{\lambda-\mu}\delta_1\varphi(\mu) \ ,$$

and take the commutator of the second variations, $2\pi i[\delta_1\delta\varphi(\mu) - \delta\delta_1\varphi(\mu)]$. Writing $\varphi_0(\lambda)\Gamma_1(\lambda)\varphi_0^{-1}(\lambda) \equiv \Phi_1(\lambda)$, we have

$$\delta_1\delta\varphi(\mu) - \delta_1\delta\varphi(\mu)$$
$$= \frac{1}{(2\pi i)^2}\left\{ \int_{\mathcal{L}}\int_{\mathcal{L}_1} [\Phi_1(\lambda')\Phi(\lambda)]\frac{d\lambda\, d\lambda'}{(\lambda'-\lambda)(\lambda-\mu)}\varphi_0(\mu)\right.$$
$$+ \int_{\mathcal{L}}\int_{\mathcal{L}_1} \frac{[\Phi(\lambda),\Phi_1(\lambda')]}{(\lambda-\mu)(\lambda'-\mu)}\, d\lambda\, d\lambda'$$
$$\left. - \int_{\mathcal{L}}\int_{\mathcal{L}_1} \frac{[\Phi(\lambda')\Phi_1(\lambda)]}{(\lambda'-\lambda)(\lambda-\mu)}\, d\lambda\, d\lambda' \varphi_0(\mu)\right\}$$
$$= \frac{1}{2\pi i}\int_{\mathcal{L}} [\Phi(\lambda)\Phi_1(\lambda)]\frac{d\lambda}{\lambda-\mu}\varphi_0(\mu)$$
$$= \frac{1}{2\pi i}\int_{\mathcal{L}} \varphi_0(\lambda)[\Gamma(\lambda)\Gamma_1(\lambda)]\varphi_0^{-1}(\lambda)\frac{d\lambda}{\lambda-\mu}\varphi_0(\mu) \ .$$

The algebra of solutions is infinite-dimensional, since the matrices $\Gamma(\lambda)$ in the general case depend on a countable number of parameters, which can be taken to be the coefficients of the Laurent series in the neighborhood of the point at infinity. In many cases, the algebra of the infinitesimal solutions (3.4.13) will be isomorphic to the so-called algebra of currents associated with a certain matrix Lie algebra. The algebra of currents is constructed from the formal Laurent series

[7] The expressions (3.4.13) determine two functions $\delta\varphi(\mu)\varphi_0^{-1}(\mu)$ holomorphic inside and outside $\mathcal{L}$, respectively, which represent an explicit solution of the linearized Riemann problem. Unlike (3.4.13), the expressions (3.4.13′), which are continuous on the contour $\mathcal{L}$ determine $\delta U(\lambda)$ and $\delta V(\lambda)$ throughout the complex λ plane.

with coefficients belonging to the matrix algebra, and the series in negative powers of λ is truncated. The variable λ introduces a natural grading in the algebra of currents.

If reductions are applied to the solutions of the systems (3.4.2–10), the $\Gamma(\lambda)$ matrix algebras have the same additional algebraic properties as the matrices $U(\lambda)$ and $V(\lambda)$. However, in contrast to the connections of U and V, the analytic structure of matrices $\Gamma(\lambda)$ depending only on λ can be arbitrary. We shall use an asterisk to denote Hermitian conjugation of the matrices $\Gamma(\lambda)$, without applying the symbol of Hermitian conjugation to the argument λ.

In the case (3.4.2), if $R \in \mathrm{SU}(2)$, then $\Gamma^*(\lambda) = -\Gamma(\lambda)$; but if $R \in \mathrm{SU}(1,1)$, then $\Gamma^*(\lambda) = -\sigma_3\Gamma(\lambda)\sigma_3$. In the case (3.4.3), in order to obtain the infinitesimal solutions of the equation for the reduction $\Gamma^*(\lambda) = -\Gamma(\lambda)$, we must also require that $\Gamma(-\lambda) = \sigma_1\Gamma(\lambda)\sigma_1$, i.e., we have a $\mathbb{Z}_2$ reduction. Similarly, in the case of the two-dimensional Toda model (3.4.8), we have a $\mathbb{Z}_n$ reduction

$$\Gamma(\zeta\lambda) = Z^{-1}\Gamma(\lambda)Z \quad ,$$

where Z is the $n \times n$ diagonal matrix of the successive powers of the root of unity.

Unusual reductions arise in the case of the electrovacuum (see Sect. 3.2). In this case, the 3×3 matrix $\Gamma(\lambda)$ is the product of an arbitrary matrix with the property $T^*(\lambda) = T(-\lambda)$ and the special matrix Ω whose nonzero components are $\Omega_{12} = -\Omega_{21} = 1$, $\Omega_{33} = 1/4\lambda$.

The Lie-Bäcklund matrix group $\{T\}$ for which the Lie-Bäcklund matrix algebra $\{\Gamma(\lambda)\}$ is the tangent space in the unit of the group can be obtained as the union of one-parameter subgroups of the form $\exp[\alpha\Gamma(\lambda)]$, where α is a parameter.

We now consider how this group $\{T\}$ acts in the space of solutions of a given functional class of a specific integrable system, converting one solution of this class into another.

By Theorem 2, (3.4.13) can be written in the form

$$2\pi \mathrm{i}\frac{d\varphi(\mu)}{d\alpha} = \int\limits_{\mathcal{C}} \varphi(\lambda)\Gamma(\lambda)\varphi^{-1}(\lambda)\frac{d\lambda}{\lambda-\mu}\varphi(\mu) \quad , \tag{3.4.14}$$

where we have omitted the dependence of the matrices φ on α, ξ, and η. The initial condition for (3.4.14) is given by the "bare" solution $\varphi(\mu)|_{\alpha=0} = \varphi_0(\mu)$. Equation (3.4.14) is an ordinary differential equation in the space of functions $\{\varphi(\mu)\}$. The intermediate "points" of the integral curve of this equation, which are exact solutions of (3.4.1), join the initial solution $\varphi_0(\mu)$ to the final solution $\varphi(\mu)$. On the space of solutions $\varphi(\lambda)$ there act one-parameter groups of diffeomorphisms generated by the elements of the algebra $\{\Gamma(\lambda)\}$. To show this, we introduce a new unknown function $S(\mu,\gamma)$:

$$S(\mu,\gamma) = (\varphi^{-1}(\mu)\varphi(\gamma) - 1)/(\mu-\gamma) \quad .$$

It is easy to see that (3.4.4) can then be rewritten as follows:

$$2\pi i \frac{d}{d\alpha} S(\mu,\gamma) = \int_{\mathcal{L}} \frac{\Gamma(\lambda)\,d\lambda}{(\lambda-\mu)(\lambda-\gamma)} + \int_{\mathcal{L}} \frac{S(\mu,\lambda)\Gamma(\lambda)\,d\lambda}{(\lambda-\gamma)} - \int_{\mathcal{L}} \frac{\Gamma(\lambda)S(\lambda,\gamma)\,d\lambda}{(\lambda-\mu)} + \int_{\mathcal{L}} S(\mu,\lambda)\Gamma(\lambda)S(\lambda,\gamma)\,d\lambda \quad (3.4.15)$$

We now introduce two important restrictions on the contour $\mathcal{L}$. We assume that: (a) inside the contour $\mathcal{L}$, the matrix function $\varphi_0(\lambda)$ is holomorphic in λ and nonsingular (i.e., its determinant vanishes nowhere inside $\mathcal{L}$): (b) outside the contour $\mathcal{L}$, the matrix $\Gamma(\lambda)$ is a holomorphic function of λ.

On the basis of property (a), we introduce a "Heisenberg" representation of the functions $S(\mu,\lambda)$ in the form of matrices of infinite order, in which at row m and column n we have the coefficient S_{mn} in the expansion of $S(\mu,\lambda)$ in positive powers of λ and μ:

$$S(\mu,\lambda) = \sum_{m,n=0}^{\infty} S_{mn}\mu^m\lambda^n \quad . \qquad (3.4.16)$$

If $\lambda = 0$ does not lie inside $\mathcal{L}$, then instead of (3.4.16) we must take an expansion in powers of $\mu - \varrho$ and $\lambda - \varrho$, where ϱ lies inside the contour $\mathcal{L}$. On the basis of property (b), on the contour $\mathcal{L}$ we write $\Gamma(\lambda)$ in the form of a Laurent series

$$\Gamma(\lambda) = \sum_{n=0}^{\infty} \frac{\Gamma_n}{\lambda^n} \quad . \qquad (3.4.16')$$

Thus, all the singularities of $\Gamma(\lambda)$ lie inside the contour $\mathcal{L}$. In view of property (b), the first term on the right-hand side of (3.4.15) vanishes and, using the residue theorem, (3.4.15) can be written in the form of an equation for the infinite-dimensional matrix $S = (S_{mn})$ $(m,n,=0,1,2,\ldots)$:

$$\frac{dS}{d\alpha} = S\Gamma_1 - \Gamma_2 S + S\Gamma_3 S \quad . \qquad (3.4.17)$$

In (3.4.16), $\Gamma_1 = (\Gamma_{m-n})$ denotes a triangular matrix in which all the elements above the main diagonal are equal to zero: $\Gamma_{m-n} = 0$ for $m < n$. The matrix Γ_2 is obtained from Γ_1 by transposition, $\Gamma_2 \equiv (\Gamma_{n-m})$, and is also triangular, with all the elements below the main diagonal equal to zero. The matrix Γ_3 is constructed from the coefficients of the Laurent series for the matrix Γ by the rule $(\Gamma_3)_{m,n} = \Gamma_{m+n+1}$. In (3.4.17) the product of matrices should be understood as the ordinary matrix product:

$$(AB)_{m,n} = \sum_{s=0}^{\infty} A_{ms}B_{sn} \quad .$$

We now multiply (3.4.17) on the right by the matrix

$$\exp(-\Gamma_1\alpha) \equiv \sum_{k=0}^{\infty}(-\alpha)^k \frac{\Gamma_1^k}{k!}$$

and on the left by the matrix

$$\exp(\Gamma_2\alpha) \equiv \sum_{k=0}^{\infty}(\alpha)^k \frac{\Gamma_2^k}{k!} \quad .$$

Then (3.4.6) can be rewritten in the form

$$\frac{d}{d\alpha}(\exp(\alpha\Gamma_2)S\exp(-\alpha\Gamma_1)) = \exp(\alpha\Gamma_2)S\Gamma_3S\exp(-\alpha\Gamma_1) \quad . \tag{3.4.18}$$

We introduce the new unknown matrix

$$T \equiv \exp(\alpha\Gamma_2)S\exp(-\alpha\Gamma_1) \quad . \tag{3.4.19}$$

Equation (3.4.18) can be rewritten for this matrix as

$$\frac{dT}{d\alpha} = TQT \quad , \quad Q \equiv \exp(\alpha\Gamma_1)\Gamma_3\exp(-\alpha\Gamma_2) \quad . \tag{3.4.18$'$}$$

Equation (3.4.18′) has the solution

$$T^{-1} = -\int_0^\alpha Q(\alpha')\,d\alpha' + T_0^{-1} \quad , \quad T_0 \equiv T|_{\alpha=0} = S_0 \quad ,$$

which (in order to avoid complications associated with the inverses of infinite-dimensional matrices T) can be written in the form

$$T_0 = -T\int_0^\alpha Q(\alpha')\,d\alpha' T_0 + T \tag{3.4.20}$$

or in the form

$$T_0 = -T_0\int_0^\alpha Q(\alpha')\,d\alpha' T + T \quad . \tag{3.4.20$'$}$$

Multiplying both sides of (3.4.20) by $\exp(-\alpha\Gamma_2)$ on the left and returning to the function S, from (3.4.20) we have

$$\exp(-\alpha\Gamma_2)S_0 = -S\exp(-\alpha\Gamma_1)\int_0^\alpha Q(\alpha')\,d\alpha' S_0 + S\exp(-\alpha\Gamma_1) \quad . \tag{3.4.21}$$

We note that $\exp(-\alpha\Gamma_1)$ and $\exp(-\alpha\Gamma_2)$ are triangular matrices with zero elements above and below the main diagonal, respectively, and they are obtained from each other by transposition. The element of $\exp(-\alpha\Gamma_1)$ in row m and column n is equal to the coefficient φ_{m-n} in the expansion of the matrix $\exp(-\alpha\Gamma(\lambda)) = \sum_{k=0}^{\infty}\varphi_k/\lambda^k$.

We shall now consider the infinite-dimensional matrix

$$\Omega \equiv -\exp(-\alpha\Gamma_1)\int_0^{\alpha} Q(\alpha')\,d\alpha'$$

and show that $\Omega_{mn} = \varphi_{m+n+1}$.

Indeed, we have

$$\begin{aligned}\Omega_{mn} &= -\left(\exp(-\alpha\Gamma_1)\int_0^{\alpha}\exp(\alpha'\Gamma_1)\Gamma_3\exp(-\alpha'\Gamma_2)\right)_{m,n}\\ &= \sum_{i=0}^{\infty}\sum_{j=0}^{\infty}\int_0^{\alpha}\frac{(\alpha'-\alpha)^i\alpha'^j}{i!j!}\,d\alpha'(\Gamma_1^i\Gamma_3\Gamma_2^j)_{m,n}(-1)^{i+j+1}\\ &= \sum_{k=1}^{\infty}(-1)^{k+1}\frac{\alpha^{k+1}}{(k+1)!}\sum_{i=0}^{k}(\Gamma_1^i\Gamma_3\Gamma_2^{k-i})_{m,n} \ .\end{aligned}$$

The result to be established follows from the identity

$$\sum_{i=0}^{k}(\Gamma_1^i\Gamma_3\Gamma_2^{k-i})_{m,n} = \sum_{i_1+\ldots+i_{k+1}=m+n+1}\Gamma_{i_1}\ldots\Gamma_{i_{k+1}} \ ,$$

where on the right we have the coefficient with index $m+n+1$ in the Laurent expansion of $[\Gamma(\lambda)]^k$.

We now multiply the components in row m and column n of the matrix equality (3.4.21) by $\mu^m\gamma^n$ and sum over m and n from 0 to ∞. Using (3.4.16, 16′), we then have

$$\begin{aligned}-\int_{\mathcal{L}}\frac{\exp(-\alpha\Gamma(\lambda))S_0(\lambda,\gamma)}{(\lambda-\mu)}\,d\lambda &= \int_{\mathcal{L}}S(\mu,\lambda)\frac{\exp(-\alpha\Gamma(\lambda))}{\lambda-\gamma}\\ &+\int_{\mathcal{L}}S(\mu,\lambda)\exp(-\alpha\Gamma(\lambda))S_0(\lambda,\gamma)\,d\lambda \ . \qquad (3.4.22)\end{aligned}$$

Using the fact that $\exp[-\alpha\Gamma(\lambda)]$ is holomorphic outside the contour $\mathcal{L}$ and returning from the function of two variables $S(\mu,\gamma)$ to the function of one variable

$$\varphi^{-1}(\mu)\varphi(\gamma) = 1 + (\mu-\gamma)S(\mu,\gamma) \ ,$$

from (3.4.22) we finally obtain the integral equation

$$\int_{\mathcal{L}}\varphi(\lambda)\exp(-\alpha\Gamma(\lambda))\frac{\varphi_0^{-1}(\lambda)\,d\lambda}{(\lambda-\mu)} = 0 \ ,$$

from which it also follows that

$$0 = \chi \equiv \frac{\mu}{2\pi i} \int \varphi(\lambda) \exp(-\alpha \Gamma(\lambda)) \frac{\varphi_0^{-1}(\lambda)\, d\lambda}{(\lambda - c)(\lambda - \mu)} , \tag{3.4.23}$$

where the points μ and c lie inside the contour $\mathcal{L}$. This relation was first obtained in the particular case of integrable equations (for the stationary axially symmetric electrovacuum in general relativity) by Ernst and Hauser. If we start from (3.4.20′), we can also write (3.4.23) in the form

$$0 = \int \varphi_0(\lambda) \exp(\alpha \Gamma(\lambda)) \frac{\varphi^{-1}(\lambda)\, d\lambda}{(\lambda - c)(\lambda - \mu)} . \tag{3.4.23′}$$

We now consider the integral (3.4.23) when μ varies outside the contour $\mathcal{L}$. In this case, $\chi = \chi_-(\mu)$ determines a function which is holomorphic outside the contour $\mathcal{L}$ and which, owing to (3.4.23), has the following limiting values as μ tends to a point on the contour $\mathcal{L}$:

$$\chi_-(\lambda) = \varphi(\lambda) \exp(-\alpha \Gamma(\lambda)) \varphi_0^{-1}(\lambda) . \tag{3.4.24}$$

The condition (3.4.24) relates the two functions $\chi_-(\lambda)$ and $\varphi(\lambda)$, which are holomorphic and nonsingular matrices outside and inside the contour $\mathcal{L}$, respectively. Instead of $\varphi(\lambda)$, it is convenient to introduce the function $\chi_+(\lambda) = \varphi(\lambda)\varphi_0^{-1}(\lambda)$. Then (3.4.24) can be written in the form of the traditional Riemann problem of determining the functions $\chi_\pm$ which are holomorphic outside and inside the contour $\mathcal{L}$, respectively, and linearly related on the contour:

$$\chi_-(\lambda) = \chi_+(\lambda) G(\lambda) , \quad G(\lambda) \equiv \varphi_0(\lambda) \exp(-\alpha \Gamma(\lambda)) \varphi_0^{-1}(\lambda) .$$

Thus, we have obtained a complete classification of the integrable systems of mathematical physics on the basis of the dividers of the connections and the Lie-Bäcklund algebras.

Out result permits new refinements of the above-mentioned Zakharov-Shabat method of "dressing the bare solution U_0, V_0". In the investigation outlined above, the following natural restrictions arise: (a) the matrices $\varphi_0(\lambda)$ and $\varphi(\lambda)$ are holomorphic inside the contour $\mathcal{L}$; (b) the matrices $\Gamma(\lambda)$ are holomorphic outside the contour $\mathcal{L}$ in the extended complex plane; (c) the matrices $\Gamma(\lambda)$ belong to some matrix algebra. We stress that the algebras of currents associated with finite matrix algebras do not exhaust all the possibilities. The matrix algebras of $\Gamma(\lambda)$ are isomorphic to the algebras of infinite-dimensional triangular matrices having the special form $\Gamma_1 = (\Gamma_{i-j})$, where $\Gamma_{i-j} = 0$ for $i < j$ and $\sum_{k=0}^{\infty} \Gamma_k/\lambda^k$ is the Laurent series for $\Gamma(\lambda)$ on the contour $\mathcal{L}$.

Finally, we note that a study of the unique solvability of the Riemann problem for solutions of the given functional classes of certain integrable systems was given in a recent monograph [3.24]. In the case of dividers of U and V consisting of a unique singular point, it is convenient to choose the contour $\mathcal{L}$ to be a contour passing through this point.

4. Propagation of Waves in the Gravitational Fields of Black Holes

In this chapter, we consider the propagation of perturbations on the background of the gravitational fields described by the Schwarzschild, Nordström-Reissner, and Kerr-Newman solutions (see Sect. 2.3); like the Kerr solution, these solutions are representatives of asymptotically flat stationary spaces possessing a nondegenerate event horizon.

The external fields of a collapsing star (assuming stability of the process of contraction) must become asymptotically stationary at large times. If, for an external observer, the boundary of a body eventually "freezes" on a nonsingular horizon, then no physical fields (other than the electromagnetic and gravitational fields) — scalar, vector, etc. — can be present outside this "frozen" star[1] [4.1, 2]. There now exist a number of arguments which indicate that the final state of the external field of a collapsar must be characterized by only three parameters: the mass, charge, and angular momentum of the body; i.e., in the general case it is described by the exact Kerr-Newman solution [4.4] (see Sect. 2.3).

For the case of static Einstein-Maxwell vacuum fields, there exists a theorem which asserts that the Nordström-Reissner solution is the only asymptotically flat solution possessing a nonsingular and simply connected event horizon [4.5] (see Chap. 2).

However, in the evolutionary stages preceding and during the collapse, complex physical processes take place inside the star, which give rise to the existence of various physical fields in the exterior space, and these fields (including the electromagnetic and gravitational fields) can have multipole components. Therefore the formation of a black hole without rotation is accompanied by the attenuation of the multipole components of the electromagnetic and gravitational fields in the surrounding space and by the vanishing of all possible fields of different types.

Particularly detailed studies have been made of wave fields outside a collapsing, weakly nonspherical star. The system of Einstein's equations, linearized near the Schwarzschild solution, has been reduced [4.6, 7] to two second-order equations, which have no eigenfrequencies for spherical harmonics of order $l \geq 2$. The stability of the black-hole state has been proved [4.6].

The gravitational collapse of a perfectly conducting, spherically symmetric mass possessing a magnetic dipole moment was studied [4.8], and it was shown

[1] An interesting approach to this problem was developed by *Markov* [4.3].

that in the quasistationary approximation the total magnetic moment of a collapsing star decreases with the time. It was shown [4.9] that when a weakly nonspherical uncharged star collapses, the resulting external field is described by the Kerr solution with a small angular momentum.

Under the assumption that the "primary" radiation (emanating from a collapsar) is effectively switched off, the laws of attenuation of the "secondary" (scattered) radiation have been studied [4.10, 11]. The quantitative details of the asymptotic law of attenuation of the *tails* of the radiation from a collapsing body were found [4.11]. We mention a preprint [4.12] in which the results of intensive investigations of gravitational collapse were summarized. A major contribution to the study of wave fields in the neighborhood of black holes was made by *Chandrasekhar* [4.13].

Various aspects of the propagation of (scalar, electromagnetic, and gravitational) waves in the Schwarzschild field have also been investigated [4.11, 14–18].

The emission of particles, either rotating in circular orbits or falling radially in the field of black holes (including charged particles), has been studied [4.19–23]. It was shown [4.14] that a collapsing dust cloud can focus, in itself, gravitational radiation from external sources, and the canonical equations of *Regge* and *Wheeler* [4.6] and *Zerilli* [4.7] were interpreted in an invariant manner as equations for the linearized invariants of the Riemann tensor. The idea of a logarithmic branch point of the scattering matrix at frequency $\omega = 0$ for Regge-Wheeler and Zerilli potentials was put forward [4.24], and it was shown that there is an intermediate asymptotic behavior, which relaxes with the time towards the asymptotic behavior found by *Price* [4.11]. A study was made [4.25] of the attenuation of the tails of scalar waves emitted during the collapse of a charged star with limiting charge.

In Sect. 4.3 we outline the results of an investigation [4.26] of the external wave fields in the case of weakly nonspherical collapse of a star which is, in general, charged. Using a new method of finding the singularities of the scattering matrix at $\omega = 0$, we study the asymptotic attenuation of the tails of the radiation from a collapsing body for various values of its charge, with allowance for the interaction of the electromagnetic and gravitational fields. The investigation is based on four independent canonical second-order equations with separable variables, which we have been able to obtain from the combined system of Einstein-Maxwell equations, linearized near the Nordström-Reissner solution, for both even and odd perturbations[2]. We note that in the case of charged rotating black holes it has hitherto been possible to study only short-wave perturbations (see Sect. 4.1).

When a black hole occurs in a binary system (in a pair with an ordinary star), the black hole becomes a "source" of scattered radiation. This effect is

[2] It is worthwhile to note that this result was obtained independently of authors [4.26] also by Moncrief [Phys. Rev. 1974, v. D9, p. 2707; ib. 1974, v. 10, p. 1057; ib. 1975, v. 12, p. 1526].

considered in Sect. 4.2. In Sect. 4.1 we show that charged black holes, if they exist, must mix the electromagnetic and gravitational radiation in the universe.

4.1 Propagation of Short Waves in the Field of a Charged Black Hole

In this section, the general theory of the behavior of short waves in strong external electromagnetic fields, developed in Sect. 1.2, is applied to the case of the field of a charged black hole, see [4.27]. In accordance with this theory, we must first find the trajectories of the light rays[3] along which short-wave perturbations propagate and then determine the tetrad component Φ_0 of the external electromagnetic field, which governs the mutual conversion of high-frequency gravitational and electromagnetic waves.

4.1.1 Short Waves in the Nordström-Reissner Field

The eikonal equation $g^{ij}u_{,i}u_{,j} = 0$ admits, in the Nordström-Reissner field, the complete integral

$$\begin{aligned} u &= t \pm R(r) \pm \psi(\theta) + N\varphi \ , \quad R(r) = \int A^{-1}\sqrt{1-\lambda^2 A/r^2}\,dr \ , \\ \psi(\theta) &= \int \sqrt{\lambda^2 - N^2/\sin^2\theta}\,d\theta \ , \quad A \equiv 1 - r_g/r + GQ^2/r^2c^4 \ . \end{aligned} \tag{4.1.1}$$

Here λ and N are arbitrary constants.

The trajectories of the light rays are described by the equations

$$\partial u/\partial\lambda = \text{const} \ , \quad \partial u/\partial N = \text{const} \ .$$

In the case of the Nordström-Reissner metric (see Sect. 2.3), the component Φ_0 in the tetrad field associated with a given normal congruence of the null geodesics has the form

$$\Phi_0 = \lambda Q/r^3 \ . \tag{4.1.2}$$

It is convenient to replace the affine parameter α by the radial coordinate r:

$$d\alpha = dr \Big/ \sqrt{1 - A\lambda^2/r^2} \ .$$

According to the general theory developed in Sect. 1.2, the period of the sinusoid which modulates mutually converted gravitational and electromagnetic packets can be found from the equation

[3] For this, it is sufficient to find the complete integral of the Hamilton-Jacobi equation for the isotropic geodesics (the eikonal equation).

$$2\pi = \sqrt{G} \int \Phi_0 \, d\alpha / c^2 = Q\lambda\sqrt{G}\, c^{-2} \int dr / r^3 \sqrt{1 - A\lambda^2/r^2} \ . \qquad (4.1.3)$$

It follows from this expression that the effect of mutual conversion on a given light ray depends on the constant λ, called the *impact parameter*. Waves trapped by a black hole have impact parameters for which the equation[4] $1 - A\lambda^2/r^2 = 0$ has no real roots. The corresponding values of λ satisfy the inequality $\lambda < \lambda_{cr}$, where

$$\lambda_{cr}^2 = r_g^2 \left[x^2 + \tfrac{5}{2} + \sqrt{1+8x^2} + (8x^2)^{-1} \left(\sqrt{1+8x^2} - 1 \right) \right] \ ,$$

$$x^2 \equiv 1 - Q^2/(GM) \ ,$$

$$\lambda_{cr} = \sqrt{27} r_g / 2 \quad \text{for} \quad Q = 0 \ ,$$

$$\lambda_{cr} = 2r_g \quad \text{for} \quad Q = M\sqrt{G} \ .$$

For $\lambda > \lambda_{cr}$, short waves incident on a black hole approach it up to a minimum radius r_2, which is the larger root of the equation $1 - A\lambda^2/r^2 = 0$. In the case in which an electromagnetic wave with amplitude B is incident on a charged black hole without being trapped, this wave, after experiencing several conversions into a gravitational wave in the field of the black hole, again escapes to infinity in the form of an electromagnetic wave and an emergent gravitational wave, whose amplitudes are

$$f = B \cos \left[2Q\lambda\sqrt{G}\, c^{-2} \int_{r_2}^{\infty} \frac{dr}{r^3} \sqrt{1 - A\lambda^2/r^2} \right] \ , \qquad (4.1.4)$$

$$P = 2c^{-2} B\sqrt{G} \sin \left[2Q\lambda\sqrt{G}\, c^{-2} \int_{r_2}^{\infty} \frac{dr}{r^3} \sqrt{1 - A\lambda^2/r^2} \right] \qquad (4.1.5)$$

(see the notation in Sect. 1.2).

The integrals in (4.1.4, 5) diverge whenever the equation $1 - A\lambda^2/r^2 = 0$ has a multiple root $\lambda = \lambda_{cr}$. The corresponding value of the impact parameter belongs to a ray winding around an unstable limit cycle — a closed circular orbit of light rays with radius

$$r_{cr} = r_g \left(3 + \sqrt{1+8x^2} \right) / 4$$

(the radius is $r_{cr} = 3r_g/2$ for $Q = 0$ and $r_{cr} = r_g$ for $Q = M\sqrt{G}$).

In the neighborhood of the limit cycle, the approximation of geometrical optics is not appropriate, and we have here leakage of short waves through a

[4] For a given λ, this equation reduces to a fourth-order equation for r, which for $\lambda > \lambda_{cr}$ has two real roots r_1 and r_2 outside the horizon. For $\lambda < \lambda_{cr}$, this equation has no roots outside the horizon. The value λ_{cr} corresponds to the case of a multiple root $r_1 = r_2$.

potential barrier (see Sect. 4.3.1) with comparable values of the reflection and transmission coefficients.

The relations (1.2.48) for the Nordström-Reissner field, after expansion of the solution in spherical harmonics $Y_{lm}(\theta, \varphi)$ and in a Fourier integral with respect to the time, take the form

$$A\frac{\partial}{\partial r}\left(A\frac{\partial}{\partial r}r\chi_{\pm}\right) + r\chi_{\pm}\left[\frac{\omega^2}{c^2} - A\left(\frac{(l+1)l}{r^2} \mp \sqrt{\frac{G}{c^4}}\frac{Q\lambda\omega}{cr^2}\right)\right] = 0 \ . \quad (4.1.6)$$

The ratio $c\sqrt{(l+1)l}/\omega$ has the meaning of the impact parameter λ.

For rays whose impact parameters are close to the critical value $|\lambda - \lambda_{cr}| \sim O(1/\omega)$, near a closed ray $|r - r_{cr}| \sim O(1/\omega)$ the relation (4.1.6) reduces to the parabolic-cylinder equation

$$\frac{r_{cr}^4}{\lambda_{cr}^4}\frac{d^2}{dr^2}\chi_{\pm} + \frac{\omega^2}{c^2}\chi_{\pm}\left[(r - r_{cr})^2\frac{6r_{cr}^2 - \lambda_{cr}^2}{r_{cr}^4}\right.$$
$$\left. + \frac{2}{\lambda_{cr}}(\lambda_{cr} - \lambda) \pm \sqrt{\frac{G}{c^4}}\frac{Qc}{\lambda_{cr}\omega r_{cr}}\right] = 0 \ . \quad (4.1.7)$$

The Weber (parabolic-cylinder) functions can be expressed in terms of the confluent hypergeometric function, the theory of which makes it possible to prove [4.28] that the amplitudes of the incident, reflected, and transmitted waves are in the ratio $1 : |R_{\pm}| : |T_{\pm}|$, where the reflection and transmission coefficients $|R_{\pm}|$ and $|T_{\pm}|$ are given by

$$|R_{\pm}| = \frac{1}{\sqrt{1 + \exp(-\pi a_{\pm})}} \ ,$$
$$|T_{\pm}| = \frac{1}{\sqrt{1 + \exp(\pi a_{\pm})}} \ , \quad (4.1.8)$$
$$a_{\pm} = \frac{\omega}{c}\lambda_{cr}\left[2(\lambda - \lambda_{cr}) \pm \sqrt{\frac{G}{c^4}}\frac{Qc}{\lambda_{cr}\omega r_{cr}}\right] \ .$$

It follows from the expansion of the hypergeometric function that at small values of the argument the amplitudes of the waves near a limit cycle become $\omega^{1/4}$ times larger than at normal points, where the expansion of geometrical optics holds.

Suppose that an electromagnetic wave is incident on a black hole, with impact parameters of the rays close to the critical value and with an asymptotic behavior at infinity given by

$$f = Br^{-1}\exp[i\omega(t + r)] \ .$$

According to (4.1.7), the gravitational wave which appears as a result of the reflection is given by the expression

$$P = \frac{B\sqrt{G}}{2ric^2}\exp[i\omega(t-r)]$$
$$\times\left\{|R_\pm|\exp\left[2i\int\limits_{\mathrm{Re}\{r_2\}}^{\infty}\left(\sqrt{\Phi_+}-\omega\sqrt{1-\frac{A\lambda^2}{r^2}}\right)dr^* + \frac{\gamma_+}{2}\right]\right.$$
$$\left.-|R_-|\exp\left[2i\int\limits_{\mathrm{Re}\{r_2\}}^{\infty}\left(\sqrt{\Phi_-}-\omega\sqrt{1-\frac{A\lambda^2}{r^2}}\right)dr^* + \frac{\gamma_-}{2}\right]\right\} .$$

Here r_2 is the root of the equation $1 - A\lambda^2/r^2 = 0$ with

$$\mathrm{Re}\{r_2\} > \frac{r_g}{2}\left(1-\sqrt{1-\frac{Q^2}{GM^2}}\right) ,$$
$$\Phi_\pm = \omega^2\left(1-\frac{A\lambda^2}{r^2}\right) \pm \sqrt{\frac{G}{c^4}}\frac{Q\lambda\omega A}{r^3 c} ,$$
$$\gamma_\pm = \arg\Gamma\left(\frac{1}{2}+\frac{ia_\pm}{2}\right) + \frac{a_\pm}{2} - \frac{a_\pm}{2}[\ln(|a_\pm|) - \ln 2] ,$$

and $R_\pm$ and $a_\pm$ are defined by (4.1.8).

For a finite difference of λ from λ_{cr}, with $\lambda > \lambda_{cr}$, the solution of (4.1.7) in the region $r < r_1$ for nontrapped short waves becomes exponentially small, since, according to the definition (4.1.8), $a_\pm$ becomes of order ω. In this case,

$$\gamma_\pm \to 0 , \quad \exp(-\pi a_\pm) \to 0 ,$$
$$\sqrt{\Psi_\pm} - \omega\sqrt{1-\frac{A\lambda^2}{r^2}} \approx \sqrt{\frac{G}{c^4}}\frac{AQ\lambda}{r^3\sqrt{1-A\lambda^2/r^2}} ,$$

and we again obtain (4.1.5).

If $|\lambda - \lambda_{cr}| \leq O(1/\omega)$, the total intensity of the reflected electromagnetic wave and the resulting gravitational wave will form a finite part of the initial intensity:

$$I_{out} = B^2(|R_+|^2 + |R_-|^2)/2 .$$

The number of mutual conversions of the waves is comparable with unity if $Q \sim \sqrt{G}M$.

Thus, charged black holes might in principle control the equilibrium of the relic gravitational and electromagnetic radiation in the universe.

4.1.2 Short Waves in the Neighborhood of a Rotating Charged Black Hole

The eikonal equation for the Kerr-Newman solution (see Sect. 2.3) admits a total integral in the form

$$u = t \pm R(r) \pm \psi(\theta) + N\varphi \ , \qquad (4.1.9)$$

where

$$R(r) = \int [(r^2 + a^2)^2 + 2aN(r_{g}r - GQ^2c^{-4}) + a^2N^2 - \lambda^2\Delta]^{1/2}\frac{dr}{\Delta} \ , \qquad (4.1.10)$$

$$\psi(\theta) = \int \left[\frac{\lambda^2 - N^2}{\sin^2\theta - a^2\sin^2\theta}\right]^{1/2} d\theta \ . \qquad (4.1.11)$$

For the Kerr-Newman solution, the principal axes of the bivector of the electromagnetic field and of the Weyl tensor in the coordinate system (2.3.67) are given by the components (2.3.69).

In Chap. 1 we derived the expression (1.2.28′) for the derivative of the argument of the tetrad component Φ_0 (which controls the mutual conversion of the waves) with respect to the affine parameter.

For the Kerr-Newman field, which has a Weyl tensor of Petrov type D, the rotation coefficients k, ν, λ, and σ of the tetrad are equal to zero. Therefore the expression (1.2.28′) for $\arg\Phi_0$ takes the form

$$\begin{aligned}\frac{d(\arg\Phi_0)}{d\alpha} = \frac{3}{2\mathrm{i}}\big[&u_{,(m)}(\pi + \tau^*) - u_{,(m^*)}(\pi^* + \tau) \\ &-u_{,(n)}(\varrho - \varrho^*) - u_{,(l)}(\mu - \mu^*)\big] \ .\end{aligned} \qquad (4.1.12)$$

We now use the equations

$$\frac{d\theta}{d\alpha} = -\Sigma^{-1}\frac{d\psi}{d\theta} \ , \quad \frac{dr}{d\alpha} = -\Delta\frac{dR}{dr} \ , \qquad (4.1.13)$$

substitute (2.3.70) into (4.1.12), and integrate. We then obtain

$$\arg\Phi_0 = -3\arctan(r/a\cos\theta) \ . \qquad (4.1.14)$$

For a rotating charged black hole, the electromagnetic field of the background in the tetrad (2.3.69) has the form (2.3.68).

Using the formulas for the transformation of the tetrad components of the electromagnetic field (see Sect. 1.1), we obtain

$$|\Phi_0| = \Sigma^{-3/2}Q\sqrt{\lambda^2 + 2aN} \ .$$

Finally, making use of the expression for the argument, we obtain for the tetrad component Φ_0 the elegant expression

$$\Phi_0 = Q\sqrt{\lambda^2 + 2aN}(r - \mathrm{i}a\cos\theta)^{-3} \ . \qquad (4.1.15)$$

In the absence of rotation, (4.1.15) reduces to (4.1.2).

A new effect for short waves, which occurs specifically in the case of their propagation in the field of a rotating charged black hole, in contrast to the case of the Nordström-Reissner field, is the rotation of the plane of polarization for waves with initial linear polarization (with respect to a tetrad which undergoes parallel transport) [4.27].

In the general case, at normal points of a congruence of null geodesics, the equation for the amplitude $\mathcal{F}$ has the form

$$\frac{d^2\mathcal{F}}{dm^2} + \mathcal{F} - \mathrm{i}\frac{d(\arg \Phi_0)}{dm}\frac{d\mathcal{F}}{dm} = 0 \qquad (4.1.16)$$

(see the notation in Sect. 1.2).

The variable m for the Kerr-Newman field is related to the radius as follows:

$$dm = Q\sqrt{\frac{G}{c^4}}\sqrt{\lambda^2 + 2aN}\frac{dr}{\Delta\sqrt{\Sigma}\, dR/dr} \quad . \qquad (4.1.17)$$

Therefore the function $d(\arg \Phi_0)/dm$ which appears in (4.1.16) is given implicitly in terms of m by the relation

$$\frac{d(\arg \Phi_0)}{dm} = \frac{3a}{Q\sqrt{\Sigma}\sqrt{\lambda^2 + 2aN}}\left(\Delta\frac{dR}{dr}\cos\theta + r\sin\theta\frac{d\psi}{d\theta}\right)\frac{c^2}{\sqrt{G}} \quad . \qquad (4.1.18)$$

The largest numbers of mutual conversions and rotations of the plane of polarization occur for waves on null geodesics which wind onto a limit cycle whose radius is a multiple root of the equation

$$(r^2 + a^2)^2 + 2aN(r_g r - GQ^2/c^4) + a^2N^2 - \lambda^2\Delta = 0 \quad .$$

This root gives the radius of a closed trajectory of massless particles.

For trajectories lying in the equatorial plane $\theta = \pi/2$, there is no rotation of the plane of polarization. In this case, the period of the modulating frequency can be found from (4.1.17). Mutual conversion of waves does not occur for rays with $N = -a$ lying in the equatorial plane.

Another interesting case is the case of trajectories winding around the cone $\theta = \arcsin(|N|/a)$ (with $|N| < a$, $\lambda^2 = -2aN$).

For these trajectories of isotropic geodesics, the tetrad component Φ_0 is equal to zero, so that the effect of mutual conversion of gravitational and electromagnetic waves does not occur for the corresponding wave packets. The general solution of (4.1.16) for large r admits the following series expansion:

$$\begin{aligned}\mathcal{F} = C_1&\left\{1 - \frac{GQ^2(\lambda^2 + 2aN)}{8r^4c^4} + \frac{\mathrm{i}aGQ^2(\lambda^2 + 2aN)\cos\theta_0}{10r^5c^4} + \ldots\right\}\\ &+ C_2\left\{\frac{1}{r^2} - \frac{2\mathrm{i}a\cos\theta_0}{r^3} + \frac{\lambda^2 - 2a^2 - 10a^2\cos^2\theta_0}{4r^4}\right.\\ &\left. + \frac{\mathrm{i}a\cos\theta_0}{5r^5}(14a^2\cos^2\theta_0 + 6a^2 - 3\lambda^2) + \ldots\right\} \quad .\end{aligned}$$

Therefore, if a ray approaches a charged rotating body at the closest distance $R \approx \lambda \gg r_g$, the argument of $\mathcal{F}$ (for $C_2 = 0$) changes by an amount $\delta \ll 1$:

$$\delta \sim aGQ^2 \cos\theta_0/(c^4 R^3) \quad .$$

The change in the argument of $\mathcal{F}$ is equal to the angle of rotation of the plane of polarization with respect to a tetrad which undergoes parallel transport along a null geodesic.

Finally, we note that for rays with large impact parameters λ, or in the case of a black hole with small charge, the rotation of the plane of polarization with respect to the tetrad will be much smaller than the rotation of the parallel-transport tetrad itself (a possible measure of the latter is given by parallel transport with respect to the pseudo-Euclidean infinity from the "point of entry" of the ray into the gravitational field to its "point of exit"). The ratio of the corresponding angles of rotation is of the order of Q^2/Mc^2R. The two effects of rotation of the plane of polarization become comparable with each other when $\lambda \sim \lambda_{cr}$ and $Q^2 \sim GM^2$, i.e., when the mutual conversion of the waves is not small.

4.2 Asymptotic Theory of Scattering of Wave Packets in the Gravitational Field of a Black Hole

The problem of small perturbations of the Einstein-Maxwell equations on the background of the Schwarzschild solution reduces to the analysis of equations of the type

$$\frac{\partial^2 Qr}{\partial r^{*2}} - \frac{\partial^2 Qr}{\partial t^2} + \left[\left(1 - \frac{r_g}{r}\right)\frac{\tilde{\Delta}}{r^2} + V(r)\right] Qr = 0 \quad , \tag{4.2.1}$$

where the quantities Q are certain combinations of small perturbations of either the metric or the bivector of the electromagnetic field, $r^* = r + r_g \ln(r/r_g - 1)$, and $\tilde{\Delta}$ is the Laplacian operator on a sphere of unit radius. *Regge* and *Wheeler* [4.6] have shown that $V(r) = 3r_g(1 - r_g/r)/r^3$ in the case of perturbations of the metric of odd type. For even perturbations of the metric (see the beginning of Sect. 4.3.1), the potential V was found by *Zerilli* [4.7, 23]; for electromagnetic waves of odd type, $V = 0$. Nevertheless, the differences between the potentials $V(r)$ for describing the propagation and scattering of short waves are unimportant asymptotically.

Following [4.29], we shall describe here the effects of scattering by the field of a black hole of radiation from the second component in a binary system. Of course, there is also a contribution to the total radiation from the radiation of the stellar wind and interstellar gas accreting on the black hole, which are not considered here. However, the specific feature of the scattering is that the black hole becomes a source of secondary radiation of a wave having the same wavelength as that of the second component, and the radiation comes from a sphere of radius $3r_g/2$.

Before investigating this problem, we recall that a homogeneous linear ordinary differential equation of the second order with variable coefficients can always be reduced, by means of a transformation of the unknown function, to the canonical form $d^2y/dx^2 + \Phi(x)y = 0$. This equation retains its canonical form under a transformation $x = f(\xi)$ of the independent variable if the unknown function is transformed as $y = \sqrt{f'}z(\xi)$ ($f' \equiv dx/d\xi$). Then $z(\xi)$ obeys the equation

$$\frac{d^2z}{d\xi^2} + \tilde{\Phi}(\xi)z = 0 \ , \quad \text{where}$$

$$\tilde{\Phi}(\xi) \equiv \Phi(x)f'^2 + \sqrt{f'}\frac{d^2}{d\xi^2}\left(\frac{1}{\sqrt{f'}}\right) \ . \tag{4.2.2}$$

In the case in which the scale L of the characteristic variation of the positive function $\Phi(x)$ is much greater than $1/\sqrt{\Phi}$, the asymptotic solution of the second-order equation (the WKB approximation) has the form

$$y \approx \text{const} \cdot \Phi^{-1/4} \exp\left(\pm \mathrm{i} \int \sqrt{\Phi}\, dx\right) \ .$$

If the function $\Phi(x)$ is positive only in a certain region, then near the zeros of $\Phi(x)$ with $\Phi' \neq 0$ (the so-called *turning points*) a more accurate asymptotic form for the solution is given by

$$y = \frac{\text{const}}{\sqrt[6]{|\Phi'|}} \mathrm{Ai}(z) \exp(2z^{3/2}/3) \exp\left(\pm \mathrm{i} \int \sqrt{\Phi(\tau)}\, d\tau\right) \ ,$$

$$z \equiv -\Phi|\Phi'|^{-2/3} \ , \quad \Phi' \equiv d\Phi/dx \ ,$$

where Ai(z) is the Airy function. A wave incident on a turning point is ideally reflected from it with a phase change $\pi/2$. The situation here is completely analogous to the approximation of geometrical optics at normal points and to the behavior of waves near caustics described in Chap. 1.

At certain isolated points, the scale L can become of order equal to or less than $\Phi^{-1/2}$, although the function Φ remains positive. In other cases, the distance between two turning points beyond which $\Phi(x) > 0$ can be of order equal to or less than $|\min \Phi(x)|^{-1/2}$. In such cases, there is a partial reflection and partial "leakage" of the waves through the potential barrier characterizing the nonhomogeneity of the background [the function $\Phi(x)$ carries information about the characteristic features of the background around which the linearization was made in obtaining the linear equation $y'' + \Phi(x)y = 0$].

We write the scattering potential $\Phi(x)$ in the form $A^2\psi(x)$, where A is a large parameter. Near a double turning point x_0, the analytic function $\psi(x)$ can be represented in the form

$$\psi(x) = (x - x_0)^2 + m_0/A + \alpha_1(x - x_0)^3 + \alpha_2(x - x_0)^4 + \dots \quad . \tag{4.2.3}$$

We shall assume that the radius of convergence of this series is of order 1. The constant m_0 can have either sign.

In the equation $y'' + \Phi(x)y = 0$, we transform to the variable $t = \sqrt{A}(x - x_0)$. Then this equation takes the form

$$\frac{d^2y}{dt^2} + \varphi(t)y = 0 \quad , \quad \varphi(t) = m_0 + t^2 + \frac{\alpha_1 t^3}{\sqrt{A}} + \frac{\alpha_2 t^4}{A} + \dots \quad . \tag{4.2.3$'$}$$

The equation for the unknown function $y(t)$ with this potential $\varphi(t)$ can be reduced by means of a substitution $y = \sqrt{dt/d\xi}\, z(\xi)$ to Weber's equation, whose general solution can be expressed in terms of confluent hypergeometric functions.

In order to find this substitution $\xi(t)$, i.e., the mapping $t \to \xi$, we must solve the equation

$$\varphi(t) = (\xi^2 + m)\left(\frac{d\xi}{dt}\right)^2 - \sqrt{\frac{d\xi}{dt}}\frac{d^2}{dt^2}\sqrt{\frac{dt}{d\xi}} \quad . \tag{4.2.2$'$}$$

We shall seek the unknown constant m and the function $\xi(t)$ in the form of asymptotic expansions

$$\begin{aligned} m &= m_0 + \frac{m_1}{A} + \frac{m_2}{A^2} + \dots = \sum_{k=0}^{\infty} \frac{m_k}{A^k} \quad , \\ \xi(t) &= t + \frac{f_1(t)}{\sqrt{A}} + \frac{f_2(t)}{A} + \dots = \sum_{k=1}^{\infty} \frac{f_k(t)}{(\sqrt{A})^k} + t \quad . \end{aligned} \tag{4.2.4}$$

Substituting the expansions (4.2.4) into (4.2.2$'$), where the function $\varphi(t)$ is represented in the form of the series (4.2.3$'$), and equating the coefficients of identical powers of $A^{-1/2}$, we obtain, in turn, the results

$$\begin{aligned} \alpha_1 t^3 &= 2tf_1 + 2f_1'(t^2 + m_0) + f_1'''/2 \quad , \\ \alpha_2 t^4 &= 2tf_2 + 2f_2'(t^2 + m_0) + f_2'''/2 + f_1^2 + m_1 + 4tf_1f_1' \\ &\quad + (f_1')^2(t^2 + m_0) - 3[(f_1')^2]''/8 + f_1'f_1'''/2, \ \dots \quad . \end{aligned} \tag{4.2.5}$$

It follows from the structure of the recurrence relations (4.2.5) that the solution for the functions f_k can be found in the form of polynomials of degree $k+1$, where the polynomials f_{2k} contain only odd powers of t, while f_{2k-1} contain only even powers. We write the required solution for f_1, f_2, and m_1:

$$\begin{aligned} m_1 &= -\frac{3}{8}\alpha_2 + \frac{13}{96}\alpha_1^2 - \frac{m_0}{8}\left[\frac{15}{32}\alpha_1^2 - 3\alpha_2\right] \quad , \\ f_1 &= \frac{\alpha_1}{6}(t^2 - m_0) \quad , \\ f_2 &= \frac{\alpha_2 - 13\alpha_1^2/36}{8}t^3 + \frac{m_0}{16}t\left[\frac{103}{36}\alpha_1^2 - 3\alpha_2\right] \quad . \end{aligned} \tag{4.2.6}$$

To analyze the character of the propagation of short waves in the Schwarzschild field, it is sufficient to know only the principal part of the differential wave operator, which, after expanding in spherical harmonics, takes the following form for a monochromatic wave with frequency ω:

$$\frac{d^2}{dr^{*2}}(Qr) + \left[\frac{\omega^2}{c^2} - \frac{(l+1)l}{r^2}\left(1 - \frac{r_g}{r}\right)\right] Qr = 0 \ . \tag{4.2.7}$$

Solutions of the Cauchy problem for short-wave perturbations which differ from zero at the initial time only in a sufficiently small region are subsequently localized in the neighborhood of the corresponding bicharacteristic (of the ray). Such solutions represent the classical transition from a wave to a particle [4.30], and the ratio lc/ω has the meaning of the impact parameter p of the ray along which the wave packet propagates.

Trapped wave packets correspond to impact parameters such that the expression

$$\Phi(r) = \frac{\omega^2}{c^2} - \frac{l(l+1)(1 - r_g/r)}{r^2}$$

remains positive up to the horizon. This requires that the impact parameter p be less than $r_g\sqrt{27}/2$.

Wave packets incident from infinity are not trapped by a black hole if their impact parameters are greater than $r_g\sqrt{27}/2$: in this case, there is a reflection of the solution from the first turning point which is encountered on the path of the wave [the root of the equation $\Phi(r) = 0$].

Finally, wave packets with near-critical impact parameters in the neighborhood of the sphere $r = 3r_g/2$ undergo strong scattering, and a certain fraction of the radiation is trapped by the black hole. In fact, when $p \approx r_g\sqrt{27}/2$ we have the situation described above, since for $p = r_g\sqrt{27}/2$ the equation $\Phi(r) = 0$ has a double root $r = 3r_g/2$. Therefore when $|p - r_g\sqrt{27}/2| \lesssim 1/\omega$ near the multiple turning point $r = 3r_g/2$ the differential equation (4.2.7) reduces asymptotically to Weber's equation

$$\frac{d^2Q}{dr^2} + \frac{\omega^2}{c^2 r_g^2}\left[12\left(\frac{r - 3r_g}{2}\right)^2 + \frac{4}{3}\left(\frac{27}{4}r_g^2 - p^2\right)\right] Q = 0 \ .$$

Using the known asymptotic form of the solution of this equation, we can determine the coefficients in the WKB solutions of (4.2.7) corresponding to the reflected and transmitted waves.

Thus, making use of the properties of the confluent hypergeometric function [4.28], we obtain for a nontrapped wave packet incident on a black hole from infinity the following asymptotic solution of (4.2.7):

$$Q = B(p)\Phi^{-1/4}\frac{1}{r}\left\{ i(1+\exp(-\pi a^2))^{-1/2}\exp\left[-i\left(\gamma + \int_{r_2^*}^{r^*}\sqrt{\Phi}\,dr^*\right)\right]\right.$$

$$\left. +\exp\left(i\int_{r_2^*}^{r^*}\sqrt{\Phi}\,dr^*\right)\right\} \quad \text{for} \quad r_2^* < r^* < \infty \ , \qquad (4.2.8)$$

$$Q = B(p)\Phi^{-1/4}\frac{1}{r}\left\{ (1+\exp\pi a^2)^{-1/2}\exp\left[i\left(\gamma + \int_{r_1^*}^{r^*}\sqrt{\Phi}\,dr^*\right)\right]\right\}$$

$$\text{for} \quad r^* < r_1^* \ . \qquad (4.2.9)$$

Here $B(p)$ is the amplitude of the incident wave, and

$$\Phi \equiv \frac{\omega^2}{c^2} - \frac{(l+1)l(1-r_g/r)}{r^2} \ ,$$

$$\gamma \equiv \arg\Gamma\left(\frac{1}{2}+i\frac{a^2}{2}\right) + \frac{a^2}{2} - a^2\ln\left(\frac{a}{2}\right) \ ,$$

$$\frac{\pi a^2}{2} = \int_{r_1}^{r_2}\frac{\sqrt{-\Phi(r)}\,dr}{1-r_g/r} \approx \frac{\pi}{\sqrt{27}}\left(p^2 - \frac{27}{4}r_g^2\right) \ .$$

In (4.2.8, 9), r_1^* and r_2^* are, respectively, the smaller and larger roots of the equation $\Phi(r) = 0$, in which r is expressed in terms of r^*. The solution (4.2.8) consists of the incident and reflected waves, and the solution (4.2.9) corresponds to the part of the wave passing into the region $r < r_1$. It is easy to see from the structure of the solution that an "energy" balance holds for the waves: the square of the modulus of the amplitude of the incident wave is equal to the sum of the squares of the moduli of the reflected and trapped waves. In order to go over to the coordinate representation of the solution, we multiply $Q(r)$ by the spherical harmonic $p_l(\cos\theta)$ and sum the result over l. If the main contribution to the sum comes from the coefficients of large l, then for normalized Legendre functions in the range $0 < \theta < \pi$ we can make use of their WKB approximation $[\pi\sin\theta/2]^{-1/2}\cos[(l+1/2)\theta+\pi/4]$ and replace the summation by an integration with respect to l from $\sqrt{27}\,\omega r_g/(2c)$ to infinity.

In order to simplify the resulting integral, we use the method of stationary phase for the calculation of an integral:

$$\int_a^b e^{iAf(x)}\psi(x)\,dx \approx \sqrt{\frac{2\pi}{A|f''(x_0)|}}\left\{\exp\left[iAf(x_0) + i\frac{\pi}{4}\operatorname{sgn} f''(x_0)\right]\right\}\psi(x_0) \ ,$$

where A is a sufficiently large number, and x_0 is the unique root of the equation $f'(x) = 0$ in the interval (a, b). Carrying out the calculations for the solution $Q(t,\theta,r)$ corresponding to the incident wave, we obtain

$$Q = \sqrt{\frac{\omega}{c}}\left[\left(\sin\theta\frac{\partial\theta}{\partial p}\right)_{p=\tilde{p}}\sqrt{\Phi}\right]^{-1/2}\frac{B(p)}{r}$$
$$\times\exp\left[-\mathrm{i}\omega\left(t+\frac{p\theta}{c}\right)+\mathrm{i}\int\limits_{r_2^*}^{r^*}\sqrt{\Phi}\,dr^*+\mathrm{i}\frac{\pi}{4}\right] \quad . \tag{4.2.10}$$

Here we can replace $\tilde{p}$ by its value from the equation

$$\theta = \int\limits_{x_0}^{x}\tilde{p}\left[r_g^2-\tilde{p}^2(x^2-x^3)\right]^{-1/2}dx \quad , \quad x \equiv r_g/r \quad . \tag{4.2.11}$$

In (4.2.10) it is assumed that all the rays emanate from the point $\theta = 0$, $x = x_0$.

We shall calculate the flux of energy arriving at a black hole as a result of scattering of untrapped wave packets in the case in which the black hole is one of the components of a binary system, in which the other component is a source of electromagnetic or gravitational radiation. We shall assume that the period of rotation of the system is much greater than the time required for a light ray to reach a distance $3r_g/2$ from the radiating component. The angle θ will be measured from the axis joining the two components.

Suppose that near the source $r = r_0$, $\theta = 0$ the function Q has the asymptotic form

$$Q = \frac{\sigma}{\sqrt{4\pi}}\frac{\exp[-\mathrm{i}\omega(t-f(r,r_0)/c)]}{f(r,r_0)} \quad . \tag{4.2.12}$$

Here

$$f(r,r_0) = \sqrt{r^2+r_0^2-2rr_0\cos\theta} \approx \sqrt{(r-r_0)^2+r_0^2\theta^2}$$

is the distance between the points with coordinates r and θ and the source with intensity $\sigma =$ const at the point $r = r_0$, $\theta = 0$, and the effects of the curvature of space near the source are neglected.

Comparing the solutions (4.2.10, 12) near the source, we can determine the function $B(p)$ in the solution (4.2.10). Indeed, according to (4.2.11), we have for θ the estimate

$$\theta \approx \frac{p}{r_0^2}\left(1-\frac{p^2}{r_0^2}\right)^{-1/2}(r-r_0) \quad .$$

Therefore, near the source, $f(r,r_0) \approx (r-r_0)/\sqrt{1-p^2/r_0^2}$. Thus, the expression (4.2.12) has the following asymptotic behavior near $r = r_0$, $\theta = 0$:

$$Q = \frac{\sigma\sqrt{1-p^2/r_0^2}}{\sqrt{4\pi}\,(r-r_0)}\exp\left[\mathrm{i}\omega\left(-t+\frac{r-r_0}{c\sqrt{1-p^2/r_0^2}}\right)\right] \quad . \tag{4.2.12'}$$

On the other hand, the expression (4.2.10) near the source has the form

$$Q = \sqrt{\mathrm{i}}\frac{\sqrt[4]{(1-p^2/r_0^2)^3}}{\sqrt{p}(r-r_0)} r_0 B(p) \times \exp\left[\mathrm{i}\int_{r_2^*}^{r_0^*}\sqrt{\Phi}\,dr^* + \mathrm{i}\omega\left(-t+\frac{r-r_0}{c\sqrt{1-p^2/r_0^2}}\right)\right] . \tag{4.2.10'}$$

Comparing (4.2.10′, 12′), we obtain

$$B(p) = \frac{\sigma\sqrt{p}\exp\left[-\mathrm{i}\int_{r_2^*}^{r_0^*}\sqrt{\Phi}\,dr^*\right]}{r_0\sqrt{4\pi}\sqrt[4]{1-p^2/r_0^2}}\mathrm{e}^{-\mathrm{i}\pi/4} . \tag{4.2.13}$$

For the complete system of spherical harmonics, we have

$$\int |Q^2|\sin\theta\,d\theta = \sum_{l=1}^{\infty}|Q_l^2| .$$

In the case under consideration, the main contribution to this sum comes from harmonics with large l, for which the summation can be replaced by an integration. For the flux of "energy" arriving at the black hole as a result of scattering of untrapped wave packets, we obtain, using (4.2.9),

$$I_1 = 2\pi\mathrm{Re}\left(\int\frac{\partial Q}{\partial t}\frac{\partial Q^*}{\partial r^*}r^2\sin\theta\,d\theta\right) = -\frac{2\pi\omega^2}{c}\int_{\sqrt{27}r_g/2}\frac{|B^2(p)|\,dp}{1+\exp(\pi a^2)} . \tag{4.2.14}$$

For impact parameters p close to the critical value, $a^2 \approx 2\omega(p^2-27r_g^2/4)/\sqrt{27}cr_g$. Substituting the expression (4.2.13) for $B(p)$ into (4.2.14) and calculating the integral in (4.2.14), we obtain, apart from small quantities of higher order in the small parameter $c/\omega r_g$,

$$I_1 = -\frac{\omega^2\sigma^2}{4cr_0^2}\int_{27r_g^2/4}^{\infty}d\gamma\left\{\sqrt{1-\frac{\gamma}{r_0^2}}\left[1+\exp\left(\frac{2\omega(\gamma-27r_g^2/4)}{r_g c\sqrt{27}}\right)\right]\right\}^{-2}$$
$$\approx -\frac{\omega\sigma^2 r_g}{8r_0^2}\frac{\sqrt{27}\ln 2}{\sqrt{1-27r_g^2/(4r_0^2)}} \approx -\frac{\omega\sigma^2\sqrt{27}r_g\ln 2}{8r_0^2} . \tag{4.2.15}$$

In the derivation of (4.2.15) we have made use of an asymptotic method of calculating an integral of the form $\int_0^{\infty} f(x)\,dx/(\mathrm{e}^{Ax}+1)$, where A is a large parameter. Expanding the function $f(x)$ in a Maclaurin series in the interval $(0, R)$, where R is the radius of convergence of the series, we obtain

$$\int_0^\infty \frac{f(x)\,dx}{e^{Ax}+1} = \sum_{k=0}^\infty \frac{f^{(k)}(0)}{k!} \int_0^{R-\varepsilon} \frac{x^k\,dx}{e^{Ax}+1} + O(e^{-AR})$$

$$= \frac{1}{A}\sum_{k=0}^\infty \frac{f^{(k)}(0)}{A^k k!} \int_0^\infty \frac{t^k\,dt}{e^t+1} + O(e^{-AR})$$

$$= \frac{\ln 2}{A} f(0) + \frac{1}{A}\sum_{k=1}^\infty \frac{f^{(k)}(0)}{A^k}\zeta(k+1)(1-2^{-k}) + O(e^{-AR}) \ ,$$

where $0 < \varepsilon \ll R$, and $\zeta(k)$ is the Riemann zeta function

$$\zeta(k) = \sum_{n=1}^\infty \frac{1}{n^k} \ .$$

In (4.2.15) we have retained only the first and most important term. We now make use of (4.2.13) to calculate the flux of "energy" trapped by the black hole, neglecting scattering:

$$I = -\frac{2\pi\omega^2}{c} \int_0^{\sqrt{27}r_g/2} |B(p)|^2\,dp$$

$$= -\frac{\omega^2\sigma^2}{4cr_0^2} \int_0^{27r_g^2/4} \frac{d\gamma}{\sqrt{1-\gamma/r_0^2}} \approx -\frac{\omega^2\sigma^2}{16c}\frac{27r_g^2}{r_0^2} \ . \tag{4.2.16}$$

If we calculate the part of the flux of "energy" scattered by the black hole from the rays trapped by it, we find a value of magnitude I_1 with the opposite sign.

Thus, we have obtained the final result that a black hole becomes a source of secondary radiation, whose output [as can be seen by comparing (4.2.15, 16)] is Ω times smaller ($\Omega \equiv \omega r_g\sqrt{27}/2c\ln 2$) than the flux of "energy" trapped by the black hole.

We shall now give estimates. For a black hole with the mass of the Earth, $r_g = 0.44$ cm. For a wavelength $2\pi c/\omega = 10^{-2}$ cm (the infrared range of radiation of the companion of a black hole), Ω has the value 165. This means that the part of the radiation scattered by the black hole is 165 times smaller than the part trapped by it. However, for the microwave range of radio emission with wavelength 1 cm, this ratio is of order unity. Consequently, the phenomenon of diffraction becomes important for wavelengths greater than 1 cm.

The effect of scattering of optical radiation will manifest itself for black holes with mass greater than about 10^{22} or 10^{23} g.

Let us now consider the effects of scattering for radiation emitted by a collapsing body. Gravitational and electromagnetic harmonics with $l \gg 1$ emitted by a collapsing body with $r_B < 3r_g/2$ (here r_B is the limiting radius of the body) can reach a distant observer with frequencies $\omega \gtrsim 2lc/\sqrt{27}r_g$ (the part of the spectrum of the radiation with $\omega \ll 2lc/\sqrt{27}r_g$ reaches a distant observer with an exponentially small amplitude).

The scattered part of the radiation of a collapsing body with $r_B < 3r_g/2$ reaches a distant observer with an asymptotic time dependence determined by the factor $\exp(2ilct/\sqrt{27}r_g)$. This result agrees with the conclusions of *Press* [4.31], obtained by means of computer calculations.

4.3 Wave Fields Outside a Collapsing Star

In Sect. 4.3.1 we make use of the Einstein-Maxwell equations, linearized around the Nordström-Reissner solution, to obtain a set of four independent second-order equations describing the behavior of arbitrary multipole perturbations. In the case of zero charge, two of these equations reduce to the equations obtained by *Regge* and *Wheeler* [4.6] and *Zerilli* [4.7] for perturbations of the gravitational field on the background of the Schwarzschild solution, and the other two reduce to equations equivalent to Maxwell's equations on this background. Each of our equations [4.26] for monochromatic waves with a fixed spherical harmonic has the form of a one-dimensional Schrödinger equation with a potential function which vanishes at the pseudo-Euclidean infinity and at the horizon. The WKB approximation for our equations, and also their short-wave asymptotic behavior taking into account the scattering of waves on an unstable closed circular photon orbit, give the same results as those obtained previously [4.27] (see Sect. 4.1).

In Sect. 4.3.2 we write down the asymptotic form of the boundary conditions on the surface of a star near the horizon for radiated waves in the case of different charge, and we describe the general features of the process of scattering of monochromatic waves.

In Sect. 4.3.3 the asymptotic properties of the "tails" of the radiation are described in terms of a certain integral along the edges of the cut for functions having a logarithmic branch point at the origin.

In Sect. 4.3.4 we obtain the asymptotic behavior of normalized monochromatic waves, as well as the transmission coefficient for monochromatic waves with a large period.

In Sect. 4.3.5 we make use of the above-mentioned integral and the results of Sect. 4.3.4 to establish the laws of attenuation of the "tails" of the radiation with time, and we find the intermediate asymptotic behavior, which, with time, goes over into the behavior given in this section.

4.3.1 Derivation of the Basis Equations

As is well known, the external field of a nonrotating, charged, spherically symmetric black hole is described by the Nordström-Reissner electrovacuum solution

$$\begin{aligned} &ds^2 = Ac^2dt^2 - A^{-1}dr^2 - r^2(d\theta^2 + \sin^2\theta\, d\varphi^2) \ , \\ &F_{4r} = -E \ , \quad E \equiv \frac{Q}{r^2} \ , \quad A \equiv 1 - \frac{2m}{r} + \frac{q^2}{r^2} \ , \\ &m \equiv \frac{GM}{c^2} \ , \quad q = \frac{\sqrt{G}\,Q}{c^2} \ , \end{aligned} \qquad (4.3.1)$$

where Q is the charge and M is the mass of the black hole.

Notation and Relations Used Here. The indices take the values $\alpha, \beta, \gamma, \ldots = 3,4$; $a, b, c, \ldots = 1,2$; $i, j, k, \ldots = 1,2,3,4$; the coordinates are $x^1 \equiv \theta$, $x^2 \equiv \varphi$, $x^3 \equiv r$, $x^4 \equiv ct$; g_{ab} and $g_{\alpha\beta}$ are the metric tensors on the coordinate surfaces (x^1, x^2) and (x^3, x^4), respectively, induced by the metric g_{ij} of the solution (4.3.1). The indices a, b and α, β are raised and lowered by means of the metrics g_{ab} and $g_{\alpha\beta}$, respectively; ∇_a and ∇_α are the operators of covariant differentiation on the coordinate surfaces (x^1, x^2) and (x^3, x^4) constructed with the metrics g_{ab} and $g_{\alpha\beta}$; and $\varepsilon_{\alpha\beta}$ and e_{ab} are the Levi-Civita tensors on these surfaces. The nonzero components of the Maxwell tensor for the Nordström-Reissner solution are $F_{\alpha\beta} = -\varepsilon_{\alpha\beta}E$; $\mu = \ln r^2$, $\mu_\alpha = \nabla_\alpha \mu$, $\nabla_\alpha E = -\mu_\alpha E$, $\Box \equiv g^{\alpha\beta}\nabla_\alpha \nabla_\beta$, $\Delta = -r^2 g^{ab}\nabla_a \nabla_b$, where Δ is the Laplacian operator on the two-dimensional sphere of unit radius.

Regge and *Wheeler* [4.6] noted that arbitrary small perturbations h_{ik} of the metric on the background of the Schwarzschild metric can be divided into two independent types: even and odd (in accordance with the different behavior of these constituents under inversion of the coordinates on the sphere). Perturbations on the background of the Nordström-Reissner metric can be divided in exactly the same way according to their parity.

Coordinate Conditions. We impose the following coordinate conditions on arbitrary small perturbations h_{ij} of the metric:

$$h_{ab} = 0 \ , \quad h^\alpha_\alpha = h^3_3 + h^4_4 = 0 \ . \qquad (4.3.2)$$

These conditions can be satisfied by choosing a particular infinitesimal transformation of the coordinates, $y^i = x^i + \xi^i(x^1, x^2, x^3, x^4)$, and the required ξ^i can be determined from h_{ik} by quadratures. The details can be found in [4.32].

A) Odd Perturbations. If the coordinate conditions are satisfied, the odd perturbations of the components of the metric tensor have the form

$$h_{ab} = 0 \ , \quad h_{\alpha\beta} = 0 \ , \quad h_{\alpha a} = e_{ab}\nabla^b h_\alpha \ . \qquad (4.3.3)$$

The perturbations of the components of the tensor F_{ik} and of its dual tensor $F^*_{ik} = \varepsilon_{iklm}F^{lm}/2$ have the form

$$\begin{aligned} &\delta F^{\alpha\beta} = 0 \ , \quad \delta F^{\alpha a} = e^{ab}\nabla_b F^\alpha \ , \quad \delta F^{ab} = e^{ab}\delta H \ , \\ &\delta F^{*\alpha\beta} = \varepsilon^{\alpha\beta}\delta H \ , \quad \delta F^{*\alpha a} = \nabla^a(\varepsilon^{\alpha\beta}F_\beta - Eh^\alpha) \ , \quad \delta F^{*ab} = 0 \ . \end{aligned} \tag{4.3.4}$$

Maxwell's Equations for the Perturbations. Under the conditions (4.3.2), the determinant g of the metric tensor remains unperturbed and Maxwell's equations for the perturbations take the form

$$(\sqrt{-g}\,\delta F^{ij})_{,i} = 0 \ , \quad (\sqrt{-g}\,\delta F^{*ij})_{,i} = 0 \ . \tag{4.3.5}$$

Using the notation introduced above, these equations can be written in the form

$$\begin{aligned} &\nabla_\beta F^\beta = \delta H \ , \quad \varepsilon^{\alpha\beta}\nabla_\alpha F_\beta + E\mu^\alpha h_\alpha = 0 \ , \\ &\nabla_\beta \delta H + \mu_\beta \delta H - \Delta(F_\beta - E\varepsilon_{\beta\gamma}h^\gamma)/r^2 = 0 \ . \end{aligned} \tag{4.3.6}$$

Applying the operator ∇_β to the last equation in (4.3.6) and eliminating F^β by means of the first two equations, we obtain

$$\Box\delta H/E - \Delta(\delta H/(Er^2)) = -\Delta[\varepsilon^{\alpha\beta}\nabla_\alpha(h_\beta/r^2)] \ . \tag{4.3.7}$$

Perturbations of the Components of the Energy-Momentum Tensor. The perturbations (4.3.4) of the electromagnetic field lead to perturbations of the components of the energy-momentum tensor:

$$\begin{aligned} &\delta T_{\alpha\beta} = 0 \ , \quad \delta T_{ab} = 0 \ , \\ &\delta T_{\alpha a} = e_{ab}\nabla^b[E^2 h_\alpha/8\pi - (E/4\pi)\varepsilon_{\alpha\beta}F^\beta] \ . \end{aligned} \tag{4.3.8}$$

Linearized Einstein Equations. The perturbations of the components of the Ricci tensor due to the perturbations (4.3.3) of the metric have the form

$$\begin{aligned} 2\delta R_{a\alpha} &= -e_{ab}\nabla^b[\Box h_\alpha - \Delta h_\alpha/r^2 - \nabla_\alpha\nabla_\beta h^\beta + \mu_\alpha\nabla_\beta h^\beta \\ &\qquad - R_* h_\alpha/2 - \nabla_\alpha(\mu_\beta h^\beta) + (2\nabla_\alpha\mu_\beta + \mu_\alpha\mu_\beta)h^\beta] \ , \\ \delta R_{\alpha\beta} &= 0 \ , \\ 2\delta R_{ab} &= e_{ac}\nabla_b\nabla^c(\nabla_\gamma h^\gamma) + e_{bc}\nabla_a\nabla^c(\nabla_\gamma h^\gamma) \ , \end{aligned} \tag{4.3.9}$$

where $R_* \equiv d^2A/dr^2$ is the curvature scalar for the metric $g_{\alpha\beta}$. By virtue of Einstein's equations, it follows from (4.3.8, 9) that

$$\nabla_\alpha h^\alpha = 0 \ , \tag{4.3.10}$$

$$\begin{aligned} &\Box h_\alpha - \Delta h_\alpha/r^2 - \nabla_\alpha(\mu_\beta h^\beta) + (2\nabla_\alpha\mu_\beta + \mu_\alpha\mu_\beta)h^\beta - R_* h_\alpha/2 \\ &\qquad = \kappa E\varepsilon_{\alpha\beta}F^\beta/2\pi - \kappa E^2 h_\alpha/4\pi \ . \end{aligned} \tag{4.3.11}$$

In (4.3.11) we have omitted terms which vanish by virtue of (4.3.10). We perform the operation of contracting (4.3.11) with μ^α and apply the Laplacian operator to both sides of the equation:

$$\Box h - \Delta h/r^2 + (3R_*/2 - \kappa E^2/8\pi)h = \kappa E \varepsilon_{\alpha\beta}\mu^\alpha \Delta F^\beta/2\pi \ ,$$
$$h \equiv \Delta\mu_\alpha h^\alpha \ . \tag{4.3.12}$$

Using (4.3.6) and eliminating F^β from the right-hand side of (4.3.12), we obtain

$$\Box h - \Delta h/r^2 + (3R_*/2 - 5\kappa E^2/8\pi)h = \kappa E^2 \varepsilon^{\alpha\beta}\mu_\alpha \nabla_\beta (\delta H/(2\pi E)) \ . \tag{4.3.13}$$

A Closed System of Equations. Using (4.3.10, 11, 13), we can express the right-hand side of (4.3.7) in terms of h. For this, we apply to (4.3.7) the operator $r\varepsilon^{\alpha\beta}\mu_\alpha\nabla_\beta$, which in the coordinates θ, φ, r, ct reduces to $-2\partial/c\,\partial t$ and hence commutes with all the other operators of differentiation. We obtain

$$\Box\Phi - \Delta\Phi/r^2 + \kappa E^2\Phi/2\pi + (\Delta + 2)h/r^3 = 0 \ , \tag{4.3.14}$$

where $\Phi = r\varepsilon^{\alpha\beta}\mu_\alpha\nabla_\beta(\delta H/E)$. Then (4.3.13) takes the form

$$\Box h - \Delta h/r^2 + (3R_*/2 - 5\kappa E^2/8\pi)h = \kappa E^2 r\Phi/2\pi \ . \tag{4.3.15}$$

Equations (4.3.14, 15) form a closed system. On the basis of their solutions, the components of the perturbations of the electromagnetic and gravitational fields can be determined from the remaining Einstein and Maxwell equations in the form of series in spherical harmonics, and the coefficients of these expansions can be calculated by quadratures. In the coordinates θ, φ, r, ct, these equations take the form

$$\frac{1}{A}\left(\frac{\partial^2}{\partial r^{*2}} - \frac{\partial^2}{c^2\partial t^2}\right)h + \frac{\Delta}{r^2}h - \frac{4q^2}{r^4}h = -6\frac{m}{r^3}h - \frac{4q^2}{r^3}\Phi \ , \tag{4.3.16}$$

$$\frac{1}{A}\left(\frac{\partial^2}{\partial r^{*2}} - \frac{\partial^2}{c^2\partial t^2}\right)\Phi + \frac{\Delta}{r^2}\Phi - \frac{4q^2}{r^4}\Phi = \frac{\Delta+2}{r^3}h \ ,$$
$$r^* = \int\frac{dr}{A} \ , \quad \Delta = -l(l+1) \ , \tag{4.3.17}$$

where $l \geq 2$, since for spherical harmonics with the index l equal to 0 or 1 the derivation of the equations becomes invalid.

It turns out to be possible to introduce new variables η_+ and η_- in the form

$$\eta_\pm = C_\pm h + 4q^2\Phi \ , \quad C_\pm = 3m \pm \sqrt{9m^2 - 4q^2(\Delta + 2)} \tag{4.3.18}$$

such that the system (4.3.16, 17) decomposes into two independent second-order equations, each of which contains only one unknown (the sign + or − is chosen according to the condition corresponding to the unknown η_+ or η_-):

$$\left(\frac{\partial^2}{\partial r^{*2}} - \frac{\partial^2}{c^2\partial t^2}\right)\eta_\pm + \left(\frac{\Delta}{r^2} + \frac{C_\pm}{r^3} - \frac{4q^2}{r^4}\right)A\eta_\pm = 0 \ . \tag{4.3.19}$$

B) Even Perturbations. An analogous procedure is possible for the even perturbations. Under the coordinate conditions (4.3.2), the nonzero components of the even perturbations of the metric are

$$h_{\alpha\beta}(h_\alpha^\alpha = 0) \ , \quad h_{\alpha a} = \nabla_a H_\alpha \ ,$$

and those of the tensor F_{ij} and its dual tensor F^*_{ij} are

$$\delta F^{\alpha\beta} = -\varepsilon^{\alpha\beta}\delta E \ , \quad \delta F^{\alpha a} = \nabla^a f^\alpha \ , \quad \delta F^{ab} = 0 \ ,$$
$$\delta F^{*\alpha\beta} = 0 \ , \quad \delta F^{*\alpha a} = -e^{ab}(\nabla_b \varepsilon^{\alpha\beta} f_\beta - EH^\alpha) \ , \quad \delta F^{*ab} = e^{ab}\delta E \ .$$

Maxwell's Equations for the Perturbations. For even perturbations, we write (4.3.5) in the form

$$\begin{aligned} &\nabla_\beta f^\beta = 0 \ , \quad \Delta f^\alpha + \varepsilon^{\alpha\beta}\nabla_\beta(r^2\delta E) = 0 \ , \\ &\varepsilon^{\alpha\beta}\nabla_\alpha f_\beta + E\mu_\alpha H^\alpha + \delta E = 0 \ . \end{aligned} \tag{4.3.20}$$

By analogy with the case of odd perturbations, by eliminating f_β we obtain from (4.3.20) the equation

$$\Box\psi - \Delta\psi/r^2 - \Delta(\mu_\alpha H^\alpha/r^2) = 0 \ , \quad \psi \equiv \delta E/E \ . \tag{4.3.21}$$

In addition, we shall use below the following equality, which is a consequence of (4.3.20):

$$\Delta\varepsilon^{\alpha\beta}\mu_\alpha f_\beta = -Q\mu^\alpha\nabla_\alpha\psi \ .$$

Perturbations of the Components of the Energy-Momentum Tensor. In the case of even perturbations, we have

$$\begin{aligned} &\delta T_{\alpha\beta} = Eg_{\alpha\beta}\delta E/4\pi + E^2 h_{\alpha\beta}/8\pi \ , \\ &\delta T_{ab} = -E\delta E g_{ab}/4\pi \ , \\ &\delta T_{\alpha a} = \nabla_a[-E\varepsilon_{\alpha\beta}f^\beta/4\pi + E^2 H_\alpha/8\pi] \ . \end{aligned}$$

Linearized Einstein Equations. From Einstein's equations for the even perturbations, using a method analogous to that given in [4.14] for obtaining a single equation describing the behavior of gravitational waves on the background of the Schwarzschild metric, it is possible to obtain an equation which, however, contains on its right-hand side a function ψ characterizing the perturbation of the electromagnetic field:

$$\begin{aligned} &\left(\frac{\partial^2}{\partial r^{*2}} - \frac{\partial^2}{c^2\partial t^2}\right)\vartheta + \left[\frac{\Delta}{r^2} + \left(\frac{6m}{r} - \frac{4q^2}{r^2}\right)V(r) + \frac{8q^2A}{r^4 p(r)}\right]A\vartheta \\ &\qquad + \kappa E^2 r^3 AV(r)\psi/4\pi = 0 \ , \\ &p(r) \equiv \Delta + 2 - 6m/r + 4q^2/r^2 \ , \\ &\vartheta \equiv (rh_{44} - \Delta H^1)/p(r) \ , \\ &V(r) = [\Delta^2 - 4 + 12m/r - 12m^2/r^2 + 4mq^2/r^3]/(r^2p^2(r)) \ . \end{aligned} \tag{4.3.22}$$

A Closed System of Equations. Equation (4.3.22) together with (4.3.21), taken after the transformation to the variable ϑ in the form

$$\left(\frac{\partial^2}{\partial r^{*2}} - \frac{\partial^2}{c^2\partial t^2}\right)\psi + \left[\frac{\Delta}{r^2} + 8\frac{q^2}{r^4}\frac{A}{p(r)}\right]A\psi$$
$$+\frac{4A}{r^2}\frac{\partial}{\partial r^*}\vartheta + \left[4A\left(\frac{3m}{r} - \frac{4q^2}{r^2}\right)\Big/ p(r) - \Delta\right]\frac{2A}{r^3}\vartheta = 0 \ , \qquad (4.3.23)$$

form a closed system, whose solutions can be used to determine all the components of the perturbations by quadratures.

Like the system (4.3.16, 17) for the odd perturbations, the system (4.3.22, 23) decomposes into two independent equations when we introduce the new variables

$$\chi_\pm = (C_\pm - 4q^2/r)\vartheta + 2q^2\psi \ , \qquad (4.3.24)$$

where $C_\pm$ are defined by (4.3.18). The variables $\chi_\pm$ satisfy the equations

$$\left(\frac{\partial^2}{\partial r^{*2}} - \frac{\partial^2}{c^2\partial t^2}\right)\chi_\pm + \left[\frac{\Delta}{r^2} + \left(C_\pm - \frac{4q^2}{r}\right)\frac{V(r)}{r} + \frac{8q^2A}{r^4p(r)}\right]A\chi_\pm = 0 \ . \qquad (4.3.25)$$

4.3.2 Boundary Conditions and General Properties of the Reflection and Transmission Coefficients of Waves

Inside a collapsing body, there can be various processes which influence the waves emitted by it into the surrounding space. The external wave fields are determined by the boundary conditions on the surface of the star. We shall not solve the complex interior problem of obtaining the values of the unknown functions on the surface of the star, but shall simply assume that they are finite (nonsingular) at the instant at which the boundary of the body crosses the exterior event horizon (according to a clock of a comoving observer).

The law of motion of the boundary of a charged sphere (without surface charge), neglecting pressure forces, can be found by calculating the law of motion of a free uncharged particle in the Nordström-Reissner field. We denote the charge and mass of the star by q and m, respectively. In what follows, we shall use a system of units in which $G = c = 1$. As in the calculations for an uncharged sphere [4.33], we find that the rate of contraction of the boundary is

$$(dr/dt)_R = -A(1 - A/\varepsilon^2)^{1/2} \ , \qquad (4.3.26)$$

where, as before, $A \equiv 1 - 2m/r + q^2/r^2$, and ε distinguishes the elliptic ($0 < \varepsilon < 1$), parabolic ($\varepsilon = 1$), and hyperbolic ($\varepsilon > 1$) laws of motion. The proper time τ of a comoving observer is related to the coordinates t and r of an observer at infinity as follows:

$$\tau = \varepsilon\left(t + \int A^{-1}\left(\frac{1-A}{\varepsilon^2}\right)^{1/2} dr\right) \ . \qquad (4.3.27)$$

Substituting t as a function of r from (4.3.26) into (4.3.27), we obtain the relation between the proper time and the radius of the surface of the star:

$$\tau - \tau_1 = -\int_{r_1}^{r} (\varepsilon^2 - A)^{-1/2} dr \ ,$$

where τ_1 is the time at which the boundary of the body crosses the exterior horizon r_1.

Suppose that the function $\psi(\tau)$ determines the law of radiation at the surface of a star for waves receding from the surface. Near the exterior horizon r_1 [the largest of the roots of the equation $A(r) = 0$], we can seek solutions of (4.3.19, 25) in the wave zone in the form of expansions in powers of $r - r_1$. In the zeroth approximation, we consider waves receding from the horizon; then the approximate solutions (4.3.19, 25) have the form $\varphi = \varphi(t - r^*)$. However, it follows from (4.3.26) that on the surface of the body

$$t = -\int \left(\frac{1-A}{\varepsilon^2}\right)^{-1/2} dr^* \approx -r^* \ .$$

(By means of a displacement of the origin from which t is measured, it is always possible to ensure that the trajectory of the boundary of the body for $t \to \infty$ is tangent to the line $t + r^* = 0$.) Therefore we determine the form of the function $\varphi(t - r^*)$ by the equality

$$\varphi(-2r^*) \approx \psi(\tau_1 - (r - r_1)/\varepsilon) \ .$$

For $q^2 < m^2$ the coordinate r^* near the horizon is related to r by $r^* \approx \gamma \ln(r - r_1)$, and for $q^2 = m^2$ it has the form $r^* \approx -m^2/(r - r_1)$. Hence for $t \gg r^*$ we have

$$\varphi(t - r^*) \approx \psi(\tau_1) - \left(\frac{\partial \psi}{\partial \tau}\right)_{\tau_1} \frac{m^2}{\varepsilon(r^* - t)} \quad \text{for} \quad q^2 = m^2 \ , \tag{4.3.28}$$

$$\varphi(t - r^*) \approx \psi(\tau_1) - \left(\frac{\partial \psi}{\partial \tau}\right)_{\tau_1} \varepsilon^{-1} \exp\left(\frac{r^* - t}{2\gamma}\right) \quad \text{for} \quad q^2 < m^2 \ ,$$

$$2\gamma \equiv \frac{r_1^2}{\sqrt{m^2 - q^2}} \ . \tag{4.3.29}$$

It is convenient to carry out the analysis of the scattering properties of the field of a charged black hole for the Fourier transforms of the unknown functions with respect to the time

$$\tilde{\chi}(\omega, r) = \int_0^\infty e^{i\omega t} \chi(t, r)\, dt \ .$$

Then the differential equations (4.3.19, 25) for the perturbations become ordinary second-order differential equations of the form

$$d^2\tilde{\chi}/dx^2 + [\omega^2 - U(x)]\tilde{\chi} = \mathrm{i}\omega\chi|_{t=0} - \partial\chi/\partial t|_{t=0} \quad , \quad x \equiv r^* \quad . \tag{4.3.30}$$

We shall call the real function $U(x)$ the *potential barrier of the curvature.*

The differential equations (4.3.30) have, for $x \to \pm\infty$, wave zones in which the homogeneous solutions possess asymptotic expansions of the form

$$\exp(\pm\mathrm{i}\omega x)\sum_{n=0}^{\infty} a_n r^{-n} \qquad \text{for} \quad r \to +\infty \quad ,$$

$$\exp(\pm\mathrm{i}\omega x)\sum_{n=0}^{\infty} b_n (r - r_1)^n \quad \text{for} \quad r \to r_1 \quad .$$

The domains of convergence of these series depend on ω. The smaller ω, the further these domains recede to $+\infty$ and $-\infty$, respectively, along the x axis. Suppose that for $x \to +\infty$ the solution has the asymptotic form

$$a_{\mathrm{out}}(\omega)T_{\mathrm{out}}(\omega)\exp(\mathrm{i}\omega x)$$

(the subscript "out" refers to a wave receding from the surface of the body to $+\infty$ along the x axis). Analytically continuing this solution to the entire x axis near the horizon in the wave zone, we obtain an expression of the form

$$a_{\mathrm{out}}(\omega)[\exp(\mathrm{i}\omega x) + R_{\mathrm{out}}(\omega)\exp(-\mathrm{i}\omega x)] \quad .$$

For a second-order equation in canonical form and with a real potential barrier, we have the equality

$$yy^{*\prime} - y'y^* = \mathrm{const} \quad .$$

Here the asterisk signifies complex conjugation, and $y(x)$ is an arbitrary solution of the equation $y'' + \Phi(x)y = 0$, in which $\mathrm{Im}\{\Phi(x)\} = 0$. Using this equality, we obtain the relation

$$|T_{\mathrm{out}}(\omega)|^2 + |R_{\mathrm{out}}(\omega)|^2 = 1 \quad ,$$

which indicates that the flux of "energy" is conserved.

The term $a_{\mathrm{out}}(\omega)\exp(\mathrm{i}\omega x)$ indicates a wave emitted by the body near the horizon, and $a_{\mathrm{out}}(\omega)R_{\mathrm{out}}(\omega)\exp(-\mathrm{i}\omega x)$ is the part of this wave reflected by the potential barrier of the curvature back to the horizon; $a_{\mathrm{out}}(\omega)T_{\mathrm{out}}(\omega)\exp(\mathrm{i}\omega x)$ is the other part of the wave, which penetrates through the barrier and reaches an observer at $x \to +\infty$. The determination of the functions $T(\omega)$ and $R(\omega)$ is the direct problem of scattering theory. [Equations of the type (4.3.30) containing potential barriers have only a continuous spectrum.]

For a wave incident on a black hole from $+\infty$, the boundary condition at $x \to -\infty$ should consist in the absence of a wave coming from the horizon. Therefore, for $x \to -\infty$ the solution has the form

$$a_{\mathrm{in}}(\omega)T_{\mathrm{in}}(\omega)\exp(-\mathrm{i}\omega x) \quad ,$$

while for $x \to +\infty$ it has the form

$$a_{\text{in}}(\omega)[\exp(-\mathrm{i}\omega x) + R_{\text{in}}(\omega)\exp(\mathrm{i}\omega x)]$$

(the subscript "in" refers to a wave coming from $+\infty$).

The transmission coefficient $T_{\text{in}}(\omega)$ characterizes the frequency distribution of the part of the energy of the radiation trapped by the black hole; the reflection coefficient $R_{\text{in}}(\omega)$ characterizes the part scattered by its field. The function $T_{\text{in}}(\omega)$ can be analytically continued into the lower half of the complex plane of the variable ω, and the function $T_{\text{out}}(\omega)$ into the upper half of this plane. We adopt the notation $y_{\text{out}}(\omega, x)$ for the homogeneous solution of (4.3.30) which for $x \to +\infty$ has the asymptotic behavior $\exp(\mathrm{i}\omega x)$. This function can be analytically continued into the upper half-plane.

The equations with potential barriers $U(x)$ obtained from (4.3.19, 25) do not admit bound states, and therefore the operators of (4.3.30) have only a continuous spectrum. It follows from this (see [4.34]) that for $\mathrm{Im}\{\omega\} \geq 0$ the function $T_{\text{out}}(\omega)$ is bounded.

4.3.3 Properties of Radiation Emitted by a Collapsing Body Near the Horizon

We shall restrict ourselves to the problem of finding the asymptotic behavior of the radiation at large t only as a result of radiation of waves by a body near the horizon, i.e., we shall set the initial perturbations equal to zero.

Thus, the properties of the coefficients of the scattering described above are required only for finding the asymptotic behavior for $t - r^* \gg m$ of the following integral:

$$\chi(t, x) = (2\pi)^{-1} \int_{-\infty}^{+\infty} a_{\text{out}}(\omega) T_{\text{out}}(\omega) y_{\text{out}}(\omega, x) \exp(-\mathrm{i}\omega t)\, d\omega \;. \tag{4.3.31}$$

We shall calculate this integral by means of a contour integration on the basis of the following two assumptions.

1) For $x > t$ the integral (4.3.31) is equal to zero for a contour around the top of the singular point $\omega = 0$. This follows from the fact that the functions $a_{\text{out}}(\omega)$, $T_{\text{out}}(\omega)$, and $y_{\text{out}}(\omega, x)$ are analytic in the upper half-plane. Therefore the integral (4.3.31) is equal to the limit of the integral of the same functions around a semicircle in the upper half-plane when its radius tends to infinity. However, this limit is zero, as was to be proved.
2) For $t > x$ the integral (4.3.31) can be calculated by means of a contour integration. A cut is placed along the negative part of the imaginary axis, and the contour shown in Fig. 4.1 lies on one sheet of the Riemann surface for the integrand of (4.3.31), which has a logarithmic branch point at $\omega = 0$.

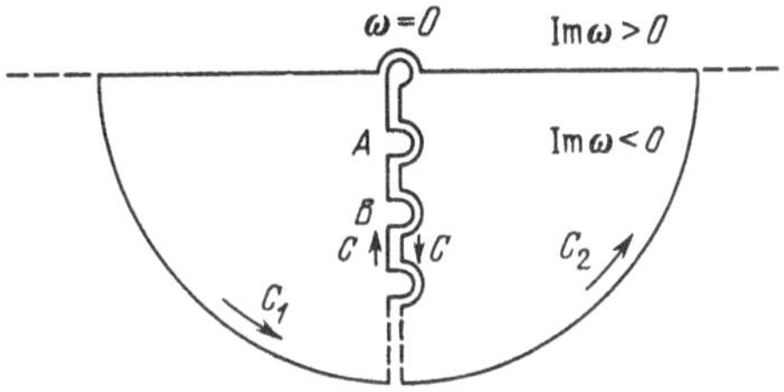

Fig. 4.1. Contour of integration in the Fourier transform

When $q^2 < m^2$, the function $a_{\rm out}(\omega)$ has poles $A, B, \ldots$ on the imaginary axis for $\mathrm{Im}\,\omega \leq 0$ [which follows from (4.3.29)] (see Fig. 4.1); when $q^2 = m^2$, the function $a_{\rm out}(\omega)$ has a logarithmic branch point at $\omega = 0$:

$$a_{\rm out}(\omega) \approx \frac{\mathrm{i}\psi(\tau_1)}{\omega} + \left(\frac{\partial\psi}{\partial\tau}\right)_{\tau_1} m^2 \ln\omega$$

[see (4.3.28)]. If the radius of the semicircle tends to infinity, the integrals over C_1 and C_2 vanish. Therefore the integral (4.3.31) is equal to the integral along the edges of the cut, i.e., along the contour C which goes around the top of the singularity at $\omega = 0$ [apart from terms which fall off exponentially with the time and arise because of the poles $A, B, \ldots$ of the integrand of (4.3.31) inside the contour of integration]. If the function $T_{\rm out}(\omega)y_{\rm out}(\omega)$ were analytic at $\omega = 0$, the integral along C would be equal to zero, and for $q^2 < m^2$ the asymptotic form of (4.3.31) would be an exponential decrease.

4.3.4 Behavior of the Transmission Coefficient for Small ω

The differential equations (4.3.30) are equations with two turning points, which for $\omega \to 0$ come arbitrarily close to the exterior event horizon $r = r_1$ and $r = \infty$. Since for sufficiently small ω the first turning point lies in the region $(r - r_1)/r_1 \ll 1$ (the region Ω_1), the solution in this region is able to change from a wave solution to a static solution.

In the region $\omega^2 r^2 \ll Al(l+1)$ (the region Ω_2, the sub-barrier region, which is separated from both turning points), the solutions can be approximated by static solutions, i.e., expanded in series in powers of ω^2.

In the region $r \gg m$ (the region Ω_3), the solutions change from monotonic to oscillating solutions.

It is easy to see that the regions of the different asymptotic expansions overlap, and therefore they can be joined in the intersections of the regions Ω_1 and Ω_2 and of the regions Ω_2 and Ω_3. In the region Ω_3, we rewrite (4.3.30) in integral form, assuming that for $x \to \infty$ there is only an outgoing wave $\exp(\mathrm{i}\omega x)$ (as a rule, we shall omit the argument ω of functions depending on ω and x):

$$y(x) = H(x) + \int_x^\infty G_3(x, x')y(x')\,dx' \;, \tag{4.3.32}$$

$$G_3(x, x') \equiv \mathrm{i}[H(x)H^*(x') - H(x')H^*(x)]\tilde{U}(x')/2\omega \;,$$

where

$$H(x) \equiv e^{i\omega x} \sum_{k=0}^{l} (-1)^k \frac{(l+k)!}{(l-k)!k!(2i\omega x)^k}$$

$$\tilde{U}(x') \equiv U(x') - l(l+1)/x'^2 \ .$$

For large x, we can express r asymptotically in terms of x:

$$r = x - 2m \ln x + \sum_{n=1}^{\infty} x^{-n} \varphi_n(\ln x) \ ,$$

where $\varphi_n(z)$ are polynomials of degree n, which can be calculated recursively: $\varphi_1(z) = m^2 - q^2 - 4m^2 z$. Therefore, for $x \gg m$ the perturbing potential $\tilde{U}(x)$ admits the expansion

$$\tilde{U}(x) = x^{-3}(4ml(l+1)\ln x - 2m - C_\pm) + \sum_{n=1}^{\infty} x^{-n-3} \tilde{\varphi}_n(\ln x) \ .$$

In the region Ω_3, we can seek a solution of (4.3.30) by the method of successive approximations:

$$y(x) = \sum_{k=0}^{\infty} Z_k(x) \ , \quad Z_0(x) = H(x) \ ,$$
$$Z_k(x) = \int_x^{\infty} G_3(x, x') Z_{k-1}(x')\, dx' \ . \tag{4.3.33}$$

In the region Ω_2, an arbitrary solution of (4.3.30) is a solution of the integral equation

$$y(x) = \alpha y_1(x) + \beta y_2(x) + \omega^2 \int_a^x G_2(x, x') y(x')\, dx' \ ,$$
$$G_2(x, x') \equiv y_1(x) y_2(x') - y_1(x') y_2(x) \ , \tag{4.3.34}$$

where $y_1(x)$ and $y_2(x)$ are the solutions of the static equation (4.3.30) (with $\omega = 0$) which are bounded for $x \to -\infty$ and $x \to +\infty$, respectively, and x is an arbitrary point in Ω_2.

The method of successive approximations leads to a formal series in powers of ω^2:

$$y(\omega, x) = \alpha W_a(\omega, x) + \beta V_a(\omega, x) \ ,$$
$$W_a(\omega, x) = \sum_{n=0}^{\infty} \omega^{2n} w_n(x) \ ,$$
$$V_a(\omega, x) = \sum_{n=0}^{\infty} \omega^{2n} v_n(x) \ , \qquad (4.3.35)$$
$$\begin{Bmatrix} w_n(x) \\ v_n(x) \end{Bmatrix} = \int_a^x G_2(x, x') \begin{Bmatrix} w_{n-1}(x') \\ v_{n-1}(x') \end{Bmatrix} dx' \ , \qquad \begin{matrix} w_0(x) \equiv y_1(x) \ , \\ v_0(x) \equiv y_2(x) \ . \end{matrix}$$

We determine the coefficients α and β by requiring continuity of $y(\omega, x)$ and $y'(\omega, x)$ at $x = a$:

$$\alpha = y_2'(a) y(\omega, a) - y_2(a) y'(\omega, a) \ , \quad \beta = y'(\omega, a) y_1(a) - y(\omega, a) y_1'(a) \ ,$$

and $y(\omega, a)$ is calculated by (4.3.33).

In the region Ω_1, after the substitution

$$y = rD \ , \quad t = \frac{r - m}{\sqrt{m^2 - q^2}} \ , \quad l_\pm(l_\pm + 1) = l(l+1) - \frac{C_\pm + 2m}{r_1} - 6\frac{q^2}{r_1^2}$$

we rewrite (4.3.30) in the form

$$(t^2 - 1)\frac{d}{dt}(t^2 - 1)\frac{d}{dt}D + [\tilde{\omega}^2 - (t^2 - 1)\tilde{l}(\tilde{l} + 1)]D = (t^2 - 1)D\tilde{\Phi}(t) \ , \quad (4.3.36)$$

where $\tilde{\omega}^2 \equiv \omega^2 r_1^2/(m^2 - q^2)$.

The perturbing potential $\tilde{\Phi}(t)$ is small throughout the region Ω_1. Linearly independent solutions of the left-hand side of (4.3.36) are given by the Legendre functions $P_{\tilde{l}}^{\mathrm{i}\tilde{\omega}}(t)$ and $Q_{\tilde{l}}^{\mathrm{i}\tilde{\omega}}(t)$. We normalize them so that for $t \to 1$ (for $x \to -\infty$) we obtain $\exp(-\mathrm{i}\omega x)$ and $\exp(\mathrm{i}\omega x)$, respectively, with

$$x = m + t\sqrt{m^2 - q^2} + \sqrt{m^2 - q^2}\left[r_1^2 \ln\sqrt{t - 1} - r_2^2 \ln\sqrt{t + 1}\right] \ ,$$
$$p_0(t) = 2^{-\mathrm{i}\tilde{\omega}/2}\Gamma(1 - \mathrm{i}\tilde{\omega})P_{\tilde{l}}^{\mathrm{i}\tilde{\omega}}(t) \ ,$$
$$q_0(t) = 2^{1+\mathrm{i}\tilde{\omega}/2}\left[Q_{\tilde{l}}^{\mathrm{i}\tilde{\omega}}(t) - \frac{\Gamma(\mathrm{i}\tilde{\omega})}{2\Gamma(1 - \mathrm{i}\tilde{\omega})}P_{\tilde{l}}^{\mathrm{i}\tilde{\omega}}(t)\right]\Gamma(1 + \tilde{l} - \mathrm{i}\tilde{\omega})$$
$$\times \Gamma(1 + \tilde{l} + \mathrm{i}\tilde{\omega})[\Gamma(1 - \mathrm{i}\tilde{\omega})]^{-1} \ ,$$

and the Wronskian of $p_0(t)$ and $q_0(t)$ is

$$-2\mathrm{i}\omega r_1^2(t^2 - 1)^{-1}\left(\sqrt{m^2 - q^2}\right)^{-1} \ .$$

An arbitrary solution of (4.3.36) in the region Ω_1 can be represented in the form

$$q(t)/[T(\tilde{\omega})] + p(t)R(\tilde{\omega})/T(\tilde{\omega}) \ , \qquad (4.3.37)$$

where $p(t)$ and $q(t)$ satisfy the integral equation

$$\begin{Bmatrix} p(t) \\ q(t) \end{Bmatrix} = \begin{Bmatrix} p_0(t) \\ q_0(t) \end{Bmatrix} + \int_1^t G_1(t,t') \begin{Bmatrix} p(t') \\ q(t') \end{Bmatrix} dt' \quad , \tag{4.3.38}$$

$$G_1(t,t') = [p_0(t)q_0(t') - q_0(t)p_0(t')]\tilde{\Phi}(t')\sqrt{m^2 - q^2}/2\mathrm{i}\omega r_1^2 \quad .$$

In the region Ω_1 we can find solutions of (4.3.38) for these functions in the usual way by taking $p_0(t)$ and $q_0(t)$, respectively, as the zeroth approximation and calculating the following approximations in terms of the preceding ones by means of the integral operator in (4.3.38).

We determine the scattering coefficients $T(\omega)$ and $R(\omega)$ by matching the solutions of (4.3.30) of the form (4.3.35) with the solutions of (4.3.30) in the form (4.3.37), using the conditions of continuity of the solution and its derivative at an arbitrary point $b \in \Omega_1 \cap \Omega_2$.

Finally, for the transmission coefficient $T(\omega)$ we obtain the expression

$$T(\omega) = \left\{ \begin{vmatrix} y(\omega,a) & y_2(a) \\ y'(\omega,a) & y_2'(a) \end{vmatrix} \begin{vmatrix} W_a(\omega,b) & p(\omega,b) \\ [W_a(\omega,b)]' & p'(\omega,b) \end{vmatrix} \right. \tag{4.3.39}$$

$$\left. - \begin{vmatrix} y(\omega,a) & y_1(a) \\ y'(\omega,a) & y_1'(a) \end{vmatrix} \begin{vmatrix} V_a(\omega,b) & p(\omega,b) \\ [V_a(\omega,b)]' & p'(\omega,b) \end{vmatrix} \right\}^{-1} 2\mathrm{i}\omega(y_1y_2' - y_2y_1') \, .$$

The primes indicate derivatives with respect to the variable x. Of course, the scattering coefficients do not depend on the choice of the points a and b in the regions $\Omega_1 \cap \Omega_2$ and $\Omega_2 \cap \Omega_3$. In the asymptotic construction of $T(\omega)$ according to (4.3.39), $y(\omega,a)$ is constructed according to (4.3.33), $V_a(\omega,b)$ and $W_a(\omega,b)$ according to (4.3.35), and $p(\omega,b)$ according to (4.3.38). We note that in the region $\Omega_1 \cap \Omega_2$ the functions $V_a(\omega,b)$, $W_a(\omega,b)$, and $p(\omega,b)$ are analytic in ω.

We shall now show that for $a \gg m$, $a\omega \ll 1$ the function $y(\omega,a)$ can be represented in the form

$$y(\omega,a) = [1 + \varphi(\omega)]\omega^{l+1}X(\omega,a) + \mathrm{i}[1 + \tilde{\varphi}(\omega)]\omega^{-l}\tilde{X}(\omega,a) \quad , \tag{4.3.40}$$

where $X(\omega,a)$ and $\tilde{X}(\omega,a)$ are formal expansions in powers of ω^2, and $\varphi(\omega)$ and $\tilde{\varphi}(\omega)$ are functions which have a logarithmic branch point at $\omega = 0$. We use for this purpose the asymptotic expansions (4.3.33). For $Z_0(\omega,x)$ with $x\omega \ll 1$, we have

$$Z_0(\omega,x) = \omega^{l+1}X_0(\omega,x) + \mathrm{i}\omega^{-l}\tilde{X}_0(\omega,x) \quad ,$$

where

$$X_0(\omega,x) = \mathrm{e}^{\mathrm{i}\pi(l+1)/2}\sqrt{\pi}\left(\frac{x}{2}\right)^{l+1} \sum_{k=0}^{\infty} \frac{(-1)^k(\omega x)^{2k}}{2^{2k}\Gamma(l+k+1/2)k!} \quad ,$$

$$\tilde{X}_0(\omega,x) = \mathrm{e}^{\mathrm{i}\pi(l+1)/2}\sqrt{\pi}\left(\frac{x}{2}\right)^{-l} \sum_{k=0}^{\infty} \frac{(-1)^k(\omega x)^{2k}}{2^{2k}\Gamma(k-l-1/2)k!} \quad .$$

The part of $Z_1(\omega)$ nonanalytic in ω can be contained only in the integral

$$(2i\omega)^{-1}\int\limits_x^\infty H^2(x')\tilde{U}(x')\,dx' \ . \qquad (4.3.41)$$

This integral consists of terms of the form

$$\text{const}\int\limits_x^\infty (x')^{-s}\ln^k(x')\exp(2i\omega x')\,dx' \ ,$$

where $s \geq 3$. By means of an integration by parts, assuming that $\mathrm{Im}\{\omega\} > 0$, these integrals can be reduced to a sum of terms of the form R^{ms} with

$$\text{const}\cdot R^{ms} \equiv (i\omega)^s\int\limits_x^\infty \ln^m(x')\exp(2i\omega x')\,dx' \quad (m = 1, \ldots, k+1)$$

and terms analytic in ω. However, the part of this integral which is nonanalytic in ω does not depend on x and has the value

$$\sum_{n=1}^{m}[\ln(-i\omega)]^n C_n \ ,$$

$$C_n = (i\omega)^s(-1)^{m-n}\int\limits_0^\infty e^{-2\mu}(\ln\mu)^{m-n}\,d\mu \ .$$

In fact, we represent the integral R^{ms} from x to ∞ as the difference of the integrals from 0 to ∞ and from 0 to x. The first of these integrals is equal to the expression written above (for the nonanalytic part). In the second integral, we may make a series expansion of $\exp(2i\omega x')$ and integrate term by term, and it follows from this that it is analytic in ω.

According to (4.3.33), the part of $Z_1(\omega, x)$ which is analytic in ω is given by

$$\frac{iH(x)}{2\omega}\int\limits_x^\infty H(x')H^*(x')\tilde{U}(x')\,dx' + \text{a.p.}\left\{\frac{iH^*(x)}{2\omega}\int\limits_\infty^x H^2(x')\tilde{U}(x')\,dx'\right\} \ ,$$

where a.p.$\{f(x,\omega)\}$ denotes the part of the function $f(x,\omega)$ which is analytic in ω. As a result, we obtain

$$Z_1(\omega, x) = \omega^{l+1}X_1 + i\omega^{-l}\tilde{X}_1 + (\omega^{l+1}X_0 - i\omega^{-l}\tilde{X}_0)\varphi_0(\omega) \ ,$$

where $\varphi_0(\omega)$ is the part of the expression (4.3.41) which is nonanalytic in ω. Repeating this process, it can be shown that $Z_k(\omega, x)$ consists of terms of the following form:

$$Z_k(\omega, x) = \omega^{l+1}[X_k + \varphi_0(\omega)X_{k-1} + \ldots + \varphi_k(\omega)X_0] + \mathrm{i}\omega^{-l}[\tilde{X}_k + \tilde{\varphi}_0(\omega)\tilde{X}_{k-1} + \ldots + \tilde{\varphi}_k(\omega)\tilde{X}_0] \ .$$

Here X_i and $\tilde{X}_i$ are formal series in powers of ω^2 for $x\omega \ll 1$, $x \gg m$, and, since the perturbing potential $\tilde{U}$ is small, we have

$$|X_i| \ll |X_{i-1}| \ , \quad |\varphi_i| \ll |\varphi_{i-1}| \ , \quad |\tilde{X}_i| \ll |\tilde{X}_{i-1}| \ , \quad |\tilde{\varphi}_i| \ll |\tilde{\varphi}_{i-1}| \ .$$

Hence we find that

$$y(\omega, x) = \sum_{k=1}^{\infty} Z_k(\omega, x)$$

has the form (4.3.40), where

$$X = \sum_{k=0}^{\infty} X_k \ , \quad \tilde{X} = \sum_{k=0}^{\infty} \tilde{X}_k \ ,$$

$$\varphi(\omega) = \sum_{k=0}^{\infty} \varphi_k(\omega) \ , \quad \tilde{\varphi}(\omega) = \sum_{k=0}^{\infty} \tilde{\varphi}_k(\omega) \ ,$$

with

$$\varphi_0(\omega) = -\tilde{\varphi}_0(\omega) \approx 8\mathrm{i}l(l+1)m\mathcal{A}\omega \ln\omega \ ,$$

$$\mathcal{A} = \sum_{k,n=0}^{l} \frac{(l+k)!(l+n)!(-1)^{k+n+1}}{k!n!(l-k)!(l-n)!(k+n+2)!} \sum_{p=1}^{k+n+2} \frac{1}{p} \ .$$

In fact, the principal terms of the nonanalytic part of (4.3.41) have the form

$$8\mathcal{B}ml(l+1)\mathrm{i}\omega \ln^2\omega - 2\mathrm{i}\omega\ln\omega\left\{\left[(8l+1)l\int_0^{\infty} \mathrm{e}^{-2\mu}\ln\mu\, d\mu - 2m - C_{\pm}\right]\mathcal{B} - 4\mathcal{A}m(l+1)l\right\} ,$$

where the value of $\mathcal{A}$ is given above, and $\mathcal{B}$ is given by the expression

$$\mathcal{B} = \sum_{k,n=0}^{l} \frac{(l+k)!(l+n)!(-1)^{k+n}}{k!(l-k)!n!(l-n)!(k+n+2)!} \ .$$

By means of cumbersome manipulations it can be shown that the expression for $\mathcal{B}$ is identically equal to zero.

If $q^2 = m^2$, the relation (4.3.31) in the region Ω_1 reduces to Bessel's equation, with the general solution

$$\chi_{\pm}(\omega, x) = \sqrt{x}\left\{C_1 H^{(1)}_{\nu\pm}(\omega x) + C_2 H^{(2)}_{\nu\pm}(\omega x)\right\} \ ,$$

where $\nu^+ = l - 1/2$, $\nu^- = l + 3/2$. In this case, the perturbing potential for $x \to -\infty$ has the form

$$\sum_{n=1}^{\infty} x^{-n-2} \varphi_n(\ln |x|) \ .$$

For $|q| \le m$, the transmission coefficient $T(\omega)$ has a branch point at $\omega = 0$ because of the logarithmic power decrease of the perturbing potential at $+\infty$. For $|q| = m$, an additional nonanalyticity in $T(\omega)$ comes from the region $-x \gg m$.

To find the asymptotic expressions for the transmission coefficients $T(\omega)$, it remains to find the static solutions $y_1(x)$ and $y_2(x)$ of (4.3.30) (for $\omega = 0$). These solutions for odd perturbations can be expressed in terms of hypergeometric functions in two limiting cases:

$$1)\ q^2 \ll m^2 \ , \quad 2)\ \alpha^2 \ll 1 \ , \quad \text{where} \quad \alpha^2 \equiv 1 - q^2/m^2 \ .$$

The expression $1 - q^2/m^2$ is non-negative because of the assumption that the horizon is nondegenerate. Finally, from (4.3.33, 35, 39), using the asymptotic behavior of the Bessel, Gauss, and Legendre functions, we obtain the following asymptotic expressions for the transmission coefficients $T(\omega)$ of the odd perturbations: for $q^2 \ll m^2$

$$T_\pm(\omega) \approx (\omega m)^{l+1} \frac{\Gamma(l+2+C_\pm/2m)\Gamma(l - C_\pm/(2m))\sqrt{\pi}\exp[-\mathrm{i}\pi(l+1)/2]}{2^{l+1}(2l)!\Gamma(l+3/2)[1-\varphi_0(\omega,l)]} \tag{4.3.42}$$

for $\alpha^2 \ll 1$

$$T_\pm(\omega) \approx (\omega m)^{l+1} \alpha^{l\mp1} \frac{(2l+3\mp1)\pi\Gamma(l+1/2\mp1/2)\exp[-\mathrm{i}\pi(l+1)/2]}{(2l-1\mp1)2^{2l+1\mp1}[\Gamma(l+1\mp1/2)]^2[1-\varphi_0(\omega,l)]} \ , \tag{4.3.43}$$

and for $q^2 = m^2$

$$T_\pm(\omega) \approx -(\omega m)^{2l+1\mp1} \frac{(2l+3\mp1)}{(2l-1\mp1)} \frac{\pi[1+\varphi_0(\omega,l)+\varphi_0(\omega,l\mp1)]}{2^{2l}|\Gamma(l+1\mp1/2)|^2} \ . \tag{4.3.44}$$

4.3.5 Laws of Attenuation of the "Tails" of the Multipole Radiation

As we demonstrated in Sect. 4.3.3, the asymptotic behavior of the radiation emitted from under the potential barrier is given by the expression

$$\int_C a(\omega)T(\omega)\exp(-\mathrm{i}\omega t)y(\omega,x)\,d\omega \ , \tag{4.3.45}$$

where C is a contour along the edges of the cut on the negative part of the imaginary axis (see Fig. 4.1). For $t \gg x$, the main contribution to (4.3.45) comes from low frequencies $\omega \lesssim 1/t$, so that for $y(\omega, x)$ it is necessary to use precisely the asymptotic form (4.3.40) for $\omega x \ll 1$. The principal term of the nonanalytic part in (4.3.45) has the form

$$2\varphi_0(\omega)y_1(x)T_0\omega^{2l+2}$$

for $|q| < m$, where for $T(\omega)$ we have used the asymptotic behavior (4.3.42, 43):

$$T(\omega) \approx T_0\omega^{l+1}[1 + \varphi_0(\omega)] \ .$$

The analytic part of $a(\omega)T(\omega)y(\omega, x)$ gives zero when the integration is made around the contour C. We shall take into account the fact that, according to (4.3.28), the function $a(\omega)$ has a simple pole at $\omega = 0$, being single-valued in the neighborhood of this point. Finally, when $q^2 < m^2$, for the attenuation of the perturbations in the case $t \gg r^* \equiv x$ we find, using (4.3.45), the law

$$\chi_\pm(t, x) \approx \frac{1}{t^{2l+3}} y_1(x)\psi_0 f_\pm(q, m) \ , \tag{4.3.46}$$

where

$$\begin{array}{llll} f_\pm(q, m) \sim 1 & \text{for} & q^2 \ll m^2 \ , & l \sim 2 \ , \\ f_\pm(q, m) \sim \alpha^{l\mp 1} & \text{for} & \alpha^2 \ll 1 \ , & l \sim 2 \ . \end{array}$$

When $q^2 = m^2$, the attenuation of the perturbations acquires a qualitatively different character:

$$\chi_\pm(t, x) \sim \frac{\psi_0}{t^{l+2\mp 1}} y_2(x) \ . \tag{4.3.47}$$

Finally, we note that the laws of attenuation (4.3.46, 47) are established not directly, but through a certain intermediate asymptotic behavior, in the region beyond the wave front $t = r^*$: $t - r^* \lesssim r^*$. For the function $y(\omega, x)$ in (4.3.45), we must take the asymptotic form in the wave zone for $x \gg m$, i.e., $\exp(i\omega x)$. In the intermediate region, it is the wave zone for monochromatic waves that forms the structure of the radiation immediately after crossing the forward front. It follows then from (4.3.42–44, 45) that for $q^2 \ll m^2$ we have

$$\chi_\pm(t, x) \sim \frac{1}{(t - x)^{l+2}} f_\pm(q, m) \ . \tag{4.3.48}$$

For $q^2 = m^2$, we obtain

$$\chi_\pm(t, x) \sim \frac{1}{(t - x)^{2l+2\mp 1}} \ . \tag{4.3.49}$$

All the expressions (4.3.46–49) include, as a proportionality factor, the quantity $\psi_0 \equiv \psi(\tau_1)$, which determines the radiation of the body at the proper time at which its boundary crosses the event horizon.

From (4.3.46–49) we deduce the interesting fact that for $\alpha^2 \ll 1$ ($q^2 \approx m^2$) the component χ_- in the radiation which has passed through the barrier can be neglected in comparison with χ_+, since χ_- is attenuated more rapidly. This indicates that in the tails of the radiation for $\alpha^2 \ll 1$ the ratio of the amplitudes of the electromagnetic and gravitational waves tends to a fixed value, which is independent of their initial ratio near the horizon:

$$\frac{h}{\Phi} \approx \frac{2\theta}{\psi} \approx \frac{2m}{l-1}$$

[see (4.3.18, 24)].

Our results concerning the attenuation of the multipole perturbations are valid only for $l \geq 2$. For $l = 1$, the linearized Einstein-Maxwell system admits a stationary solution without attenuation for a slow rotation, corresponding to the Kerr-Newman solution in the case of a small angular momentum. From (4.3.17), on the basis of the results of Sects. 4.3.2–5, we can deduce that there is attenuation of the odd dipole electromagnetic radiation: $\Phi \to 0$ for $t \to \infty$ according to the law (4.3.46) or (4.3.47), so that there remains only the stationary part of δH, in terms of which it is possible to express the other components of the electromagnetic field tensor and the components of the metric of the Kerr-Newman solution, linearized in the angular momentum. For $l = 1$, there is also attenuation of the even perturbations.

5. Relativistic Hydrodynamics

In this chapter, we consider problems of the wave dynamics of a relativistic gas. Relativistic hydrodynamics is distinguished from Newtonian hydrodynamics in the following respects: (a) it gives a correct description of the motion of particles of a fluid having speeds comparable with the speed of light; (b) it uses, of necessity, equations of state which are qualitatively different from those of the Newtonian theory, either in regions of high densities (at low temperatures) or at very high temperatures.

In the first two sections of this chapter, which have an introductory character, we make a qualitative analysis of the evolution of the composition of matter when its density increases up to the nuclear density in the case of relatively low temperatures, and we note the difficulties in obtaining the corresponding equations of state. In the case of high temperatures, it is pointed out that matter is described asymptotically by the so-called *ultrarelativistic equation of state*.

In Sect. 5.3 we consider adiabatic motions of an ultrarelativistic gas. In the particular case of isentropic motions, it is shown that the vortices are frozen in the fluid particles and that there exists a class of potential motions (when a curvilinear shock wave occurs, the isentropic and potential character of the flow is, in general, destroyed). In Sect. 5.3.3 we derive equations of nonlinear acoustics for short sound waves propagating in an arbitrary potential flow of a relativistic gas. Using model equations obtained here, an analysis is made of the attenuation of weak spherical, cylindrical, and plane shock waves in the case of the Friedman-Lemaître cosmological model for various equations of state of the matter. In Sect. 5.4 we present the basic principles of relativistic magnetohydrodynamics. In Sect. 5.5 we analyze self-similar motions of an ultrarelativistic gas with spherical and cylindrical symmetries in the framework of the special theory of relativity.

In Sect. 5.6 we describe the stationary flow of an ultrarelativistic gas in the Schwarzschild field, resulting from the production of particles near the event horizon.

5.1 Relativistic Dynamics of a Point and Gas of Free Particles

A material point on which forces act describes, in its motion in the manifold M_4, a curve called the *world line* of the particle. For arbitrary motion of the particle, one can associate with it a natural orthonormalized tetrad, one of whose vectors coincides with the unit vector $\boldsymbol{u}$ tangential to the world line and having components dx/ds, dy/ds, dz/ds, dt/ds. The vector $\boldsymbol{u}$ is called the *4-velocity* of the particle. A second vector coincides in direction with the derivative of the vector $\boldsymbol{u}$ with respect to the canonical parameter s, equal to the so-called *4-acceleration* of the particle; the 4-acceleration of a particle characterizes the presence of forces which deflect the motion of the particle from rectilinear and uniform motion. By definition, an increment of the canonical parameter is equal to the interval between two neighboring points on the world line. It is proportional to the proper time of the particle, $d\tau = ds/c$.

It is natural to reformulate Newton's second law in the theory of relativity in the form $d(mc\boldsymbol{u})/d\tau = \boldsymbol{F}$. The constant m characterizes the inertial properties of the particle and is called its *rest mass*. If m = const, the *force 4-vector* $\boldsymbol{F}$ must be orthogonal to the 4-velocity $\boldsymbol{u}$ in the sense of the geometry of pseudo-Euclidean space. The vector $mc\boldsymbol{u}$, proportional to the 4-velocity, is called the *4-momentum* of the particle.

Kinetic Description of a Gas of Free Particles. We shall now consider a system of *many* particles[1] and introduce the particle distribution function. We consider at the point x the set T_x of vectors tangential to all possible world lines, directed to the future, and passing through the point x. In the tangent space T_x at the point x, we identify the vectors obtained from one another by multiplication by a positive number.

Consider two world lines passing through a point x, and suppose that on each of them there passes an infinitesimal time interval $d\tau$. After the time $d\tau$, these curves are separated by an interval $ds^2 = (\boldsymbol{u}_1 - \boldsymbol{u}_2)^2 c^2 d\tau^2$. In the tangent space T_x of unit vectors in a locally Lorentzian coordinate system, the natural interval $ds^2 = \eta_{ij} du^i du^j$ is induced, and the space T_x becomes a Riemannian space. Here $\eta_{\alpha\beta} = -\delta_{\alpha\beta}$, where $\delta_{\alpha\beta}$ is the Kronecker delta symbol, and $\alpha, \beta = 1, 2, 3$, $\eta_{\alpha 4} = 0$, $\eta_{44} = c^2$. The geometry of this manifold is the same as the geometry of a pseudosphere (or of Lobachevsky space). If for the coordinates we choose the first three components of the 4-velocity vector in a Cartesian coordinate system, the metric of the three-dimensional pseudosphere will have the form

$$ds_T^2 = \left(\delta_{\alpha\beta} - \frac{u_\alpha u_\beta}{1 + (u^1)^2 + (u^2)^2 + (u^3)^2}\right) du^\alpha du^\beta \quad ,$$

[1] The reader can become acquainted with the problems of describing a system of interacting particles in the theory of relativity from the monograph of [5.1], for example.

where the indices α and β take the values[2] 1, 2, 3, and the volume element in this space has the form

$$du = \frac{du^1 du^2 du^3}{(1 + (u^1)^2 + (u^2)^2 + (u^3)^2)^{1/2}} \ .$$

The invariant volume element on the pseudosphere in the tangent space at the point x is given by $du = \sqrt{-g}\, du^1 du^2 du^3 / u_4$, where g is the value of the determinant of the metric tensor at the point x.

Fourier integral expansions of arbitrary functions in Lobachevsky space in the form of superpositions of spherical (or cylindrical) waves, and also expansions in terms of different systems of eigenfunctions, were obtained in [5.2, 3].

Suppose that the world lines of particles cross a hyperarea dV with projections onto the coordinate hyperplanes equal to ($dx^2 dx^3 dt$, $dx^3 dt\, dx^1$, $dt\, dx^1 dx^2$, $dx^1 dx^2 dx^3$). Then the number dN of particles crossing the hyperarea with velocity components in the intervals $[u^1, u^1 + du^1;\ u^2, u^2 + du^2;\ u^3, u^3 + du^3]$ can be written in the form

$$dN = (n_i u^i) f(u, x)\, dV du \ , \tag{5.1.1}$$

where n_i are the components of the 4-normal to the area, determined by $n_1 dV \equiv dx^2 dx^3 dt, \ldots, n_4 dV \equiv dx^1 dx^2 dx^3$, and $f(u, x)$ has the meaning of the distribution function of the particles, given on the set of tangent spaces at all the points of the manifold M_4. The total number of particles crossing the hyperarea dV is given by the expression

$$dN = dV\, n_i \int_{-\infty}^{+\infty} \int_{-\infty}^{+\infty} \int_{-\infty}^{+\infty} u^i f(u, x)\, du^1 du^2 du^3 \sqrt{-g} / u_4 \ . \tag{5.1.2}$$

Consider an elementary volume $dx^1 dx^2 dx^3 dt$ at a point x. Suppose that at this point all the Christoffel symbols vanish inside this volume. If the particle world lines do not terminate in this volume, the total flux of particles through its surface must be equal to zero, and by Gauss's theorem we obtain the equation of continuity[3] $\partial N^i / \partial x^i = 0$, where

$$N^i = \int_{-\infty}^{+\infty} \int_{-\infty}^{+\infty} \int_{-\infty}^{+\infty} u^i f(u, x)\, du^1 du^2 du^3 / u_4 \ . \tag{5.1.3}$$

When they cross the hyperarea, the particles carry energy and momentum. The 4-momentum of a particle is defined as the vector $mc\boldsymbol{u}$. In the nonrelativistic approximation, the first three components of the 4-momentum form the momentum 3-vector $m\boldsymbol{v}$, and the fourth physical component of the 4-momentum

---

[2] For what follows, we adopt the convention that Greek indices take the values 1, 2, 3, and Latin indices take the values 1, 2, 3, 4.

[3] In an arbitrary coordinate system, this equation can be rewritten in the form $\nabla_i N^i = 0$.

has the value $(mc^2 + mv^2/2)/c$, i.e., it is proportional to the sum of the kinetic energy and the *rest energy* of the particle.

In elastic collisions, the total 4-momentum of the particles, determined by the motion of their center of mass, is conserved: $m_1 \boldsymbol{u}_1' + m_2 \boldsymbol{u}_2' = m_1 \boldsymbol{u}_1'' + m_2 \boldsymbol{u}_2''$. The flux of 4-momentum through the area dV is $cm\boldsymbol{u}\, dN$, where dN is determined by (5.1.2).

The ith component of the flux of 4-momentum through the hyperarea dV is $n_j T_i^j \, dV$, where

$$T_{ij} = m \int u_i u_j f(u, x)\, du \quad . \tag{5.1.4}$$

The tensor T_{ij} which we have introduced in this way is called the *energy-momentum tensor* of the relativistic gas. We note that the transition from the discrete particle-number function to a continuous distribution function implies a transition to a continuous medium moving in the phase space [5.4].

If no external forces act on the particles, and their momentum is changed only in collisions, then the total flux of energy and momentum through the surface of the elementary 4-volume $dx^1 dx^2 dx^3 dt$ is equal to zero, and we obtain the equations of conservation of energy and momentum[4]:

$$\frac{\partial}{\partial x^i} T^{ij} = 0 \quad .$$

The average macroscopic velocity in the gas can be defined in two ways: either as an eigenvector of the tensor T^{ij} [5.5] or as the normalized particle flux vector N_i (in this case, the normalization factor obviously has the meaning of the particle number density [5.6]).

5.2 Thermodynamic Equilibrium in an Ideal Gas

Owing to collisions of the particles in the gas, thermodynamic equilibrium is established in each volume whose dimensions are much greater than the mean free path of the particles. The state of the gas in thermodynamic equilibrium is determined by the following macroscopic parameters: the velocity 4-vector u^i, the temperature T, and the pressure p.

The analysis which follows is carried out in a coordinate system in which the given small volume V is at rest. We adopt the notation Φ for the value of the thermodynamic potential $\Phi \equiv E + pV - TS$ of the volume of gas V (E is its internal energy, and S is the entropy), and the notation μ for the value of Φ per particle, $\mu = \Phi/N$, where N is the (variable) number of particles in the volume V.

4 In an arbitrary coordinate system, these equations can be rewritten in the form $\nabla_i T^{ij} = 0$.

The equilibrium state of the gas can be determined by specifying the temperature T and the chemical potential μ.

In order to find the equation of state of the gas, we must use the formula of the Gibbs distribution for a system with a variable number of particles. According to the general Gibbs formula, the thermodynamic function pV is given by the expression

$$pV = -kT \sum_i \ln \sum_k \exp \left[\left(\frac{\mu - e_i}{kT} \right) n_{ik} \right] , \tag{5.2.1}$$

where e_i and n_{ik} are, respectively, the energy and occupation numbers of the state i.

According to the Pauli principle, for particles with half-integral spin the occupation numbers of each state can take the values 0 and 1. Therefore from (5.2.1) we have in this case

$$pV = -kT \sum_i \ln \left[1 + \exp \left(\frac{\mu - e_i}{kT} \right) \right] . \tag{5.2.2}$$

For a gas of Bose particles (particles with integer spin), the occupation numbers n_{ik} are not restricted in any way and can have arbitrary values $(0, 1, 2, \ldots)$. Summing the geometric progression in the argument of the logarithmic function in (5.2.1), with the condition that the chemical potential is negative, we obtain

$$pV = kT \sum_i \ln \left[1 - \exp \left(\frac{\mu - e_i}{kT} \right) \right] . \tag{5.2.3}$$

The number of particles N in the system is related to the function $\Omega \equiv pV$ by the equation

$$N = \left(\frac{\partial \Omega}{\partial \mu} \right)_{T,V} . \tag{5.2.4}$$

By "particles"we shall mean those having only translational degrees of freedom. The quantum state of a particle is determined, for a given value of its momentum, by the direction of its spin and by its charge state (we neglect completely the interaction between the particles). Therefore, for a given e_i in the sums (5.2.2, 3), there are s identical terms. For electrons, $s = 2$ (two spin directions); for nucleons, $s = 4$ (two spin directions and two charge states); and for pions, $s = 3$, since there are three states π^+, π^-, π^0.

The energy of a free particle is related to its momentum $\boldsymbol{P}$ and its rest mass m by the equation

$$\frac{e^2}{c^2} = \boldsymbol{P}^2 + m^2 c^2 . \tag{5.2.5}$$

In a small volume V of a continuous medium, the number of particles N is, by definition, quite large. Using the law of large numbers, we can convert the summations in the distributions (5.2.2, 3) to integrations (this replacement is not valid for a degenerate Bose gas, in which there is an accumulation of particles in the state $e = 0$).

According to quantum statistics, the number of quantum states of the translational motion of the particles with momentum components in the intervals $[P_x, P_x + dP_x]$, $[P_y, P_y + dP_y]$, $[P_z, P_z + dP_z]$ in a stationary volume V is $V dP_x dP_y dP_z/(2\pi\hbar)^3$ ($\hbar$ is Planck's constant). Let n be the number of particles per unit volume: $n = N/V$. Then from (5.2.2–5) we have, in a locally Lorentzian coordinate system,

$$
\begin{aligned}
n &= \frac{s}{(2\pi\hbar)^3}\int \frac{dP_x dP_y dP_z}{\exp\left((e-\mu)/kT\right)\pm 1} \ , \\
\varepsilon &= \frac{s}{(2\pi\hbar)^3}\int \frac{e\, dP_x dP_y dP_z}{\exp\left((e-\mu)/kT\right)\pm 1} \ ,
\end{aligned} \tag{5.2.6}
$$

$$
\begin{aligned}
p &= \frac{s}{(2\pi\hbar)^3 3m}\int \frac{\boldsymbol{P}^2 dP_x dP_y dP_z}{\left[\exp\left((e-\mu)/kT\right)\pm 1\right]\sqrt{1+P^2(mc)^{-2}}} \ , \\
\boldsymbol{P}^2 &= P_x^2 + P_y^2 + P_z^2 \ .
\end{aligned} \tag{5.2.7}
$$

Here and in what follows, the upper and lower signs correspond to Fermi and Bose statistics, respectively.

For macroscopic motions of a gas in local equilibrium, the particle-flux 4-vector is determine by (5.1.3), and the energy-momentum tensor by (5.1.4), where the invariant distribution function is given by

$$
\begin{aligned}
&f(\boldsymbol{P}, T(x^k), \mu(x^k), u^i(x^k)) \\
&\quad = s\left(\frac{mc}{2\pi\hbar}\right)^3\left[\exp\left(-\frac{\mu}{kT}+\frac{c}{kT}P_i u^i\right)\pm 1\right]^{-1} \ .
\end{aligned} \tag{5.2.8}
$$

Cold Fermi Gas. At low temperatures, $kT/(mc^2) \to 0$, we have quantum degeneracy of the gas. For a Fermi gas, the distribution function tends to a step function:

$$
f(\boldsymbol{P}) = s\left(\frac{mc}{2\pi\hbar}\right)^3\left[\exp\left(\frac{e-\mu}{kT}\right)+1\right]^{-1} \to \begin{cases} s\left(\frac{mc}{2\pi\hbar}\right)^3 & \text{for } e<\mu \ , \\ 0 & \text{for } e>\mu \ . \end{cases}
$$

Therefore the chemical potential μ for a degenerate Fermi gas is equal to the limiting value of the energy e, called the *limiting Fermi energy* e_F. In this case, (5.2.6, 7) give, after integration,

$$
\begin{aligned}
n &= \frac{s}{(2\pi\hbar)^3} \int\limits_{e<\mu} dP_x dP_y dP_z = \frac{s(mc)^3}{(2\pi\hbar)^3} \frac{4\pi}{3} (\xi^2 - 1)^{3/2} \ , \\
p &= \frac{s}{3(2\pi\hbar)^3 m} \int\limits_{e<\mu} \frac{\boldsymbol{P}^2 dP_x dP_y dP_z}{\sqrt{1 + \boldsymbol{P}^2/m^2c^2}} \\
&= \frac{sm^4c^5}{(2\pi\hbar)^3} \frac{\pi}{6} \left[2(\xi^2 - 1)^{3/2}\xi - 3(\xi^2 - 1)^{1/2}\xi + 3\ln(\xi + \sqrt{\xi^2 - 1}) \right] \ , \\
\varepsilon &= \frac{sm^4c^5}{(2\pi\hbar)^3} \pi \left[(\xi^2 - 1)^{1/2}\xi^3 - \frac{1}{2}(\xi^2 - 1)^{1/2}\xi - \frac{1}{2}\ln(\xi + \sqrt{\xi^2 - 1}) \right] \ ,
\end{aligned}
\tag{5.2.9}
$$

where $\xi \equiv \mu/mc^2 = e_F/mc^2$. The expressions (5.2.9) become particularly simple in the ultrarelativistic limit, when the limiting Fermi energy is much greater than mc^2: $\xi \gg 1$; in this case, from (5.2.9) we obtain

$$
n \approx \frac{s(mc)^3}{6\pi^2\hbar^3} \left(\frac{e_F}{mc^2} \right)^3 \ , \quad p \approx \frac{sm^4c^5}{24\pi^2\hbar^3} \left(\frac{e_F}{mc^2} \right)^4 \ , \quad \varepsilon \approx 3p \ .
$$

Eliminating e_F from these relations, we have

$$
\frac{\varepsilon}{3} = p = \frac{1}{4} \left(\frac{6\pi^2}{s} \right)^{1/3} \hbar c n^{4/3} \ . \tag{5.2.10}
$$

In the nonrelativistic case, $|e_F - mc^2| \ll mc^2$, i.e., $|\xi - 1| \ll 1$, and from (5.2.9) we readily obtain

$$
\begin{aligned}
n &\approx s \frac{(mc)^3}{(2\pi\hbar)^3} \frac{4\pi}{3} 2^{3/2} (\xi - 1)^{3/2} \ , \\
p &\approx \frac{1}{5} \left(\frac{6\pi^2}{s} \right)^{2/3} \frac{\hbar^2}{m} n^{5/3} \ , \\
\varepsilon &\approx mnc^2 + 3p/2 \ .
\end{aligned}
\tag{5.2.11}
$$

It follows from (5.2.9) that the limiting Fermi energy for a cold gas is related to the density of the gas by the equation

$$
\frac{e_F}{mc^2} = \sqrt{1 + \left(\frac{2\pi\hbar}{mc} \right)^2 \left(\frac{3n}{4\pi s} \right)^{2/3}} \ . \tag{5.2.12}
$$

The ratio of the limiting Fermi energy (minus the rest energy) and Boltzmann's constant is called the *degeneracy temperature*: $e_F - mc^2 = kT_0$. A Fermi gas at temperatures $T \gg T_0$ much higher than the degeneracy temperature has a Maxwell distribution, while at temperatures $T < T_0$ the gas is degenerate. We note that the temperature of a degenerate gas can be relatively high at high densities.

In a cold gas, the condition for the transition to the ultrarelativistic equation of state $\varepsilon = 3p$ is the condition $e_F \gg mc^2$. According to (5.2.12), this is equivalent to the condition that the average distance between the particles is much less than their Compton wavelength $\lambda_C \equiv \hbar/(m_e c)$ (for an electron the Compton wavelength is 3.8×10^{-11} cm, and for a neutron it is 2.10×10^{-14} cm).

Formation of a Gas of Free Electrons. Owing to the compression in stars which have exhausted their supplies of thermonuclear energy, the density increases. Therefore it becomes possible for electrons to escape from one atom to another (the tunneling effect). The matter becomes ionized. The tunneling of electrons is particularly effective when the average distances between the atoms, $l \sim n^{-1/3}$, become comparable with the radius of the lowest energy level for the electrons in an atom (the radius of the K shell), i.e. , for $n^{-1/3} \sim \hbar^2/m_e e^2 Z$, where m_e is the electron mass and Z is the number of protons in the nucleus. Therefore the density of particles required for complete ionization is of order

$$n \sim (m_e e^2 Z/\hbar^2)^3 \quad .$$

The corresponding mass density is of order

$$\varrho \approx A m_n n \sim 10 Z^4 \, \text{g/cm}^3 \quad ,$$

where A is the number of nucleons in a nucleus and m_n is the neutron mass.

At sufficiently high densities, the bare nuclei of the atoms find themselves in the environment of a degenerate electron gas (we have the "electron-nucleus phase of matter").

Neutronization of Matter. The condition of neutrality of matter reduces to the condition of equality of the numbers of protons and electrons per unit volume. For matter with mass densities $\varrho \gtrsim 10^7 \, \text{g/cm}^3$ and for a limiting Fermi energy $e_F > 1$ MeV (1 MeV, i.e., one million electron volts, is equal to 1.783×10^{-27} g), neutronization of matter sets in, since the nuclei become unstable with respect to the process of inverse β decay, in which one of the protons in a nucleus absorbs an electron from the electron gas surrounding the nucleus, emitting an electronic neutrino ν_e:

$$(A, Z) + e \rightarrow (A, Z-1) + \nu_e \quad .$$

The limiting Fermi energy e_F of the electrons in the electron gas, corresponding to this reaction, is

$$e_F = (M(A, Z-1) - M(A, Z))c^2 \quad .$$

Here $M(A, Z)$ is the mass of the nucleus with Z protons and A nucleons. We can calculate the threshold values of e_F [and the density of particles related to e_F by (5.2.12)] on the basis of the nuclear mass formula

$$M(A, Z) = (A - Z)m_n + Z m_p + E_b/c^2 \quad , \tag{5.2.13}$$

where m_p is the proton mass and the binding energy E_b is given by the semiempirical formula [5.7, 8]

$$E_b = f(A, Z) \quad . \tag{5.2.13'}$$

Formation of a Gas of Free Neutrons. When the limiting Fermi energy e_F of the electrons exceeds the binding energy E_b of a nucleus, the electron-nucleus plasma must contain free neutrons as well. Thermodynamic equilibrium between different components of the matter is maintained through the reactions

$$(A, Z) + Ze \rightleftharpoons (A, n) + Z\nu_e \quad .$$

Besides processes of β decay, in cold but dense matter there can also occur so-called *picnonuclear reactions*, in which light nuclei fuse and form heavier nuclei. Therefore, in cold matter in which the neutrons form a superfluid, heavy nuclei forming a crystal lattice exist up to extremely high densities [5.9].

Qualitatively, the thermodynamic equilibrium can be described by means of the principle of the minimum of the internal energy. Let n be the number of nucleons per unit volume, and n_n be the number of free neutrons per unit volume. Then there will be $(n - n_n)/A$ nuclei per unit volume.

The energy density will consist of the energy of the nonrelativistic degenerate gas of neutrons [see (5.2.11)], the energy of the relativistic degenerate gas of electrons [see (5.2.10)], and the energy of the nuclei [5.10]:

$$\begin{aligned} \varepsilon = {} & M(A, Z)c^2(n - n_n)/A + n_n m_n c^2 + 3a^2 n_n^{5/3}/(10 m_n)c^2 \\ & + 3a[Z(n - n_n)/A]^{4/3}/4 \quad , \end{aligned} \tag{5.2.14}$$

where $a \equiv (3\pi^2)^{1/3}\hbar c$, and we have taken into account the condition that the plasma is electrically neutral:

$$Z(n - n_n) = A n_e \quad .$$

For a fixed number of nucleons n, the energy ε as a function of n_n, A, and Z has a minimum for the most stable nuclei. Therefore, to determine n_n, A, and Z for fixed n as functions of n, we have the three equations

$$\frac{\partial \varepsilon}{\partial A} = 0 \ , \quad \frac{\partial \varepsilon}{\partial Z} = 0 \ , \quad \frac{\partial \varepsilon}{\partial n_n} = 0 \quad . \tag{5.2.14'}$$

At the threshold for production of the electron-nucleus phase with free neutrons, the limiting Fermi energy of the electrons is equal to the binding energy E_b of the particles in the nucleus:

$$Z e_F = -E_b \quad .$$

We stress that at high densities, owing to the reaction of β decay, the presence of a gas of free protons is energetically forbidden.

Although the extrapolation of the empirical formula (5.2.13) to the case of nuclei in which the number of neutrons is much larger than the number of protons is not entirely accurate, the relations (5.2.14′) together with (5.2.13′, 14) nevertheless provide a means of gaining a qualitative idea of the rate of increase of the number of free neutrons with increasing density of the total number of nucleons.

At $\varrho \approx 2 \times 10^{13}$ g/cm^3, a continuous nuclear medium is formed, consisting mainly of free neutrons, with a small number of protons and electrons.

By means of scattering of high-energy electrons on heavy nuclei, the density distribution in these nuclei has been determined. It has been found [5.7] that the density at the center of the nucleus does not depend on the number of nucleons in the nucleus. Experiments have revealed an attraction of the nucleons in the nucleus and a strong repulsion of nucleons at distances of order 2×10^{-14} cm (at supernuclear densities). At the same time, the nuclear forces at very small distances cannot be determined from scattering experiments. It is doubtful whether they can be described at all by a static potential [5,7, 11].

So far, the only theoretical method which makes it possible to obtain equations of state of nuclear matter with equal numbers of protons and neutrons is the method of Brueckner and Goldstone [5.7, 12–14]. Semiempirical equations of state have also been given in the literature (see the references in [5.7, 10]). At supernuclear densities, neither the theory nor experiments give as yet reliable direct indications of how to construct equations of state.

Neglecting the strong interactions between the particles, the equilibrium in a cold mixture of various free fermions and bosons with allowance for all possible nuclear-reaction channels was considered by *Ambartsumyan* and *Saakyan* [5.10, 15] under the assumption that the heavy bosons which occur are condensed onto the lowest energy level $e_\pi = m_\pi c^2$.

Equilibrium in Hot Matter in the Relativistic Region. By "hot matter" we mean matter whose temperature is much greater than the degeneracy temperature [see the definition after (5.2.12)] at the density in question.

For particles whose number in the system is determined by the condition of thermal equilibrium, the chemical potential is equal to zero. Therefore the integrals in (5.2.6, 7) must be calculated with $\mu = 0$. For particles with nonzero mass, the energy and momentum are related by (5.2.5). Examples of a hot gas are a gas of pions (Bose particles) produced by collisions of superfast nucleons [5.16–19] or gases of photons (Bose particles) and neutrinos (Fermi particles) [in this case, the rest mass of the particles is zero, and calculations based on (5.2.7) are greatly simplified]. The masses of the π^-, π^+, and π^0 mesons differ on account of the contribution to the mass from the Coulomb energy of the charged pions. These mesons can therefore be regarded as three different states of the pions.

At high temperatures $kT > m_\pi c^2$, i.e., large collision energies, for a pion gas we obtain from (5.2.7) the following equation of state:

$$\varepsilon \approx \frac{\pi^2}{10} kT \left(\frac{kT}{\hbar c}\right)^3 \quad , \quad \varepsilon \approx 3p \ . \tag{5.2.15}$$

We note that in these expressions there is no dependence on the pion mass m_π.

In the case of a photon gas, we obtain for all T the well-known Stefan-Boltzmann formulas

$$\varepsilon = \frac{\pi^2}{15} kT \left(\frac{kT}{\hbar c}\right)^3 \quad , \quad \varepsilon = 3p \quad , \quad n = 0.244 \left(\frac{kT}{\hbar c}\right)^3 \ . \tag{5.2.16}$$

At high temperatures, matter contains electron-positron pairs, and the equilibrium distribution functions of the electrons and positrons are described by (5.2.8). The condition for equilibrium with respect to pair production has the form $\mu_+ + \mu_- = 0$, where μ_+ and μ_- are the chemical potentials of the positrons and electrons, respectively. The actual value of μ_+ is determined by the condition of electrical neutrality of the gas, $n_- - n_+ = \sum_k Z_k n_k$, where Z_k is the number of protons in the nucleus of type k, and n_k is the concentration of nuclei of type k. If Z_k, n_k, and T are specified, we can completely determine μ_+ and μ_-, and thereby determine the parameters in the distribution function (5.2.8). At temperatures $kT \gg m_e c^2$ the concentrations of positrons and electrons are practically the same, $(n_+ - n_-)/n_+ \ll 1$, and in a first approximation we have $\mu_+ = \mu_- = 0$. Then for a gas of electron-positron pairs we obtain from (5.2.6, 7) the equation of state

$$\varepsilon = \frac{7\pi^2}{60} kT \left(\frac{kT}{\hbar c}\right)^3 \quad , \quad \varepsilon = 3p \quad , \quad n = 0.366 \left(\frac{kT}{\hbar c}\right)^3 \ . \tag{5.2.17}$$

5.3 Relativistic Dynamics and Acoustics of an Ideal Gas

The equations of the relativistic dynamics of an ideal gas in the absence of external forces and heat exchange between the particles reduce to the five equations

$$\nabla_i \left[(p+\varepsilon) u^i u_j - p\delta^i_j\right] = 0 \quad , \quad j = 1,2,3,4 \quad , \quad \nabla_i (\varrho u^i) = 0 \ . \tag{5.3.1}$$

Here u_i denotes the macroscopic 4-velocity of a fluid particle, and ε is the internal energy per unit volume of the gas. These equations are written in covariant form, and they hold also in the Riemannian space determined by the equations of general relativity.

The continuous field of time-like vectors u_i determines the world lines of the fluid particles. We shall write $d\tau$ for a small interval of length on a world line of a fluid particle. Then the derivative of an arbitrary function φ with respect to the proper time of a particle (called the *total derivative* in the nonrelativistic limit) can be written as

$$cu^i\varphi_{,i} \equiv \frac{d\varphi}{d\tau} \ .$$

From the thermodynamic identity (S is the entropy, and w is the enthalpy per unit mass)

$$\delta w = \frac{\delta p}{\varrho} + T\,\delta S \ , \quad w \equiv \frac{(p+\varepsilon)}{\varrho} \ ,$$

it follows that

$$\nabla_i\, p = \varrho(\nabla_i\, w - T\nabla_i\, S) \ . \tag{5.3.2}$$

Using (5.3.2) and the equation of continuity, we rewrite (5.3.1) in the form [5.20]

$$u^i\Omega_{ij} = -T\nabla_j\, S \ , \quad \Omega_{ij} = -\Omega_{ji} = \nabla_i\,(wu_j) - \nabla_j\,(wu_i) \ . \tag{5.3.3}$$

Contracting (5.3.3) with the components u^j, we find that the motions of the gas determined by (5.3.1) are adiabatic: $dS/d\tau = 0$.

In the particular case of isentropic motions of the gas, when $p = p(\varrho)$, the determinant of the antisymmetric tensor Ω_{ij} is equal to zero, and in the comoving coordinate system it follows from (5.3.3) that

$$\Omega_{4\alpha} = \partial_4(w\hat{u}_\alpha) - \partial_\alpha(w\hat{u}_4) = 0 \ , \quad \alpha = 1,2,3 \ . \tag{5.3.4}$$

The components of the velocity in the comoving coordinate system can be expressed in terms of the metric coefficients as follows:

$$\hat{u}^\alpha = 0 \ , \quad \hat{u}^4 = \frac{1}{\sqrt{\hat{g}_{44}}} \ , \quad \hat{u}_\alpha = \frac{\hat{g}_{\alpha 4}}{\sqrt{\hat{g}_{44}}} \ , \quad \hat{u}_4 = \sqrt{\hat{g}_{44}} \ .$$

By differentiation, from (5.3.4) we readily obtain the equations

$$\partial_4\left[\partial_\beta(w\hat{u}_\alpha) - \partial_\alpha(wu_\beta)\right] = 0 \ , \tag{5.3.5}$$

from which we deduce the relativistic integral of freezing of the vortices (Thomson's theorem):

$$\partial_\beta(w\hat{u}_\alpha) - \partial_\alpha(w\hat{u}_\beta) = \Omega_{\alpha\beta} \ , \quad \partial_4\Omega_{\alpha\beta} = 0 \ .$$

In the Newtonian limit, the relations (5.3.5) also have the form (5.3.4), where $\hat{u}_\alpha$ are equal to the velocity components in a Lagrangian coordinate system, and the function w must be set equal to c^2.

5.3.1 Shock Waves and Vortex Motions of an Ideal Gas

According to (5.3.3), the motion of a gas cannot have a potential character if the entropy distribution varies with the particles.

If the flow of a gas involves a shock wave with variable intensity, then the flow beyond this wave has a variable entropy distribution in the particles and will have a vortex character, even if the flow had an isentropic and potential character before the break. This fact, which is well known in classical gas dynamics, is equally true in the relativistic case. The corresponding analysis is given below.

The geometry of the surface of discontinuity is characterized by the first and second quadratic forms. The first quadratic form is nondegenerate, since shock waves propagate with speeds less than the speed of light. The first quadratic form can be reduced at each point to the form $-(dx^2)^2 - (dx^3)^2 + (dx^4)^2$.

In the neighborhood of a shock wave, let us transform to a coordinate system constructed as follows. From each point of the shock wave we construct, along the normal to it, a space-like geodesic characterized by the parameter l which measures the arc length from the shock wave.

Let χ^α $(\alpha = 1,2,3)$ be the internal parameters of the surface of discontinuity. The square of the interval in some neighborhood of the discontinuity then takes the form

$$ds^2 = -dl^2 + g_{\alpha\beta} d\chi^\alpha d\chi^\beta \quad .$$

The coefficients of the first quadratic form of the surface of discontinuity are $g_{\alpha\beta}$ (at $l = 0$), and the coefficients of the second quadratic form $b_{\alpha\beta}$ are $(\partial g_{\alpha\beta}/\partial l)_{l=0}$.

The projection of the equation of motion (5.3.1) of an ideal gas onto the plane tangential to the shock wave in the system of coordinates χ^1, χ^2, χ^3, l at a fixed point has the form

$$j\partial(wu_\alpha)/\partial l + \varrho u^\beta \nabla_\beta (wu_\alpha) - \nabla_\alpha p = 0 \quad . \tag{5.3.6}$$

Here u_α is the component of the tangential part of the velocity, and the covariant differentiation is carried out by means of the three-dimensional metric $g_{\alpha\beta}|_{l=0}$, $j \equiv \varrho u_{(n)}$, $u_{(n)} = u_l = u_i n^i$, in which n is the normal to the shock wave.

The conditions on shock waves for a relativistic gas in an arbitrary coordinate system have the form

$$[(p+\varepsilon)u_{(n)}u_i - pn_i] = 0 \quad , \tag{5.3.7}$$

$$[j] = 0 \quad , \quad j \equiv \varrho u_{(n)} \quad . \tag{5.3.8}$$

The square brackets denote the difference of the corresponding quantities "before" and "after" the break.

Writing the projections of (5.3.7) onto the normal and onto the plane tangential to the surface of discontinuity at a given point, using (5.3.8), we obtain, respectively,

$$[jwu_{(n)} - p] = 0 \quad , \tag{5.3.9}$$

$$[wu_\alpha] = 0 \ , \quad \alpha = 1,2,3 \ . \tag{5.3.10}$$

Differentiating the condition (5.3.9) with respect to χ_α and subtracting the result from (5.3.6) individually from each of the two sides of the surface of discontinuity, after using (5.3.10) to separate the terms continuous at the break, we obtain

$$j[\Omega_{l\alpha}] + \left[\frac{\varrho}{w}\right]\left(wu^\beta \Omega_{\beta\alpha} + \nabla_\alpha \frac{w^2 u_\beta u^\beta}{2}\right) - j^{-1}[p]\nabla_\alpha j = 0 \ . \tag{5.3.11}$$

We stress that in deriving this relation we did not make use of the thermodynamic identity (5.3.2).

Taking advantage of (5.3.3), in which we have made use of (5.3.2), from (5.3.11) we have

$$[\varrho T \nabla_\alpha S] = -\left[\frac{\varrho}{w}\right]\nabla_\alpha \frac{w^2 u_\beta u^\beta}{2} + [p]\nabla_\alpha \ln j = 0 \ . \tag{5.3.12}$$

According to the relations (5.3.11, 12), there is a break in the component of the vortex bivector and in the tangential component of the entropy gradient if at least one of the following two situations occurs: (a) nonuniformity of the distribution of the square of the tangential component (with respect to the surface of a strong break) of the pseudovelocity q_β, related to the 4-velocity components by the equation $q_i \equiv wu_i$; (b) nonuniformity of the mass flux density, $\{\nabla_\alpha j\} \neq 0$. If the flow before the break has a potential character, these circumstances lead to a nonuniformity of the particle entropy distribution and to a vortex character of the flow beyond the break.

5.3.2 Potential Motions

In the important special case of isentropic motions of a gas, there exists a pseudovelocity potential, $wu_j = \varphi_{,j}$, so that the relations (5.3.3) are satisfied identically. For potential motions of a gas, the equation of continuity leads to an equation for the potential φ:

$$\begin{aligned}\nabla_i \varrho u^i &= \nabla_i ((\nabla^i \varphi)\varrho^2/(p+\varepsilon)) \\ &= \varrho^2/(p+\varepsilon)[\nabla_i \nabla^i \varphi + \nabla^i \varphi \nabla_i \ln(\varrho^2/(p+\varepsilon))] = 0 \ .\end{aligned} \tag{5.3.13}$$

From the relation (5.3.2) with $S = \text{const}$, we have

$$d[\ln((p+\varepsilon)\varrho^{-2})] = -\frac{1}{2}\left(\frac{d\varepsilon}{dp} - 1\right) d(\ln w^2) \ , \tag{5.3.14}$$

while $\nabla_i \varphi \nabla^i \varphi = w^2$ by the definition of φ, so that from (5.3.13), using (5.3.14), we deduce the equation

$$2(\nabla_i \nabla^i \varphi)(\nabla_k \varphi \nabla^k \varphi) + \left(\frac{d\varepsilon}{dp} - 1\right)(\nabla^i \varphi)(\nabla_i(\nabla_k \varphi \nabla^k \varphi)) = 0 \ , \tag{5.3.15}$$

which can be written in the form [5.20]

$$\frac{dp}{d\varepsilon}(g^{ij}-u^iu^j)\nabla_i\nabla_j\varphi+u^iu^j\nabla_i\nabla_j\varphi=0 \ . \tag{5.3.16}$$

According to the theory of equations of hyperbolic type, this equation has real characteristics $\psi=\text{const}$:

$$\frac{dp}{d\varepsilon}\psi_{,i}\psi_{,j}(g^{ij}-u^iu^j)+u^iu^j\psi_{,i}\psi_{,j}=0 \ . \tag{5.3.17}$$

Equation (5.3.17) gives rise to a system of ordinary differential equations for sound rays:

$$\begin{aligned}\frac{dx^i}{d\alpha}&=\left(u^iu^j+\frac{dp}{d\varepsilon}(g^{ij}-u^iu^j)\right)k_j \ ,\\ \frac{dk_i}{d\alpha}&=-k_lk_m\frac{\partial}{\partial x^i}\left(u^lu^m+\frac{dp}{d\varepsilon}(g^{lm}-u^lu^m)\right) \ ,\\ k_i&\equiv\partial\psi/\partial x^i \ ,\end{aligned} \tag{5.3.18}$$

where α is a parameter on a sound ray, related to the proper time on the sound ray by the equation

$$c\,d\tau=|u_ik^i|\sqrt{1-dp/d\varepsilon}\,d\alpha \ .$$

5.3.3 Acoustic Waves in a Relativistic Gas

We shall derive equations for rapidly varying perturbations propagating in an arbitrary potential flow of a relativistic gas [5.21].

Suppose that the sound characteristics ψ = const stratify some region of 4-space in accordance with (5.3.17) and that we can introduce a "comoving" coordinate system for the sound rays: ψ, α, ξ^1, ξ^2, where ξ^1 and ξ^2 are Lagrangian coordinates of a sound ray on the sound wave front ψ = const; the transition to such a coordinate system is determined by the solution of (5.3.18).

In acoustic waves, the derivatives of hydrodynamical quantities with respect to the coordinate ψ will always be much greater than their derivatives with respect to the coordinates α, ξ^1, ξ^2, except in the neighborhood of caustic surfaces. Therefore, in the equation for the perturbations we shall neglect the second derivatives with respect to the slow variables, but we shall retain the quadratic terms involving products of derivatives with respect to the fast variable ψ. We note that such an approach to arbitrary nonlinear hyperbolic systems was developed by *Choquet–Bruhat* [5.22]. The acoustic equation has the form

$$2\frac{dx^i}{d\alpha}\nabla_i\Phi + \Phi\left[\nabla_i\left(\frac{dx^i}{d\alpha}\right) + \left(1 - \frac{dp}{d\varepsilon}\right)\frac{dx^i}{d\alpha}\nabla_i \ln\varrho\right]$$
$$-2\left(\frac{d\varepsilon}{dp} - 1\right)\frac{(k_i u^i)^3}{w}\Phi\frac{\partial\Phi}{\partial\psi} = 0 \ ,$$
$$\Phi \equiv \frac{\partial\varphi}{\partial\psi} \ , \quad \frac{dx^i}{d\alpha} = k_j\left(\frac{dp}{d\varepsilon}(g^{ij} - u^i u^j) + u^i u^j\right) \ . \tag{5.3.19}$$

We denote by dV the 4-volume element $dV = +\sqrt{-g}\, dx^1 dx^2 dx^3 dx^4$. If each point of this volume is displaced along the sound lines, then a variation of the parameter α on each sound ray by an amount $\Delta\alpha$ leads to a change in the volume by an amount

$$\Delta\, dV = dV\, \nabla_i\, (dx^i/d\alpha)\Delta\alpha \ . \tag{5.3.20}$$

This relation is analogous to the kinematic interpretation, well known in the mechanics of continuous media [5.23], of the divergence of the velocity as the rate of relative change of the volume. On the other hand, the element of volume can be expressed in terms of the Lagrangian coordinates as

$$dV = \sqrt{-\hat{g}}\, d\psi\, d\alpha\, d\xi^1 d\xi^2 \ , \quad \sqrt{-\hat{g}} = \frac{D(x^1, x^2, x^3, x^4)}{D(\psi, \alpha, \xi^1, \xi^2)}\sqrt{-g} \ .$$

Therefore (5.3.19) can be rewritten in the form

$$2\frac{\partial\Phi}{\partial\alpha} + \Phi\frac{\partial}{\partial\alpha}\ln\sqrt{-\hat{g}} + \Phi\left(1 - \frac{dp}{d\varepsilon}\right)\frac{\partial}{\partial\alpha}\ln\varrho$$
$$2\left(\frac{d\varepsilon}{dp} - 1\right)\frac{(k_i u^i)^3}{w}\Phi\frac{\partial\Phi}{\partial\psi} = 0 \tag{5.3.21}$$

We shall assume that the gas has the equation of state $p = \lambda\varepsilon$. We introduce the notation

$$v \equiv \Phi\sqrt[4]{-\hat{g}}\,\frac{\varrho}{\sqrt{p+\varepsilon}} \ , \quad dx = -\frac{(1-\lambda)(k_i u^i)^3 d\alpha}{\sqrt[4]{-\hat{g}}\sqrt{p+\varepsilon}\,\lambda} \ .$$

Then (5.3.21) reduces to the equation $v_{,x} + vv_{,\psi} = 0$. It can be seen that sound waves in relativistic hydrodynamics have the same qualitative behavior as in classical gas dynamics (see Sect. 4.1). The only change is in how far the initial sinusoidal wave varies up to the point at which it deviates from the parameters of the unperturbed solution, and in the characteristic dissipation interval of the wave. In the case of the most rigid equation of state $p = \varepsilon$, we obtain, according to (5.3.15), a linear equation for the pseudovelocity potential, so that this effect does not occur in such a gas.

5.3.4 Nonlinear Acoustics of an Expanding Universe: Relativistic Theory

We shall consider an important example of the application of our equation (5.3.21) when there is a background consisting of a homogeneous isotropic universe with a flat comoving space. For matter in the early stage of expansion at high temperatures, we adopt the relativistic equation of state $p = \lambda\varepsilon$ with $\lambda = \text{const}$. In the case of the model commonly known as the *Friedman-Lemaître model*, it follows from Einstein's equation $R_4^4 - R/2 = \kappa\varepsilon$ that the scale factor a in the metric $ds^2 = -a^2(t)(dx^2 + dy^2 + dz^2) + dt^2$ is related to the comoving time t by the law $a = a_0 t^{2/(3\lambda+3)}$, where $a_0 = \text{const}$.

We introduce a variable η such that $a\,d\eta = dt$, $\eta \sim t^{(3\lambda+1)/(3\lambda+3)}$. The equation of the characteristics (5.3.17) for one-dimensional motions (spherical, cylindrical, and planar) on the background of the Friedman-Lemaître model can be written in the form $\lambda(\psi_{,r})^2 - (\psi_{,\eta})^2 = 0$, from which we have $\psi = r \mp \sqrt{\lambda}\,\eta$. In what follows, we shall consider waves moving in the direction of increasing r, i.e., we shall take the upper sign.

It follows from the equation of continuity that $\varrho \sim a^{-3}$ and hence $\varepsilon \sim \varrho^{\lambda+1} \sim a^{-3(\lambda+1)} \sim t^{-2}$.

We now calculate the determinant of the metric tensor constructed on the sound rays. The components of the vector $\boldsymbol{k}(k_r, k_\eta)$ are $k_r = 1$, $k_\eta = -\sqrt{\lambda}$. Therefore, according to (5.3.18), the components of the vector tangential to the sound rays are

$$\frac{dr}{d\alpha} = -\frac{\lambda}{a^2}\,, \quad \frac{d\eta}{d\alpha} = -\frac{\sqrt{\lambda}}{a^2}\;.$$

According to (5.3.20), the elementary volume in the coordinate system comoving with the sound rays can be calculated from the equation

$$\frac{\partial}{\partial\alpha}\ln\sqrt{-\hat{g}} = \nabla_i\left(\frac{\partial x^i}{\partial\alpha}\right) = -\frac{\sqrt{\lambda}}{a^4}\frac{d}{d\eta}a^2 - \frac{\lambda}{r^\nu a^2}\frac{d}{dr}r^\nu\;,$$

and therefore $\sqrt[4]{-\hat{g}} = ar^{\nu/2}$ const. Here $\nu = 2$ in the spherical case, $\nu = 1$ in the cylindrical case, and $\nu = 0$ in the planar case.

Substituting these data into (5.3.21), we obtain

$$2r^{\nu/2}a^2\sqrt{p+\varepsilon}\frac{\partial}{\partial\eta}\left(vr^{\nu/2}a^2\sqrt{(p+\varepsilon)}\right) + \frac{\partial}{\partial\psi}[v^2a^4(p+\varepsilon)r^\nu](1-\lambda) = 0\;. \tag{5.3.22}$$

In this equation, v is the perturbation of the velocity:

$$\Phi \equiv \frac{\partial\varphi}{\partial\psi} = -\frac{(p+\varepsilon)va}{\varrho}\;.$$

Substituting the dependence of the background quantities on η, we have

$$\begin{aligned}&\left(\psi+\eta\sqrt{\lambda}\right)^{\nu/2}\eta^{(1-3\lambda)/(3\lambda+1)}X_{,\eta}+(1-\lambda)XX_{,\psi}=0\ ,\\&X\equiv v\left(\psi+\eta\sqrt{\lambda}\right)^{\nu/2}\eta^{(1-3\lambda)/(3\lambda+1)}\ .\end{aligned}\tag{5.3.23}$$

Hence we obtain, for the curves along which $X=\text{const}$, the differential equation

$$(1-\lambda)X\left.\frac{d\eta}{d\psi}\right|_{X=\text{const}}=\left(\psi+\eta\sqrt{\lambda}\right)^{\nu/2}\eta^{(1-3\lambda)/(3\lambda+1)}\ .\tag{5.3.24}$$

In what follows, we shall consider separately the cases of plane, spherical, and cylindrical waves. In the case $\nu=0$, we introduce the variable τ by the equation

$$\tau=\frac{3\lambda+1}{6\lambda}\eta^{6\lambda/(3\lambda+1)}(1-\lambda)\ .$$

Then (5.3.23) can be written in the form

$$X_{,\tau}+XX_{,\psi}=0\ .\tag{5.3.25}$$

This equation admits discontinuous solutions, in which the discontinuity moves with speed[5]

$$\frac{d\psi}{d\tau}=\frac{1}{2}(X_1+X_2)\ .$$

Here X_1 and X_2 are, respectively, the values of X before and after the discontinuity. This follows from the fact that at the discontinuity $f_{,\tau}[X]+f_{,\psi}[X^2]/2=0$, where $f(\psi,\tau)=0$ is the equation of the surface of discontinuity.

Let us calculate the asymptotic behavior of weak shock waves propagating through an unperturbed background. In this case, $X_2=0$. The intensity X_1 of the shock wave is attenuated in time, since the Riemannian wave comes onto the surface of discontinuity: the discontinuity moves with speed $X_1/2$, while the particles move with speed X_1. Therefore the shock wave is overtaken by the particles with a smaller and smaller speed.

The general solution of (5.3.25) has the form

$$\psi-X\tau=C(X)\ .\tag{5.3.26}$$

The function $C(X)$ can be expanded at small X in a series in powers of X: $C(X)\approx C_1X+\dots$.

Differentiating (5.3.26) with respect to τ for particles in front of the shock wave, and using the fact that $d\psi/d\tau=X_1/2$, we obtain for $X_1(\tau)$ the equation

$$X_1/2+\tau\dot{X}_1=-C_1\dot{X}_1\ .\tag{5.3.27}$$

[5] Thus, weak shock waves move through the gas with a speed which depends strongly on the size of the perturbation. In other words, the sound characteristics of the unperturbed solution do not coincide with the fronts of the weak shock waves.

By integrating, we readily obtain the asymptotic law of attenuation of plane shock waves:

$$X_1 = \frac{\text{const}}{\sqrt{\tau + C_1}} \ .$$

For a plane wave obtained from an inverted periodic wave, we have $X_1 + X_2 = 0$ at the discontinuity, so that from (5.3.26), by differentiating with respect to τ, using the fact that $d\psi/d\tau = 0$ for points on the shock wave, and then integrating, we obtain

$$X \sim 1/\tau \ .$$

Returning to the quantity v, the perturbation of the speed, we obtain the law of attenuation of plane shock waves and periodic waves in an expanding universe:

$$v \sim \eta^{-1/(3\lambda+1)} \sim 1/\sqrt{a} \quad \text{for shock waves} \ ,$$
$$v \sim \eta^{-1} \quad \text{for periodic waves} \ .$$

We note that for a photon gas ($\lambda = 1/3$) these expressions are completely analogous to the well-known formulas for plane waves in a gas at rest, provided that η is replaced by the time t.

We now turn to the case of cylindrical and spherical waves. Using (5.3.23), we find that the discontinuity moves with a speed given by

$$\frac{d\psi}{d\eta} = \frac{X_1 + X_2}{2} \left(\psi + \eta\sqrt{\lambda}\right)^{-\nu/2} \eta^{(3\lambda-1)/(3\lambda+1)} \ .$$

We introduce the notation $\nu/2 + (1 - 3\lambda)/(1 + 3\lambda) \equiv b$. The effects of inversion and dissipation occur only when $b \leq 1$, i.e., only for $\lambda \geq 1/3$ in the spherical case, and only for $\lambda \geq 1/9$ in the cylindrical case.

If $b > 1$, i.e., if $\lambda < 1/3$ for $\nu = 2$ and $\lambda < 1/9$ for $\nu = 1$, then the intensity X becomes asymptotically constant, and this leads to a change in the amplitude of periodic waves and in the intensity of shock waves according to the laws of geometrical optics: $v \sim \eta^{-b}$. The effect of wave inversion is absent, owing to the expansion of the universe and the geometrical expansion of the wave itself; the nonlinearity in these cases is unimportant.

The nonlinear asymptotic behavior of the attenuation in the case $b = 1$ ($\lambda = 1/3$ for $\nu = 2$ and $\lambda = 1/9$ for $\nu = 1$) is as follows:

$$v \sim (\eta \ln \eta)^{-1} \quad \text{for periodic waves} \ ,$$
$$v \sim \left(\eta\sqrt{\ln \eta}\right)^{-1} \quad \text{for shock waves} \ .$$

For $\lambda > 1/3$ the law of attenuation of the intensity of shock waves in the spherical case is

$$v \sim \eta^{(3\lambda+3)/(6\lambda+2)} \sim t^{-1/2} \quad ,$$

while in the cylindrical case for $\lambda > 1/9$ we have

$$v \sim \eta^{-1/4-1/(3\lambda+1)} \sim (a^2\eta)^{-1/4} \quad .$$

When $\lambda = 1/3$ in the cylindrical case, we have $v = \eta^{-3/4}$.

In the case of periodic sawtooth-shaped waves, independently of the symmetry and the equation of state, an essentially nonlinear law of attenuation is established when $b < 1$, namely, $v \sim 1/\eta$.

We note that the importance of nonlinear acoustic phenomena for an expanding universe was pointed out by *Peebles* [5.24]; the treatment given above is due to the present author.

5.4 Relativistic Magnetohydrodynamics

Various aspects of the relativistic electrodynamics of a continuous medium are also considered in other books which should be consulted for further details[6]. In the theory of relativity, the electromagnetic field is described by an antisymmetric tensor F_{ij}. The properties of molecules of matter in changing their electromagnetic characteristics under the influence of fields external to the molecules are characterized macroscopically by a polarization and magnetization tensor M_{ij}. In a medium, electric currents can flow and charges can accumulate. In the theory of relativity, one introduces the electric-current 4-vector j^i, whose first three components in the Cartesian coordinates x, y, z, t are the components of the electric-current density vector, while the fourth component gives the density of charge.

If in some coordinate system the current 4-vector lies inside the light cone in the tangent space, the fourth component of the current 4-vector cannot be transformed to zero by means of a transformation of the coordinates. In this case, it is possible to introduce the proper density of electric charge, $c^2\varrho_e^2 = j_i j^i$. It is convenient to introduce the tensor F^*_{ij} which is dual to the tensor F_{ij} by the relation $F^*_{ik} = \frac{1}{2}\varepsilon_{iklm}F^{lm}$. Then Maxwell's equations can be rewritten in the form

$$\nabla_i F^{*ij} = 0 \quad , \quad \nabla_i (F^{ik} + M^{ik}) = \frac{4\pi}{c} j^k \quad .$$

Ohm's law in the special theory of relativity can be reformulated as

$$j^i = c\varrho_e u^i + \sigma F^{ij} u_j \quad .$$

Magnetohydrodynamics (MHD) is the study of the dynamical properties of a fluid with infinite conductivity, $\sigma = \infty$, and a space-like electric-current 4-vector. In this case, it follows from Ohm's law that $F^{ij}u_j = 0$. We shall restrict

[6] See [5.23, 25–28] and the references to journal publications cited therein.

ourselves to the case $M_{ij} = 0$. It is convenient to introduce [5.27] a space-like magnetic-field 4-vector $h^i = F^{*ij}u_j$. Then the energy-momentum tensor of the electromagnetic field, $\tau_{ij} = [F_{ik}F_j^k - g_{ij}F_{kl}F^{kl}/4]/4\pi$, can be represented in the form

$$\tau_{ij} = \frac{1}{4\pi}\left\{h_k h^k\left(-u_i u_j + \frac{1}{2}g_{ij}\right) - h_i h_j\right\} .$$

The combined energy-momentum tensor of the matter and the field has the form

$$T_{ij} = \left(p + \varepsilon - \frac{1}{4\pi}h_k h^k\right)u_i u_j - \left(p - \frac{1}{8\pi}h_k h^k\right)g_{ij} - \frac{1}{4\pi}h_i h_j . \quad (5.4.1)$$

Therefore the closed system of equations of relativistic MHD consists of the equations of energy-momentum conservation $\nabla_i T^i_j = 0$, where T_{ij} is given by (5.4.1), the equation of continuity, and the magnetic-induction equations

$$\nabla_i (h^i u^j - h^j u^i) = 0 .$$

In the comoving system of coordinates ξ^1, ξ^2, ξ^3, τ, this equation can be readily integrated, since

$$\frac{\partial}{\partial\tau}\left[h^i \frac{\partial \xi^\alpha}{\partial x^i}\sqrt{-\hat{g}/\hat{g}_{44}}\right]_{\xi^\alpha} = 0 , \quad \alpha = 1,2,3 ; \quad i = 1,2,3,4 ,$$

from which, using the equation of continuity

$$\frac{\partial}{\partial\tau}\left[\varrho\sqrt{-\hat{g}/\hat{g}_{44}}\right]_{\xi^\alpha} = 0$$

for an arbitrary coordinate system of the observer, we have

$$h^i \frac{\partial \xi^\alpha}{\partial x^i} = \varrho\chi^\alpha(\chi^1, \chi^2, \chi^3) . \quad (5.4.2)$$

Analogous first integrals of the magnetic-induction equation hold also in the nonrelativistic theory. After transforming the equations

$$\frac{\partial}{\partial t}\boldsymbol{B} = \mathrm{curl}[\boldsymbol{v} \times \boldsymbol{B}]$$

to the comoving coordinate system, using the equation of continuity, we have

$$\frac{\partial}{\partial t}\left[\frac{B^\gamma}{\varrho}\frac{\partial \xi^\alpha}{\partial x^\gamma}\right]_{\xi^\alpha} = 0 , \quad \text{whence} \quad B^\gamma \frac{\partial \xi^\alpha}{\partial x^\gamma} = \varrho\chi^\alpha ,$$
$$\chi^\alpha = \chi^\alpha(\xi^1, \xi^2, \xi^3) , \quad \alpha, \gamma = 1,2,3 . \quad (5.4.2')$$

The equation $\nabla_i F^{ik} = 4\pi j^k/c$ makes it possible to use the resulting solution to determine the electric currents flowing in the matter.

Let $\psi(x,y,z,t)$ = const be the equation of a characteristic surface of the equations of relativistic MHD. Then the characteristics of the fast and slow magneto-sound waves satisfy the equation

$$P(\boldsymbol{k}) \equiv (p+\varepsilon)\left(\frac{d\varepsilon}{dp}-1\right)(u_i k^i)^4 + \left(p+\varepsilon-\frac{1}{4\pi}\frac{d\varepsilon}{dp}h_k h^k\right)(u^i k_i)^2 k_m k^m$$
$$-(4\pi)^{-1}(h^i k_i)^2 k_m k^m = 0 \ ,$$
$$k_m = \partial\psi/\partial x^m \ , \qquad (5.4.3)$$

and that of the Alfvén wave satisfies the equation

$$\left(p+\varepsilon-\frac{1}{4\pi}h_k h^k\right)(u_i k^i)^2 - \frac{1}{4\pi}(h_i k^i)^2 = 0 \ . \qquad (5.4.3')$$

The velocity of motion of the surface ψ = const relative to the gas is determined by the equation

$$v_\psi^2/c^2 = (u^i k_i)^2/[(u^i k_i)^2 - k_i k^i] \ . \qquad (5.4.4)$$

From the requirement that the speed of sound in the gas does not exceed the speed of light, $dp/d\varepsilon < 1$, it follows that the speeds of waves in relativistic MHD are always less than the speed of light. Moreover, (5.4.3, 3′) imply the inequalities

$$v_\mathrm{s} < v_\mathrm{A} < v_\mathrm{f} \ ,$$

where v_s, v_f, and v_A are the speeds of the slow and fast MHD waves and the Alfvén wave, respectively.

5.4.1 Shock Waves in Magnetohydrodynamics and the Hugoniot Adiabat

The conditions of continuity of the flux of energy-momentum and of the rest mass in crossing the surface of a break in the theory of relativity have the form

$$[T_{ij}]n^j = 0 \ , \quad [\varrho u_i]n^i = 0 \ ,$$

where n_i is the 4-vector of the normal to the surface of the shock wave. For the energy-momentum tensor in the case of an ideally conducting gas, we must substitute the expression (5.4.1). Maxwell's equations lead to the condition of freezing of the magnetic field:

$$f^{ij}n_j = 0 \ , \quad f^{ij} \equiv h^i u^j - h^j u^i \ . \qquad (5.4.5)$$

The corresponding algebraic equations in the nonrelativistic case are analyzed, for example, in the book of [5.29]. In the relativistic case, the analysis of the conditions on MHD breaks as shown by *Lichnerowitz* [5.27] also has an inherent elegance.

From the conditions of conservation of the flux of energy-momentum and of the magnetic induction on the break, we can form four scalars:

$$[T_{ik}]n^i n^k = 0 \ , \quad [T_{ik}T^i_l]n^k n^l = 0 \ ,$$
$$[f_{ij}T^i_k]n^j n^k = 0 \ , \quad [f_{ij}f^j_k]n^i n^k = 0 \ , \qquad \text{or}$$

$$\left[(p+\varepsilon+h^2)u_n^2 + (p+h^2/2) - h_n^2/4\pi\right] = 0 \ , \qquad h^2 \equiv -h^k h_k/4\pi \ , \quad h_n \equiv h^i n_i \ , \tag{5.4.6}$$

$$\left[(p+\varepsilon+h^2)^2 u_n^2 - (p+h^2/2)^2 - 2(p+\varepsilon+h^2)(p+h^2/2)u_n^2 + p h_n^2/2\pi\right] = 0 \ , \tag{5.4.7}$$

$$[(p+\varepsilon)u_n h_n] = 0 \ , \tag{5.4.8}$$

$$[H] = 0 \ , \quad H \equiv (h^2 u_n^2 - h_n^2/4\pi) \ . \tag{5.4.9}$$

In (5.4.6–9) the subscript n is used to denote the normal components of vectors, for example, $u_n \equiv u_i n^i$. We have made use of the fact that, according to its definition, h_i has the property $h_i u^i = 0$. We add to (5.4.6–9) the condition of conservation of the flux of mass across the surface of the break:

$$[j] = 0 \ , \quad j \equiv \varrho u_n \ . \tag{5.4.10}$$

We introduce the notation $w \equiv (p+\varepsilon)/\varrho$ for the specific enthalpy, and also the function $W \equiv (p+\varepsilon)/\varrho^2$. If we form the expression $T_{ik}T^k_j n^i n^j + (T_{ik}n^i n^k)^2$, which is continuous on the shock wave, then from the conditions (5.4.6, 7), using (5.4.9, 10), we obtain the relation

$$[w^2 + W^2 j^2 + 2W(h^2+H) + (h^2+H)H/j^2] = 0 \ . \tag{5.4.11}$$

After squaring (5.4.8) and eliminating h_n and u_n, we obtain

$$[W^2 j^2 h^2 - w^2 H] = 0 \ . \tag{5.4.12}$$

Using (5.4.9, 10), we rewrite the condition (5.4.6) in the form

$$[E] = 0 \ , \quad E \equiv Wj^2 + p + (h^2+H)/2 \ . \tag{5.4.13}$$

Using (5.4.12) to eliminate the function w from (5.4.11), we obtain

$$[F] = 0 \ , \quad F \equiv (h^2+H)(Wj^2+H)^2 \ . \tag{5.4.14}$$

We note that, according to the definition of H in (5.4.9), the expression h^2+H cannot be negative. We therefore rewrite (5.4.14) in the form

$$\left[\sqrt{h^2+H}\,(Wj^2+H)\right] = 0 \ . \tag{5.4.15}$$

According to this relation, the function Wj^2+H remains unchanged in sign on crossing the shock wave. Fast and slow MHD breaks correspond to $Wj^2+H > 0$

and $Wj^2+H<0$, respectively. On Alfvén breaks, the function Wj^2+H is equal to zero.

Using (5.4.13, 14), we write the relation (5.4.11) in the form [5.27]

$$\mathcal{H}(Z_1,Z_2)\equiv w_2^2-w_1^2+(p_1-p_2)(W_2+W_1)$$
$$+\tfrac{1}{2}(W_2-W_1)\left(\sqrt{h_2^2+H}-\sqrt{h_1^2+H}\right)^2=0\ , \qquad (5.4.16)$$

where the subscripts 1 and 2 are used to denote quantities before and after the break, respectively, and Z_1 and Z_2 denote the states before and after the break.

Equation (5.4.16) is called the equation of the *relativistic Hugoniot adiabat*. If we fix the constants E, F, and H, the state Z_2 for (5.4.13, 14) is determined by specifying two parameters, which can be taken to be j^2 and the entropy S, or W and the effective pressure $\overline{p}=p+(h^2+H)/2$.

5.4.2 Properties of MHD Breaks

In Newtonian gas dynamics, the equations of state of many gases satisfy the conditions

$$\text{I)}\ \partial V/\partial p|_s<0\ ,\quad \text{II)}\ \partial V/\partial S|_p>0\ ,\quad \text{III)}\ \partial^2 V/\partial p^2|_s>0\ , \qquad (5.4.17)$$

where V is the specific volume and S is the entropy per unit mass.

If the function W is substituted into these conditions in place of the function V, we obtain following *Lichnerowitz* [5.27] the corresponding conditions for a relativistic gas.

From the thermodynamic identity

$$dw=dp/\varrho+T\,dS \qquad (5.4.18)$$

it follows that $\partial(w^2)/\partial p=2W$.

Corollary I. We fix the state Z_1 and consider the state Z_2+dZ, where Z_2 is a point on the Hugoniot adiabat, for given j^2, E, F, and H. From (5.4.13, 14, 16) we find that

$$d\mathcal{H}=2wT\,dS\ . \qquad (5.4.19)$$

Corollary II. The propagation speed of small adiabatic perturbations $c\sqrt{\partial p/\partial\varepsilon|_s}$ is less than the speed of light c. Indeed, from the definition of W and from the condition I of (5.4.17), $\partial W/\partial p|_s<0$, we have $\partial W/\partial p|_s=\varrho^{-2}(1-\partial\varepsilon/\partial p|_s)<0$ and hence $\partial p/\partial\varepsilon|_s<1$.

Corollary III. The Poisson adiabat $S=$ const in the $\overline{p}$, W plane is convex: $\partial^2\overline{p}/\partial W^2|_s>0$. Indeed, from (5.4.14) it follows that

$$h^2+H=F(Wj^2+H)^{-2}\ .$$

Using this, by differentiating $\overline{p} \equiv p+(h^2+H)/2$ with respect to W with S = const, we obtain

$$\partial^2\overline{p}/\partial W^2|_s = -Q(\partial W/\partial p|_s)^{-3} \quad ,$$
$$Q \equiv \partial^2 W/\partial p^2|_s - 3\frac{(h^2+H)(\partial W/\partial p|_s)^3 j^4}{(Wj^2+H)^2} \quad .$$

This expression is greater than zero as a consequence of the conditions I and III of (5.4.17).

Corollary IV. On a line Δ in the $\overline{p}$, W plane, any extremum of the entropy must be a maximum. If we differentiate S with respect to W twice along the line $(\overline{p}-\overline{p}_1)/(W_1-W)$ = const, at a point at which $\partial S/\partial W|_\Delta = 0$ we obtain

$$\left.\frac{\partial^2 S}{\partial W^2}\right|_\Delta = -\frac{Q}{\partial W/\partial S|_p(\partial W/\partial p|_s)^2} \quad .$$

This expression is less than zero, in view of the conditions I–III of (5.4.17).

Corollary V. On weak shock waves, the discontinuity in the entropy and the discontinuity in the pressure are related by the equations

$$12wT[S] = Q[p]^3 \quad . \tag{5.4.20}$$

In view of the conditions I and III of (5.4.17), $Q > 0$, and a positive discontinuity in the pressure causes a positive discontinuity in the entropy of the third order. In order to prove (5.4.20), we expand w_2^2 in a series in powers of $[p]$ up to the term of third order and retain only the first power of $[S]$ in the expansion. We represent the term $(W_2+W_1)(p_2-p_1)$ in the form $\{2W_1+(\partial W/\partial p)[p]+2^{-1}(\partial^2 W/\partial p^2)[p]^2\}[p]$. The last term in (5.4.16) is of third order in $[p]$, and we transform it by means of a linearization of the condition (5.4.14). After this, the relation (4.20) is obtained by means of thermodynamic identities when collecting similar terms in (5.4.16).

Corollary VI. If the conditions I and III of (5.4.17) are satisfied on breaks of finite intensity, it follows from the law of growth of the entropy on crossing a break, $S_2 > S_1$, that $p_2 > p_1$ and $W_2 < W_1$. We shall prove that $p_2 > p_1$ for $S_2 > S_1$. Let us assume the contrary, i.e., that $p_2 < p_1$. Then along the Poisson adiabat S_1 = const, using its convexity [condition III of (5.4.17)], we have

$$w_1^2 - w^2(p_2, S_1) = 2\int_{p_2}^{p_1} W\,dp < (p_1-p_2)(W(p_2,S_1)+W_1) \quad . \tag{5.4.21}$$

Making use of the inequalities $\partial w^2/\partial S|_p = 2wT > 0$, $\partial W/\partial S > 0$, from (5.4.21) we obtain

$$\begin{aligned} &w_1^2 - w_2^2 < (p_1 - p_2)(W_2 + W_1) \ , \\ &W_1 = W(p_1, S_1) \ , \quad W_2 = W(p_2, S_2) \ . \end{aligned} \tag{5.4.22}$$

However, it now follows from this and from (5.4.16) that $W_2 < W_1$. This contradicts the assumptions $p_1 > p_2$, $S_2 > S_1$ and the conditions I and II of (5.4.17).

We shall now prove that $W_2 < W_1$. If we assume the contrary, i.e., that $W_2 > W_1$, then from (5.4.16), using $p_2 > p_1$, we obtain $I \equiv w_2^2 - w_1^2 - 2W_2(p_2 - p_1) < 0$. However, the expression I cannot be negative, since it can be represented in the form

$$I = 2 \int_{p_1}^{p_2} [W(p', S_2) - W(p_2, S_2)]\, dp' + w^2(p_1, S_2) - w^2(p_1, S_1) \ ,$$

where $W(p', S_2) > W(p_2, S_2)$, since $p_2 > p'$ and $\partial W/\partial p < 0$. The expression $w^2(p_1, S_2) - w^2(p_1, S_1)$ is greater than zero because $\partial w^2/\partial S|_p > 0$.

Corollary VII. On slow MHD breaks, the magnetic field intensity h^2 decreases on crossing the break; in the case of fast breaks, it increases. This property of MHD breaks is readily obtained from (5.4.14) by using the condition $W_2 < W_1$, which holds according to Corollary VI.

5.4.3 Relative Positions of the Poisson and Hugoniot Adiabats

It follows from Corollary V that the Poisson and Hugoniot adiabats passing through a point $\overline{p}_1$, W_1 in the $\overline{p}$, W plane have a second-order tangency at this point.

Corollary VIII. The Hugoniot function $\mathcal{H}(Z, Z_1)$ on the Poisson adiabat passing through the point Z_1 is negative for $W < W_1$. Calculating the second derivative of the Hugoniot function $\mathcal{H}(Z, Z_1)$ along the Poisson adiabat, we obtain $\partial^2\mathcal{H}/\partial W^2|_s = (W - W_1)\partial^2\overline{p}/\partial W^2|_s$. For $W < W_1$, it follows from condition III of (5.4.17) that the function $\partial^2\mathcal{H}/\partial W^2|_s$ is negative. Using the fact that $\partial\mathcal{H}/\partial W|_s = 0$ for $W = W_1$, we conclude from this that $\partial\mathcal{H}/\partial W|_s > 0$ for $W < W_1$ and, in turn, $\mathcal{H}(Z, Z_1) < 0$ for $W < W_1$.

Corollary IX. In the $\overline{p}$, W plane, let us draw an arbitrary straight line Δ with slope $-j^2 = (\overline{p} - \overline{p}_1)/(W - W_1)$ which is less than the slope of the tangent to the Poisson adiabat at the point $Z = Z_1$. The line Δ intersects the Poisson adiabat at two points Z_1 and Z_A, since the latter is convex. We shall prove that on the line Δ there is a unique Z_2 belonging to the Hugoniot adiabat $\mathcal{H}(Z_2, Z_1) = 0$.

The entropy has the same value at the points Z_1 and Z_A: $S = S_1$. Therefore the entropy has at least one extremum on the line Δ between these points. It follows from Corollary IV that this extremum can be only a maximum, and therefore the entropy has no other extrema. Consequently, $\partial S/\partial W|_\Delta$ is greater than zero at $Z = Z_A$ and less than zero at $Z = Z_1$. According to Corollary

I, on the line Δ the function $\mathcal{H}(Z, Z_1)$ has a maximum at the same point as the entropy, and $\partial\mathcal{H}/\partial W|_\Delta > 0$ at $Z = Z_A$ and $\partial\mathcal{H}/\partial W|_\Delta < 0$ at $Z = Z_1$. According to Corollary VIII, at the point Z_A we have $\mathcal{H}(Z_A, Z_1) < 0$, while at the point $Z = Z_1$ this function is equal to zero. Consequently, between the points Z_A and Z_1 there is a unique point Z_2 at which $\mathcal{H}(Z_2, Z_1) = 0$.

Corollary X. In relativistic MHD, for a given initial state Z_1 and mass flux j^2 the state Z_2 after a shock wave is uniquely determined.

Corollary XI. For a small change of the state along the line Δ, it follows from (5.4.13, 14) that

$$(\partial W/\partial S)_p(Wj^2 + H)(\partial S/\partial W)|_\Delta = P(\boldsymbol{n}) \quad , \tag{5.4.23}$$

where $P(\boldsymbol{n})$ is defined as $P(\boldsymbol{k})$ according to (5.4.3): $P(\boldsymbol{n})$ vanishes on the fast and slow MHD characteristics, and $Wj^2 + H$ vanishes on the Alfvén characteristics (5.4.3′). We now make use of Corollary IX: $\partial S/\partial W|_\Delta > 0$ at $Z = Z_2$ and $\partial S/\partial W|_\Delta < 0$ at $Z = Z_1$. Then for the speeds of propagation of the waves calculated according to (5.4.4), we readily obtain from (5.4.23) the following inequalities: for fast MHD breaks,

$$\begin{aligned} &v_{1s} < v_{1A} < v_{1f} < \mathcal{D}_1 \quad , \\ &v_{2s} < v_{2A} < \mathcal{D}_2 < v_{2f} \quad ; \end{aligned} \tag{5.4.24}$$

for slow MHD breaks,

$$\begin{aligned} &v_{1s} < \mathcal{D}_1 < v_{1A} < v_{1f} \quad , \\ &\mathcal{D}_2 < v_{2s} < v_{2A} < v_{2f} \quad , \end{aligned}$$

where $\mathcal{D}_1$ and $\mathcal{D}_2$ are the speeds of propagation of a strong break in the particles before and after the break, and v_s, v_A, and v_f are the characteristic speeds of propagation of the slow, Alfvén, and fast MHD waves. In the derivation of (5.4.24), we have taken into account the fact that on fast MHD breaks $Wj^2+H > 0$, while on slow breaks $Wj^2 + H < 0$.

We note that Corollaries I, IV–VI, VIII, and IX are ingenious generalization, to the case of relativistic MHD [5.27], of well-known results (in particular, results due to Weyl [5.30]) in Newtonian gas dynamics.

5.5 Hydrodynamical Flow Resulting from Production of Ultrarelativistic Particles in the Field of a Black Hole

In the case of quantum production of particles near a black hole, there is a gas-like dynamical flow which carries the particles to the pseudo-Euclidean infinity.

As a quantitative characteristic of this process, we cite its parameters for various black-hole masses M [5.31–34]: for example, for $M \sim 10^{17}$ g, the main component of the radiation consists of electronic and muonic neutrinos and antineutrinos (~81 %), about 17 % of the energy is radiated in the form of photons, and about 2 % in the form of gravitons. The total power of the radiation is

$$3.5 \times 10^{12} \frac{\text{erg}}{\text{s}} \left(\frac{10^{17}\,\text{g}}{M} \right)^2 .$$

With decreasing mass of the black hole, the output of energy increases: for example, for $M = 10^{15}$ g the power of the radiation is 6.3×10^{16} erg/s, and in this case ~45 % of the energy is radiated in the form of electron-positron pairs, ~45 % in the form of neutrinos and antineutrinos, ~9 % in the form of photons, and about 1 % in the form of gravitons.

The smaller the mass of the black hole, the heavier the particles which appear in its radiation, and this leads to a quantum explosion of black holes [5.31, 35, 36]. To calculate the mass distribution of the produced particles at a given temperature, we can use semiempirical theories.

We shall confine ourselves to black-hole masses of order 10^{15} g and above, when the main contribution to the radiation comes from light particles and we can use the ultrarelativistic equation of state $p = \varepsilon/3$. For such black-hole masses the rate of loss of mass is comparatively low, and we shall therefore neglect the changes in the gravitational field due to the decrease of the mass of the black hole.

We shall assume that the resulting spherically symmetric hydrodynamical flows have been established. In static gravitational fields, the equations of hydrodynamics admit the relativistic Bernoulli integral $Tu_4 = T_\infty = \text{const}$ [5.5]. It is assumed that at infinity the gas has finite temperature and zero velocity. From the condition of constancy of the flux of energy, we have

$$4\pi(p + \varepsilon)u_4 u^r r^2 = \dot{M}c^2 . \qquad (5.5.1)$$

Let v be the three-dimensional radial velocity of the gas in units of the speed of light, in terms of which the components of the 4-velocity can be expressed as follows:

$$u^r = \sqrt{1-x}\, v \Big/ \sqrt{1-v^2} ,$$

$$u_4 = \sqrt{1-x} \Big/ \left(c\sqrt{1-v^2} \right) ,$$

$$x \equiv r_g/r , \quad r_g = 2GM/c^2 .$$

Then Bernoulli's equation and (5.5.1) can be written in the form of the system

$$\begin{aligned} &\tau(1-x) = (1-v^2) \ , \quad \tau^2(1-x)v = \delta(1-v^2)x^2 \ , \\ &\tau \equiv (T/T_\infty)^2 \ , \quad \delta \equiv \dot{M}c/\left[4\pi(p+\varepsilon)_\infty r_g^2\right] \ . \end{aligned} \tag{5.5.2}$$

Eliminating either v or τ from the system (5.5.2), we obtain

$$\tau^2[1-\tau(1-x)] = x^4\delta^2 \ , \quad v(1-v^2) = \delta(x^2-x^3) \ . \tag{5.5.3}$$

The case of accretion corresponds to $\dot{M} < 0$ or $\delta < 0$, and in the case of outflow of the gas to infinity we have $\dot{M} > 0$ or $\delta > 0$. The equations (5.5.3) describe both cases. It is interesting to note that the gravitational field of a black hole acts as a huge Laval nozzle with the minimal cross section[7] for $r = (3/2)r_g$.

We pose the problem of finding the stationary regimes of the hydrodynamical flow of the particles produced by quantum processes in the region $[r_g, (3/2)r_g]$ with a finite temperature corresponding to the temperature of a black hole. The only curve from the family of curves described by (5.5.3) for which the temperature is finite and nonzero for $x \to 0$ ($r \to \infty$) and for $x \to 1$ ($r \to r_g$) is the curve with a continuous transition through the speed of sound. This solution corresponds to $\delta = \sqrt{27}/2$.

The solution in question can be given in parametric form as follows:

$$\begin{aligned} &\frac{r}{r_g} = \frac{1}{2y}\left[3y^2 + (y-1)\sqrt{3(3y^2+2y+1)}\right] \ , \\ &\tau = 3yr_g/r \ , \quad v = \sqrt{3}\, r_g/(2yr) \ . \end{aligned}$$

As y varies from $\sqrt{3}/2$ to infinity, r varies from r_g to ∞, τ varies form $\sqrt{27}/2$ to 1, and the speed decreases from the speed of light to zero at infinity. The point $y = 1$ corresponds to a transition through the speed of sound. In this case, $x = 2/3$, $\tau = 2$, $v = \sqrt{3}/3$.

Thus, a stationary outflow from a black hole is possible only when the temperature of the black hole is $\sqrt{\sqrt{27}/2}$ greater than the temperature of the gas at infinity. Stationary outflow regimes with $\delta > \sqrt{27}/2$ are impossible, and here, as in the case of the Laval nozzle in the nondesign regime, shock waves occur.

For $\delta < \sqrt{27}/2$, stationary subsonic regimes are possible only in the case of unbounded growth of the temperature at the horizon. The asymptotic behavior of the temperature for small accretion numbers δ, but for all x, is as follows:

$$(T/T_\infty)^2 \approx (1-x)^{-1} - \delta^2(1-x)x^4 \ , \quad v \approx \delta(1-x)x^2 \ .$$

In the case of a finite temperature at the horizon, subsonic regimes must entail rarefaction waves violating the stationarity condition.

[7] It follows from (5.5.3) that the families of curves $\tau = \tau(\delta, x)$, $v = v(\delta, x)$ can be continuously deformed into families of curves for the distribution of temperature and Mach number along a current tube whose area has a minimum at some interior point and is maximal at the ends of the current tube, for adiabatic flow of an ideal gas in classical gas dynamics.

In supersonic outflow regimes with $\delta < \sqrt{27}/2$, the speed of the gas tends to the speed of light at large r and near the horizon, and the temperature tends to zero at large r.

We note that accretion onto dense objects has been studied in [5.37, 38], and the stationary flow of a gas in the field of a black hole has also been studied in [5.39].

5.6 Self-Similar Motions of an Ultrarelativistic Gas with Spherical or Cylindrical Symmetry

In Sect. 5.2 we described the physical conditions under which matter can have the ultrarelativistic equation of state $p = \varepsilon/3$. We shall distinguish the cases of a cold superdense gas [see (5.2.10)] and a hot gas whose particle-number density is determined by the condition of thermal equilibrium [see (5.2.16, 17)]. In a hot gas and in a nuclear fluid [5.11, 16, 40], the flux of the number of particles on a shock wave is not conserved, owing to the production of particles on a strong break.

In an ultrarelativistic gas, the conditions of conservation of the flux of energy and momentum on a shock wave are sufficient to describe p_2 and v_2 when p_1, v_1, and the speed of the wave are known. The condition of continuity of the flux of the number of particles serves to determine the jump in the entropy in crossing the shock wave.

The conditions of continuity of the flux of energy and momentum on a direct jump in the coordinate system in which the jump is at rest have the form

$$[(p + \varepsilon\hat{v}^2/c^2)/(1 - \hat{v}^2/c^2)] = 0 \ , \quad [\hat{v}(p + \varepsilon)/(1 - \hat{v}^2/c^2)] = 0 \ ,$$

where $\hat{v}$ is the speed in this coordinate system. These conditions can be solved [5.5] for the square of the speed before and after the jump:

$$\begin{aligned} \frac{\hat{v}_1^2}{c^2} &= \frac{(p_1 - p_2)(p_1 + \varepsilon_2)}{(\varepsilon_1 - \varepsilon_2)(p_2 + \varepsilon_1)} \ , \\ \frac{\hat{v}_2^2}{c^2} &= \frac{(p_2 - p_1)(p_2 + \varepsilon_1)}{(\varepsilon_2 - \varepsilon_1)(p_1 + \varepsilon_2)} \ . \end{aligned} \tag{5.6.1}$$

Multiplying the relations (5.6.1), we readily obtain

$$\frac{\hat{v}_1 \hat{v}_2}{c^2} = \frac{p_1 - p_2}{\varepsilon_1 - \varepsilon_2} \ . \tag{5.6.2}$$

In a coordinate system in which the speed of the shock wave $\mathcal{D}$ is $c\lambda$, we obtain for the speeds $\hat{v}_1$ and $\hat{v}_2$, according to the relativistic rule for the composition of velocities, the values

$$\hat{v}_{1,2} = (v_{1,2} - c\lambda)/(1 + v_{1,2}\lambda/c) \ . \tag{5.6.3}$$

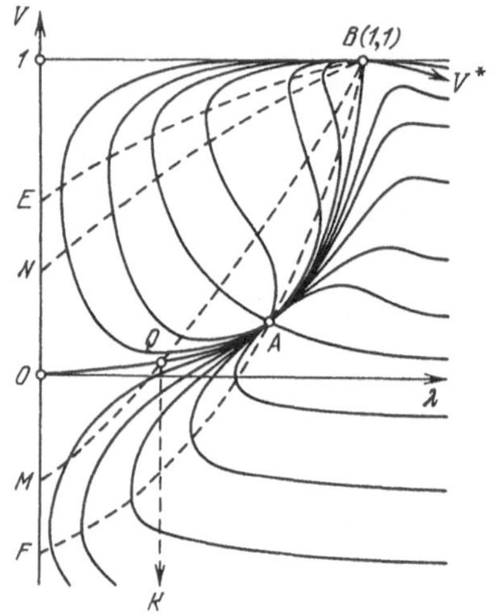

Fig. 5.1. The integral curves (5.6.5) for $0 < l < 1 - 1/\sqrt{3}$, $\nu = 2$. The point A corresponds to a weak break. A transition onto the separatrix segment OA is possible through a strong break

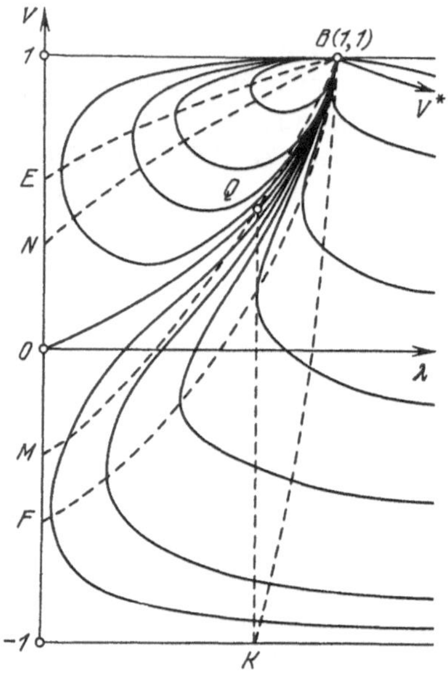

Fig. 5.2. The integral curves for $1 - 1/\sqrt{3} \leq l \leq 1 + 1/\sqrt{3}$, $\nu = 2$. There are no solutions with a weak break. A transition from the curve KB onto the curve QB is possible through a strong break

For an ultrarelativistic gas, $p = \varepsilon/3$, we find from the relations (5.6.2, 3) that

$$V_2 = \frac{3\lambda^2 - 1 - 2V_1\lambda}{V_1(\lambda^2 - 3) + 2\lambda} \ , \quad cV_1 \equiv v_1 \ , \quad cV_2 \equiv v_2 \ . \tag{5.6.4}$$

It is convenient to study the transformation (5.6.4) in the λ, V plane; it carries the straight line $V_1 = 1$ into the curve NB, $V_2 = (3\lambda + 1)/(3 + \lambda)$ (see Figs. 5.1–5). The image of the line $V_1 = -1$ is the curve MB, $V_2 = (3\lambda - 1)/(3 - \lambda)$. On the curves EB, $V = (\lambda\sqrt{3} + 1)/(\sqrt{3} + \lambda)$ and FB, $V = (\lambda\sqrt{3} - 1)/(\sqrt{3} - \lambda)$ there are points which, after the break, transform into themselves. Since the speed of a shock wave is bounded by the speed of light, breaks are possible only for $\lambda < 1$. No breaks are possible in the region $NBMON$.

The system of equations representing the conservation of energy and momentum for self-similar motions of an ultrarelativistic gas can be reduced to the equation

$$\lambda\frac{dV}{d\lambda} = \frac{(1 - V^2)[3l\lambda + (\nu - 3l)V^2\lambda - \nu V]}{(1 - V\lambda^2) - 3(\lambda - V)^2} \ , \tag{5.6.5}$$

where $\lambda \equiv r/ct$, and $\nu = 2$ in the spherical case and $\nu = 1$ in the cylindrical case.

In self-similar solutions, $p = P(\lambda)r^{-4l}$. From the known function $V(\lambda)$, the function $P(\lambda)$ can be found by quadrature:

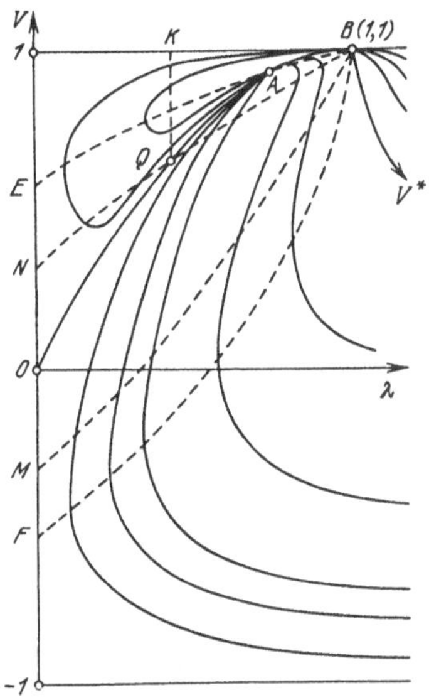

Fig. 5.3. The integral curves for $1 + 1/\sqrt{3} < l < 2$, $\nu = 2$. The point A is a node. Continuous solutions for $\lambda < 1$ coincide with the separatrix OB. Discontinuous solutions for $\lambda < 1$ coincide with the piecewise-smooth curve $OQKB$ with a strong break at the point Q

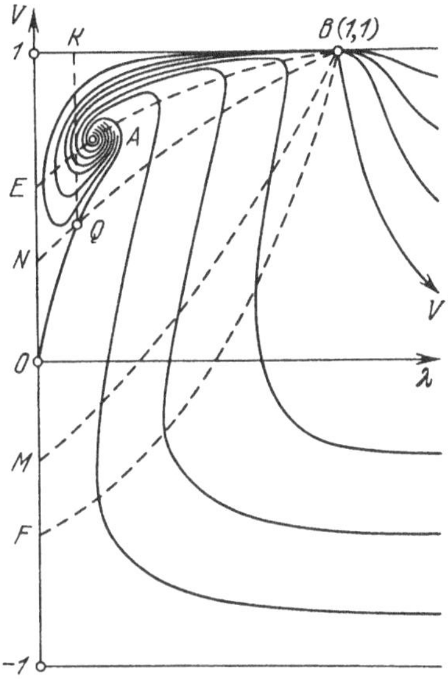

Fig. 5.4. The integral curves for $l > 2$, $\nu = 2$. The point A is a focus. The solution of the Cauchy problem for all initial speeds of dispersion is given by the piecewise-smooth curve $OQKB\ldots$ with a strong break at the point Q

$$\frac{d(\ln P(\lambda))}{d\lambda} = \frac{4\lambda(\lambda - V)dV/d\lambda + 4l(1 - V^2)}{\lambda(1 - V^2)(1 - \lambda V)} . \tag{5.6.6}$$

In what follows, we summarize the results of a previously published investigation and physical interpretation of self-similar motions of an ultrarelativistic gas [5.40, 41]. The problem of finding self-similar solutions in the theory of relativity was raised in [5.42, 43]. Self-similar solutions in the planar case ($\nu = 0$) were considered in [5.44, 45].

5.6.1 Qualitative Investigation of (5.6.5)

We shall list the properties of the integral curves of (5.6.5), which are qualitatively identical for the spherical and cylindrical cases. If $0 < l < (1 - 1/\sqrt{3})\nu/2$ (Fig. 5.1), there exists a singular point A (a node singularity) with coordinates $\nu/[(\nu - 2l)\sqrt{3}]$, $\sqrt{3}l/(\nu - 3l)$. All the integral curves reach the point A, being tangent to the separatrix of the node A with asymptote

$$V - \frac{\sqrt{3}l}{\nu - 3l}$$
$$= \left(\lambda - \frac{\nu}{\sqrt{3}(\nu - 2l)}\right) \frac{(2l - \nu)^2}{8\nu(3\nu - l)^2} \left(-a + \sqrt{a^2 - 144\nu l(3l - \nu)}\right) ,$$

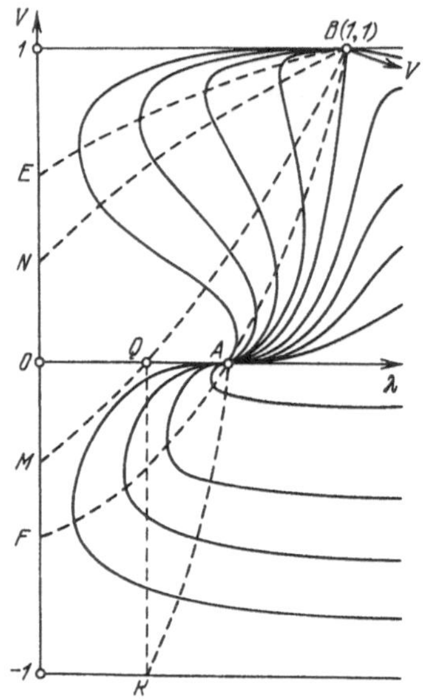

Fig. 5.5. The integral curves for $l = 0$, $\nu = 2$. The point A, a degenerate node, corresponds to a weak break. A transition from a point on the curve KA onto the segment QA is possible through a strong break

where $a = 3\nu^2 - 12l\nu - 6\nu$. The asymptote of the other separatrix of the point A differs from this expression only in the sign in front of the square root. As l increases, the singular point A moves along the curve FB, $V = (\lambda\sqrt{3}-1)/(\sqrt{3}-\lambda)$ from the point $(1/\sqrt{3}, 0)$ (for $l = 0$) to the point $B(1,1)$ (for $l = (1 - 1/\sqrt{3})\nu/2$).

In the interval $(1 - 1/\sqrt{3})\nu/2 \leq l \leq (1 + 1/\sqrt{3})\nu/2$ (Fig. 5.2), the singular point A is absent. When $l > (1 + 1/\sqrt{3})\nu/2$ but $a^2 > 144\nu l(3l - \nu)$, a singular point $A((2l-\nu)\nu/\sqrt{3}, \sqrt{3}l/(3l-\nu))$ reappears. As l increases, the node A moves from the point $B(1,1)$ (Fig. 5.3) along the curve $V = (\lambda\sqrt{3}+1)/(\sqrt{3}+\lambda)$ up to some critical point $\tilde{A}$ for $l = \tilde{l}$, where $\tilde{l}$ is a root of the equation $a^2 = 144\nu\tilde{l}(3\tilde{l}-\nu)$. If l varies in the interval $((1 + 1/\sqrt{3})\nu/2, \tilde{l})$, then the curves reach the point A being tangent to the separatrix OB at the point A (a node singularity) with asymptote

$$\left(V - \frac{\sqrt{3}l}{3l-\nu}\right) = \left(\lambda + \frac{\nu}{\sqrt{3}(\nu - 2l)}\right)\frac{(2l-\nu)^2}{8\nu(3l-\nu)^2}\left(-a - \sqrt{a^2 - 144\nu l(3l-\nu)}\right) .$$

For the other separatrix of the node A, the asymptote differs only in the sign in front of the square root. For all $l \in (0, \tilde{l})$, the separatrix of the saddle point $O(0,0)$ reaches B with asymptote[8] $V - 1 = (\lambda - 1)[2 + \nu/2 - 3l +$

[8] The asymptote of the other separatrix of the degenerate singular point B, on which $V \to V^*$ as $\lambda \to \infty$, differs from the one given here in the sign in front of the square root. Curves on which $V(\infty) > V^*$ reach the point B, being tangent to the line $V = 1$. When $V(\infty) < V^*$, the corresponding curves do not reach the point B.

$\sqrt{(2+\nu/2-3l)^2+\nu-1}$]. In the intervals $0 < l < (1-1/\sqrt{3})\nu/2$ and $(1+1/\sqrt{3})\nu/2 < l < \bar{l}$, the separatrix of the saddle point O, passing through the node A, coincides with the separatrix of the point A to which the integral curves are tangent. For $l \in ((1-1/\sqrt{3})\nu/2,\ (1+1/\sqrt{3})\nu/2)$, some of the integral curves reach the point B, being tangent to the separatrix OB.

The node A degenerates when $l = \bar{l}$; we shall write $\tilde{A}$ for the point A when $l = \bar{l}$. For $l > \bar{l}$, the singular point A (Fig. 5.4) becomes a focus, and as l increases we move along the curve $V = (\lambda\sqrt{3}+1)/(\sqrt{3}+\lambda)$ from $\tilde{A}$ to $(0, 1/\sqrt{3})$ as $l \to \infty$. We now note the difference between the cases $\nu = 2$ and $\nu = 1$. In the spherical case ($\nu = 2$), practically all the integral curves at the point B for all l are tangent to the line $V = 1$ with the asymptote $V - 1 \approx \text{const} \cdot (\lambda - 1)^2$. The asymptotes of the separatrices at the point B are $V - 1 = (\lambda - 1)[3(1-l) \pm \sqrt{9(1-l)^2+1}]$.

In the cylindrical case ($\nu = 1$), the integral curves (with the exception of the separatrix and the line $V = 1$) reach the point B, being tangent to the line $V = 1$ on only one side (the asymptote is

$$V - 1 = (\lambda - 1)/[(6\lambda - 5)(\text{const} + \ln(\lambda - 1))]) \ .$$

When $l < 5/6$, the tangency is only on the left; when $l > 5/6$, it is only on the right. When $l = 5/6$, the integral curves do not reach the point B. The point B has, for all l, only one separatrix, and its asymptote at B is $V-1 = (5-6l)(\lambda-1)$.

The character of the point A for $l = 0$ is different in the spherical and cylindrical cases: for $\nu = 2$ the point A (Fig. 5.5) is a degenerate node with asymptote of the integral curves given by

$$V = (3\lambda - \sqrt{3})/[\ln(\lambda - 1/\sqrt{3}) + \text{const}] \ ,$$

while for $\nu = 1$ it is an ordinary node. All the integral curves for $\nu = 1$ reach the point A, being tangent to the line $V = (3\lambda - \sqrt{3})/4$. The foregoing qualitative investigation of (5.6.5) admits a double physical interpretation (Sects. 5.6.2, 3).

5.6.2 Ejection of Matter from a Singular Point (Axis) at the Instant $t = 0$

At a fixed instant t, the matter ejected from a point can be contained only within a sphere (cylinder) with radius ct and with center at the point (on the axis) of symmetry. The difference between the methods of ejection of matter is controlled by variation of the parameter l. The separatrix OB gives the velocity distribution at a fixed instant for $l \in (0, \bar{l})$.

If $l > \bar{l}$ ($\bar{l} = 2$ for $\nu = 2$ and $\bar{l} = (3+\sqrt{11})/8$ for $\nu = 1$), the motion takes place with a shock wave: the point in question jumps from one separatrix to the other and, along it, reaches the point $B(1,1)$. For the pressure, we have for all l near the center (axis) of symmetry, $\lambda \ll 1$, the expansion $p(r,t) = \text{const}[1 + \beta r/(ct) + \ldots]/t^{4l}$, where $\beta = 12l(l\nu - 2l + \nu + 1)/(\nu+1)^2$. If $l < 5/6$ for axial symmetry, and for all l in the case of spherical symmetry, near $B(1,1)$ the pressure along the separatrix has the asymptote

$$p = \frac{\text{const}}{r^{4l}}(1-\lambda)^{\sigma} \ , \quad \sigma = \frac{2(3-4l)}{3l-1-\sqrt{9(l-1)^2+1}} \quad (\text{for } \nu = 2) \ ,$$

$$p = \frac{\text{const}}{r^{4l}}(1-\lambda)^{-4/3} \quad (\text{for } \nu = 1) \ .$$

It can be seen from these expressions that the matter is concentrated mainly near a sphere (cylinder) which expands with the speed of light.

For $l > 5/6$ in the cylindrical case, outflow from an axis is also impossible without a shock wave, but the nature of the motion becomes different: beyond the shock wave (the point K), all the matter moves with the speed of light. Outside the core restricted by the shock wave, where the velocity distribution is represented by the segment of the separatrix of the saddle from the point $O(0,0)$ to the point Q, the intersection with the image $V = 1$ of the curve $V = (3\lambda + 1)/(3 + \lambda)$, the pressure distribution is given by the expression

$$p = \frac{\text{const}}{r^{4l}} \left(\frac{r}{ct - r} \right)^{2-4l}$$

5.6.3 Solution of the Cauchy Problem

We pose the Cauchy problem for an ideal gas with the equation of state $p = \varepsilon/3$ and with initial distributions of the speed $v_0(x)$ and pressure $p_0(x)$ given by $v_0(x) = \beta \nabla r$, $|\beta| < c$, $p_0(x) = \alpha / r^{4l}$, where c is the speed of light, and α and β are characteristic dimensional constants. We shall assign the argument r the meaning of the distance to the center or to the axis of symmetry. For such initial conditions, the subsequent motion will be self-similar and will possess spherical or axial symmetry.

The Problem of Focusing. For $l \in (0, (\nu+1)/4)$, there exists a solution to the problem of focusing onto a point. There is a spherical (cylindrical) core restricted by the shock wave, which expands with constant speed (for $l = 0$, with speed less than $c/\sqrt{3}$). Inside the core, the velocity distribution is given by the segment of the separatrix of the saddle $(0,0)$. For $l = 0$, the matter inside the core is at rest[9] [5.46]. For $l \geq (\nu+1)/4$, there does not exist a self-similar solution of the problem of focusing with a subcritical speed, since in this case the separatrix of the point O does not intersect the curve $V = (3\lambda - 1)/(3 - \lambda)$.

The Problem of Dispersion. When $\nu = 2$, there exists for all l a certain critical dispersion speed V^*. When $\nu = 1$, a critical speed does not exist for l less than 5/6. Dispersive motion (with subcritical initial speeds for $\nu = 2$ and arbitrary initial speeds of dispersion for $\nu = 1$) for $l \in (0, (1 - 1/\sqrt{3})\nu/2)$ take place with a weak break.

[9] We have cited the monograph [5.46] in order to emphasize the qualitative analogy between the conclusions of this section and the corresponding results of Sedov in nonrelativistic gas dynamics.

In going from infinity to the center, the speed first increases up to some value $\tilde{V} = \max V(\lambda)$ on the curve $3l\lambda + (\nu - 3l)V^2\lambda - \nu V = 0$ and then decreases down to $V = \sqrt{3}l/(\nu - 3l)$ at $r = \nu\sqrt{3}ct/(3\nu - 6l)$, corresponding to the point A. At this value of r, there is a weak break. Inside the spherical core $r < \nu\sqrt{3}ct/(3\nu - 6l)$ for fixed t, the motion is standard for all initial speeds. It is given by the segment of the separatrix of the saddle from the point $O(0,0)$ to the point $A\ (\nu\sqrt{3}/(3\nu - 6l),\ \sqrt{3}l(\nu - 3l))$.

For $l \in ((1 - 1/\sqrt{3})\nu/2,\ (\nu + 1)/4)$ in the case of dispersion of the gas (with subcritical initial speeds when $\nu = 2$ and arbitrary initial speeds when $\nu = 1$), a shock wave is formed: the image point comes onto the separatrix of the saddle $O(0,0)$ discontinuously.

In all the solutions with $l \geq (\nu + 1)/4$, the pressure becomes infinite on approaching the sphere (cylinder) $r = ct$ from either the outside or the inside. We shall consider this case separately for spherical and axial symmetry.

For $\nu = 2$ and $l \in (3/4,\ 5/4)$, the solution of the Cauchy problem for any supercritical initial speed inside the light sphere has a unique continuation. It is given by the separatrix OB, onto which it is possible to come only at the point B (on the light sphere), since for $l \in (3/4,\ 5/4)$ it is impossible to jump onto the curve OB through a shock wave.

For $l > 5/4$, the continuation of the solution for $r < ct$ becomes nonunique. In the interval $5/4 < l < 2$, one (smooth) solution is given by the separatrix, and another solution is discontinuous. For $l > 2$, only discontinuous solutions are possible.

The cosmological solution of *Milne* [5.47] is a particular solution for $l = 1$. It coincides with the separatrix OB with the equation $V = \lambda$. The other separatrix of the point B has the equation $V = 1/\lambda$.

Thus, the analysis of the solution of the Cauchy problem with the initial pressure distribution $p = \alpha/r^4$ for a stationary gas leads to the conclusion that inside the light sphere the solution is given by the known cosmological solution of Milne, while outside the sphere the motion of the fluid particles is uniformly accelerated and is given by the expressions

$$v = c^2t/r \ , \quad p = \alpha r^{-8/3}(r^2 - c^2t^2)^{-2/3} \ .$$

In the case of axial symmetry $\nu = 1$ with $1/2 < l \leq 5/6$, there are solutions with initial speed of outflow from the axis equal to the speed of light, and only such solutions. When $l \in (1/2,\ 3/4)$, the curve OB is the only continuation of the solution $V = 1$ into the light cylinder. The continuation into the light cylinder is nonunique if $l \in (3/4,\ 5/6)$, but shock waves occur for all continuations through $\lambda = 1$. In particular, the solution $V = 1$ can be analytically continued through the singular cylinder $\lambda = 1$ up to a certain value $\tilde{\lambda}$, at which the speed decreases discontinuously from the speed of light to the speed $c(3\tilde{\lambda} + 1)/(\tilde{\lambda} + 3)$.

If $l > 5/6$, there exists a certain critical initial speed. Only motions with initial speeds not less than the critical speed V^* can be physically realized. The continuation of the solution into the light cylinder becomes unique but nonanalytic. For $r < ct$, the matter moves with the speed of light up to a certain value $r = ct\tilde{\lambda}$ (the point K), at which the speed decreases discontinuously to the value $c(3\tilde{\lambda} + 1)/(\tilde{\lambda} + 3)$ (corresponding to the point Q).

6. Some Problems of the Dynamics of Waves in Relativistic Cosmology

In spite of the striking successes of modern observational astronomy — the discovery and investigation, by means of artificial satellites, of x-ray sources, quasistellar objects (quasars), radio pulsars, and achievements in the study of individual stars, galaxies, and clusters of galaxies — comparatively few facts are known at the present time about the global structure of the universe, and these must be extracted from observations for the most part by means of a statistical analysis. We shall enumerate the most important of these facts.

1) Hubble's Law. The celebrated law of Hubble states that the remote galaxies are receding from us with a speed proportional, on the average, to the distance.

This law was established in two stages. From observations of Cepheids in the Magellanic Clouds situated at approximately the same distances from the Earth, Shapley concluded that the period of oscillations of the apparent magnitude (the flux density of radiated energy from a star per unit time at the surface of the Earth) is related to the total Cepheid luminosity (the energy radiated by the star per unit time). This made it possible to determine the relative distance to a Cepheid on the basis of the period of oscillations and the apparent magnitude. The method of statistical parallax was used to find the absolute distance.

By observing Cepheids in other galaxies of the local system (in particular, in the Andromeda Nebula M-31), Shapley estimated the distance to neighboring galaxies for the first time[1].

The second stage consisted in the use by Hubble of the brightest stars as indicators of the distances to remote galaxies, assuming that they are distributed uniformly on the average over the galaxies. It is possible to calibrate the total luminosity of a "typical" bright star in a galaxy on the basis of the distance determined by means of Cepheids in the same galaxy.

Using data obtained by Slipher on the red shift of spectral lines and comparing them with estimates of the distances to remote galaxies, Hubble concluded that

[1] Baade subsequently discovered an error in the period-luminosity calibration: this relationship gives for the classical Cepheids (stellar population I) a luminosity four times larger than for variables like RR Lyrae (stellar population II). Population I consists of stars like the Sun with a relatively high content of metals. Population II includes stars with a lower content of metals. These are weakly luminous, slowly evolving stars, which represent the first generation of the galaxy.

$v = Hl$, where v is the speed of recession of a galaxy from the Earth, l is its distance from the Earth, and H is a constant.

2) Age of the Universe. There is some uncertainty in the value of the Hubble constant H. From 1936, when Hubble obtained the estimate $H = 530$ km/s Mpc [Mpc (one million parsecs) = 3.086×10^{24} cm], to the present time, the value of this constant has dropped by almost a factor of 10 (see the references in [6.1]), and we can only say with confidence that the true value of the Hubble constant can hardly fall outside the interval 50–100 km/s Mpc. The quantity H^{-1} has the dimension of time and is of the order of 10^{10} yr; for the very simple Friedman-Lemaître cosmological model, H^{-1} is equal to the time which has elapsed since the infinitely dense and hot mass (the primordial fireball) began to expand.

It is remarkable that the value $H^{-1} = 10^{10}$ yr agrees with estimates obtained from data on the radioactive decay of the isotopes of uranium and from the theory of stellar evolution. We shall cite such estimates.

a) In ore of uranium U^{238}, the content of lead Pb^{206} is related to the geological age of the ore, since the lead is the final decay product of the uranium. Therefore, knowing the rate of decay of uranium, it is possible to determine the age of the ore from the ratio of the contents of lead and uranium. On the basis of this method, the estimate 1.3×10^9 yr is obtained for the maximum age of the ore.
b) A more delicate method of estimating the age is based on the abundance of the uranium isotope U^{235}, whose final decay product is the lead isotope Pb^{207}. The ratio of the abundances of the uranium isotopes U^{235} and U^{238} is 0.003. The relative content of the isotope Pb^{207} in lead ore is 0.07. Knowing the rate of decay of uranium U^{238}, we can calculate from this what interval of time is required for ore with the same content of U^{235} and U^{238} to be converted into ore with the ratio of the contents equal to 0.003. This interval of time is 3×10^9 yr.
c) Estimates of the age of stars on the basis of the theory of stellar evolution with thermonuclear sources of hydrogen combustion lead to a characteristic lifetime of a star on the main sequence in the range $(3\text{–}25) \times 10^9$ yr.

Thus, it follows from independent data that an extraordinary event took place about 10^{10} yr ago in the life of the metagalaxy surrounding us.

3) Homogeneity of the Universe. The positions of the bright galaxies in the night sky indicate that the distribution of galaxies is far from uniform; for example, very many objects are concentrated in the Virgo cluster and in a band emanating from it. Therefore a statistical analysis of the distribution of galaxies must be made over characteristic scales much larger than the distance to the Virgo cluster (equal to approximately 10 Mpc).

To test the hypothesis of a uniform distribution of galaxies, it is customary to construct a curve from the data of observations of distant galactic radio sources by plotting the number N of sources with magnitude f greater than a fixed value as a function of the apparent magnitude for a given wavelength[2]. Counts of radio sources by different groups of observers have given mutually consistent results, in spite of the statistical scatter and different wavelengths at which the observations were made. The experimental data show [6.1] that the dependence of $\ln N$ on $\ln f$ is practically linear, with the tangent of the slope angle equal to -1.8 ($\ln N \approx -1.8 \ln f + \text{const}$); in the region of weak sources, the tangent of the slope angle is -0.8. Allowance for the expansion of the universe makes it easier to obtain good agreement between the theory and the observations. The mean density in structures with a scale of 1000 Mpc varies in any case by not more than a factor of two. It has not yet been possible to give a definitive proof that the universe does not have a hierarchical structure. The microwave and x-ray radiations are relic phenomena not associated with a possible hierarchy: galaxies, clusters of galaxies, metagalaxies, etc.

4) Cosmic Microwave Radiation. The most important of the observations from which we can learn about the state of the universe in the distant past is the cosmic radiation in the microwave range, discovered by Penzias and Wilson in 1965. The spectrum of radiation with wavelengths $\lambda \gtrsim 0.3$ cm is close to the Planck spectrum with temperature 2.83 K. This radiation cannot be created as a result of scattering of radiation from sources in interstellar or intergalactic gas or dust. However, if we assume that in the distant past the universe was dense and hot, we can conclude that the radiation interacted with the matter as a result of scattering by the hot plasma. After recombination of the hydrogen, the radiation ceases to interact with the matter, retaining a "memory" of the thermal spectrum. The isotropy of the background radiation has been established with a very high degree of accuracy. The interpretation of the "cosmic" noise in the radiometers of Penzias and Wilson as relic black-body radiation and the modern scenario of the evolution of the chemical composition of the universe with time were given for the first time by *Dicke* et al. [6.2].

A theory of an expanding universe with a high temperature of matter and radiation in the early stages of the evolution — the model of a hot universe — was put forward by Gamow (see [6.1]). Gamow and his coworkers were responsible for the first consistent theory of the formation of the chemical elements in the early stages of expansion of the universe, in particular, for the theory of the formation of helium.

5) Uniform Distribution of Helium. The observational data on the presence of helium in the cosmos indicate [6.3] a surprising uniform distribution of helium in the universe. The relative helium content Y in the Orion Nebula is approximately

---

2 In a statistical homogeneous Euclidean model with a uniform distribution of sources, the number of them inside a sphere of radius r is proportional to r^3, and the radiation flux f from a source falls off in inverse proportion to the square of the distance. Therefore the number N of sources with magnitude greater than a fixed f depends on the magnitude as $f^{-3/2}$: $N(>f) \sim f^{-3/2}$.

the same as in the Sun. In regions of ionized hydrogen H_{II} in our galaxy and in other galaxies of the local system, $Y = 0.28$–0.29. Models of the stellar populations of classes I and II are best fitted to the observational data if one takes a relative helium concentration $Y = 0.30$–0.32. Such a helium concentration is estimated in the case of the planetary nebula of the spherical cluster M15, which belongs to the oldest stellar population II.

It is quite possible that the reason for the uniform distribution of helium lies in the universality of its origin in the period of high temperature in the history of the universe.

6) Mean Density of Matter. A basic tool for the calculation of the mean density of matter in the universe is the luminosity function $N(> f)$ for galaxies, in terms of which one can calculate the luminosity of galaxies per unit volume, as well as the mass-luminosity $(M - L)$ relation for typical galaxies: for spiral galaxies $M/L \sim 1$–$10\, M_\odot/L_\odot$, and for elliptic galaxies $M/L \sim 50\, M_\odot/L_\odot$, where $M_\odot$ and $L_\odot$ are, respectively, the mass and luminosity of the Sun. Multiplying the luminosity $\mathcal{L}$ of galaxies per unit volume (Oort, Van der Berg, and Kiang estimate $\mathcal{L}$ as $3 \times 10^8 L_\odot\, \mathrm{Mpc}^{-3}$) by the mean mass-luminosity ratio $M/L \approx 20\, M_\odot/L_\odot$, we obtain the value 4×10^{-31} g/cm^3 for the mean density of matter. A discussion of the systematic errors in finding the mean density of matter was given, for example, in Chap. IV of the book by *Peebles* [6.1]. The neglected forms of matter (apart from galaxies) can strongly alter the estimate of the mean density.

7) Cosmological Magnetic Field. Hoyle was the first to indicate the difficulties in explaining the origin of the magnetic field 10^{-6} G of our galaxy. Estimates of the lifetime of the galactic magnetic field on the basis of ohmic dissipation give a time much longer than the age of the universe.

If we assume that the magnetic field of our galaxy has a relic origin, then for the cosmological magnetic field we can take the value 10^{-10} G; it is this value that leads to the appearance of a field of intensity 10^{-6} G in our galaxy if the matter with a frozen magnetic field and with an assumed mean density 10^{-30} g/cm^3 in the universe is compressed to the mean density of matter in the galaxy, 10^{-24} g/cm^3, with conservation of the magnetic flux.

Cosmology is characterized by a large number of models of the expanding universe, special attention to the various possible physical and chemical processes taking place during the expansion, and bold assertions about the distant past of the universe (going right back to the time of the singular state). Nevertheless, a reasonable interpretation of the accumulated facts is possible only on the basis of definite theoretical constructions, which must be sufficiently flexible to encompass as many as possible of the observationally accessible manifestations of the part of the universe around us. For a detailed exposition of problems only mentioned here, see the monographs of [6.1, 4–10]. At present the concept of the inflationary universe is popular, its mathematical model was first introduced by Starobinsky [Phys. Lett. B **91**, 99 (1980)].

6.1 Development of Inhomogeneities in Models of the Universe with a Cosmological Magnetic Field

Solutions of Einstein's equations which admit three-parameter groups of motions on three-dimensional space-like surfaces of transitivity (Bianchi groups) have been studied intensively in recent years in connection with the problem of choosing a general relativistic homogeneous model of the universe (*Zel'manov* [6.11] first introduced a definition of homogeneity, not related to the concept of a group of motions, on the basis of his theory of chronometric invariants).

Particularly interesting properties hold for the corresponding dynamical system of ordinary differential equations in the case of the nonsolvable Bianchi groups VIII and IX [6.12–14]. The accumulated observational data (see the introduction to this chapter) do not enable us to make a categorical choice of a definite theoretical cosmological model from the class of models with isotropization[3] in the time or with initial isotropic conditions.

Secondly, the theoretical model must ensure a rate of development of arbitrary small perturbations after the instant of recombination[4] which is capable of explaining the observed strong concentration of matter in the galaxies (this being six orders of magnitude greater than the cosmological value). The current status of this problem is described in several books [6.6, 7, 9]. Investigations have been made of the development of inhomogeneities on the background of expanding Friedman models [6.16, 17], and also of the development of perturbations in axially symmetric Bianchi-I models [6.18].

Below, we present the results of investigations [6.19, 20] of the development of small perturbations in cosmological models with a magnetic field (the corresponding homogeneous solutions were found in [6.21–24]). Such an approach offers the attractive possibility of explaining the appearance of a magnetic field in our galaxy as a result of the curvature of the lines of force of a cosmological magnetic field frozen into the large-scale perturbations. Upper estimates have been obtained [6.25] for the intergalactic magnetic field on the basis of the Faraday effect of rotation of the polarization. We note also that some authors [6.1] believe that the presence of a magnetic field in the early stages of expansion would facilitate the production of primordial helium.

---

[3] By "isotropization" we mean equalization of the velocities of the cosmological expansion in all directions.

[4] Prior to this instant, small perturbations in the plasma are resorbed under the action of the radiation, which, owing to the Compton effect, smoothes the inhomogeneities [6.15].

6.1.1 Unperturbed Solution

Suppose that the group properties (axial symmetry and spatial homogeneity) of some solution of the Einstein-Maxwell equations make it possible to write the interval in the synchronous coordinate system, which is the comoving system for a gas, in the form

$$ds^2 = dt^2 - a^2(t)\left[dr^2 + k^{-1}\sin^2\left(\sqrt{k}\,r\right)d\varphi^2\right] - b^2(t)\,dz^2 \quad ,$$

where $k = +1, 0, -1$ for closed, flat, and open models, respectively (the system of units is such that the speed of light is $c = 1$).

Einstein's equations for spaces with such a metric reduce to the following system of ordinary differential equations:

$$\begin{aligned} R_1^1 - R/2 &= \ddot{a}/a + \dot{a}\dot{b}/ab + \ddot{b}/b = -\kappa(p + W) \quad , \\ R_3^3 - R/2 &= 2\ddot{a}/a + (\dot{a}/a)^2 - k/a^2 = -\kappa(p - W) \quad , \\ R_4^4 - R/2 &= (\dot{a}/a)^2 + 2\dot{a}\dot{b}/ab - k/a^2 = \kappa(\varepsilon + W) \quad . \end{aligned} \tag{6.1.1}$$

It is assumed here that only the z component of the magnetic field F_{12} is nonzero, $W = F_{12}F^{12}/8\pi = B_0^2/8\pi a^4(t)$, where $B_0 = \text{const}$, and the dots always indicate differentiation with respect to t. We assume that all space is filled with an ideal gas with the magnetic field frozen in it.

It follows from (6.1.1) that when $k = 1$ this system describes oscillations of the comoving ("fluid") ellipsoids of rotation between the "pancake" configuration ($b \approx b_0 t \to 0$, $a \to a_0 = \text{const}$, $\varrho \approx \varrho_0/t \to \infty$) and a certain limiting ellipsoid (these solutions are qualitatively similar to the Dirichlet solutions [6.26] for oscillations of the ellipsoid of an incompressible gravitating fluid). The role of the magnetic field is to prevent the ellipsoid consisting of any particular fluid particles from being drawn out into a filament.

Flat models for certain equations of state of the gas undergo isotropization for $t \to \infty$, i.e., they tend asymptotically to the Friedman-Lemaître solution (a solution with a flat comoving space). Here the magnetic field forbids the appearance of filament-like singularities.

For $k = -1$, the solutions tend in the limit $t \to \infty$ to the well-known axially symmetric solution of Milne (uniform outflow from the axis) in the special theory of relativity.

6.1.2 Notation for Small Perturbations and Coordinate Restrictions

We shall assume that one of the solutions of (6.1.1) is the basic homogeneous cosmological background, and superimpose on it arbitrary small inhomogeneous perturbations of the density and velocity of an ideal conducting medium and of the gravitational and electromagnetic fields. The electric field in the comoving coordinate system is equal to zero. In the perturbed Riemannian space with the metric $g_{ij} + h_{ij}$, we also use the synchronous reference frame:

$$h_{44} = h_{4\mu} = h_{43} = 0 \quad , \quad \mu = 1, 2 \quad .$$

Here and below, the indices μ, η, ν take the values 1, 2. Covariant differentiation will be carried out in what follows by means of the metric

$$a^2(t)\left[dr^2 + k^{-1}\sin^2\left(\sqrt{k}\,r\right)d\varphi^2\right] \quad . \tag{6.1.2}$$

We adopt the notation δA for a perturbation of the quantity A. In the linear approximation, $\delta u_4 = 0$.

We introduce the following notation for perturbations of the magnetic field components:

$$H^3 = \frac{\delta F_{12}}{F_{12}} \, , \quad \frac{\partial H^1}{\partial z} = \frac{\delta F_{23}}{F_{12}} \, , \quad \frac{\partial H^2}{\partial z} = \frac{\delta F_{31}}{F_{12}} \quad . \tag{6.1.3}$$

The perturbation of the total energy-momentum tensor of an ideal gas and of the magnetic field frozen in it, in the linear approximation, has the form

$$\begin{aligned} \delta T^\mu_\eta &= -[\delta p + W(2H^3 + h_\perp)]\delta^\mu_\eta \; , & \delta T^\mu_3 &= 2W\partial(H^\mu)/\partial z \; , \\ \delta T^3_3 &= -\delta p + W(2H^3 + h_\perp) \; , & \delta T^\mu_4 &= (p + \varepsilon + 2W)\delta u^\mu \; , \\ \delta T^3_4 &= (p + \varepsilon)\delta u^3 \; , & \delta T^4_4 &= \delta\varepsilon + W(2H^3 + h_\perp) \; . \end{aligned} \tag{6.1.4}$$

Here and in what follows, $h_\perp \equiv h^1_1 + h^2_2$, $h_\parallel \equiv h^3_3$.

The required system of equations for small perturbations is obtained by linearization of the combined system of relativistic magnetohydrodynamics (see Sect. 4.4) and Einstein's equations near the solution (6.1.1).

It can be shown that in the case under consideration the system of linear equations for small perturbations decomposes into two independent subsystems for even and odd perturbations (for the definition of even and odd perturbations for this case, see the footnote below).

Let $\varepsilon_{\mu\nu}$ be the Levi-Civita tensor in the two-dimensional subspace of the coordinates r, φ with the metric (6.1.2). Instead of the three components of the tensor h^μ_ν in the two-dimensional space (6.1.2), we can introduce three scalars $h_\perp$, Q, K:

$$\begin{aligned} h^\mu_\nu &= \delta^\mu_\nu h_\perp/2 + (\nabla^\mu\nabla_\nu Q - \delta^\mu_\nu \Delta Q/2) \\ &\quad + a^2(\varepsilon_{\nu\eta}\nabla^\mu\nabla^\eta + \varepsilon^{\mu\eta}\nabla_\nu\nabla_\eta)K \quad . \end{aligned} \tag{6.1.5}$$

Instead of the two-dimensional vectors $h_{3\mu}$, δu_μ, H_μ, we can introduce scalars L, M, Φ, φ, H, G in accordance with the relations[5]

$$\begin{aligned} h_{3\mu} &= \partial[\nabla_\mu L + a^2\varepsilon_{\mu\nu}\nabla^\nu M]/\partial z \; , \\ \delta u_\mu &= \nabla_\mu \Phi + a^2\varepsilon_{\mu\nu}\nabla^\nu\varphi \; , \end{aligned} \tag{6.1.6}$$

[5] In (6.1.5–7) the terms containing $\varepsilon_{\mu\nu}$ correspond to odd perturbations, and the remaining terms to even perturbations. The independence of the even and odd perturbations follows from the fact that Laplace's equation does not admit nonconstant solutions which are everywhere bounded and regular.

$$H_\mu = \nabla_\mu H + a^2 \varepsilon_{\mu\nu} \nabla^\nu G \ . \tag{6.1.7}$$

Rotational perturbations (see the definition in [6.16] for Friedman models) in anisotropic models in a "pure" form do not exist; the transverse components of the curl of the velocity are coupled in the differential equations to the perturbation of the density.

Gravitational waves in anisotropic models can be distinguished only for short-wave perturbations; for perturbations of sufficiently long length, gravitational waves interact with the matter.

We now expand all the scalar perturbations in a Fourier integral with respect to the coordinate z and in terms of cylindrical waves, i.e., the eigenfunctions of the Laplacian operator $\nabla_\mu \nabla^\mu$ in the plane of x^1 and x^2. For $k = 0$ (flat model) and for $k = -1$ (open model), the eigenvalues of "cylindrical" waves form a continuous spectrum (for a discussion of cylindrical waves in an open model, see [6.27]) and take the values $-n^2/a^2$ and $-(n^2 + 4^{-1})/a^2$, respectively. For $k = 1$, the eigenfunctions of the Laplacian operator are spherical harmonics with the discrete eigenvalues $-n(n+1)/a^2$ ($n = 0, 1, 2, \ldots$). In all three cases, we shall denote the eigenvalues of the operator $\nabla_\mu \nabla^\mu$ by $-\alpha^2(t)$, and those of the operator $\nabla_3 \nabla^3$ by $-\beta^2(t)$.

6.1.3 Equations of Conservation of Energy-Momentum and of the Magnetic Induction

If we substitute the perturbation of (6.1.4) into the equations for the conservation of momentum, $\delta(\nabla_i T^i_\mu) = 0$, we obtain equations for the even perturbations,

$$\begin{aligned} &2W\beta^2(H + L) + [(p + \varepsilon + 2W)a^2 b\Phi]^\cdot/a^2 b + \delta p \\ &\quad + W(2H^3 + h_\perp) + W h_\parallel = 0 \ , \\ &\delta p + [(p + \varepsilon)a^2 bU]^\cdot/a^2 b - 2W h_\perp = 0 \ , \quad \delta u_3 = \partial U/\partial z \ , \end{aligned}$$

and for the odd perturbations,

$$2W\beta^2 a^2 (G + M) + [(p + \varepsilon + 2W)a^4 b\varphi]^\cdot/a^2 b = 0 \ .$$

Here a dot signifies differentiation with respect to the time.

It follows from these equations that in the absence of a magnetic field the curl of the velocity is frozen in the fluid particles, so that

$$[(p + \varepsilon)a^2 b(\Phi - U)]^\cdot = [(p + \varepsilon)a^4 b\varphi]^\cdot = 0 \ . \tag{6.1.8}$$

Eliminating the electric field from Maxwell's equations $\nabla_{[i} F_{kl]} = 0$ by means of the condition of freezing, we obtain for the even perturbations

$$(H^3)^\cdot = \alpha^2 \Phi \ , \quad (H/a^2)^\cdot = \Phi/a^2 \ , \quad H^3 = \alpha^2 H \ , \tag{6.1.9}$$

and for the odd perturbations

$$\dot G = \varphi \ . \tag{6.1.10}$$

6.1.4 A Closed System for the Odd Perturbations

Each of the expressions $\delta R^3_\eta - \kappa\delta T^3_\eta$ and $\delta R^4_\eta - \kappa\delta T^4_\eta$, after substitution of the perturbations of the metric and of the matter, taken in the form (6.1.4–7), can be regrouped and represented in the symbolic form

$$\nabla_\eta A + \varepsilon_{\eta\nu}\nabla^\nu B \ . \tag{6.1.11}$$

Equating to zero an expression of this kind, by virtue of Einstein's equations we obtain for the even perturbations the symbolic equation $A = 0$, and for the odd perturbations the equation $B = 0$.

The components of the two-dimensional tensor $\delta[R^\nu_\mu - \kappa(T^\nu_\mu - T\delta^\nu_\mu/2)]$, after substitution of the expressions (6.1.5–7) and after grouping of the even and odd terms separately, can be represented in the symbolic form

$$\delta^\nu_\mu C + \nabla^\nu\nabla\mu D + (\varepsilon^{\nu\eta}\nabla_\eta\nabla_\mu E + \varepsilon_{\mu\eta}\nabla^\eta\nabla^\nu E) \ , \tag{6.1.12}$$

from which, by virtue of Einstein's equations, we obtain for the even perturbations the symbolic equations $C = 0$ and $D = 0$, and for the odd perturbations the symbolic equation $E = 0$.

These equations for the odd perturbations [the B components of the equations $\delta(R^3_\mu - \kappa T^3_\mu) = 0$, $\delta(R^4_\mu - \kappa T^4_\mu) = 0$ and the symbolic equation $E = 0$] have, respectively, the form

$$\begin{aligned}&\ddot M + \dot M(a^4/b)^{\cdot}\, b/a^4 + 4\kappa W M \\ &\quad + (\alpha^2 + k/a^2)(M - K) + 4\kappa W G = 0 \ ,\end{aligned} \tag{6.1.13}$$

$$\beta^2\dot M + (\alpha^2 + k/a^2)\dot K - 2\kappa(p + \varepsilon + 2W)\varphi = 0 \ , \tag{6.1.14}$$

$$-\beta^2(M - K) + \ddot K + \dot K(a^2b)^{\cdot}/a^2b = 0 \ . \tag{6.1.15}$$

Eliminating the functions G and φ from (6.1.13, 14), by means of the magnetic induction equation (6.1.10) we obtain

$$\begin{aligned}&\{a^4[\ddot M + \dot M(a^4/b)^{\cdot}\, b/a^4 + 4\kappa W M + (\alpha^2 + k/a^2)(M - K)]\}^{\cdot} \\ &\quad + 2Wa^4(p + \varepsilon + 2W)[\beta^2\dot M + (\alpha^2 + k/a^2)\dot K] = 0 \ .\end{aligned} \tag{6.1.16}$$

Equations (6.1.15, 16) form the required closed system for the functions M and K, and the function M can be eliminated. It is interesting to note that we then obtain a fourth-order equation for the single function K, which we have written down elsewhere [6.19]. The exact solution $K = \text{const}$ corresponds to a perturbation of the metric in going over to a perturbed coordinate system and is therefore physically meaningless.

6.1.5 A Closed System for the Even Perturbations

The C component of the equation $\delta(R^\mu_\nu - \kappa(T^\mu_\nu - \delta^\mu_\nu T/2)) = 0$ [see (6.1.12)] has the form

$$\begin{aligned} &-2k(h_\perp + \alpha^2 Q) + \alpha^2(h_\perp + \alpha^2 Q) + 2^{-1}[(h_\perp + \alpha^2 Q)^{\cdot\cdot} \\ &\quad + 2(\dot{h}_\| + \dot{h}_\perp)\dot{a}/a + (h_\perp + \alpha^2 Q)^\cdot (a^2 b)^\cdot / a^2 b] \\ &\quad - \kappa[\delta\varepsilon - \delta p + W(2\alpha^2 H + h_\perp)] = 0 \ , \end{aligned} \tag{I}$$

and the D component of this same equation has the form

$$\beta^2(Q - 2L) - h_\| + a^2[(Q/a^2)^{\cdot\cdot} + (Q/a^2)^\cdot (a^2 b)^\cdot / a^2 b] = 0 \ . \tag{II}$$

We now separate the A components of Einstein's equations $\delta\left[R^\eta_3 - \kappa T^\eta_3\right] = 0$ and $\delta\left[R^\eta_4 - \kappa T^\eta_4\right] = 0$ [see the definition (6.1.11)]:

$$\begin{aligned} &-(h_\perp + \alpha^2 Q)/2 + kQ/a^2 - \ddot{L} + \dot{b}\dot{L}/b - 2\kappa W H \\ &\quad - L[\kappa(p + \varepsilon) + 4\dot{a}\dot{b}/ab] = 0 \ , \end{aligned} \tag{III}$$

$$\begin{aligned} &-\dot{h}_\perp + \dot{h}_\| + (h_\| + \beta^2 L)(b/a)^\cdot\, a/b - 2^{-1}[2^{-1}(h_\perp + \alpha^2 Q)^\cdot \\ &\quad - kQ/a^2 - \alpha^2 a^2 (Q/a^2)^\cdot] - \kappa(p + \varepsilon + 2W)(H/a^2)^\cdot\, a^2 = 0 \ . \end{aligned} \tag{IV}$$

Einstein's equations $\delta(R^4_4 - R/2) = \kappa\delta T^4_4$ and $\delta R^3_3 = \kappa(\delta T^2_3 - \delta T/2)$ have, respectively, the form

$$\begin{aligned} &\dot{h}_\perp (ab)^\cdot/ab + 2\dot{h}_\|\dot{a}/a + 2\alpha^2\beta^2 L + \alpha^2 h_\| - \alpha^2\beta^2 Q \\ &\quad + (h_\perp + \alpha^2 Q)(\beta^2 + \alpha^2/2 + k/a^2) \\ &\quad = 2\kappa(\delta\varepsilon + 2W\alpha^2 H + W h_\perp) \ ; \end{aligned} \tag{V}$$

$$\begin{aligned} &2\alpha^2\beta^2 L + \beta^2 h_\perp + \alpha^2 h_\| + \ddot{h}_\| + 2\dot{h}_\|(ab)^\cdot/ab + \dot{h}_\perp\dot{b}/b \\ &\quad = \kappa[(\delta\varepsilon - \delta p)/2 - W(2\alpha^2 H + h_\perp)] \ . \end{aligned} \tag{VI}$$

In deriving the system (I–VI), we have used Maxwell's equations (6.1.9) to eliminate Φ and H^3.

6.1.6 Exact Solutions of the Linearized Equations Corresponding to Perturbations of the Coordinate System

Not all solutions of the linearized combined system of Einstein's equations and equations of magnetohydrodynamics are physically meaningful. The point is that if in the unperturbed homogeneous model we go over to a perturbed coordinate system $\tilde{x}^i = x^i + \xi^i(x)$, where the functions $\xi^i(x)$ are regarded as small displacements, then in the perturbed coordinates all the quantities at the point x acquire increments, although the pseudo-Riemannian space itself remains the same.

The metric tensor acquires increments $h_{ij} = \nabla_i \xi_j + \nabla_j \xi_i$. If the "new" coordinate system is to be synchronous, the components of the vector ξ^i will have the special form

$$\xi^4 = f(x^\alpha) \ , \quad \xi^\mu = \nabla^\mu f(x^\alpha) a^2 \int a^{-2} dt + f^\mu(x^\alpha) \ ,$$
$$\xi^3 = \nabla^3 f(x^\alpha) b^2 \int b^{-2} dt + f^3(x^\alpha) \ , \quad \mu = 1,2 \ ; \quad \alpha = 1,2,3 \ . \tag{6.1.17}$$

We stress that the functions f, f^μ, and f^3 depend only on the spatial coordinates x^1, x^2, and x^3.

By means of the expressions (6.1.17) for the components $\xi^\imath$, we obtain the following exact solution of the system (I–VI) for the even perturbations, defined in accordance with (6.1.5,6):

$$h_\perp = -2\alpha^2 f_1 a^2 \int a^{-2} dt + 4 f_1 \dot{a}/a - 2\alpha^2 a^2 f_2 \ ,$$
$$h_\parallel = -2\beta^2 f_1 b^2 \int b^{-2} dt + 2 f_1 \dot{b}/b - 2\beta^2 b^2 f_3 \ ,$$
$$Q = 2 f_1 a^2 \int a^{-2} dt + 2a^2 f_2 \ , \tag{6.1.18}$$
$$L = f_1 \left(b^2 \int b^{-2} dt + a^2 \int a^{-2} dt \right) + f_2 a^2 + f_3 b^2 \ .$$

Here f_1, f_2, and f_3 do not depend on the time t. We make use of our "false" solution of the system (I–VI) for the even perturbations in order to reduce its order.

For this, we make a linear replacement of the unknown functions: instead of $h_\perp$, $h_\parallel$, L, and Q, we introduce unknown functions $\tilde{f}_1$, $\tilde{f}_2$, $\tilde{f}_3$, and g according to the relations

$$h_\perp = -2\alpha^2 a^2 \left(\int a^{-2} dt \right) \tilde{f}_1 + 4 \tilde{f}_1 \dot{a}/a - 2\alpha^2 a^2 \tilde{f}_2 \ ,$$
$$h_\parallel = -2\beta^2 b^2 \left(\int b^{-2} dt \right) \tilde{f}_1 + 2 \tilde{f}_1 \dot{b}/b - 2\beta^2 b^2 \tilde{f}_3 + 2g \ ,$$
$$Q = 2 \tilde{f}_1 a^2 \left(\int a^{-2} dt \right) + 2a^2 \tilde{f}_2 \ , \tag{6.1.19}$$
$$L = \tilde{f}_1 \left(a^2 \int a^{-2} dt + b^2 \int b^{-2} dt \right) + \tilde{f}_2 a^2 + \tilde{f}_3 b^2 \ .$$

After the substitutions (6.1.19), the system of equations (I–VI) for the even perturbations contains only the derivatives of the functions $\tilde{f}_1$, $\tilde{f}_2$, and $\tilde{f}_3$ with respect to the time. It is therefore convenient to seek not these functions directly, but the following combinations of them:

$$f \equiv \dot{\tilde{f}}_1 \ , \quad m \equiv \dot{\tilde{f}}_1 a^2 \int a^{-2} dt + a^2 \dot{\tilde{f}}_2 \ ,$$
$$n \equiv \dot{\tilde{f}}_1 b^2 \int b^{-2} dt + b^2 \dot{\tilde{f}}_3 \ . \tag{6.1.20}$$

This linear substitution enables us to reduce the order of the system for the even perturbations by three units.

6.1.7 Closed System of Equations for the Even Perturbations

We shall assume that the perturbations are adiabatic and eliminate from (I–VI) the perturbations associated with the magnetic field and with the matter, H, $\delta\varepsilon$, and δp [see the definitions (6.1.3–5)]. We then obtain a closed system of four equations for the functions $h_{\|}$, $h_{\perp}$, L, and Q.

To obtain the first equation, we eliminate H from (III) and (IV). We obtain the second and third equations of the system by substituting in place of $\delta\varepsilon$ and H in (I) and (IV) their values calculated by means of (III) and (V). We take (II) to close the system.

We now replace $h_{\perp}$, $h_{\|}$, L, and Q in this system by the expressions (6.1.19) and make use of the notation f, m, and n defined in (6.1.20).

We obtain the following closed system of equations for the functions f, g, m, and n:

$$\begin{aligned} &a^{-2}\{a^2[\dot{m}+\dot{n}+(a^2/b)^{\cdot}(b/a^2)m+\dot{b}n/b+2f]\}^{\cdot}+4\kappa Wm \\ &\quad +2W(p+\varepsilon+2W)^{-1}\times[-2km/a^2+\beta^2(n-m)+2f(ab)^{\cdot}/ab \\ &\quad +2\dot{g}+2g(b/a)^{\cdot}\,a/b]=0\ , \end{aligned} \tag{6.1.21}$$

$$\begin{aligned} &\dot{a}(-\alpha^2m-\beta^2n+\dot{f}+\dot{g})+\kappa(1+dp/d\varepsilon)(W+k/a^2)f \\ &\quad =(1-dp/d\varepsilon)[-\alpha^2(ab)^{\cdot}/2ab+(\dot{g}-\beta^2n)\dot{a}/a+\alpha^2g/2] \\ &\qquad -(1+dp/d\varepsilon)4^{-1} \\ &\qquad \times\alpha^2[\dot{m}+\dot{n}+(a^2/b)^{\cdot}(b/a^2)m+n\dot{b}/b+2f]\ , \end{aligned} \tag{6.1.22}$$

$$\begin{aligned} &\dot{g}+2(ab)^{\cdot}\,\dot{g}/ab+\alpha^2g-\beta^2\dot{n}-2\beta^2n\dot{a}/a-\alpha^2m\dot{b}/b \\ &\quad -\beta^2f+\dot{b}\dot{f}/b+2f(\ddot{b}/b+2\dot{a}\dot{b}/ab) \\ &\quad =2^{-1}(1-dp/d\varepsilon)[\alpha^2g+(ab)^{\cdot}/ab(-\alpha^2m+2\dot{a}f/a) \\ &\qquad +2\dot{a}/a(-\beta^2n+\dot{g}+\dot{b}f/b)] \\ &\qquad +(3-dp/d\varepsilon)4^{-1} \\ &\qquad \times\alpha^2[\dot{m}+\dot{n}+(a^2/b)^{\cdot}(b/a^2)m+n\dot{b}/b+2f]\ , \end{aligned} \tag{6.1.23}$$

$$-g+m\dot{b}/b+\dot{m}+f=0\ . \tag{6.1.24}$$

In the absence of a cosmological magnetic field, the system is of fourth order, since it admits an integral corresponding to conservation of the even component of the curl of the velocity [see (6.1.8)]:

$$\begin{aligned} &(\alpha^2+\beta^2)(m-n)+2\dot{g}+2(g+f)(b/a)^{\cdot}\,a/b-2km/a^2 \\ &\quad =L(a^2b)^{-1}\ ,\quad L=\text{const}\ . \end{aligned} \tag{6.1.25}$$

Moreover, for $W=0$ we can integrate (6.1.21).

6.1.8 Analysis of the Closed System of Equations for Odd Perturbations

The system of equations (6.1.15, 16) for M and K with odd perturbations having a scale of variation much smaller than the characteristic scale of variation of the background contain solutions for gravitational and Alfvén waves. In fact, if we seek solutions of (6.1.15, 16) in the form of rapidly oscillating expressions

$$M = \mu \exp\left(\mathrm{i}\int \Omega\, dt\right) \quad , \quad K = \chi \exp\left(\mathrm{i}\int \Omega\, dt\right) \quad ,$$

where $\Omega \gg \max(\dot{a}/a,\ \dot{b}/b)$, we obtain for Ω the two values

$$\Omega_1 = \sqrt{\alpha^2 + \beta^2} \quad , \quad \Omega_2 = \sqrt{2W/(p+\varepsilon+2W)}\,\beta \quad ,$$

of which Ω_1 corresponds to a gravitational wave, and Ω_2 to an Alfvén wave.

For the amplitudes $\mu(t)$ and $\chi(t)$ in the gravitational wave, we obtain the expressions

$$\mu(t) = \mu_0 \alpha^2 b \Big/ \sqrt{a^2 b \Omega_1^3} \quad , \quad \chi(t) = -\mu_0 \beta^2 b \Big/ \sqrt{a^2 b \Omega_1^3} \quad , \quad \mu_0 = \text{const} \quad .$$

The rate of change of the amplitude of gravitational waves in anisotropic models depends on the direction of their propagation. Therefore in anisotropic models the spectrum of gravitational radiation cannot be isotropic, provided that the gravitational radiation is not thermalized by some process, such as the interaction with the electromagnetic radiation (see Sect. 1.3), which has an equilibrium character at a high temperature of the matter.

For $2W(p+\varepsilon+2W)^{-1}\beta^2 \lesssim 1$, the Alfvén perturbations have a characteristic time of variation of the order of the characteristic time of variation of the scale factors $a(t)$ and $b(t)$, which, according to the system (6.1.1), is of order $(\sqrt{\kappa(\varepsilon+W)})^{-1}$. It is meaningless to speak of gravitational and Alfvén waves individually if the characteristic time of their variation is greater than or of the order of $(\sqrt{\kappa(\varepsilon+W)})^{-1}$, since for such scales the odd subsystem cannot be split into purely Alfvén and purely gravitational perturbations.

6.1.9 Analysis of the Closed System for Even Short-Wave Perturbations

For short-wave perturbations, we shall seek solutions for f, g, m, and n (6.1.21–24) in the form of amplitudes $\tilde{f}$, $\tilde{g}$, $\tilde{m}$, and $\tilde{n}$, multiplied by a rapidly oscillating factor $\exp(\int \mathrm{i}\Omega\, dt)$. Each of the amplitudes is represented as a formal series in powers of $1/\Omega$, whose principal terms we denote by f_0, g_0, m_0, and n_0.

Retaining the principal terms in (6.1.21–24), we obtain

$$
\begin{aligned}
&- \Omega^2(m_0+n_0)+2\mathrm{i}\Omega f_0 - 2W(p+\varepsilon+2W)^{-1} \\
&\qquad \times [\beta^2(n_0-m_0)-2\mathrm{i}\Omega g_0]=0 \ , \\
&-(1-dp/d\varepsilon)g_0/2+4^{-1}(1+dp/d\varepsilon)[\mathrm{i}\Omega m_0+\mathrm{i}\Omega n_0+2f_0]=0 \ , \\
&(\Omega^2+\alpha^2)g_0+\mathrm{i}\Omega\beta^2 n_0+\beta^2 f_0=0 \ , \quad -g_0+\mathrm{i}\Omega m_0+f_0=0 \ , \\
&\tilde{f}=f_0+f_1/\Omega+\dots \ , \quad \tilde{g}=g_0+g_1/\Omega+\dots \ , \\
&\tilde{m}=m_0+m_1/\Omega+\dots \ , \quad \tilde{n}=n_0+n_1/\Omega+\dots \ .
\end{aligned}
$$

This system has the solution $m_0=n_0$, $g_0=0$, $f_0=-\mathrm{i}\Omega n_0$. In order to obtain the modes of the short-wave oscillations in the equations, we must retain, after the principal terms, the following terms:

$$
\begin{aligned}
&(m_1-n_1)(\Omega^2+2\beta^2 w)+\mathrm{i}\Omega m_0[2(a/b)^{\cdot}\, b/a-4w(ab)^{\cdot}/(ab)] \\
&\qquad +2\mathrm{i}\Omega g_1(1+2w)=0 \ , \\
&w=W/(p+\varepsilon+2W) \ , \\
&\Omega/4(1+dp/d\varepsilon)\alpha^2(m_1-n_1)-\alpha^2(dp/d\varepsilon)g_1 \\
&\qquad +m_0[\dot{a}/a(\Delta^2 dp/d\varepsilon-\Omega^2)+\alpha^2(b/a)^{\cdot}\,(a/b)\,dp/d\varepsilon]=0 \ , \\
&\mathrm{i}\Omega\Delta^2(m_1-n_1)+(\Delta^2+\Omega^2)m_0(\dot{b}/b-\dot{a}/a)-g_1(\Delta^2+\Omega^2)=0 \ , \\
&\mathrm{i}\Omega m_1+\dot{b}m_0/b+f_1-g_1=0 \ , \quad \Delta^2\equiv\alpha^2+\beta^2 \ .
\end{aligned}
\tag{6.1.26}
$$

Equating the determinant of the system (6.1.26) to zero, we obtain

$$
\begin{aligned}
&(\Omega^2-\Delta^2)\{\Omega^4-\Omega^2[\Delta^2 dp/d\varepsilon+2w(\Delta^2-\alpha^2 dp/d\varepsilon)] \\
&\qquad +2w\beta^2\Delta^2 dp/d\varepsilon\}=0 \ .
\end{aligned}
\tag{6.1.27}
$$

The roots $\Omega=\pm\Delta$ refer to a gravitational wave. The corresponding solution of (6.1.26) does not contribute to the perturbation of the density and velocity of the matter, and also does not lead to the appearance of an electromagnetic wave.

The expression in the curly brackets in (6.1.27) is the dispersion relation for the fast and slow magnetohydrodynamic waves (see Chap. 4). Substituting the magnetohydrodynamic modes for Ω into the system (6.1.26), we can calculate the perturbations of the metric, of the magnetic field and of the density of the matter in these waves (the corresponding expressions were given in [6.19]).

Short-wave even perturbations diffuse in magnetohydrodynamic and gravitational waves. However, if the inhomogeneities have a sufficiently large characteristic scale, they will be sustained by the forces of gravity. We assume that for such scales the solution for the gravitational waves is, as before, rapidly oscillating, i.e., $\Delta \gg \max(|\dot{a}/a|, |\dot{b}/b|)$. Then, using the gravitational-wave solution, we can reduce the order of the system (6.1.21–24) by two units.

After this, with accuracy up to terms of order $\dot{a}/a\Delta$ and $\dot{b}/b\Delta$, we obtain the equations of general relativistic magnetohydrodynamics with gravity in a cosmological model with an unperturbed metric tensor satisfying the system (6.1.1). An analysis of the resulting system of equations leads to the following conclusions. In the presence of a cosmological magnetic field, the minimal dimensions of the inhomogeneities of the matter which are sustained by gravity are different from those in isotropic models, where one can use the Jeans criterion: if along the field the critical size of the inhomogeneities is of the Jeans order $\sqrt{[\kappa(\varepsilon + W)]^{-1} dp/d\varepsilon}$, then across the magnetic field the critical size of the clustering of matter is of order

$$\sqrt{[dp/d\varepsilon + 2w(1 - dp/d\varepsilon)][\kappa(\varepsilon + W)]^{-1}} \quad , \quad w = W/(p + \varepsilon + 2W) \ .$$

The evolution of a cluster of matter with dimensions much greater than these critical dimensions takes place according to the linear theory during the characteristic time of development of the unperturbed model. The rate of increase of the density of matter in these clusters depends significantly on the configuration of the initial perturbation (i.e., on the ratio a/β) if the dimensions of the perturbations are much smaller than $(\sqrt{\kappa(\varepsilon + W)})^{-1}$.

6.1.10 Evolution of Perturbations of Arbitrary Finite Scales Near a "Pancake" Singularity

Let us consider the character of the development of inhomogeneities in a sufficiently small interval of time near the singular instant at which a "pancake" is produced. For a gas with the equation of state $0 \le dp/d\varepsilon \equiv \lambda = \text{const} < 1$, we can deduce from the system (6.1.1) the following asymptotic behavior for the energy density and the scale factors:

$$\begin{aligned} &a = a_0[1 + (2 - 2\lambda)^{-1} \kappa \varepsilon_0 t^{(1-\lambda)} + \ldots] \ , \quad b \approx b_0 t \ , \\ &\varepsilon = \varepsilon_0 t^{-(1+\lambda)} + \ldots \ , \quad \lambda \equiv dp/d\varepsilon \ , \\ &a_0 = \text{const} \ , \quad b_0 = \text{const} \ , \quad \varepsilon_0 = \text{const} \ . \end{aligned}$$

Because of the contraction of the scales along the z axis, the moduli of the gradients of the functions with respect to z, for a sufficiently small interval of time, will be much greater than the moduli of the gradients in the plane of x^1 and x^2. We shall now assume that $\alpha^2 \ll \beta^2$. For any scale of inhomogeneity, this inequality is satisfied for sufficiently small t, since β has the asymptotic behavior $\beta \approx \beta_0/t$.

Equations (6.1.21–24) for a gas with the equation of state indicated above take the following asymptotic form:

$$\begin{aligned} &[\dot{m} + \dot{n} - m/t + n/t + 2f]^{\cdot} = 0 \ , \\ &\dot{f} + (\dot{g} - n/t^2)\lambda = 0 \ , \\ &\ddot{g} + 2\dot{g}/t - \beta_0^2 \dot{n} + \dot{f}/t - f\beta_0^2/t^2 = 0 \ , \\ &- g + m/t + \dot{m} + f = 0 \ . \end{aligned} \tag{6.1.28}$$

A solution of the system (6.1.28) can be sought in the form

$$m = m_0 t^{\gamma+1} \ , \quad n = n_0 t^{\gamma+1} \ , \quad f = f_0 t^{\gamma} \ ,$$
$$g = g_0 t^{\gamma} \ , \quad \gamma = \text{const} \ ; \tag{6.1.29}$$

substituting the expression (6.1.29) into (6.1.28) and equating to zero the determinant of the resulting system for m_0, n_0, f_0, and g_0, we obtain the "dispersion equation"

$$\gamma^2 \left[\beta_0^2 + (\gamma + 2)^2\right] \left[(\gamma + 1 - \lambda)\gamma + \lambda\beta_0^2\right] = 0 \ . \tag{6.1.30}$$

The relative perturbation of the density, $\delta\varepsilon/\varepsilon$, then has the asymptotic behavior t^{γ}.

Thus, the magnetic field near a "pancake" singularity is of little importance, so that the angular momentum in an ideal gas is conserved. We recall that if there were no magnetic field at all, the system (6.1.21–24) would admit the first integral (6.1.25).

It should be stressed that the dispersion relation (6.1.30) indicates the existence of a wave zone for perturbations of arbitrary scales in an ideal gas and in a gravitational field near a "pancake" singularity.

The root $\gamma = 0$ corresponds to rotational perturbations which are frozen in the matter. The vanishing of the second factor in (6.1.30) corresponds to the modes of the gravitational waves. A "pancake" singularity is unstable with respect to such perturbations, which become infinitely large in the limit $t \to 0$.

The third factor corresponds to adiabatic perturbations in an ideal gas. The dispersion relation for adiabatic perturbations, $\gamma(\gamma + 1 - \lambda) + \beta_0^2\lambda = 0$, indicates that there are wave properties only in the case of perturbations for which $\beta_0^2 > (1 - \lambda)^2/4\lambda$ (analog of the Jeans criterion). The amplitude of "sound" perturbations falls off with time as $t^{-(1-\lambda)/2}$. Long-wave perturbations, for which the inequality sign is reversed, do not have wave properties. The larger the scale of these inhomogeneities, the slower the falloff of the relative perturbations of the density. Finally, for the largest scales, the relative perturbations become frozen (see below).

As t increases, the boundary between the scales of the perturbations which diffuse in sound waves and those which are sustained by gravity is shifted toward smaller scales. Therefore the perturbations of a fixed scale with $\beta_0^2 > (1 - \lambda)^2/4\lambda$ first diffuse with a decrease of the amplitude $\delta\varepsilon/\varepsilon \sim t^{-(1-\lambda)/2}$, but after a certain time the perturbations become frozen, as it were, losing their oscillatory properties. This means that after that time the inhomogeneities fall in the long-wave part of the spectrum.

After isotropization (equalization of the rates of expansion in all directions) of anisotropic models, the inverse process begins: the growth of the inhomogeneities leads to an increase of the gradients of the density. In a certain interval of scales, the growth of the gradient of the density causes the inhomogeneities to "burst", and they then diffuse in sound waves.

As to the rotational perturbations, for sufficiently small scales, some time after the singular instant they begin to diffuse in slow magnetohydrodynamic waves.

By the time of isotropization, the strength of the cosmological magnetic field falls so much that slow MHD waves are again converted into frozen vortices.

6.2 Self-Similar Motions of a Photon Gas in the Friedman-Lemaître Model

In the early stages of expansion of the universe, the dominant contribution to the energy of the matter comes from the electromagnetic radiation, which is related to the cosmic radiation of a black hole with temperature 2.83 K [6.2]. In the simplest model of a universe filled with radiation — the Friedman-Lemaître model with a flat comoving space — the analytic expressions for the metric and the energy density in Lagrangian coordinates have the form

$$\begin{aligned} ds^2 &= c^2 d\tau^2 - a_0 \tau [d\xi^2 + \xi^2 (d\theta^2 + \sin^2\theta \, d\varphi^2)] \ , \\ \varepsilon &= 3c^2/32\pi G\tau^2 \ . \end{aligned} \tag{6.2.1}$$

Here a_0 is a constant, τ is the "world time", and ξ is the Lagrangian radial coordinate.

The solution (6.2.1) of Einstein's equations is a particular self-similar solution. Therefore it is natural to consider the position of the Friedman-Lemaître solution in the class of self-similar spherically symmetric solutions and to study the physical and analytic properties of these solutions, including solutions with shock waves.

In this section, spherically symmetric self-similar motions are studied in an orthogonal spherical coordinate system of an observer:

$$ds^2 = c^2 e^{\nu} dt^2 - e^{\gamma} dr^2 - r^2 (d\theta^2 + \sin^2\theta \, d\varphi^2) \ . \tag{6.2.2}$$

Here $\nu(r,t)$ and $\gamma(r,t)$ are unknown functions.

The problem of self-similar spherically symmetric motions of a gravitating gas in the general theory of relativity was first formulated in [6.28]. Owing to a choice of variables which was not entirely successful, the correctly written conditions on gas-dynamical shock waves in general relativity lacked the simplicity that is inherent in these conditions in the special theory of relativity [6.29]; however, it was possible to derive a condition on a break after which the gas goes into a state of rest.

Einstein's equation was reduced [6.28] to a second-order equation with radicals. For a fixed velocity of the shock wave, the parameters of the gas were calculated behind the shock wave, and from these data an integral curve outside the stationary core was constructed numerically. A system of ordinary differential equations for self-similar motions was written down [6.30] and was later greatly simplified [6.31] by choosing a new scaling variable. In the comoving coordinate system, a study was made [6.32] of the self-similar problem in the case of the maximally rigid equation of state $p = \varepsilon$. In the theory of small perturbations, self-similar perturbations of the Friedman-Lemaître model were considered in [6.33]. A qualitative analysis of a self-similar system in a conformally static coordinate system was given in [6.34]. An extended concept of self-similarity in general relativity was analyzed from a group-theoretical point of view in [6.35].

The system of ordinary differential equations for self-similar motions of a gravitating gas in general relativity was studied in the comoving coordinate system in [6.36], where it was shown, in particular, that it is possible to match the known solutions by means of the conditions on the breaks. A study was made [6.37] of the self-similar formation of nonstationary black holes in cosmological models which at large distances tend to the flat Friedman-Lemaître model. Use was made of the comoving coordinate system.

This section is based on [6.38]. Here we derive a closed system of two first-order ordinary differential equations, which is convenient for a qualitative investigation. It is shown that the conditions on shock waves for a gas with the equation of state $p = \varepsilon/3$ have, in special variables, the same form in general relativity as in special relativity. For certain solutions, we demonstrate the existence of a limiting sphere (a nonsingular horizon) outside which the coordinate system (6.2.2) becomes meaningless. This happens because, for these cosmological solutions, particles outside the light horizon have a speed with respect to the center of symmetry exceeding the speed of light.

The integral curves on which the speed of the gas is equal to zero at the center of symmetry form a one-parameter family containing the Friedman-Lemaître solution. Therefore all solutions not containing a source or empty space at the center of symmetry must be converted into this family by means of either a shock transition or a weak break. It is interesting to note that all solutions with weak breaks have a light horizon on which the speed of the gas with respect to the system (6.2.2) is equal to the speed of light but the pressure is finite. In these solutions, the whole of space-time cannot be covered by the coordinate grid (6.2.2).

The situation with regard to solutions possessing shock waves is quite different. It turns out that there exists a critical wave intensity. If a solution exhibits a break with a subcritical intensity, then the solution possesses a light horizon. Solutions containing a break with a supercritical intensity do not possess a horizon and have all the qualitative properties of the self-similar Cauchy problem of focusing of a gas to the center, as described in [6.39].

In the case of formation of a stationary core, none of the solutions with a shock wave in the system (6.2.2) have a light horizon and, conversely, all those with a weak break do have one. We establish for what initial data the problems of outflow and focusing are physically meaningful. We describe the phase portraits of the integral curves near the light horizon, the sound line, and the coordinate origin. We present the results of the following numerical calculations: (1) solutions with a spherical shock wave of subcritical or supercritical intensity, inside which there is either a "segment" of a Friedman-Lemaître universe or a stationary core; (2) solutions with a weak break on a sound surface, inside which there is either a Friedman solution or a stationary core; (3) inhomogeneous cosmological models possessing a light horizon.

6.2.1 Derivation of a Closed System of Ordinary Differential Equations and Conditions on Shock Waves

For our purposes, it is convenient to introduce the radial velocity V, measured in terms of the speed of light, which is defined at a point as the ordinary velocity with respect to a local Lorentz coordinate system with basis vectors directed along the coordinate lines of (6.2.2). In terms of the components of the 4-velocity u^i and the metric coefficients $\exp\gamma$ and $\exp\nu$, the velocity V can be expressed as follows:

$$V \equiv u^r \exp[(\gamma - \nu)/2]/u^4 \ .$$

For self-similar solutions, the unknown functions $\nu(r,t)$, $\gamma(r,t)$, $V(r,t)$, and $p(r,t)$ have the form

$$\nu = \nu(\lambda) \ , \quad \gamma = \gamma(\lambda) \ , \quad \kappa p = P(\lambda)/r^2 \ , \quad \lambda \equiv r/(ct) \ .$$

We introduce a new scaling variable ζ having the meaning of the velocity of the surface $\lambda = \text{const}$ with respect to the local Lorentz coordinate system:

$$\zeta = \lambda \exp[(\gamma - \nu)/2] \ . \tag{6.2.3}$$

We introduce the notation $x \equiv \exp\gamma$, and we shall find the dependence of the functions ν, x, P, and V on the variable ζ.

Einstein's equations in this case have the form

$$\kappa T_4^r = R_4^r \ , \quad 4PV = x^{-2}\zeta^2(1-L)(1-V^2)\,dx/d\zeta \ , \tag{6.2.4}$$

$$\kappa T_r^r = R_r^r - R/2 \ ,$$
$$-P(3V^2+1) = \left[1 - \frac{1}{x} - \frac{\zeta}{x}(1-L)\frac{d\nu}{d\zeta}\right](1-V^2) \ , \tag{6.2.5}$$

$$\kappa T_4^4 = R_4^4 - R/2 \ ,$$
$$P(3+V^2) = \left[1 - \frac{1}{x} + \frac{\zeta}{x^2}\frac{dx}{d\zeta}\right](1-V^2) \ , \tag{6.2.6}$$

$$L \equiv \frac{2(x-1)(V^2\zeta + \zeta - 2V)}{\zeta(3+V^2) - 4V} \ . \tag{6.2.7}$$

In the expressions on the left, we have indicated the corresponding components.

The function $P(\zeta)$ can be calculated explicitly from (6.2.4, 6) if the solutions $x(\zeta)$ and $V(\zeta)$ are known:

$$P(\zeta) = (x-1)(1-V^2)\zeta x^{-1}[(3+V^2)\zeta - 4V)]^{-1} \ . \tag{6.2.8}$$

The coefficient $\nu(\zeta)$ can be found by quadrature from (6.2.5) in terms of the known $x(\zeta)$ and $V(\zeta)$ if $P(\zeta)$ is replaced by the expression (6.2.8):

$$\zeta\frac{d\nu}{d\zeta}(1-L)=\frac{4V(x-1)}{\zeta(3+V^2)-4V} .$$

Eliminating the function $P(\zeta)$ from (6.2.4, 6), we obtain

$$(1-L)\zeta\frac{dx}{d\zeta}=\frac{4V(x-1)x}{\zeta(3+V^2)-4V} . \tag{6.2.9}$$

Eliminating the pressure from the conservation equations $\nabla_i T^i_r = 0$ and $\nabla_i T^i_4 = 0$, we have

$$\begin{aligned}\zeta\frac{dV}{d\zeta}(1-L)&=\frac{(1-V^2)[2V-\zeta V^2/2-3\zeta/2+\Omega]}{3(V-\zeta)^2-(1-V\zeta)^2} ,\\ \Omega&\equiv 4(x-1)[V^3\zeta-3V\zeta+3\zeta^2/2+V^2-V^4\zeta^2/2]\\ &\quad\times[\zeta(3+V^2)-4V]^{-1} .\end{aligned} \tag{6.2.10}$$

Equations (6.2.9, 10) constitute a system of equations for $x(\zeta)$ and $V(\zeta)$. The relation between the variables ζ and λ can be obtained by integration of the equation

$$\zeta(1-L)\,d\ln\lambda/d\zeta=1 , \tag{6.2.11}$$

which follows from the relation (6.2.3).

Conditions on Breaks. The definition of differentiable manifolds involves a fixed class of local coordinate systems, within which the transition from one coordinate system to another must satisfy given smoothness conditions. In order to encompass the possible appearance of gas-dynamical shock waves, the coordinate transformations in the distinguished class of coordinate systems must be twice differentiable and must have piecewise-smooth third derivatives. In this case, in the absence of a medium, breaks in the second derivatives of the metric on nonisotropic surfaces can be eliminated by an appropriate choice of breaks in the third derivatives of the coordinate transformations.

The differential operator $(R_{ik}-g_{ik}R/2)n^i$ contains only first derivatives of the metric along the normal to the surface of the break and is therefore continuous. As a consequence of Einstein's equations, we can then deduce continuity of the flux of energy and momentum across a shock wave, $[T_{ik}]n^k = 0$.

When a coordinate system is distinguished by means of some auxiliary constraints on the form of the metric of the four-dimensional space, it may happen that the distinguished coordinate grid does not belong to the class of coordinate systems indicated above. In such coordinate systems, there can be breaks in both the metric itself and its first derivatives. In this case (see, for example, [6.40]), on a nonisotropic surface of a break the first and second quadratic forms of the surface of the break must both be continuous. For the coordinate system (6.2.2), this implies continuity of the metric and of the flux of energy and momentum across the surface of the break.

Suppose that the equation of the surface of a shock wave has the form $f(r,t) = 0$. We denote by $c\tilde{\zeta}$ the velocity of the shock wave with respect to an orthonormalized tetrad of the local coordinate basis (6.2.2):

$$\tilde{\zeta} = -\exp[(\gamma - \nu)/2] f_{,t}/f_{,r} \quad .$$

We write down the components of the 4-normal to the shock wave:

$$n_r = -\exp(\gamma/2)(1-\tilde{\zeta}^2)^{-1/2} \quad , \quad n_4 = \exp(\nu/2)\tilde{\zeta}(1-\tilde{\zeta}^2)^{-1/2} \quad .$$

From the conditions $n_i[T^i_r] = n_i[T^i_4] = 0$ it follows that

$$[(p+\varepsilon)\tilde{\zeta}V - p(1-V^2) - (p+\varepsilon)V^2/(1-V^2)] = 0 \quad , \tag{6.2.12}$$

$$[(p+\varepsilon)(\tilde{\zeta} - V)/(1-V^2) - p\tilde{\zeta}] = 0 \tag{6.2.13}$$

respectively. It is remarkable that these conditions have the same form as in special relativity [6.41].

Eliminating the pressure from the conditions (6.2.12, 13), we obtain for an ultrarelativistic gas the expression

$$V_2 = (3\tilde{\zeta}^2 - 1 - 2V_1\tilde{\zeta})/(2\tilde{\zeta} + V_1(\tilde{\zeta}^2 - 3)) \quad , \tag{6.2.14}$$

where V_1 and V_2 are the speeds of the particles of the gas before and after the break, in units of the speed of light.

Equation (6.2.14) means that, in the photon gas in the coordinate system associated with the break, the product of the speeds before and after the break is equal to the square of the speed of sound.

6.2.2 Friedman Solution and Qualitative Investigation of the System of Equations for $x(\zeta)$ and $V(\zeta)$

The metric tensor in the Friedman solution in the Lagrangian coordinates (6.2.1) can be readily transformed to the coordinate system (6.2.2) of the observer. One transformation is obviously $r^2 = a_0\tau\xi^2$, and another can be found from the condition of orthogonality of the metric (6.2.2). It is convenient to seek it in the form

$$\tau = r/f(\lambda) \quad , \quad \lambda \equiv r/ct \quad .$$

After simple calculations, we obtain for the function $f(\lambda)$ and the speed $V(\lambda)$ the expressions

$$V = f/2 = \lambda/\left(1+\sqrt{1-\lambda^2}\right) \quad . \tag{6.2.15}$$

The metric coefficients $\exp\gamma$ and $\exp\nu$ in the Friedman solution are identical:

$$\exp\gamma = \exp\nu = \left(1+\sqrt{1-\lambda^2}\right) \Big/ 2\sqrt{1-\lambda^2} \ . \tag{6.2.16}$$

For the pressure as a function of the coordinates, we have

$$p = c^6 t^2 \left(1-\sqrt{1-\lambda^2}\right)^2 \Big/ 8\pi G r^4 \ . \tag{6.2.17}$$

The coordinates (6.2.2) can span only part of the Friedman solution in the infinite comoving space.

We shall carry out a qualitative investigation of the system (6.2.9, 10).

We first linearize the system (6.2.9, 10) near the straight line $\zeta = 0$, $V = 0$. We introduce the notation $V/\zeta \equiv q$ and reduce the system (6.2.9, 10) to the form

$$\begin{aligned} \frac{d\zeta}{\zeta} &= \frac{2\,dq[3-4q-2(x-1)(1-2q)]}{(3-4q)[3(1-2q)-4(x-1)(1-q)]} \\ &= \frac{dx[3-4q+2(x-1)(1-2q)]}{4q(x-1)x} \ . \end{aligned} \tag{6.2.18}$$

It follows from this that in the first approximation in V and ζ the integral curves near the straight line $\zeta = 0$, $V = 0$ will lie on surfaces $x = x(V/\zeta, \alpha)$, where $x(q, \alpha)$ is an integral curve of the equation

$$\frac{dx}{dq} = \frac{8q(x-1)x}{(3-4q)[3(1-2q)-4(x-1)(1-q)]} \ . \tag{6.2.19}$$

A schematic representation of the integral curves of this equation is shown in Fig. 6.1. According to (6.2.19), in order to reach the value $\zeta = 0$ along the integral curve with the direction $V/\zeta = q_0$, it is necessary that the point q_0, x_0 be a singular point of the equation (6.2.19). Further investigation shows that one of these singular points (the point E in Fig. 6.1) has coordinates $q_0 = 1/2$, $x_0 = 1$, and the other point D has coordinates $q = 0$, $x_0 = 7/4$. The family of integral curves depending on a single parameter P_1 and going into the point $E(q_0 = 1/2$, $x_0 = 1)$ has the following asymptotic behavior:

$$\begin{aligned} x - 1 &= P_1\zeta^2\left[1+\left(\frac{4}{5}-\frac{P_1}{5}\right)\zeta^2\right. \\ &\quad \left.+\left(\frac{207}{280}-\frac{17}{35}P_1+\frac{4}{35}P_1^2\right)\zeta^4+\dots\right] \ , \\ V &= \zeta\left[\frac{1}{2}+\left(\frac{7}{40}-\frac{P_1}{5}\right)\zeta^2\right. \\ &\quad \left.+\frac{1}{14}\left(\frac{61}{40}-\frac{5}{2}P_1-\frac{2}{5}P_1^2\right)\zeta^4+\dots\right] \ , \\ \kappa p &= P_1 t^{-2}[1+(1-2P_1)\zeta^2+\dots] \ . \end{aligned} \tag{6.2.20}$$

For the parameter value $P_1 = 1/4$, these expressions represent the first terms of the expansion of the Friedman solution (6.2.15–17). The exact solution $x_0 = 7/4$,

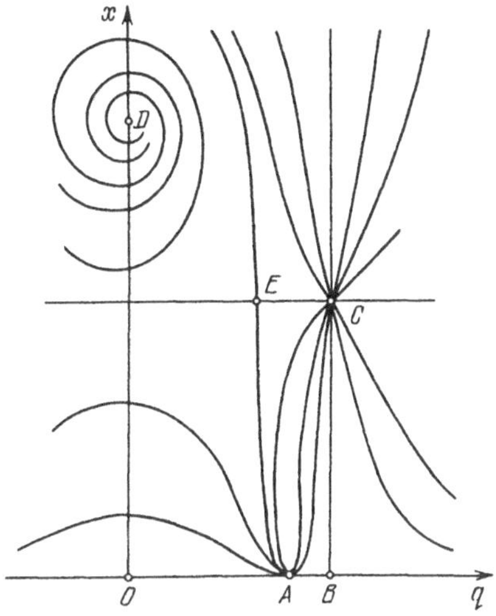

Fig. 6.1. Integral curves of (6.2.19). The singular points have the coordinates $A(7/10, 0)$, $B(3/4, 0)$, $C(3/4, 1)$, $D(0, 7/4)$, and $E(1/2, 1)$

$V_0 = 0$, $p_0 = (7\kappa r^2)^{-1}$ corresponds to a static configuration of the gas (the state of rest).

The complete set of solutions of (6.2.9, 10) depends on two parameters, while the solutions (6.2.20) depend on only one parameter, so that all the other solutions must go over into one of the curves of (6.2.20) or into a static solution, either through a weak break along a sound line or through a strong break (if, of course, there is no source or empty space at the center).

We shall now study the behavior of the integral curves of the system (6.2.9, 10) near the light cone $V = 1$, $\zeta = 1$.

If we linearize the system (6.2.9, 10) in $\zeta - 1$ and $V - 1$, then, introducing the notation $(V - 1)/(\zeta - 1) \equiv \tilde{q}$, this system can be reduced to the form

$$\frac{d\zeta}{\zeta - 1} = \frac{(4 - \tilde{q} - 2x)(\tilde{q}^2 + 1 - 4\tilde{q})\, d\tilde{q}}{\tilde{q}(2 - \tilde{q})(3\tilde{q} - \tilde{q}^2 + 1)} = \frac{(4 - \tilde{q} - 2x)\, dx}{2(x - 1)x} . \tag{6.2.21}$$

In the first approximation, the integral curves lie on the family of surfaces $x = x((V - 1)/(\zeta - 1), \alpha)$, where $x(\tilde{q}, \alpha)$ is an integral curve of the equation

$$\frac{dx}{d\tilde{q}} = \frac{2(x - 1)x(\tilde{q}^2 + 1 - 4\tilde{q})}{\tilde{q}(2 - \tilde{q})(3\tilde{q} - \tilde{q}^2 + 1)} . \tag{6.2.22}$$

A schematic representation of the integral curves of (6.2.22) is shown in Fig. 6.2, where the arrows indicate the direction of increase of ζ. A schematic representation of the integral curves of (6.2.21) for $x > 1$ and for a fixed value of the constant of integration in (6.2.21), projected onto the plane of V and ζ, is shown in Fig. 6.3. The direction $(V - 1)/(\zeta - 1) = 2$ is singular.

The asymptotic behavior of the ends of the loops (Fig. 6.3), which approach the straight line $V = \zeta = 1$ and $x \to \infty$ with a finite slope $\tilde{q}_0 \neq 0$, is as follows:

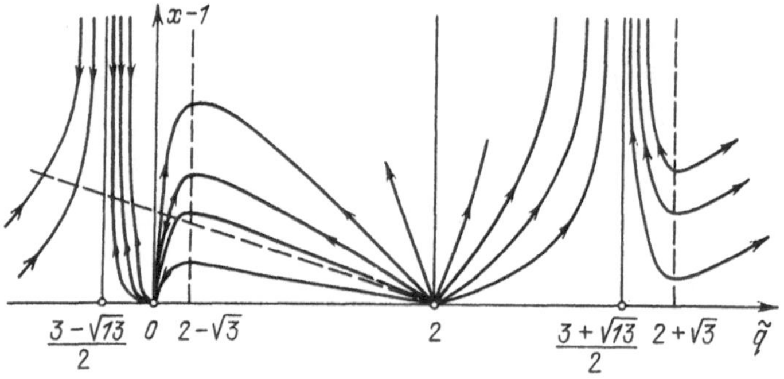

Fig. 6.2. Integral curves of (6.2.22)

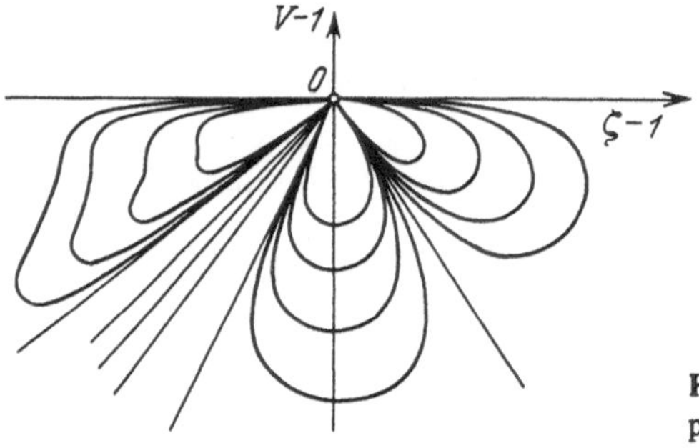

Fig. 6.3. Integral curves of the system (6.2.21), projected onto the plane of V and ζ

$$x = \frac{\alpha}{(1-\zeta)} + O(1) ,$$
$$V - 1 = q_0(\zeta - 1) + O((\zeta-1)^2) , \qquad (6.2.23)$$
$$P \approx \frac{q_0}{q_0 - 2} .$$

For the slope $q = 0$, we find that the curves have the following asymptotic behavior:

$$(x-1) = \alpha_1(\zeta - 1) + O((\zeta-1)^2) ,$$
$$(V-1) = -\beta_1(\zeta-1)^2 + O((\zeta-1)^3) ,$$
$$P = \alpha_1\beta_1(\zeta-1)^2/2 .$$

Here α_1 and β_1 are arbitrary positive constants.

It is quite essential to consider the asymptotic behavior of the curves for $q \to \infty$. An analysis shows that there exists a one-parameter family of integral curves, not contained in the two-parameter family of integral curves of (6.2.21), having the following asymptotic behavior:

$$V = 1 - \alpha\sqrt{(1-\zeta)} + (3/4 + \alpha^2/16)(1-\zeta) + \dots ,$$
$$x(\alpha^2+2) = \alpha \Big/ \sqrt{(1-\zeta)} + (\alpha^4 - 36)/16 + \dots . \qquad (6.2.24)$$

For $\alpha = \sqrt{2}$, the expressions (6.2.24) represent the asymptotic form of the Friedman solution for $\zeta \to 1$. Numerical calculations show that the one-parameter family having the asymptotic behavior (6.2.24) for $\zeta \to 1$ will have the asymptotic behavior (6.2.20) for $\zeta \to 0$.

As we shall show below, these solutions are submanifolds of inhomogeneous cosmological models bounded by a light horizon. It is interesting that for the Friedman-Lemaître solution the coefficient $\alpha/(\alpha^2+2)$ in the asymptotic behavior (6.2.24) for x has the maximum value.

We now investigate the structure of the weak breaks near the sound lines, which are determined as the zeros of the numerator and denominator of the right-hand side of (6.2.10):

$$\zeta_0 = \frac{\left(V_0\sqrt{3}+1\right)}{\left(V_0+\sqrt{3}\right)} \; , \quad x_0 - 1 = \frac{\left(V_0-\sqrt{3}\right)^2}{4\left(1-V_0^2\right)} \; . \tag{6.2.25}$$

As ζ_0 varies from $\sqrt{3}/3$ to $\sqrt{3}/2$, we obtain the sound lines in the various inhomogeneous cosmological models described above. When ζ_0 is equal to $\sqrt{3}/3$, we obtain from (6.2.25) the sound line in the static solution $x_0 = 7/4$, $V_0 = 0$. Near this line, the curves form a degenerate node at which the curves intersect the sound line, being tangential to the straight line $V = 0$, $x = 7/4$.

For $\sqrt{3}/3 < \zeta_0 < \sqrt{3}/2$, the curves near the sound line (6.2.25) form a node lying in the plane:

$$x - x_0 = \delta(\zeta - \zeta_0) \; ,$$
$$\delta = \frac{V_0\left(3-V_0^2\right)\left(\sqrt{3}+V_0\right)\left(7-2\sqrt{3}\,V_0-3V_0^2\right)}{2\left(1-V_0^2\right)^2\left(\sqrt{3}-3\sqrt{3}\,V_0^2+4V_0\right)\left(1+\sqrt{3}\,V_0\right)} \; .$$

Finally, if $\zeta_0 = \sqrt{3}/2$, we obtain from (6.2.25) the sound line in the Friedman-Lemaître solution (6.2.15–17) ($x_0 = 3/2$, $V_0 = \sqrt{3}/2$), near which the curves form a degenerate node. All the curves intersect the sound line, being tangential to the straight line

$$x - \frac{3}{2} = 2\sqrt{3}\left(\zeta - \frac{\sqrt{3}}{2}\right) \; , \quad V - \frac{\sqrt{3}}{3} = 4\left(\zeta - \frac{\sqrt{3}}{2}\right)\Big/3 \; .$$

For $\zeta_0 > \sqrt{3}/2$, the points on the sound line (6.2.25) are foci.

An investigation of the system (6.2.9, 10) gives the following asymptotic expressions for $x(\zeta)$, $V(\zeta)$, and $P(\zeta)$ in the limit $\zeta \to \infty$:

$$x = x_0 + \mathrm{O}(1/\zeta) \; , \quad V = V_0 + \mathrm{O}(1/\zeta) \; , \quad P = P_0 + \mathrm{O}(1/\zeta) \; ,$$
$$P_0 = \left(1-V_0^2\right)(x_0-1)x_0^{-1}\left(3+V_0^2\right)^{-1} \; . \tag{6.2.26}$$

(Concrete expressions for the coefficients of $1/\zeta$ were given in [6.19], and for $x_0 - 1 = (3+V_0^2)[2(1+V_0^2)]^{-1}$ the solutions were decomposed [6.19] into powers of $\zeta^{-1/2}$.)

The relation between the scaling variable λ and the variable ζ defined in (6.2.3) has the following form at large ζ:

$$d\ln\zeta/d\ln\lambda = 1 - 2(x_0 - 1)\left(V_0^2 + 1\right) \Big/ \left(3 + V_0^2\right) \ .$$

Large positive values of ζ correspond to $\lambda \gg 1$ for $1 - L > 0$ or for

$$x_0 < \left(5 + 3V_0^2\right)\left[2\left(1 + V_0^2\right)\right]^{-1} \ . \tag{6.2.27}$$

For such x_0, when $\zeta \to \infty$, we obtain a restriction on the possible value of the constant P_0 in (6.2.26):

$$P_0 < \left(1 - V_0^2\right)\left(5 + 3V_0^2\right)^{-1} \ . \tag{6.2.28}$$

Thus, the self-similar Cauchy problem of focusing or outflow for a gravitating photon gas has a solution subject to the constraints (6.2.27, 28). The actual integration of the system (6.2.9, 10) may make these inequalities even more stringent.

6.2.3 Discussion of the Results

In order to investigate the possibility of analytic continuation of the solutions beyond the light horizon, let us consider the relation between the comoving coordinate system and the coordinate system of the observer for self-similar solutions. Suppose that the metric in the comoving coordinate system has the form

$$ds^2 = a^2 d\tau^2 - b^2 d\xi^2 - r^2(d\theta^2 + \sin^2\theta\, d\varphi^2) \ , \tag{6.2.29}$$

where the metric coefficients a^2, b^2, and r^2 are unknown functions of ξ and τ. In the comoving coordinate system, the momentum equation gives $p = f(\tau)a^{-4}$, and from the energy equation it follows that $p^{3/4}r^2 b = \varphi(\xi)$, where $f(\tau)$ and $\varphi(\xi)$ are arbitrary functions of their arguments. We fix the time τ and the Lagrangian coordinate ξ by putting $f(\tau) = A^4\tau^{-2}$, $\varphi(\xi) = A^3\xi^{1/2}$, $A = \text{const.}$

The self-similar solutions in the comoving coordinate system are distinguished by the requirement

$$a = a(\mu) \ , \quad b = b(\mu) \ , \quad r = \xi R(\mu) \ , \quad p = A^4/(\tau^2 a^4(\mu)) \ ,$$

where $\mu \equiv \xi/\tau$.

By definition, the velocity 4-vector in the comoving coordinate system has only one nonzero component u^4: $u^i(0,0,0,a^{-1})$. According to the rule for transformation of the components of vectors in the coordinate system with the metric (6.2.2), the vector u^i will have the components

$$u^r = a^{-1}\partial r/\partial\tau \quad , \quad u^4 = a^{-1}\partial t/\partial\tau \quad . \tag{6.2.30}$$

Dividing the first of these equations by the second and making use of the definition of the velocity V, we obtain

$$\partial r/\partial\tau = V\exp[(\nu-\gamma)/2]\,\partial t/\partial\tau \quad . \tag{6.2.31}$$

From the condition of orthogonality of the comoving coordinate system ($g_{\tau\xi} = 0$) we obtain, using this relation,

$$V\,\partial r/\partial\xi = \exp[(\nu-\gamma)/2]/(\partial t/\partial\xi) \quad . \tag{6.2.32}$$

Substituting into (6.2.31,32) a self-similar dependence of r and t on ξ and τ in the form $r = \xi R(\mu)$, $t = \xi R(\mu)/\lambda$, from (6.2.31,32) we obtain

$$\begin{aligned} &(1-L)\,d\ln R/d\zeta = V(V-\zeta)^{-1}\zeta^{-1} \quad , \\ &(1-L)\,d\ln\mu/d\zeta = -(1-V^2)(V-\zeta)^{-1}(1-V\zeta)^{-1} \end{aligned} \tag{6.2.33}$$

[L is defined by (6.2.7)].

Using (6.2.33) in going from the system (6.2.2) to the system (6.2.29) according to the tensor law of transformation of the components of the metric, we obtain for the coefficients a and b the expressions

$$\begin{aligned} &a^2 = (\mu R/\zeta)^2 x(1-V\zeta)^2/(1-V^2) \quad , \\ &b^2 = (R/\zeta)^2 x(\zeta-V)^2/(1-V^2) \quad . \end{aligned} \tag{6.2.34}$$

Equations (6.2.33, 34) determine the dependence of R, a, and b on μ in parametric form (in terms of ζ).

Substituting into (6.2.34) the asymptotic behavior (6.2.24) of the inhomogeneous cosmological models near the horizon, we obtain expressions for $R(\zeta)$, $\mu(\zeta)$, $a^2(\zeta)$, and $b^2(\zeta)$ in the form of series in powers of $\sqrt{1-\zeta}$:

$$\begin{aligned} &R = R_0\left[1+(\alpha^2+2)\sqrt{1-\zeta}/(2\alpha^3)+\mathrm{O}(1-\zeta)\right] \quad , \\ &\mu = \mu_0\left[1-(\alpha^2+2)\sqrt{1-\zeta}/\alpha^3+\mathrm{O}(1-\zeta)\right] \quad , \\ &a^2 = \mu_0^2 R_0^2\left[\alpha^2/(2\alpha^2+4)+\mathrm{O}\left(\sqrt{1-\zeta}\right)\right] \quad , \\ &b^2 = R_0^2\left[\alpha^2/(2\alpha^2+4)+\mathrm{O}\left(\sqrt{1-\zeta}\right)\right] \quad , \end{aligned} \tag{6.2.35}$$

where μ_0, R_0, and α are arbitrary constants. Expressing $\sqrt{1-\zeta}$ in terms of $\mu-\mu_0$, we obtain for R, a^2, and b^2 analytic series in powers of $\mu-\mu_0$ which contain no singularities on the light horizon. Formally, the continuation through the horizon in this case corresponds to a reversal of the sign in front of the square root $\sqrt{1-\zeta}$, with ζ remaining less than unity outside the light horizon as well.

Solutions with the asymptotic behavior (6.2.23) can also be analytically continued through the light horizon. For the functions $R(\zeta)$, $\mu(\zeta)$, $a^2(\zeta)$, and $b^2(\zeta)$ in this case, we have series expansions in powers of $\zeta-1$:

$$
\begin{aligned}
R &= R_0\left\{1-\frac{q_0-2}{(q_0-1)\alpha}(\zeta-1)+\dots\right\} \; ,\\
\mu &= \mu_0\left\{1+\frac{2q_0(q_0-2)}{(q_0^2-1)\alpha}(\zeta-1)+\dots\right\} \; ,\\
a^2 &= (\mu_0 R_0)^2\alpha\frac{(q_0+1)^2}{2q_0}+\dots \; ,\\
b^2 &= R_0^2\alpha\frac{(q_0-1)^2}{2q_0}+\dots \; ,
\end{aligned}
\tag{6.2.36}
$$

where R_0, μ_0, α, and q_0 are arbitrary constants.

Expressing $\zeta - 1$ in terms of $\mu - \mu_0$, we obtain analytic series in powers of $\mu - \mu_0$ for the required functions. The solution outside the light horizon is given by (6.2.36) for $\zeta > 1$.

Thus, the continuous and discontinuous self-similar motions of an ultrarelativistic gas which we have described above indicate that all the continuous self-similar motions of the gas have a cosmological character, since, according to (6.2.35), they can be analytically continued beyond the light horizon. This one-parameter class of solutions, which contains the Friedman-Lemaître solution, represents a set of (in general) inhomogeneous expanding cosmological models.

The segments of the curves of this class of solutions up to the sound line (6.2.25) appear as central cores in the larger two-parameter class of solutions possessing weak breaks. [In the space (x, V, ζ), the sound spheres are represented by the points on the sound line (6.2.25).] Each central core of the sound sphere can be attached to some solution in the two-parameter class of solutions. The analytic continuation beyond the light horizon of each of these solutions with weak breaks in the comoving coordinate system is given by (6.2.36). These solutions are inhomogeneous cosmological models with a weak break along the sound characteristic. They have greater generality than the class of regular cosmological solutions (6.2.20, 24), since they depend on two parameters.

In contrast to solutions with a weak break, in solutions with shock waves of sufficiently large intensity there is no light horizon. Inside a shock wave, a solution is described by a segment of one of the solutions (6.2.20, 24) or by a static solution. Thus, the uniformity and expansion of the matter around an observer does not guarantee the uniformity of the universe as a whole. At a sufficiently large distance from the center, a shock wave can lead to a complete change in the structure of the solution, when the pressure of the matter falls off to zero at infinity. In other words, regions with a uniform outflow (segments of Friedman-Lemaître universes) can be formed as a result of focusing [$V_0 < 0$ in (6.2.26)] (accumulation) of the gas to the center, with preservation of the pseudo-Euclidean asymptotic behavior at infinity.

The results of a numerical integration of the system (6.2.9, 10) are shown in Figs. 6.4–7, in which the abscissa represents the scaling variable $\lambda \equiv r/(ct)$.

In Figs. 6.4 and 6.5 we show the solutions for the velocity and the pressure with a stationary core. The curves a with weak breaks correspond to inhomo-

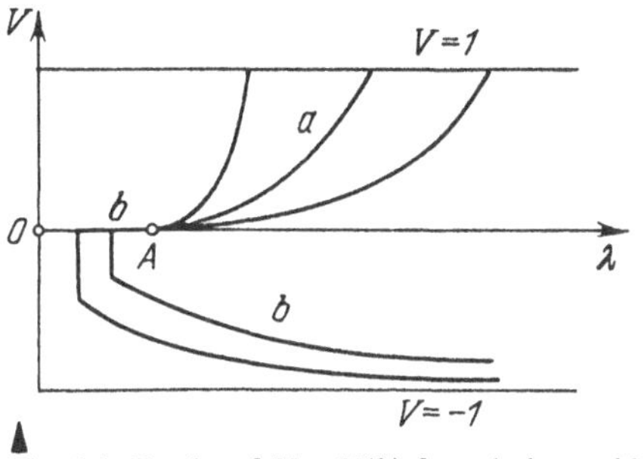

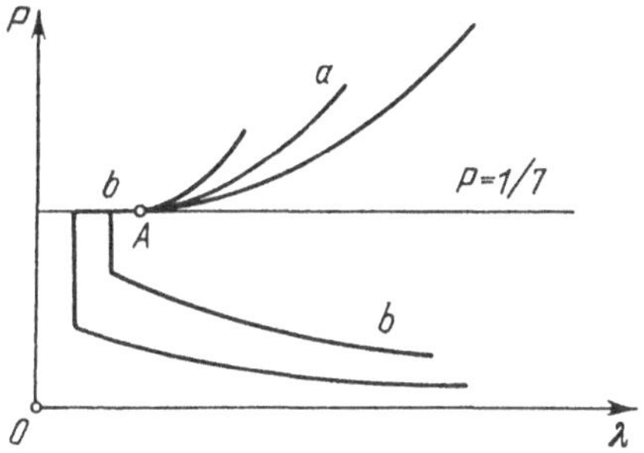

Fig. 6.4. Graphs of $V = V(\lambda)$ for solutions with weak and strong breaks in the presence of a static stationary core. Here and in Fig. 6.5, the point A corresponds to a weak break; OAa is a typical solution with a weak break, and Ob is a typical solution with a strong break

Fig. 6.5. Graphs of $P = P(\lambda)$ for solutions with weak and strong breaks in the presence of a static stationary core

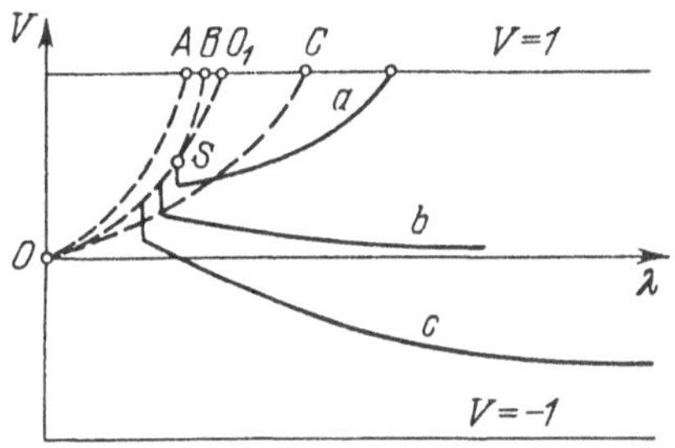

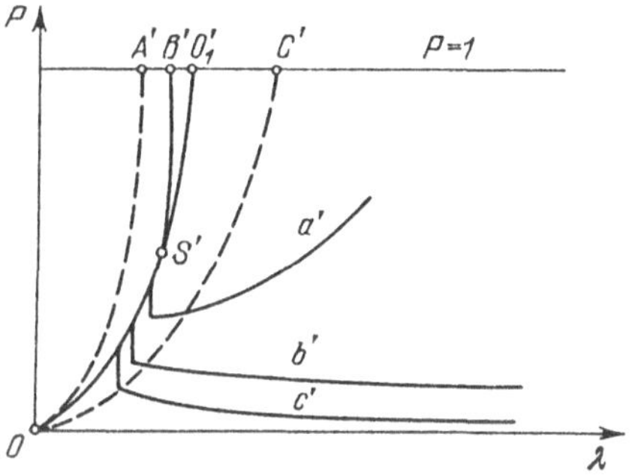

Fig. 6.6. Graphs of $V = V(\lambda)$ for solutions with weak and strong breaks in the presence of a homogeneous Friedman core. Here Oa, Ob, and Oc are typical curves with a strong break: Oa is a solution with a light horizon, on the curve Ob we have $V \to 0$ for $\lambda \to \infty$, and the curve Oc corresponds to a solution of the problem of accumulation towards the center. On the Friedman curve OSO_1 there is a point with a weak break at $\lambda = \sqrt{3}/2$, $V = \sqrt{3}/3$, $x = 3/2$, and OSB is a typical curve with a weak break. Inhomogeneous cosmological solutions are shown as OA and OC

Fig. 6.7. Graphs of $P = P(\lambda)$ corresponding to the curves in Fig. 6.6

geneous cosmological solutions possessing a light horizon. The curves a with strong breaks represent solutions of the problem of initial focusing to the center, when a stationary core is formed inside the shock wave. The point A in these two figures corresponds to the surface of a weak break with $\lambda = 1/\sqrt{3}$, $V = 0$, $P = 1/7$.

In Figs. 6.6 and 6.7 we show the solutions for the velocity and the pressure with a core of uniform outflow. The curves OSO_1 in Fig. 6.6 and, correspondingly, $OS'O_1'$ in Fig. 6.7 represent the Friedman-Lemaître solution for the velocity and the pressure inside the light sphere. The curves OSB and $OS'B'$ represent a typical solution with a weak break, for which part of the solution (the curves OS and OS') coincides with the Friedman-Lemaître solution.

The curves a and a' correspond to a typical inhomogeneous cosmological solution with a shock wave and a light horizon. The shock wave includes the Friedman-Lemaître solution. The curves b and b' give an idea of the dynamical evolution of the initially stationary gas, in which a "fragment" of a homogeneous universe is formed inside the shock wave.

The curves c and c' correspond to a typical solution of the problem of initial focusing to a point. Besides the Friedman-Lemaître solution, in which the parameter P_1 in (6.2.20) has the value 0.25, we have plotted in Figs. 6.6 and 6.7 the solutions (6.2.20) with $P_1 = 0.125$ (the curves OA and OA') and with $P_1 = 0.75$ (the curves OC and OC'). These curves correspond to smooth inhomogeneous cosmological models. The point S in these two figures corresponds to the surface of a weak break for $\lambda = \sqrt{3}/2$, $V = \sqrt{3}/2$, $x = 3/2$.

7. Acoustic Phenomena in Strong Gravitational and Magnetic Fields

The launch of the spacecraft Uhuru, Ariel, SAS–3, and Copernicus containing x-ray detectors led to the discovery of a large number of sources of pulsating x-ray emission (x-ray pulsars). The largest of the known periods of their pulsations exceeds the smallest period by three orders of magnitude. For example, the x-ray pulsars Her X-1 and Cen X-3 have periods 1.24 and 4.8 s, respectively, and the x-ray sources 3U0352 + 30 and 3U0900 – 40 according to the third Uhuru catalog have periods 835 and 284 s, respectively. The pulsations of the emission from an x-ray source can be explained by the variability of the direction of the emission from the regions of the magnetic poles, which occurs because the axis of rotation does not coincide with the axis of the magnetic dipole. If the x-ray pulsars belong to a single class of physical objects and are magnetized, rotating, accreting stars belonging to binary stellar systems, it is natural to ask why there is such a large spread in the angular speed of their rotation. (We note that at least six long-period sources belong to binary systems [7.1].)

We shall trace the main stages in the evolution of the mechanisms of retardation of a neutron star in a binary system whose normal component is characterized by strong loss of mass[1]. In its early stages, a neutron star is characterized by rapid rotation and is an ejecting pulsar (like the pulsar in the Crab Nebula) which produces intense acceleration of cosmic rays and emits radiation in all parts of the electromagnetic spectrum. The resulting radio emission may not be observed because of its absorption in the stellar wind of the normal star. The period of rotation of the pulsar grows as a result of the loss of the total angular momentum carried away by the magnetic dipole radiation. In turn, the power of the magnetic dipole radiation of the ejecting pulsar falls off rapidly with increasing period, and at a certain time the plasma pressure of the stellar wind begins to exceed the pressure of the cosmic rays. The plasma penetrates into the region of the light cylinder $r < ct$ and disrupts the operation of the pulsar mechanism.

Accretion of matter onto the neutron star during this stage is impossible. The pressure of the magnetic field stops the matter on a certain surface, called the *magnetosphere*, on which $p = H^2/8\pi$. Since the surface of the magnetosphere is not spherical, it acts like a propeller and throws off matter which is incident on it. Naturally, the propeller mechanism contributes to the retardation of the

[1] The observational data and theoretical ideas on x-ray sources have been reviewed in a number of papers [7.1–5].

rotation of the neutron star. This mechanism operates as long as the component of the rotational velocity of the magnetosphere normal to the surface of the magnetosphere is much greater than the parabolic velocity.

At lower rotational velocities, other mechanisms of retardation of the neutron star operate. The rotating neutron star generates acoustic and shock waves, which effectively heat the plasma surrounding it. There is turbulent convection, which deflects the angular momentum. An important role in the retardation of the rotation of the neutron star is played by the viscous forces acting in the gas surrounding it. The operation of these mechanisms is possible because of the subsonic (see Sect. 7.3) flow of the accreting gas.

The retardation of the neutron star continues until the angular momentum carried by the accreting matter per unit time begins to compensate the moment of the hydrodynamic forces acting on the magnetosphere. The equilibrium period of rotation of the neutron star is determined by the parameters of the binary system. The acoustic and viscous mechanisms of retardation lead to dissipation of energy in the accreting gas, heat it, and hinder the accretion. Therefore it is highly probable that accretion becomes possible only when the period of rotation of the pulsar comes close to the equilibrium period. The neutron star is then converted into an x-ray source. If allowance is made for the nonspherical form of the pulsar and the resulting moment of the gravitational forces in the presence of accretion, there will be a long-period variation of the angular velocity of rotation of the pulsar (see Sect. 7.8).

The periods of long-period x-ray pulsars are so long that they are not likely to be converted into radio pulsars that can be seen by observers after the normal star runs through its evolutionary path and is converted into a dead star. Thus, besides the mechanisms of retardation of the rotation of magnetized neutron stars which have already been considered (such as magnetic dipole radiation [7.6], acceleration of cosmic rays [7.7], and the supersonic "propeller" [7.8]), there exist also gas-dynamical mechanisms of retardation of the rotation, which become very important for long-period pulsars: (a) generation of shock and sound waves by the rotating magnetosphere in the plasma surrounding the star; (b) retardation of the rotation due to the viscosity of the plasma and its transition to turbulence at large Reynolds numbers. These processes are important in the case of a subsonic value of the normal component, with respect to the surface of the magnetosphere, of the rotational velocity of the boundary of the magnetosphere (see Sect. 7.3).

In Sect. 7.1, which is introductory in character, we present the fundamentals of the theory of short acoustic waves on the background of an arbitrary potential flow of a gas with constant entropy. In the case of variable entropy in the particles, acoustic waves and rotational perturbations influence each other. In Sect. 7.4 we investigate the behavior of arbitrary small adiabatic perturbations of radial pulsations in a well-known model of the Cepheids, in which the entropy is distributed nonuniformly over the particles and the unperturbed density depends only on the time. In Sect. 7.2 we develop analytic methods which make it possible to determine the explicit form of the magnetosphere, both in the case of supersonic rotation and in the case of negligible small rotation. In Sect. 7.5 and 7.6 we study nonlinear transverse oscillations in layers of a weakly compressible liquid or gas with ideal conductivity in strong magnetic fields in the presence of resonant oscillations of the electric current at the boundary. It is found that a characteristic feature of such oscillations is the appearance of shock waves and Alfvén breaks. We note that it is possible to write the equations of ideal magnetohydrodynamics in Lagrangian coordinates as the equations for a certain material with nonlinear elasticity.

7.1 Propagation of Nonlinear Short Acoustic Waves

In Sects. 7.1.1 and 7.1.2 we derive model equations for the description of acoustic waves at normal points on sound rays and near the envelope surfaces for sound rays on an arbitrary potential background. Formally, the problem reduces to the addition of quadratic terms in the approximation of geometrical acoustics, these having the maximum order.

In Sect. 7.1.3 the model equations are reduced to the simplest form, and known results in the one-dimensional case are used to analyze the formation and dissipation of a sawtooth wave in the general case.

In Sect. 7.1.4 we analyze the reflection of acoustic waves from a shock wave in the case of direct or oblique incidence. Some of the results of that section have been obtained by many authors, beginning with Rayleigh (of Soviet authors, we mention R. V. Khokhlov and O. S. Rizhov). For example, the second approximation in the theory of acoustic waves has been discussed in the literature since the 1930s. For detailed references, we refer the reader to various monographs and articles [7.9–13]. The references given in the text make no claim to completeness.

7.1.1 Derivation of the Model Equations

Suppose that in some region, in which the external forces have a potential U, the motion of a perfect gas is described by the potential in the form $\bar{v} = \nabla\varphi$ and is isentropic. Then the pressure p is related to the density ϱ by a power law $p = \text{const} \cdot \varrho^\gamma$, where γ is the ratio of the specific heats at constant pressure and at constant volume.

Under the stated assumptions, the equations of gas dynamics have the Cauchy–Lagrange integral

$$\frac{\partial \Phi}{\partial t} + \frac{v^2}{2} + U + \frac{a^2}{\gamma - 1} = \frac{a_0^2}{\gamma - 1} = \text{const} \ , \quad a^2 \equiv \gamma p/\varrho \ .$$

Eliminating the density from the equation of continuity, by means of the Cauchy–Lagrange integral we obtain

$$\left(\frac{\partial}{\partial t}+\boldsymbol{v}\cdot\nabla\right)\left(\frac{\partial\Phi}{\partial t}+\frac{v^2}{2}+U\right)+(\gamma-1)\left[-\frac{a_0^2}{\gamma-1}+\frac{\partial\Phi}{\partial t}+\frac{v^2}{2}+U\right]\Delta\varphi$$
$$=0 \ . \tag{7.1.1}$$

The characteristic ψ of (7.1.1) satisfies the equation

$$Q=Q(\boldsymbol{r},t,\boldsymbol{k},\omega)\equiv\frac{\varrho}{2a^2}[\boldsymbol{k}^2a^2-(\omega-\boldsymbol{k}\cdot\boldsymbol{v})^2]=0 \ , \tag{7.1.2}$$

where $\boldsymbol{k}\equiv\nabla\psi$, $\omega\equiv-\psi_{,t}$, and $\boldsymbol{r}$ is the vector with components x, y, z (the radius vector).

For the first-order partial differential equation (7.1.2), there is an associated *characteristic system* of ordinary differential equations in the Hamiltonian form:

$$\frac{d\boldsymbol{r}}{d\alpha}=\frac{\partial Q}{\partial\boldsymbol{k}} \ , \quad \frac{dt}{d\alpha}=-\frac{\partial Q}{\partial\omega} \ , \quad \frac{d\boldsymbol{k}}{d\alpha}=-\frac{\partial Q}{\partial\boldsymbol{r}} \ , \quad \frac{d\omega}{d\alpha}=\frac{\partial Q}{\partial t} \ ; \tag{7.1.3}$$

here α is the canonical parameter on the integral curve of the system (7.1.3). The system (7.1.3) determines a normal congruence of sound rays, the *bicharacteristics* of (7.1.1). We note that the sound rays lie on a surface $\psi=\text{const}$, covering it, as it were, since $\nabla\psi(d\boldsymbol{r}/d\alpha)+\psi_{,t}\,dt/d\alpha=2Q=0$.

The individuality of each sound ray on the three-dimensional surface $\psi=\text{const}$ can be characterized by means of Lagrangian coordinates ξ^1 and ξ^2, so that the solution of (7.1.3) can be written in the form of the dependence of the functions $\boldsymbol{r}$, t, $\boldsymbol{k}$, ω on the arguments α, ψ, ξ^1, ξ^2, this dependence being determined by the initial data for the system (7.1.3):

$$\boldsymbol{k}=\nabla f(x_0,y_0,z_0) \ , \quad \boldsymbol{r}=\boldsymbol{r}_0 \quad \text{for} \quad t=t_0 \ .$$

These conditions correspond to initial conditions for the solution of (7.1.2):

$$\psi=f(x,y,z) \quad \text{for} \quad t=t_0 \ .$$

Suppose that some solution of (7.1.1), which we shall call the *background* (or unperturbed solution), has a characteristic length $\tilde{L}$ and a characteristic time T of variation of the gas-dynamical quantities. We shall henceforth speak of a solution corresponding to an acoustic wave in the case of a perturbed solution of (7.1.1) which has a characteristic scale of variation l such that $l\ll L\equiv\min(\tilde{L},aT)$, where a is the characteristic speed of sound in the unperturbed solution. We shall assume that the relative amplitude of the perturbations is of an order not greater than the ratio of l^2 to L^2.

We shall denote the perturbations of the hydrodynamic quantities in an acoustic wave by a prime. Let φ be the potential of the perturbation of the velocity: $\boldsymbol{v}'=\nabla\varphi$. Quantities referring to the unperturbed solution will appear in the coefficients in the equations for the perturbations. We shall consider the terms of highest order in (7.1.1) when the solution is formally expanded in the small parameter $\varepsilon=l/L$.

In an acoustic wave, the field of the hydrodynamic quantities varies "rapidly" along the directions "extracted" from the surface $\psi=\text{const}$. If we transform from the variables x, y, z, t in (7.1.1) to the variables α, ψ, ξ^1, ξ^2, then ψ is a "fast" variable, while α, ξ^1, and ξ^2 are "slow" variables "almost everywhere"; near caustics, α also becomes a "fast" variable. The largest of the nonlinear terms in (7.1.1) will be those of the form $\varphi_{,\psi}$ and $\varphi_{,\psi\psi}$, where φ is the potential of the velocity perturbation. We shall assume that the characteristic length in an acoustic wave is of the order of $|\nabla\psi|^{-1}$, i. e., $l \sim |\nabla\psi|^{-1}$.

In what follows, the index j in the notation x^j will take the values $1, 2, 3, 4$, and $x^4 \equiv t$. We note that the affine parameter α on sound rays is determined only to the extent that an arbitrary function $f(\psi, \xi^1, \xi^2)$ can be added to it. Transforming in (7.1.1) to the variables α, ψ, ξ^1, ξ^2 and neglecting the derivatives with respect to ξ^1 and ξ^2 in comparison with the derivatives with respect to α and ψ, we find that the potential of the velocity perturbation satisfies the equation

$$2\frac{\partial^2\varphi}{\partial\psi\,\partial\alpha} + \frac{\partial}{\partial x^i}\left(\frac{\partial Q(\psi_{,j}; x^j)}{\partial\psi_{,i}}\right)\frac{\partial\varphi}{\partial\psi} + 2\frac{\partial^2\varphi}{\partial\alpha^2}Q(\alpha_{,j}; x^j)$$
$$+ \frac{\partial}{\partial x^i}\left(\frac{\partial Q(\alpha_{,j}; x^j)}{\partial\alpha_{,i}}\right)\frac{\partial\varphi}{\partial\alpha} + \frac{\varrho}{a^2}(\gamma+1)\boldsymbol{k}^2(\omega - \boldsymbol{k}\cdot\boldsymbol{v})\varphi_{,\psi}\varphi_{,\psi\psi} = 0 \ . \tag{7.1.4}$$

Far from *caustics* [envelopes for the bicharacteristics of (7.1.3)], the third and fourth terms in (7.1.4) can be dropped in comparison with the terms containing derivatives with respect to ψ. Therefore, far from caustics, we obtain from (7.1.4), after multiplication by $\varphi_{,\psi}$, the equation

$$\frac{\partial}{\partial t}\left[\varrho\left(\frac{\omega - \boldsymbol{k}\cdot\boldsymbol{v}}{a^2}\right)(\varphi_{,\psi})^2\right] + \operatorname{div}\left[\left(\boldsymbol{v} + \frac{\boldsymbol{k}a^2}{\omega - \boldsymbol{k}\cdot\boldsymbol{v}}\right)\varrho\left(\frac{\omega - \boldsymbol{k}\cdot\boldsymbol{v}}{a^2}\right)(\varphi_{,\psi})^2\right]$$
$$+ \frac{\varrho}{a^2}(\gamma+1)\boldsymbol{k}^2(\omega - \boldsymbol{k}\cdot\boldsymbol{v})\varphi^2_{,\psi}\varphi_{,\psi\psi} \sim 0 \ . \tag{7.1.5}$$

If there is an envelope for a family of rays (a caustic surface), we shall assume that it is convex. Sound rays which are tangential to a caustic remain on the same side of it. Caustics separate the "light" region from the "dark" region, i. e., the region with real bicharacteristics (constructed on the basis of the initial data) from the region with imaginary bicharacteristics. If $\sigma(x^j) = 0$ is the equation of a caustic, then, by the definition of a caustic, the normal to it is orthogonal to the direction of a ray:

$$\frac{d\boldsymbol{r}}{d\alpha}\cdot\nabla\sigma + \frac{dt}{d\alpha}\sigma_{,t} = 0 \quad \text{or} \quad \frac{\partial Q}{\partial\psi_{,j}}\frac{\partial\sigma}{\partial x^j} = 0 \ . \tag{7.1.6}$$

Taking advantage of the arbitrariness in the definition of the affine parameter on a sound ray, we shall measure the affine parameter on each ray from the point of tangency of the sound ray to the caustic. From the fact that the rays remain on the same side of the caustic, $\sigma > 0$, it follows that

$$\frac{d^2\boldsymbol{r}}{d\alpha^2}\cdot\nabla\sigma + \frac{d^2t}{d\alpha^2}\sigma_{,t} > 0 \ . \tag{7.1.7}$$

We transform in the neighborhood of the caustic to the coordinates σ, u^1, u^2, u^3, i.e., we stratify the space of the variables x, y, z, t by means of an analytic definition of the surfaces $\sigma = \text{const}$ in the neighborhood of the caustic $\sigma = 0$ (u^1, u^2, u^3 are internal variables on these surfaces). Then the conditions (7.1.6,7) take the form

$$\frac{d\sigma}{d\alpha} = 0 \ , \quad \frac{d^2\sigma}{d\alpha^2} = 2c^2 > 0 \quad \text{for} \quad \alpha = \sigma = 0 \ ,$$

where c is some function of u^1, u^2, u^3.

At their tangency, the sound rays leave a field of directions on the caustic $\sigma = 0$. From this field on the caustic, it is possible to reconstruct a normal congruence of curves which cover the caustic[2]. Let $\theta(u^1, u^2, u^3)$ be a continuous continuation of the characteristic ψ onto the caustic.

It follows from (7.1.6) that $\sigma \approx c^2\alpha^2$ and $d\sigma/d\alpha \approx 2c^2\alpha \approx 2c\sqrt{\sigma}$. On the other hand, according to (7.1.3),

$$\frac{d\sigma}{d\alpha} = -\frac{\partial Q}{\partial \psi_{,\sigma}} \ . \tag{7.1.8}$$

From the fact that $\partial Q/\partial\psi_{,\sigma} \sim \sqrt{\sigma}$ according to (7.1.8), it follows that $\psi_{,\sigma} \sim \sqrt{\sigma}$. Consequently, near a caustic the solution of the equation for the characteristics has the form

$$\psi = \theta \pm \frac{2\sigma^{3/2}}{3} \ . \tag{7.1.9}$$

[This form of the solution can always be obtained by multiplying σ by some function $f(u^1, u^2, u^3)$, with no change in the equation of the caustic.]

It follows from (7.1.8,9) that near a caustic the quadratic form Q has the structure

$$2Q(\psi_{,j};\, x^j) \approx 2c(\psi_{,\sigma})^2 + \sum_{A,B=1}^{3} c_{AB}\psi_{,A}\psi_{,B} \ ,$$

where the c_{AB} are certain functions of u^1, u^2, u^3.

We now calculate, near a caustic, the coefficients which appear in (7.1.4):

$$\frac{\partial}{\partial x^j}\left(\frac{\partial Q}{\partial \psi_{,j}}\right) \approx \frac{\partial}{\partial \sigma}\left(\frac{\partial Q}{\partial \psi_{,\sigma}}\right) \approx \frac{c}{\sqrt{\sigma}} \approx \frac{1}{\alpha} \ ,$$

$$Q(\alpha_{,j};\, x^j) \approx \frac{1}{2}\left[2c(\alpha_{,\sigma})^2 + \sum_{A,B=1}^{3} c_{AB}\alpha_{,A}\alpha_{,B}\right] \approx \frac{1}{4c\sigma} \approx \frac{1}{4c^3\alpha^2} \ ,$$

$$\frac{\partial}{\partial x^j}\left[\frac{\partial Q}{\partial \alpha_{,j}}\right] \approx -\frac{1}{2\sigma^{3/2}} \approx \frac{1}{2c^3\alpha^3} \ .$$

---

[2] These curves can also have envelopes corresponding to cusps on the caustic (see, for example, [7.14]).

Thus, near a caustic the equation for the acoustic waves takes the form

$$2\frac{\partial^2\varphi}{\partial\psi\,\partial\alpha}+\frac{\partial\varphi}{\partial\psi}\frac{1}{\alpha}+\frac{1}{2c^3\alpha}\frac{\partial}{\partial\alpha}\left(\alpha^{-1}\frac{\partial\varphi}{\partial\alpha}\right)$$
$$+\frac{(\gamma+1)\varrho}{a^2}\boldsymbol{k}^2(\omega-\boldsymbol{k}\cdot\boldsymbol{v})\varphi_{,\psi}\varphi_{,\psi\psi}=0\ . \tag{7.1.10}$$

Making the substitution

$$\psi=\theta+\frac{2\sigma^{3/2}}{3}\ ,\quad \alpha=\sqrt{\frac{\sigma}{c}}\ ,$$

from (7.1.10) we finally obtain

$$\frac{\partial^2\varphi}{\partial\sigma^2}-\left[\sigma-A\frac{\partial\varphi}{\partial\theta}\right]\frac{\partial^2\varphi}{\partial\theta^2}=0\ . \tag{7.1.11}$$

We stress that the coefficient A in (7.1.11) is a function of θ, ξ^1, and ξ^2. In the particular case in which $\partial A/\partial\theta=0$, (7.1.11) is readily reduced in the hodograph plane to the Tricomi equation (A. A. Bagdoev). Indeed, making the substitution $\varphi=(\tilde\varphi+\sigma\theta)/A$, we obtain for $\tilde\varphi$ the equation

$$\frac{\partial^2\tilde\varphi}{\partial\sigma^2}+\frac{\partial\tilde\varphi}{\partial\theta}\frac{\partial^2\tilde\varphi}{\partial\theta^2}=0\ . \tag{7.1.12}$$

We transform from the variables σ and θ to the variables $\lambda\equiv\tilde\varphi_{,\sigma}$ and $\mu\equiv\tilde\varphi_{,\theta}$. Dividing both sides of (7.1.12) by the Jacobian of the transformation $\tilde\varphi_{,\sigma\sigma}\tilde\varphi_{,\theta\theta}-(\tilde\varphi_{,\sigma\theta})^2$, we obtain $\theta_{,\mu}+\mu\sigma_{,\lambda}=0$. From the condition $\lambda_{,\theta}=\mu_{,\sigma}$, which follows from the definition of λ and μ, we obtain $\theta_{,\lambda}=\sigma_{,\mu}$. From this equation it follows that there exists a function $F(\lambda,\mu)$ such that $\theta=F_{,\mu}$ and $\sigma=F_{,\lambda}$. For this function we obtain the required Tricomi equation

$$\frac{\partial^2F}{\partial\mu^2}+\mu\frac{\partial^2F}{\partial\lambda^2}=0\ . \tag{7.1.13}$$

7.1.2 The Energy Density, Enthalpy Flux Vector, and Momentum Density in an Acoustic Wave Traveling Through an Arbitrary Background

The equation of conservation of energy in adiabatic flow of an ideal gas has the form

$$\frac{\partial}{\partial t}\left[\varrho\left(\frac{\boldsymbol{v}^2}{2}+\varepsilon\right)\right]+\operatorname{div}\left[\varrho\boldsymbol{v}\left(\frac{\boldsymbol{v}^2}{2}+\varepsilon+\frac{p}{\varrho}\right)\right]=0\ . \tag{7.1.14}$$

In the expression $\varrho(v^2/2+\varepsilon)$ for the volume energy density, we substitute $\boldsymbol{v}+\boldsymbol{v}'$ in place of $\boldsymbol{v}$ and $\varrho+\varrho'$ in place of ϱ, and we expand this expression in a series in the small quantities $\boldsymbol{v}'$ and ϱ' up to the terms of the second order inclusive. Then we obtain

$$
\begin{aligned}
E &= (\varrho+\varrho')\frac{(\boldsymbol{v}+\boldsymbol{v}')^2}{2} + \varrho\varepsilon + \frac{\partial(\varrho\varepsilon)}{\partial\varrho}\varrho' + \frac{\partial^2\varrho\varepsilon}{\partial\varrho^2}\frac{\varrho'^2}{2} \\
&= \varrho\left(\frac{\boldsymbol{v}^2}{2}+\varepsilon\right) + \varrho'\left(\frac{\boldsymbol{v}^2}{2}+\varepsilon+\frac{p}{\varrho}\right) + \varrho\boldsymbol{v}'\cdot\boldsymbol{v} \\
&\quad + \left\{\varrho'\boldsymbol{v}'\cdot\boldsymbol{v} + \frac{1}{2}\varrho'^2\frac{a^2}{\varrho} + \varrho\frac{\boldsymbol{v}'^2}{2}\right\} \quad . \qquad (7.1.15)
\end{aligned}
$$

When (7.1.15) is integrated over a volume much greater than the scale l but less than the scale L, the linear terms drop out.

It follows from the equations of motion that in an acoustic wave in the first approximation the perturbations of the velocity and of the density are related by the equation

$$
\varrho\boldsymbol{v}'(\omega-\boldsymbol{k}\cdot\boldsymbol{v}) = \boldsymbol{k}p' \quad . \qquad (7.1.16)
$$

In (7.1.15) the energy density in the acoustic wave is obviously represented by those terms in the curly brackets which are quadratic in the perturbations. Eliminating the density perturbation, we obtain

$$
E_{\mathrm{ac}} = \frac{\varrho\omega}{\omega-\boldsymbol{k}\cdot\boldsymbol{v}}\boldsymbol{v}'^2 \quad . \qquad (7.1.17)
$$

We now consider the flux vector of the total enthalpy:

$$
W = \varrho\boldsymbol{v}\left(\varepsilon+\frac{p}{\varrho}+\frac{\boldsymbol{v}^2}{2}\right) \quad .
$$

By analogy with the procedure for extracting the density of acoustic energy from (7.1.15), for the enthalpy flux vector in the acoustic wave we obtain

$$
W_{\mathrm{ac}} = \left(\frac{\varrho'}{\varrho}\boldsymbol{v}+\boldsymbol{v}'\right)(p'+\varrho\boldsymbol{v}\cdot\boldsymbol{v}') + \dots \quad .
$$

Using (7.1.16), this expression for W_{ac} can be readily transformed to

$$
W_{\mathrm{ac}} = \varrho\boldsymbol{v}'^2\left(\boldsymbol{v}+\frac{\boldsymbol{k}a^2}{\omega-\boldsymbol{k}\cdot\boldsymbol{v}}\right)\frac{\omega}{\omega-\boldsymbol{k}\cdot\boldsymbol{v}} \quad . \qquad (7.1.18)
$$

Omitting the analogous calculations, for the momentum density in the acoustic wave we obtain

$$
\boldsymbol{J}_{\mathrm{ac}} = \varrho\boldsymbol{v}'^2\boldsymbol{k}/(\omega-\boldsymbol{k}\cdot\boldsymbol{v}) \quad . \qquad (7.1.19)
$$

Taking into account the fact that $\boldsymbol{v}'\approx\boldsymbol{k}\varphi_{,\varphi}$, we can write (7.1.5) in the form

$$
\begin{aligned}
&\frac{\partial}{\partial t}\left(\frac{\varrho\boldsymbol{v}'^2}{\omega-\boldsymbol{k}\cdot\boldsymbol{v}}\right) + \operatorname{div}\left[\left(\boldsymbol{v}+\frac{\boldsymbol{k}a^2}{\omega-\boldsymbol{k}\cdot\boldsymbol{v}}\right)\frac{\varrho\boldsymbol{v}'^2}{(\omega-\boldsymbol{k}\cdot\boldsymbol{v})}\right] \\
&\quad + \frac{\varrho(\gamma+1)}{ka^2}(\omega-\boldsymbol{k}\cdot\boldsymbol{v})\boldsymbol{v}'^2\frac{\partial v'}{\partial\psi} = 0 \quad . \qquad (7.1.20)
\end{aligned}
$$

Multiplying (7.1.20) by ω, in accordance with (7.1.17, 18), we obtain an equation for the acoustic energy; multiplying (7.1.20) by $\boldsymbol{k}$, we obtain an equation for the momentum in the acoustic wave. Therefore (7.1.20) has the meaning of a kinetic equation for the acoustic "particles", where the last term describes the dissipation of the acoustic particles as a result of the effect of distortion of the acoustic waves. We note that for $\boldsymbol{v} = 0$ the expressions (7.1.17, 18) reduce to the well-known expressions for the corresponding quantities in a stationary gas. For one-dimensional motions and perturbations, when the vector $\boldsymbol{k}$ is parallel to $\boldsymbol{v}$, we find from (7.1.2) that $\omega - kv = \mp ak$. Therefore the expressions (7.1.17, 18) take the form

$$E_{\mathrm{ac}} = \varrho(v \mp a)v'^2/a \ , \quad W_{\mathrm{ac}} = \varrho(v \mp a)^2 v'^2/a \ .$$

7.1.3 Distortion of Short Acoustic Waves

We denote by D the Jacobian of the transformation from the variables x^i ($i = 1,2,3,4$) to the variables α, ψ, ξ^1, ξ^2:

$$D = \frac{D(x,y,z,t)}{D(\alpha,\psi,\xi^1,\xi^2)} \ .$$

It is easy to show that

$$\frac{d}{d\alpha} \ln D = \frac{\partial}{\partial x^i}\left(\frac{dx^i}{d\alpha}\right)$$

by analogy with the interpretation of the divergence of the velocity as the rate of relative change of the volume.

Away from caustics, we have $D \neq 0$, and therefore (7.1.4) can be rewritten in the form

$$2\frac{\partial^2 \varphi}{\partial \psi\, \partial \alpha} + \frac{1}{D}\left(\frac{\partial}{\partial \alpha} D\right)\frac{\partial \varphi}{\partial \psi} + \frac{\varrho(\gamma+1)}{a^2} k^2 (\omega - \boldsymbol{k}\cdot\boldsymbol{v})\varphi_{,\psi}\varphi_{,\psi\psi} = 0 \ . \quad (7.1.21)$$

In (7.1.21) we make the substitution $u \equiv \varphi_{,\psi}\sqrt{D}$ and take advantage of the fact that the function D is a slowly varying function of its arguments. Then (7.1.21) takes the form

$$2\sqrt{D}\frac{\partial u}{\partial \alpha} + \frac{(\gamma+1)\varrho}{a^2} k^2 (\omega - \boldsymbol{k}\cdot\boldsymbol{v}) u \frac{\partial u}{\partial \psi} = 0 \ . \quad (7.1.21')$$

Instead of the parameter α on sound rays we introduce the parameter χ given by

$$d\chi = \frac{(\gamma+1)\varrho k^2(\omega - \boldsymbol{k}\cdot\boldsymbol{v})}{2a^2\sqrt{D}}\, d\alpha \ . \quad (7.1.22)$$

Equation (7.1.21′) then takes the form

$$\frac{\partial u}{\partial \chi} + u\frac{\partial u}{\partial \psi} = 0 \ , \tag{7.1.23}$$

whose general solution is

$$u = f(\psi - u\chi) \ . \tag{7.1.24}$$

On the basis of (7.1.24), by means of a well-known analysis (see, for example, [7.10]), it can be shown that any sinusoidal profile of an acoustic wave for $\chi = 0$ leads to the formation of a sawtooth profile as χ increases.

If $u = u_0 \sin\psi$ for $\chi = 0$, then in the range $0 < \chi < 1/u_0$ there is a smooth growth of the higher harmonics, and the Bessel–Fubini solution of (7.1.23) has the form [7.10]

$$u = \sum_{n=1}^{\infty} (-1)^{n-1} 2I_n(nu_0\chi)\frac{(\sin n\psi)}{\chi} \ .$$

At $\chi u_0 = 1$, breaks begin to form, and the profile takes a sawtooth shape. The process of formation of a sawtooth profile is completed at $\chi u_0 = \pi/2$, and the required solution for $\chi u_0 > \pi/2$ acquires a sawtooth profile:

$$u = -\sum \frac{2u_0}{n(\chi u_0 - 1)} \sin n\psi \ .$$

Thus, the ampltitude u decreases in inverse proportion to χ, and the energy of the acoustic wave is converted into heat with the passage of time.

Let us consider the particular case in which the vector $\boldsymbol{k}$ is parallel to the velocity $\boldsymbol{v}$ and the unperturbed solution is planar ($\nu = 0$), cylindrical ($\nu = 1$), or spherical ($\nu = 2$) and, in addition, stationary, depending on the single coordinate r.

Then from (7.1.21) we obtain

$$\begin{aligned} &2\frac{\partial}{\partial r}\left(\varphi_{,\psi}\sqrt{\frac{r^v \varrho}{a}}\right) \mp \frac{(\gamma+1)\omega^2}{(1 \mp M)^3 a^2}\sqrt{\varrho a r^v}\varphi_{,\psi}\varphi_{,\psi\psi} = 0 \ , \\ &\frac{dr}{d\alpha} = \mp\frac{\varrho\omega}{a} \ . \end{aligned} \tag{7.1.25}$$

Here $M \equiv v/a$ and $\psi = \pm t + \int dr/(a \mp v)$. According to (7.1.22), the coordinate χ, which characterizes the distortion of the profile, is related to the coordinate r as follows:

$$\chi = \mp\frac{(\gamma+1)\omega^2}{2}\int \frac{dr}{\sqrt{\varrho r^v a}\, a^2(1 \mp M)^3} \ . \tag{7.1.25'}$$

Equation (7.1.25) leads, in particular, to the well-known results that in the case $\varrho = \text{const}$ and $M = 0$ the distortion of the profile for $\nu = 2$ increases with the distance as its logarithm, while for $\nu = 1$ the distortion of the profile grows as $\sqrt{r}$ [7.9].

7.1.4 Reflection of Acoustic Waves from Strong Breaks

The conditions on direct discontinuities in the coordinate system in which the gas after the discontinuity is at rest have the form (see, for example, [7.15])

$$\begin{gathered} \varrho(v-c)=-\varrho_1 c\ , \quad p=p_1+(1-\varrho_1/\varrho)\varrho_1 c^2\ , \\ \frac{a^2}{(\gamma-1)}+\frac{(v-c)^2}{2}=a_1^2/(\gamma-1)+c^2/2\ . \end{gathered} \tag{7.1.26}$$

Here the quantities before the discontinuity have no subscript, those after the discontinuity have the subscript 1, and c is the speed of the shock wave. In an acoustic wave incident on a shock wave from the region before the discontinuity, the state 1 cannot be perturbed, since perturbations after the discontinuity propagate with speed less than the speed of the shock wave. Therefore for small perturbations (denoted by primes) in the acoustic wave we find from (7.1.26) that

$$\begin{gathered} \varrho'(v-c)+\varrho(v'-c')=-\varrho_1 c'\ , \\ p'=2\varrho_1 cc'(1-\varrho_1/\varrho)+\frac{\varrho'\varrho_1^2 c^2}{\varrho^2}\ , \\ \frac{\gamma}{\gamma-1}\left(\frac{p'}{\varrho}-\frac{\varrho' p}{\varrho^2}\right)+(v'-c')(v-c)=cc'\ . \end{gathered} \tag{7.1.27}$$

Hence, eliminating c' and using (7.1.26), for the unperturbed solution we have

$$\frac{p'}{p}\left[(\gamma+1)-(\gamma-1)\frac{\varrho_1}{\varrho}\right]\left[(\gamma+1)\frac{\varrho_1}{\varrho}-(\gamma-1)\right]-4\gamma\frac{\varrho'}{\varrho}\frac{\varrho_1}{\rho}=0\ , \tag{7.1.28}$$

$$p'B-\varrho_1 cv'=0\ , \tag{7.1.29}$$

$$B\equiv\frac{3-\gamma+(\gamma+1)\varrho_1/\varrho}{4}\ . \tag{7.1.30}$$

It follows from (7.1.16) that in the incident wave (denoted by the subscript "in") and reflected wave (subscript "out") the perturbations of the velocity and of the pressure are related by the equations

$$p'_{\text{in}}-\varrho a v'_{\text{in}}=0\ , \quad p'_{\text{out}}+\varrho a v'_{\text{out}}=0\ .$$

From (7.1.29), after simple manipulations, putting $v'=v'_{\text{in}}+v'_{\text{out}}$ and $p'=p'_{\text{in}}+p'_{\text{out}}$, we obtain the elegant result

$$\frac{v'_{\text{out}}}{v'_{\text{in}}}=\frac{\delta B-1}{\delta B+1} \quad \text{where} \quad \delta^2\equiv\frac{\varrho^2 a^2}{\varrho_1^2 c^2}=\frac{1}{2}\left[(\gamma+1)\frac{\varrho}{\varrho_1}-(\gamma-1)\right]\ . \tag{7.1.31}$$

Therefore the coefficient of reflection of acoustic waves from a shock wave is given by the expression

$$\frac{W_{out}}{W_{in}} = \frac{(a-v)^2 v_{out}'^2}{(a+v)^2 v_{in}'^2} = \frac{(\delta + 1 - \varrho/\varrho_1)^2(\delta B - 1)^2}{(\delta - 1 + \varrho/\varrho_1)^2(\delta B + 1)^2} .$$

Here W_{in} is the flux of enthalpy in the incident wave, and W_{out} is the flux of enthalpy in the reflected wave. For strong shock waves, we have [7.16]

$$\frac{\varrho_1}{\varrho} = \frac{(\gamma-1)}{(\gamma+1)} , \quad B = \frac{1}{2} , \quad \delta^2 = \frac{2\gamma}{\gamma-1} .$$

Therefore the reflection coefficient in this case is

$$\frac{W_{out}}{W_{in}} = \frac{[\delta(\gamma-1)-2]^2}{[\delta(\gamma+1)+2]^2} \frac{(\delta-2)^2}{(\delta+2)^2} , \quad \delta = \sqrt{\frac{2\gamma}{\gamma-1}} .$$

A shock wave which has an acoustic wave incident on it in the subsonic region will radiate acoustic and entropy waves, as a consequence of (7.1.28). In the entropy wave, the perturbations are frozen in the fluid particles, and some of the energy of the incident acoustic wave is irreversibly expended on heating of the gas.

We note that since the direction of the wave vector is reversed, the frequency in the reflected wave changes discontinuously: $\omega_{out} = k(v-a)$; this can be treated as a Doppler effect, since $\omega_{in} = k(v+a)$.

Suppose now that an acoustic wave is incident on a shock wave at an angle θ, i.e., that the vector $\boldsymbol{k}$ is directed at an angle θ with respect to the normal to the shock wave. We shall perform a calculation of the local quantities – the angle of reflection θ_{out} of the acoustic wave and the reflection coefficient – in a coordinate system in which, at the point M of incidence of the acoustic wave on the shock wave, the velocity of the unperturbed flux is directed along the local normal to the shock wave (taken as the x axis). After the shock wave the gas is at rest. We take the y axis to be directed along the projection of the vector $\boldsymbol{k}$ onto the plane tangential to the shock wave at the point M. Then instead of (7.1.29) we obtain, in a similar way,

$$p'B - \varrho_1 c v_x' = 0 , \quad v_y' = 0 . \tag{7.1.32}$$

We represent the perturbations of the velocity and of the pressure on the shock wave in the form of sums of perturbations in the incident and reflected acoustic waves (the perturbation of the density also includes an entropy perturbation): $p' = p_{in}' + p_{out}'$, $\boldsymbol{v}' = \boldsymbol{v}_{in}' + \boldsymbol{v}_{out}'$. Using (7.1.16) and the dispersion equation (7.1.2), for the incident and reflected waves, respectively, we have

$$\frac{p_{in}' - \varrho a v_{x\,in}'}{\cos\theta_{in}} = 0 , \quad \frac{p_{out}' - \varrho a v_{x\,out}'}{\cos\theta_{out}} = 0 . \tag{7.1.33}$$

Here θ_{out} is the angle which the vector $\boldsymbol{k}_{out}$ makes with the normal to the surface of the shock wave.

We introduce the notation $\theta_1 \equiv \pi - \theta_{out}$, so that the angle θ_1 varies in the range from 0 to $\pi/2$. Using (7.1.33), we can then write (7.1.32) in the form

$$
\begin{aligned}
&\delta B\left[\frac{v_i}{\cos\theta} - \frac{v_o}{\cos\theta_1}\right] - (v_i + v_o) = 0 \quad , \\
&v_i \tan\theta = v_o \tan\theta_i \quad , \qquad\qquad (7.1.34)\\
&v_i \equiv v'_{x\,\mathrm{in}} \quad , \qquad v_o \equiv v'_{x\,\mathrm{out}} \quad .
\end{aligned}
$$

From the homogeneous equations (7.1.34) we obtain an equation which relates the angle of incidence θ to the angle θ_1:

$$
\delta B(\sin\theta_1 - \sin\theta) - \sin(\theta + \theta_1) = 0 \quad . \tag{7.1.35}
$$

Solving (7.1.35) for θ_1, we obtain

$$
\tan\frac{\theta_1}{2} = (\delta B + 1)\frac{\tan(\theta/2)}{\delta B - 1} \quad . \tag{7.1.36}
$$

We define the critical angle of incidence θ_{cr} as the angle at which the angle of reflection is $\theta_1 = \pi/2$. At supercritical angles of incidence $\theta > \theta_{cr}$ there is no reflection, since in this case, according to (7.1.36), the angle θ_1 becomes greater than the direct angle $\pi/2$. For $\tan(\theta_{cr}/2)$ we obtain from (7.1.36) the equation

$$
(\delta B + 1)\tan(\theta_{cr}/2) = \delta B - 1 \quad . \tag{7.1.37}
$$

The expression (7.1.37) makes sense only for $\delta B - 1 > 0$ or for $\delta^2 B^2 - 1 > 0$. According to the definition of B and δ by means of (7.1.30, 31), we have

$$
\delta^2 B^2 - 1 = \frac{1}{32}\left(\frac{\varrho_1}{\varrho}\right)^2\left(\frac{\varrho}{\varrho_1} - 1\right)^2\left[(3-\gamma)^2\frac{\varrho}{\varrho_1} - (\gamma^2 - 1)\right] \quad .
$$

Therefore the expression for $\tan(\theta_{cr}/2)$ makes sense only for $\varrho/\varrho_1 > (\gamma^2-1)(3-\gamma)^{-2}$. On the other hand, the degree of compression ϱ/ϱ_1 in shock waves cannot exceed the value $(\gamma+1)/(\gamma-1)$ [7.15]. Therefore reflection of acoustic waves from a shock wave is possible only for $\gamma < 2$, which is a consequence of the inequality

$$
\frac{\gamma+1}{\gamma-1} > \frac{\gamma^2-1}{(3-\gamma)^2} \quad .
$$

For $\gamma < 5/3$, reflection of acoustic waves is possible for any intensity of the shock wave.

For $5/3 < \gamma < 2$, reflection of acoustic waves is possible only in the interval

$$
\frac{\gamma+1}{\gamma-1} > \frac{\varrho}{\varrho_1} > \frac{\gamma^2-1}{(3-\gamma)^2} \quad .
$$

We define the reflection coefficient for an oblique acoustic wave as the ratio of the moduli of the incident and reflected fluxes of enthalpy. It follows from (7.1.34) that

$$\frac{|v'_{\text{in}}|}{|v'_{\text{out}}|} = \frac{\delta B - \cos\theta_1}{\delta B + \cos\theta} \quad .$$

According to (7.1.18), for the incident and reflected fluxes of enthalpy in an acoustic wave we have the expressions

$$|W_{\text{in}}| = \varrho\sqrt{a^2 + 2av\cos\theta + v^2}\,\frac{a + v\cos\theta}{a}(v'_{\text{in}})^2 \quad ,$$

$$|W_{\text{out}}| = \varrho\sqrt{a^2 + 2av\cos\theta_1 + v^2}\,\frac{a - v\cos\theta_1}{a}(v'_{\text{out}})^2 \quad .$$

Therefore the reflection coefficient in this case has the form

$$\left|\frac{W_{\text{out}}}{W_{\text{in}}}\right| = \left(\frac{\delta B - \cos\theta_1}{\delta B + \cos\theta}\right)^2 \left(\frac{\delta - (\varrho/\varrho_1 - 1)\cos\theta_1}{\delta + (\varrho/\varrho_1 - 1)\cos\theta}\right)$$
$$\times \sqrt{\frac{\delta^2 - 2\delta(\varrho/\varrho_1 - 1)\cos\theta_1 + (\varrho/\varrho_1 - 1)^2}{\delta^2 + 2\delta(\varrho/\varrho_1 - 1)\cos\theta + (\varrho/\varrho_1 - 1)^2}} \quad .$$

The reflection coefficient attains its largest value for strong shock waves, but this is a small value which tends to zero as $\gamma \to 2$. This implies almost complete "absorption" of the incident acoustic wave, i. e., almost complete transformation of it after the reflection into an entropy wave. For example, for $\gamma = 1.4$ and $\theta = 0$ the reflection coefficient (for a maximum degree of compression $\varrho/\varrho_1 = 6$) is 0.0018.

7.2 The Form of the Magnetosphere in a Nonuniform Plasma

7.2.1 The Form of the Static Magnetosphere in a Nonuniform Plasma

Neutron stars, owing to their high density of matter, have small dimensions (their radii are of the order of 10 km). In the process of evolution of a star, its magnetic field is frozen in the matter in a first approximation, so that the magnetic flux is conserved. Simple estimates show that if a star with the characteristic parameters of the Sum is compressed to the dimensions of a neutron star, the magnetic field strength on its surface rises to enormous values of the order of 10^{12} G.

The atmospheric plasma surrounding the neutron star then is casted by its magnetic field to distances of the order of 100–200 km from the star. The surface on which the pressure in the plasma is balanced by the pressure of the magnetic field is called the *magnetosphere*. On the magnetosphere, there are flows of surface electric currents, which screen the magnetic field of the neutron star. From the continuity of the normal component of the magnetic field, which is absent outside the magnetosphere, it follows that the latter must consist of magnetic lines of force. As the neutron star is approached, the magnetic field tends to the field of a magnetic dipole, since the characteristic dimension in the problem is the size of the magnetosphere, which is much greater than the radius of the neutron star.

The determination of the surface of the magnetosphere in the three-dimensional case is a cumbersome problem of mathematical physics. In the static case (without allowance for the rotation of the star) it was solved approximately in [7.17] for a uniform plasma, and in [7.18] for a nonuniform plasma with a pressure which falls off according to the law $p \sim r^{-5/2}$.

The two-dimensional case is attractive, above all, in that it is then possible to obtain a solution of the problem in closed form. Certain qualitative features of two-dimensional solutions make it possible to get some idea of the character of the solution in the three-dimensional case. For $p = \text{const}$ the two-dimensional problem was solved in [7.19], and for $p \sim r^{-2}$ in [7.20][3].

Below, we give an exact solution for the case of an atmosphere with an arbitrary power-law distribution of pressure $p = p_0 r^{-l}$, where p_0 and l are arbitrary constants. For the special values $l = 0$ and $l = 2$, our solution is identical to the solutions known from [7.19, 22]. We note that the form of the magnetosphere cannot be stable for $l > 4$. In this case, for a small perturbation of the magnetosphere the forces of the magnetic pressure and of the pressure in the plasma take the elements of the surface of the magnetosphere away from the position of equilibrium. In the three-dimensional case, the surface of the magnetosphere is not stable if the pressure in the plasma falls off according to the law r^{-l} with $l > 6$.

In the planar case, it follows from the equation $\operatorname{div} \boldsymbol{H} = 0$ (where $\boldsymbol{H}$ is the magnetic field intensity) that there exists a function ψ such that $H_x = \psi_{,y}$, $H_y = -\psi_{,x}$. The complex potential W of the magnetic field is $\varphi + i\psi$, where φ is the potential of the magnetic field. The function W depends on the complex variable $z = x + iy$, with $|z| = r$.

On the magnetosphere, the function ψ must be constant, and we choose it so that $\psi = 0$. If we determine $\psi = \operatorname{Im}\{W\}$ as a function of x and y, then the required surface of the magnetosphere is determined by the equation $\psi(x, y) = 0$. On the magnetosphere there are flows of surface currents, which screen the field from the point dipole situated at the point $z = 0$. The pressure of the magnetic field for $\psi = 0$ must be equal to $p_0 r^{-l}$:

$$\left| \frac{dW}{dz} \right|^2 = \frac{8\pi p_0}{r^l} \quad . \tag{7.2.1}$$

We introduce the new variable $\tilde{z} = (\alpha z^{\alpha})^{-1}$, where $\alpha \equiv l/2 - 1$. For the variable $\tilde{z}$, the boundary condition for $\psi = 0$ can be rewritten in the form

$$\left| \frac{dW}{d\tilde{z}} \right|^2 = 8\pi p_0 \quad . \tag{7.2.2}$$

[3] The generalization of this problem of the case of a quadrupole magnetic moment instead of a magnetic dipole was given in [7.21].

In the limit $z \to 0$, the function W must tend to the complex potential of a point dipole:

$$W \approx A/z = A(\alpha\tilde{z})^{1/\alpha} \ .$$

As a result, for the derivative $dW/d\tilde{z}$ near $z = 0$ we have the limiting behavior

$$dW/d\tilde{z} \approx A(\alpha\tilde{z})^{1/\alpha-1} \ .$$

Consequently, the functions W and $dW/d\tilde{z}$ near $z = 0$ are related by the equation

$$W \approx A^{\alpha/(\alpha-1)}(dW/d\tilde{z})^{1/(1-\alpha)} \ . \tag{7.2.3}$$

We direct the x axis along the axis of the dipole. Then $A^{\alpha/(\alpha-1)}$ will be a real constant. To find the required dependence of W and $dW/d\tilde{z}$, we recall that a circulation-free flow around a cylinder of radius R in the plane of the complex variable ζ is described by the complex potential

$$W = v_\infty(\zeta + R^2/\zeta) \ ,$$

where v_∞ is the speed at large ζ (the solution with arbitrary circulation is discussed below).

The role analogous to the coordinate ζ is played in our case by the variable

$$(dW/d\tilde{z})^{1/(1-\alpha)} \ ,$$

and the role of v_∞ is played by the expression $A^{\alpha/(\alpha-1)}$.

In fact, according to (7.2.3), W at large ζ behaves as $v_\infty\zeta$, and on the circle $|\zeta| = R$ (corresponding to $\psi = 0$) $(dW/d\tilde{z})^{1/(1-\alpha)}$ must be equal in modulus to the quantity $(8\pi p_0)^{1/2(1-\alpha)}$ in accordance with the boundary condition (7.2.2). Therefore the role of R^2 in our case is played by the quantity $(8\pi p_0)^{1/(1-\alpha)}$. Thus, the required relation between the functions W and $dW/d\tilde{z}$ has the form

$$W = A^{\alpha/(\alpha-1)}[(dW/d\tilde{z})^{1/(1-\alpha)} + (8\pi p_0)^{1/(1-\alpha)}(dW/d\tilde{z})^{1/(\alpha-1)}] \ .$$

Solving this equation for $dW/d\tilde{z}$, we obtain

$$dW/d\tilde{z} = \left[2^{-1}WA^{\alpha/(1-\alpha)} + \sqrt{4^{-1}W^2A^{2\alpha/(1-\alpha)} - (8\pi p_0)^{1/(1-\alpha)}}\right]^{(1-\alpha)} \ . \tag{7.2.4}$$

In the W plane, we have a cut along the real axis, where $\psi = 0$.

It can be readily shown from (7.2.4) that

$$\frac{8\pi p_0}{A^{2\alpha}}\frac{1}{\alpha}\left(\frac{A}{z}\right)^\alpha = \int dW\left[\frac{W}{2} - \sqrt{\frac{W^2}{4} - \left(\frac{8\pi p_0}{A^{2\alpha}}\right)^{1/(1-\alpha)}}\right]^{(1-\alpha)} \ . \tag{7.2.5}$$

The cut in the W plane is transformed by the conformal mapping (7.2.5) into the surface of the magnetosphere. In order to obtain the equation of the surface of the magnetosphere in parametric form, in (7.2.5) we must take points W lying on the cut:

$$W = W^* = \varphi = 2[8\pi p_0/A^{2\alpha}]^{1/[2(1-\alpha)]}\cos\theta \ .$$

The upper edge of the cut corresponds to variation of θ from 0 to π, and the lower edge corresponds to variation from π to 2π.

Then from (7.2.5) it is easy to show that

$$R/z = \mathrm{e}^{-\mathrm{i}\theta}(1 + \alpha\exp(2\mathrm{i}\theta)/(2-\alpha))^{1/\alpha} \ ,$$

where

$$R \equiv [A^{2\alpha}/8\pi p_0]^{1/[2(1-\alpha)]} \ .$$

Separating the real and imaginary parts in this equation, we obtain the equation of the surface of the magnetosphere in parametric form:

$$\begin{aligned} x &= R\frac{\cos\{\theta - \alpha^{-1}\mathrm{arctg}[\alpha\sin 2\theta(2-\alpha+\alpha\cos 2\theta)^{-1}]\}}{[1+(\alpha/(2-\alpha))^2 + 2\alpha/(2-\alpha)\cos 2\theta]^{1/(2\alpha)}} \ , \\ y &= R\frac{\sin\{\theta - \alpha^{-1}\mathrm{arctg}[\alpha\sin 2\theta(2-\alpha+\alpha\cos 2\theta)^{-1}]\}}{[1+(\alpha/(2-\alpha))^2 + 2\alpha/(2-\alpha)\cos 2\theta]^{1/(2\alpha)}} \ , \end{aligned} \tag{7.2.6}$$

These equations describe the shape of the surface of the magnetosphere as α varies from -1 to $+1$. For $\alpha = -1$ and $p = \text{const}$, the expressions (7.2.6) give a known result. Letting α tend to zero, from (7.2.6) we obtain in the limit

$$\begin{aligned} x &= \sqrt{\frac{A^2}{8\pi p_0}}\exp\left(-\frac{\cos 2\theta}{2}\right)\cos\left[\theta - \frac{\sin 2\theta}{2}\right] \ , \\ y &= \sqrt{\frac{A^2}{8\pi p_0}}\exp\left(-\frac{\cos 2\theta}{2}\right)\sin\left[\theta - \frac{\sin 2\theta}{2}\right] \ . \end{aligned}$$

It follows from (7.2.6) that the magnetosphere has a symmetry with respect to replacement of x by $-x$ and y by $-y$. In the neighborhood of the points $\theta = 0$ and $\theta = \pi$, the shape of the magnetosphere is described asymptotically by the equation of a semicubic parabola:

$$\left[\left(\frac{x}{\tilde{R}} - 1\right)\frac{2}{2-\alpha}\right]^3 = \left[\frac{3y}{\tilde{R}(\alpha^2+2-3\alpha)}\right]^2 \ , \quad \tilde{R} \equiv R\left(\frac{2-\alpha}{2}\right)^{1/\alpha} \ .$$

For $\theta = \pi/2$, we obtain on the magnetosphere a point with zero abscissa and ordinate $y_0 = R(1-\alpha)^{-1/\alpha}$. As the degree of nonuniformity increases, i. e., as α increases, the magnetosphere deviates more and more from a circle: the ratio of the distances from the center to the cusp of the semicubic parabola on the x axis and to the point on the y axis tends to zero (Fig. 7.1).

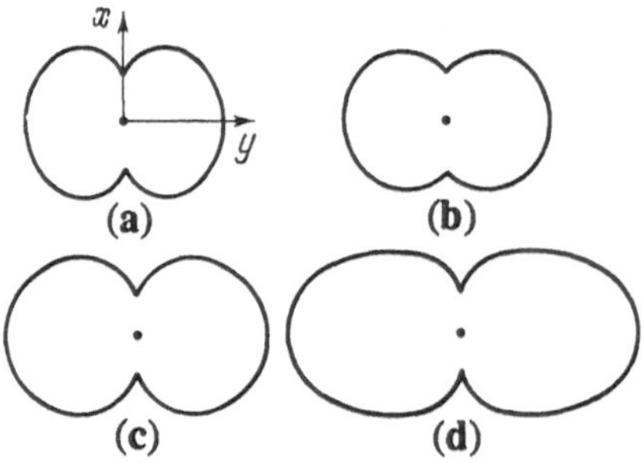

Fig. 7.1a–d. Evolution of the shape of the magnetosphere with increasing α, the degree of nonuniformity of the pressure distribution, with equal R: (a) $\alpha = -1$ (uniform case); (b) $\alpha = -1/2$, (c) $\alpha = 0$; (d) $\alpha = 1/2$

Above, we studied a special class of solutions with symmetric surfaces of the magnetosphere. We shall now consider the general case. We introduce the variable $\tilde{z} \equiv \sqrt{8\pi p_0}\,(\alpha z^\alpha)^{-1}$. Then the boundary condition (7.2.1) on the magnetosphere can be rewritten in the form

$$\left|\frac{dW}{d\tilde{z}}\right|^2 = 1 \ . \tag{7.2.7}$$

Using the asymptotic behavior $W \approx A/z$ for $z \to 0$, in the neighborhood of this point we have, for the relation between W and $dW/d\tilde{z}$, the asymptotic expression

$$W \approx v_\infty \xi \ , \quad \xi = \left(\frac{dW}{d\tilde{z}}\right)^{1/(1-\alpha)} \ , \quad v_\infty \equiv \left(\frac{A^\alpha}{\sqrt{8\pi p_0}}\right)^{1/(1-\alpha)} \ . \tag{7.2.8}$$

Let us consider the plane of the complex variable ξ. For $\xi \to \infty$, we have $W \approx v_\infty \xi$ according to (7.2.8). For $|\xi| = 1$, we have $\psi = \text{const}$ according to the boundary condition (7.2.7). A function $W(\xi)$ which is harmonic outside the circle $|\xi| = 1$ and satisfies these conditions, with allowance for the real constant v_∞, is given by

$$W = v_\infty(\xi + 1/\xi + \mathrm{i}\Gamma \ln \xi) \ , \tag{7.2.9}$$

where Γ is an arbitrary real constant.

Differentiating (7.2.9) with respect to ξ, we have

$$\frac{dW}{d\tilde{z}}\frac{d\tilde{z}}{d\xi} = \xi^{1-\alpha}\frac{d\tilde{z}}{d\xi} = v_\infty\left(1 - \frac{1}{\xi^2} + \frac{\mathrm{i}\Gamma}{\xi}\right) \ ,$$

from which

$$a\tilde{z} = \xi^\alpha\left(1 - \frac{\alpha}{(\alpha-2)}\frac{1}{\xi^2} + \frac{\alpha}{(\alpha-1)}\frac{\mathrm{i}\Gamma}{\xi}\right) \ . \tag{7.2.10}$$

Returning from the variable $\tilde{z}$ to the variable z, from (7.2.10) we obtain

$$\frac{R_0}{z} = \xi\left(1 - \frac{\alpha}{(2-\alpha)}\frac{1}{\xi^2} + \frac{\alpha}{(\alpha-1)}\frac{\Gamma \mathrm{i}}{\xi}\right)^{1/\alpha} \ ,$$
$$R_0 = \left(\frac{\sqrt{8\pi p_0}}{A}\right)^{1/(\alpha-1)} \ . \tag{7.2.11}$$

Equations (7.2.9, 11) give a parametric solution to the problem of the distribution of the magnetic field inside the magnetosphere and the configuration of the magnetic lines of force, since, according to (7.2.9, 11), we have for the complex magnetic intensity

$$\begin{aligned}\frac{dW}{dz} &= -\frac{v_\infty}{R_0}\xi^2\left(1+\frac{\alpha}{(2-\alpha)}\frac{1}{\xi^2}+\frac{\alpha}{\alpha-1}\frac{\Gamma\mathrm{i}}{\xi}\right)^{(\alpha+1)/\alpha}\\ &= -\frac{v_\infty}{z}\xi\left(1+\frac{\alpha}{(2-\alpha)}\frac{1}{\xi^2}+\frac{\alpha}{(\alpha-1)}\frac{1}{\xi}\right) .\end{aligned}$$

The equation of the magnetosphere itself can be readily obtained from (7.2.11) by putting $\xi = \exp(-\mathrm{i}\theta)$:

$$\frac{z}{R_0} = \mathrm{e}^{\mathrm{i}\theta}\left(1+\frac{\alpha\mathrm{i}}{(\alpha-1)}\Gamma\mathrm{e}^{\mathrm{i}\theta}+\frac{\alpha}{(2-\alpha)}\mathrm{e}^{2\mathrm{i}\theta}\right)^{-1/\alpha} .$$

From this it is easy to obtain a parametric form for the equation of the surface of the magnetosphere in polar coordinates $z = r\exp(\mathrm{i}\varphi)$:

$$r = R_0\left[1+\left(\frac{\alpha}{2-\alpha}\right)^2+\frac{2\alpha}{2-\alpha}\cos 2\theta+\frac{\alpha^2}{(\alpha-1)^2}\Gamma^2+\frac{4\alpha}{2-\alpha}\Gamma\sin\theta\right]^{-1/2\alpha} , \tag{7.2.12}$$

$$\varphi = \theta-\frac{1}{\alpha}\mathrm{arctg}\left\{\alpha\frac{[(2-\alpha)\Gamma\cos\theta+(\alpha-1)\sin 2\theta]}{[(2-\alpha)(\alpha-1)+\alpha(\alpha-1)\cos 2\theta-\alpha(2-\alpha)\Gamma\sin\theta]}\right\} . \tag{7.2.13}$$

For the conformal mapping of the ξ plane into the z plane to be one-to one outside the circle $|\zeta| - 1$, it is necessary (but not sufficient) that in the ξ plane for $|\xi| > 1$ there be no branch points, i.e., that the derivative $dz/d\xi$ not vanish for $|\xi| > 1$. From this it is easy to obtain a bound on the arbitrary constant Γ in the solution (7.2.9): $|\Gamma| \leq 2$.

For $\alpha > 0$ finite points of the plane with $|\xi| > 1$ cannot be mapped into the point at infinity in the z plane, in view of the compactness of the magnetosphere, and for $\alpha < 0$ the point $z = 0$ cannot have several inverse images in the ξ plane. Therefore the expression $1+\alpha\mathrm{i}\Gamma\xi^{-1}/(\alpha-1)+\alpha\xi^{-2}/(2-\alpha)$ cannot vanish for $|\xi| > 1$. This requirement also imposes a bound on Γ:

$$|\Gamma| < 2(1-\alpha)^2/[|\alpha|(2-\alpha)] . \tag{7.2.14}$$

We now calculate the derivative $d\varphi/d\theta$ of the function $\varphi(\theta)$ defined by (7.2.13):

$$\begin{aligned}\frac{d\varphi}{d\theta} &= 4\left(\frac{r}{R_0}\right)^{4\alpha}\frac{1-\alpha}{2-\alpha}(\sin\theta+a_1)(\sin\theta-a_2) ,\\ a_1 &= \frac{\Gamma\alpha(2-\alpha)}{2(1-\alpha)^2} , \quad a_2 = \frac{\Gamma}{2} .\end{aligned}$$

Owing to the bounds on Γ, the quantity φ as a function of θ in the interval $|\theta| < \pi$ has four extrema for $|\Gamma| < 2$ and three extrema for $|\Gamma| = 2$. In what follows, we shall assume that $\Gamma > 0$ (the case $\Gamma < 0$ is reduced to the case $\Gamma > 0$ by the transformation $\theta \to \theta + \pi$).

Consider the value of the function φ at the extreme points θ_1 and θ_2, where $\sin\theta_{1,2} = -a_1$, $|\theta_1| < \pi$. According to (7.2.13), the value of φ at the point $\theta = \theta_1$ is $(\alpha - 1)\theta_1/\alpha$, and $\varphi(\theta_2)$ is equal to $\pi - (\alpha - 1)\theta_1/\alpha$ for $\alpha < 0$ and $-\pi - (\alpha - 1)\theta_1/\alpha$ for $\alpha > 0$. The change of the function between these extreme points is $|\varphi(\theta_1) - \varphi(\theta_2)| = \pi + 2\theta_1(\alpha - 1)/\alpha$. If the curve $r = r(\varphi)$ is to have no self-intersections, taking into account the fact that $r(\theta_1) = r(\theta_2)$, it is necessary that $|\varphi(\theta_1) - \varphi(\theta_2)|$ be less than 2π. Hence we find that $|\theta_1| < \pi\alpha/(2\alpha - 2)|$, and therefore

$$\Gamma < -\sin\left(\frac{\pi\alpha}{2\alpha - 2}\right)\frac{2(\alpha - 1)^2}{\alpha(2 - \alpha)} \quad . \tag{7.2.15}$$

This inequality holds for $-1 \le \alpha \le 1/2$ (for example, $\Gamma < 4\sqrt{2}/3$ for $\alpha = -1$, $\Gamma < \pi/2$ for $\alpha = 0$, and $\Gamma < 2/3$ for $\alpha = 1/2$). We note that for $1/2 < \alpha < 1$ the change of the function φ between the extreme values $\varphi(\theta_1)$ and $\varphi(\theta_2)$ is always less than 2π, and therefore the constant Γ must satisfy only the inequality (7.2.14) (the inequality $\Gamma < 2$ is satisfied automatically in this case).

We now consider the extrema of the function $\varphi(\theta)$, at which $\sin\theta_{3,4} = \Gamma/2$. At these same values of θ, there are extrema of the function $r(\theta)$. Indeed, from (7.2.12) it follows that

$$\frac{dr}{d\theta} = \frac{4\cos\theta}{2 - \alpha}\left(\sin\theta - \frac{\Gamma}{2}\right)(r/R_0)^{2\alpha+1}R_0 \quad .$$

The extrema of $r(\theta)$ and $\varphi(\theta)$ for $\sin\theta = \Gamma/2$ in the case $\Gamma < 2$ correspond to cusps on the magnetosphere [near cusps, the curve $r = r(\varphi)$ has the asymptotic behavior of a semicubic parabola]. Thus, for $\alpha \le 1/2$ the unit circle in the ξ plane is mapped in a one-to-one manner into a curve without self-intersections in the z plane for Γ satisfying the inequality (7.2.15), while for $1/2 < \alpha < 1$ there is a one-to-one mapping for Γ satisfying the inequality (7.2.14)[4]. It follows from the one-to-one character of the conformal mapping on the boundaries of the region and from the absence of branch points inside the region that the conformal mapping $z(\xi)$ is single-sheeted everywhere inside the magnetosphere. In Fig. 7.2 we show the evolution of the shape of the magnetosphere for $\alpha = -1$ with increasing Γ up to the critical value $4\sqrt{2}/3$.

We now calculate the principal vector $\boldsymbol{F}$ of the forces acting on the magnetosphere. We form the complex number $F_x - \mathrm{i}F_y$, where F_x and F_y are the

[4] In a multisheeted solution for a supercritical value of the parameter Γ, there is trapping of the plasma inside the magnetosphere, which evidently corresponds to the formation of a plasma droplet with simultaneous restructuring of the magnetosphere. The subsequent severance of the droplet and its fall onto the neutron star should lead to a burst of radiation from such a star.

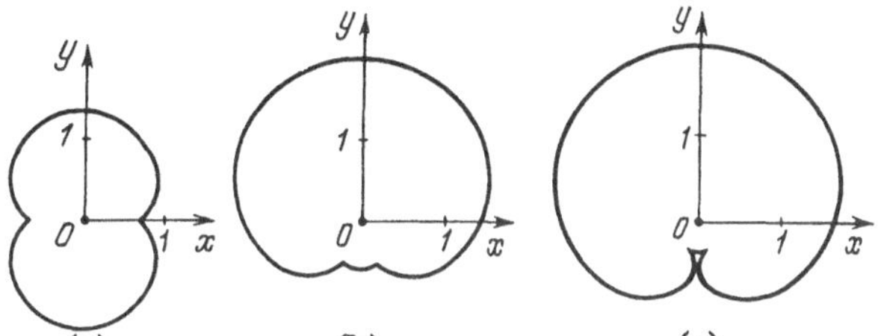

Fig. 7.2a–c. Evolution of the shape of the surface of the magnetosphere as the parameter Γ varies in the case $\alpha = -1$: (a) $\Gamma = 0$; (b) $\Gamma = \sqrt{2}$; (c) $\Gamma = 4\sqrt{2}/3$

components of the vector $\boldsymbol{F}$. According to the Blasius–Chaplygin formula (see, for example, [7.23]), we have

$$F_x - \mathrm{i}F_y = \frac{\mathrm{i}}{8\pi}\int\limits_{c_1}\left(\frac{dW}{dz}\right)^2 dz = \frac{\mathrm{i}}{8\pi}\int\limits_{c_2}\left(\frac{dW}{d\xi}\right)^2 \frac{d\xi}{dz}\, d\xi \ .$$

In order to calculate the integral on the right-hand side of this equation, we deform the contour c_2 into an infinitely remote contour c in the ξ plane. Then this integral can be calculated by means of the residue theorem if we expand the derivative $d\xi/dz$ in the ξ plane in the neighborhood of the point at infinity in a Laurent series, retaining in this series the terms up to the third order inclusive. This gives

$$F_x - \mathrm{i}F_y = \frac{\mathrm{i}v_\infty^2(\alpha+1)}{4R_0}\left(\frac{2\Gamma}{2-\alpha} - \frac{\Gamma^3(\alpha-1)^2}{3}\right) .$$

It is clear from symmetry arguments that the moment of the forces acting on the magnetosphere is equal to zero.

From the expression just obtained, we find the natural result that in the uniform case ($\alpha = -1$) the principal force vector is equal to zero.

7.2.2 Shape of the Magnetosphere of a Star Rotating in the Supersonic Regime

Let us consider the situation in which a gas falls onto a star rotating around some axis, in a uniform manner from all directions with supersonic velocity. The pressure in the gas can be neglected in this case [7.16]. We shall consider the gas as an aggregate of dust particles which, after colliding elastically on the magnetosphere, recede from it with velocities sufficient to overcome the attraction of the star. In [7.8] this model was called the propeller model. Under the action of the collisions of the dust particles, the magnetosphere acquires some unknown shape. Let v_n' be the normal component of the velocity ($\boldsymbol{n}$ is the outward normal to the magnetosphere) of a particle with respect to the magnetosphere. Then along the normal there is a force per unit area of the magnetosphere equal to $2\varrho(v_n')^2$, where ϱ is the density of the gas falling on the surface of the magnetosphere.

The calculation of v_n' presents no difficulty. In a coordinate system rotating together with the magnetosphere, its form does not depend on the time. In a stationary spherical coordinate system, the equation of the surface of the magnetosphere has the form $r = R(\theta, \varphi - \Omega t)$, where Ω is the angular speed of rotation and the z axis is coincident with the axis of rotation. The speed of motion of this surface along the normal is

$$D_n = \Omega\gamma\frac{\partial R}{\partial\varphi} , \quad \text{where}$$

$$\gamma \equiv \left[1 + \left(\partial \ln \frac{R}{\partial\theta}\right)^2 + \sin^{-2}\theta \left(\partial \ln \frac{R}{\partial\varphi}\right)^2\right]^{-1/2} .$$

In spherically symmetric accretion, only the radial component of the velocity $-v_0$ is nonzero. Its projection onto the normal to the magnetosphere is $v_n = -v_0\gamma$.

Thus, for the velocity of the gas with respect to the magnetosphere we have the expression

$$v_n' = v_n - D_n = -\left(\Omega\frac{\partial R}{\partial\varphi} + v_0\right)\gamma .$$

The component of the magnetic field intensity normal to the magnetosphere vanishes on the magnetosphere. Let Φ be the potential of the magnetic field. Then for the harmonic function Φ we have the following conditions on the magnetosphere:

$$|\nabla\Phi|^2 = 8\pi\varrho\left(\Omega\frac{\partial R}{\partial\varphi} + v_0\right)^2\gamma^2 , \qquad (7.2.16)$$

$$-\frac{\partial\Phi}{\partial r} + \frac{\partial\Phi}{\partial\theta}\frac{\partial R}{\partial\theta}\frac{1}{R^2} + \frac{\partial\Phi}{\partial\varphi}\frac{\partial R}{\partial\varphi}\frac{1}{R^2\sin^2\theta} = 0 . \qquad (7.2.17)$$

Near the neutron star, the magnetic field tends to the field of a dipole:

$$\Phi \to (\boldsymbol{A}\cdot\nabla)(1/r) \quad \text{for} \quad r \to 0 , \qquad (7.2.18)$$

where $\boldsymbol{A}$ is the magnetic moment of the neutron star.

In the three-dimensional case, the boundary-value problem (7.2.16–18) has not yet been studied. In what follows, we present our solution for the two-dimensional case, obtained in explicit form. In the two-dimensional case, the boundary conditions (7.2.16–18) can be rewritten for the complex potential $W = \Phi + \mathrm{i}\Psi$ in the form

$$\left|\frac{dW}{dz}\right|^2 = 8\pi\varrho\left(\Omega\frac{\partial R}{\partial\varphi} + v_0\right)^2\left[1 + \left(d \ln \frac{R}{\partial\varphi}\right)^2\right]^{-1} , \qquad (7.2.19)$$

$$\Psi = 0 \quad \text{for} \quad r = R(\varphi - \Omega t) \ , \tag{7.2.20}$$

$$W \to \frac{A}{z} \quad \text{for} \quad z \to 0 \ ; \quad z = x + \mathrm{i}y = r\exp(\mathrm{i}\varphi) \ . \tag{7.2.21}$$

From the boundary condition (7.2.19), for the increment of the potential along a magnetic line of force we obtain

$$d\Phi = \pm\sqrt{8\pi\varrho}\,(\Omega R\,dR + v_0 R\,d\varphi) \ . \tag{7.2.22}$$

For the magnetic pressure to balance the pressure on the magnetosphere, v'_n must be negative everywhere on the magnetosphere; otherwise, the gas would break away from the surface, and it would be impossible to satisfy the boundary condition (7.2.19). Consequently, the right-hand side of (7.2.22) does not change sign. In what follows, we shall choose the upper sign and impose on the solution the additional condition of positivity of the right-hand side of (7.2.22).

In the two-dimensional case, v_0 is the speed of axially symmetric flow toward the symmetry axis. The density ϱ and the speed v_0 depend on the radius r.

The conditions (7.2.20, 21) are identical to the conditions for plane-parallel flow of an incompressible fluid around a cylindrical body in the $1/z$ plane. Therefore the solution for W must have the form

$$W = D^*\zeta + D\zeta^{-1} + 2\Gamma \ln\zeta \ . \tag{7.2.23}$$

Here $\zeta^{-1}(z^{-1})$ is a function which gives a one-to-one conformal mapping of the exterior of the streamline body onto the exterior of the unit circle, in which the dependence of ζ on z near $z = 0$ must be linear:

$$\zeta \approx Dz/A \ . \tag{7.2.24}$$

Consider the function $F(\zeta) \equiv \mathrm{i}d(\ln z)/d(\ln\zeta)$, which is analytic inside the unit circle $|\zeta| \leq 1$. Then the condition (7.2.22) can be written as a linear relation between the real and imaginary parts of the functions $F(\zeta)$ on the contour $|\zeta| = 1$:

$$a\,\mathrm{Re}F + b\,\mathrm{Im}F = c \ , \tag{7.2.25}$$

$$a = \sqrt{8\pi\varrho}\,\Omega R^2 \ , \quad b = \sqrt{8\pi\varrho}\,v_0 R \ , \quad c = \mathrm{i}(D^*\zeta - D\zeta^{-1} + \mathrm{i}\Gamma) \ . \tag{7.2.26}$$

The condition (7.2.25) for the required function $F(\zeta)$ could be interpreted as a Hilbert boundary-value problem if the coefficients a and b in (7.2.25) were given together with $c = c(\zeta)$ as functions of $\theta = -\mathrm{i}\ln\zeta$ for $|\zeta| = 1$. From the condition $W \approx A/z$ for $z \to 0$, we have $F(0) = \mathrm{i}$.

As is well known, the solution of the Hilbert problem (7.2.25) (in the case in which it is uniquely solvable with the condition $F(0) = \mathrm{i}$) has the form [7.24, 25]

$$F(\zeta) = \frac{1}{2\pi\mathrm{i}}\varphi(\zeta)\left(\zeta \int\limits_{|\tau|=1} \frac{2c(\tau)\,d\tau}{\tau\varphi(\tau)(a(\tau) - \mathrm{i}b(\tau))(\tau - \zeta)} - \frac{2\pi}{\varphi(0)}\right) \ , \tag{7.2.27}$$

where

$$\varphi(\zeta) \equiv \exp \frac{1}{2\pi \mathrm{i}} \int\limits_{|\tau|=1} \frac{\ln G(\tau)}{\tau - \zeta}\, d\tau \quad ,$$

$$G(\tau) \equiv \frac{a(\tau) + \mathrm{i}b(\tau)}{a(\tau) - \mathrm{i}b(\tau)} \quad .$$

However, according to (7.2.26), a and b are given as functions of R, and not as functions of $\theta = -\mathrm{i} \ln \zeta$ for $|\zeta| = 1$. This makes it difficult to obtain solutions with the functions $a(R)$ and $b(R)$ specified arbitrarily.

Particular exact solutions can be constructed as follows. We shall specify a and b as real functions of θ. Using (7.2.27), we determine $z = z(\zeta)$. The magnetosphere corresponds to the circle $|\zeta| = 1$. Therefore from $z = z(\zeta)$ we obtain the expression $R = R(\theta)$. Using this expression to eliminate the parameter θ from the originally specified $a(\theta)$ and $b(\theta)$, we obtain $a(R)$ and $b(R)$.

Suppose, for example, that

$$\begin{aligned} a(\theta) &= a_0 \left(\alpha + \frac{1}{\alpha} + 2\cos\theta \right) \cos\gamma \quad , \\ b(\theta) &= a_0 \left(\alpha + \frac{1}{\alpha} + 2\cos\theta \right) \sin\gamma \quad . \end{aligned} \tag{7.2.28}$$

Here a_0, α, and γ are positive constants, with $0 < \alpha < 1$ and $0 < \gamma < 2^{-1}\pi$. Then

$$G(\zeta) = \exp(2\mathrm{i}\gamma) \quad , \quad \varphi(\zeta) = \varphi(0) = \exp(2\mathrm{i}\gamma) \quad .$$

From (7.2.27) we now obtain

$$\begin{aligned} F(\zeta) &= \mathrm{i} - \frac{\zeta}{\pi} \int \frac{(D^*\tau - D/\tau + \mathrm{i}\Gamma)\, d\tau}{(\tau + \alpha)(\tau + 1/\alpha)(\tau - \zeta)} \\ &= \mathrm{i} \left[1 - \frac{\zeta}{a_0} \frac{\exp(\mathrm{i}\gamma)}{(\zeta + 1/\alpha)} \frac{D^* - D\alpha^2 - \mathrm{i}\Gamma\alpha}{(1 - \alpha^2)} \right] \quad . \end{aligned} \tag{7.2.29}$$

From (7.2.28), using (7.2.24), we have

$$z = \zeta(1 + \alpha\zeta)^m \frac{A}{D} \quad , \quad m = -\frac{\exp(\mathrm{i}\gamma)(D^* - D\alpha^2 - \mathrm{i}\Gamma\alpha)}{a_0(1 - \alpha^2)} \quad . \tag{7.2.30}$$

Let us consider the case of a real constant m, which, as is shown below, corresponds to a power dependence of a and b on R. Then from (7.2.30) we find for $\zeta = \exp(\mathrm{i}\theta)$ that

$$R = |z| = \left| \frac{A}{D} \right| (1 + \alpha^2 + 2\alpha \cos\theta)^{m/2} \quad ,$$

from which

$$1 + \alpha^2 + 2\alpha \cos\theta = \left|\frac{RD}{A}\right|^{2/m} .$$

Eliminating θ by means of this relation, from (7.2.28) we have

$$a(R) = a_0 \cos\gamma \left|\frac{RD}{A}\right|^{2/m} , \quad b(R) = a_0 \sin\gamma \left|\frac{RD}{A}\right|^{2/m} .$$

It follows from these relations that the constants α and m cannot be equal to zero. For $m = 2$ the equation of continuity $\varrho v_0 R = \text{const}$ is satisfied, since in this case $\varrho \sim 1/R^2$ and $v_0 \sim R$. As a consequence of (7.2.30), the shape of the magnetosphere is determined with accuracy up to a similarity transformation, corresponding to the arbitrariness in the choice of the modulus of the constant D. If the circulation of Γ is equal to zero, the argument of D is determined uniquely by the requirement that m be real.

In parametric form, the shape of the magnetosphere is determined by the equations

$$R = \left|\frac{A}{D}\right| (1 + \alpha^2 + 2\alpha\cos\theta)^{m/2} , \tag{7.2.31}$$

$$\varphi - \Omega t = \theta + m \arctan\left(\frac{\alpha\sin\theta}{\alpha\cos\theta + 1}\right) + \arg\frac{A}{D} . \tag{7.2.32}$$

The relation (7.2.31) follows from (7.2.30) with $|\zeta| = 1$ and $\arg\zeta = \theta$.

A sufficient condition for the curves (7.2.31) in the plane to have no self-intersections is that the derivative $d\varphi/d\theta$ be greater than zero. Then from (7.2.32) we have the inequality

$$1 + \alpha^2(1 + m) + \alpha(2 + m)\cos\theta \geq 0 , \tag{7.2.33}$$

which must hold for all θ. From (7.2.33) we obtain a bound on the arbitrary constant α which enters into the solution (7.2.30):

$$\alpha \leq \frac{1}{|m + 1|} . \tag{7.2.34}$$

We recall that $0 < \alpha < 1$. Therefore the bound (7.2.34) is significant only for $m > 0$ and $m \leq -2$. Self-intersections of the curves for $\alpha > 1/|m+1|$ evidently correspond to drops of the magnetized plasma which become detached from the magnetosphere and then fall onto the pulsar.

A limiting case for the solution (7.2.30) is the case in which $a_0 \to 0$ and $\alpha \to 0$, with the existence of a nonzero limit of a_0/α. In this case, proceeding to the limit in (7.2.30), we have

$$z = \frac{A}{D}\zeta \exp(\mu\zeta) .$$

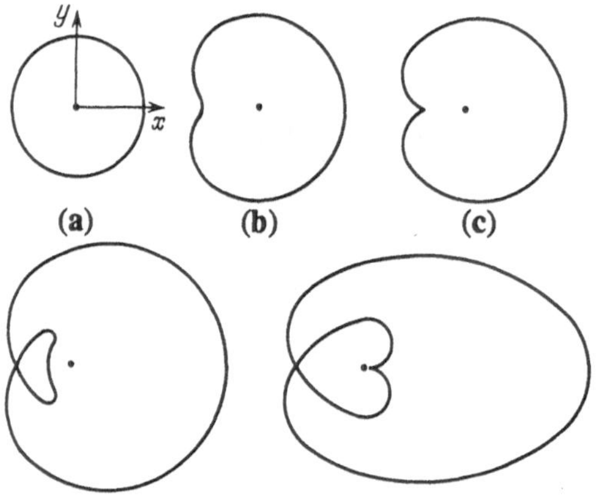

Fig. 7.3

This solution corresponds to functions $a(R)$ and $b(R)$ which degenerate into constants. Then the shape of the magnetosphere is given parametrically as follows:

$$R = \frac{A}{D}\exp(\mu\cos\theta) \ ,$$

$$\varphi - \Omega t = \theta + \mu\sin\theta + \arg\left(\frac{A}{D}\right) \ .$$

This family of curves with the arbitrary parameter μ will have no self-intersection only for $0 < \mu < 1$.

We note that each curve belonging to the family (7.2.31, 32) will be a circle for all admissible α if $m = -1$.

The equations of these circles are as follows:

$$|z^2| = \left|\frac{A}{D} - \alpha z\right|^2 \ .$$

It follows from (7.2.31, 32) that the real constants m and $a_0|D|^{m/2}/\alpha$ and the complex constant A are physically specified. Thus, there remains an arbitrariness associated with the constants Γ and α. In Fig. 7.3 we show the characteristic evolution of the shape of the magnetosphere for $m = 2$ when the parameter α evolves from 0 to 1. The curves with no self-intersections for $\alpha \le 1/3$ are physically meaningful.

We now calculate the force and the torque of the magnetic field acting on the magnetosphere. They can be determined by means of expressions of the Blasius–Chaplygin type (see, for example, [7.23]):

$$8\pi F^* = \mathrm{i}\int_C \left(\frac{dW}{dz}\right)^2 dz \ , \quad 8\pi M = \mathrm{Re}\int_C z\left(\frac{dW}{dz}\right)^2 dz \ ,$$

where C is the contour of the magnetosphere in the complex z plane, and F^* is the complex-conjugate force $F^* = F_x - \mathrm{i}F_y$. Conformally mapping the interior of the magnetosphere into the interior of the circle $|\zeta| < 1$ by means of (7.2.30), we have

$$8\pi F^* = \mathrm{i} \int\limits_{|\zeta|=1} \frac{d\zeta}{dz} \left(\frac{dW}{d\zeta}\right)^2 d\zeta \ , \quad 8\pi M = \mathrm{Re} \int z \frac{d\zeta}{dz} \left(\frac{dW}{d\zeta}\right)^2 d\zeta \ .$$

Hence we have

$$\begin{aligned} F^* = & \frac{\mathrm{i}\Gamma|D^2|}{2A} - \frac{m\alpha D}{2A}(\Gamma^2 + 2|D^2|) \\ & - \frac{(m-4)(m+1)\alpha^2}{8A}\mathrm{i}D^2\Gamma - \frac{m(m+1)(2m+1)\alpha^3}{3A}D^3 \ , \end{aligned} \tag{7.2.35}$$

$$2M = \mathrm{Re}\left[m\alpha D\Gamma + \mathrm{i}D^2\frac{m(m+1)\alpha^2}{2}\right] \ . \tag{7.2.36}$$

As the accreting matter is discarded, an amount of work $M\omega$ per unit time is done. It follows from (7.2.36) that this work is equal to zero (in the absence of circulation, $\Gamma = 0$) only if $m = -1$, i.e., if the magnetosphere has the shape of a circle (the cases with $m = 0$ and $\alpha = 0$ are degenerate). Besides the torque, according to (7.2.35) the magnetic dipole will be acted upon by a force which is constant in the rotating coordinate system. For $a(R) = \mathrm{const}$ and $b(R) = \mathrm{const}$, we must proceed to the limit $\lim_{m\to\infty} m\alpha = \mu$ in (7.2.35, 36). We then obtain

$$F^* = \frac{\mathrm{i}\Gamma D^2}{2A} - \frac{\mu D}{2A}(\Gamma^2 + 2|D^2|) - \frac{\mu^2}{8A}\mathrm{i}D^2\Gamma - \frac{\mu^3 D^3}{3A} \ ,$$

$$2M = \mathrm{Re}\left[\mu D\Gamma + \mathrm{i}\frac{\mu^* D^2}{2}\right] \ .$$

7.3 Generation of Acoustic Waves by the Rotating Magnetosphere in the Stellar Wind[5]

For slow rotations of a neutron star, the main contribution to the radiation of acoustic perturbations comes from the rotation of the magnetosphere as a solid body; if the boundary conditions are formulated with allowance for the deformation of the surface of the magnetosphere due to the rotation, there is a relative correction to the shape of the magnetosphere of the order of the square of the Mach number, $(\Omega R/a)^2$, where Ω is the angular speed, R is the radius of the

[5] The author expresses gratitude to R. A. Sunyaev for his part in the formulation of the problems of this section and for discussion of the results.

magnetosphere, and a is the speed of sound. Slowness of the rotation implies that the Mach number $\Omega R/a$ is small. In order to calculate the generation of sound by the rotating ellipsoid which approximately replaces the magnetosphere, we must know the speed of motion of the surface along the normal to it.

Suppose that in the principal axes the equation of the ellipsoid has the form

$$x_0^2 + y_0^2 + z_0^2/(1 - e^2) = R^2 \ .$$

The moving system of coordinates x_0, y_0, z_0 rotates together with the ellipsoid. The angle α is the angle between the minor axis of the ellipsoid (taken as the z_0 axis and the corresponding axis of the magnetic dipole) and the vector $\boldsymbol{\Omega}$ of the angular velocity of rotation of the neutron star (taken as the z axis of the stationary system of coordinates x, y, z). Let the relation between the two Cartesian systems of coordinates x_0, y_0, z_0 and x, y, z be given by

$$\begin{pmatrix} x \\ y \\ z \end{pmatrix} = \begin{pmatrix} \cos\Omega t & \sin\Omega t & 0 \\ -\sin\Omega t & \cos\Omega t & 0 \\ 0 & 0 & 1 \end{pmatrix} \begin{pmatrix} \cos\alpha & 0 & \sin\alpha \\ 0 & 1 & 0 \\ -\sin\alpha & 0 & \cos\alpha \end{pmatrix} \begin{pmatrix} x_0 \\ y_0 \\ z_0 \end{pmatrix} \ ,$$

from which it follows that

$$z_0 = \sin\alpha[x\cos\Omega t - y\sin\Omega t] + z\cos\alpha \ .$$

Apart from terms of order e^4, the equation of the ellipsoid can be written in the form

$$x_0^2 + y_0^2 + x_0^2 + e^2 z_0^2 = R^2 \ .$$

Using the fact that $x_0^2 + y_0^2 + z_0^2 = x^2 + y^2 + z^2$, in the stationary coordinate system we have

$$f(x, y, z, t) \equiv x^2 + y^2 + z^2 + e^2 z_0^2 - R^2 = 0 \ .$$

By definition, the speed of motion of the surface along the normal to it is

$$\begin{aligned} D_n &= -|\mathrm{grad} f|^{-1} \partial f/\partial t \approx -e^2 z_0 \partial z_0/\partial t R^{-1} \\ &= e^2 \Omega R \sin\alpha \sin(\varphi + \Omega t) \sin\theta [\sin\alpha \sin\theta \cos(\varphi + \Omega t) + \cos\alpha\cos\theta] \ . \end{aligned}$$

Here R, θ, φ are the coordinates in the spherical system associated with the stationary Cartesian system of coordinates x, y, z with center at its origin. The boundary condition that the gas does not flow across the magnetosphere obviously has the form

$$v_n' = v_n - D_n = 0 \ , \qquad (7.3.1)$$

where v_n is the component of the velocity $\boldsymbol{v}$ normal to the surface of the ellipsoid. We note that the expression for D_n is a linear combination of two independent quadrupole spherical harmonics.

In Sects. 7.3.1 and 7.3.5 we calculate the total flux of acoustic energy generated by a slowly rotating ellipsoid in various models of the atmosphere surrounding the magnetosphere: in an isothermal atmosphere (Sect. 7.3.1) and in an adiabatic atmosphere with allowance for viscosity of the gas (Sect. 7.3.5). In Sect. 7.3.2 we calculate the power of the acoustic energy carried away by short waves. In Sect. 7.3.3 we study the question of the distances from the magnetosphere at which inversion of the acoustic waves occurs as a result of nonlinear effects, and in Sect. 7.3.4 we carry out a qualitative analysis of the consequences of the heating of the plasma by the acoustic waves. In Sect. 7.3.6 we give a qualitative description of the generation of acoustic perturbations in the case of supersonic rotation of the magnetosphere. In Sect. 7.3.7 we consider a model problem of stationary accretion of gas in the presence of energy release. In Sect. 7.3.8 we show that, for the hydrodynamical mechanisms of retardation considered in Sects. 7.3.1–6, nonsymmetric pulsars have a stable long-period variation of the angular velocity of rotation.

7.3.1 Acoustic Waves in an Isothermal Atmosphere

In equilibrium of an unperturbed isothermal atmosphere filled with an ideal gas, we must have fulfillment of the condition $p = a^2\varrho$, $a^2 = \text{const}$, and of the equilibrium condition

$$\frac{dp}{dz} + \frac{GM}{r^2} = 0 \ , \quad \text{and hence} \quad \varrho = \varrho_0 \exp\left(\frac{GM}{(a^2 r)}\right) \ ;$$

here M is the mass of the star (we neglect the gravitation of the gas). The perturbations of the pressure p' and of the density ϱ' are related at constant temperature by the equation $p' = a^2\varrho'$.

From the Cauchy–Lagrange integral for potential perturbations, we have

$$\frac{\partial \Phi}{\partial t} + a^2 \frac{\varrho'}{\varrho} = 0 \ . \tag{7.3.2}$$

Here Φ is the potential of the perturbed velocity, and we neglect the term $v^2/2$ as a small quantity of the second order. From the equation of continuity it follows that

$$\frac{\partial}{\partial t}\frac{\varrho'}{\varrho} - \frac{GM}{a^2 r^2}\frac{\partial \Phi}{\partial r} + \Delta\Phi = 0 \ . \tag{7.3.3}$$

Using (7.3.2), from (7.3.3) we have

$$-\frac{1}{a^2}\frac{\partial^2}{\partial t^2}\Phi - \frac{r_a}{r^2}\frac{\partial \Phi}{\partial r} + \Delta\Phi = 0 \ , \quad r_a \equiv GM/a^2 \ . \tag{7.3.4}$$

We shall expand the function Φ in spherical harmonics and seek a solution of the form $\Phi = \exp(\mathrm{i}\Omega t) P_l^m(\cos\theta) \mathrm{e}^{\mathrm{i}m\varphi} \tilde{\Phi}(r)$. Then from (7.3.4) we obtain

$$\frac{\Omega^2}{a^2}\tilde{\Phi} - \frac{d\tilde{\Phi}}{dr}\frac{r_a}{r^2} + \frac{d^2\tilde{\Phi}}{dr^2} + \frac{2}{r}\frac{d\tilde{\Phi}}{dr} - \frac{l(l+1)}{r^2}\tilde{\Phi} = 0 \ . \tag{7.3.5}$$

Owing to the boundary condition of impenetrability (7.3.1), the acoustic emission has a quadrupole character, i. e., $l = 2$, and the boundary condition (7.3.1) contains two linearly independent angular harmonics: $l = 2$ with $m = 1, 2$.

We shall assume that the radius R of the magnetosphere is less than r_a: $R < r_a$. The condition of slowness of the rotation can be written as $\Omega r_a/a \lesssim 1$. Then the region outside the magnetosphere can be divided into a near zone [where in (7.3.5) the term $\Omega^2\tilde{\Phi}/a^2$ is small] and a wave zone. In the wave zone, for the power of the emitted energy we have (see Sect. 7.1.2)

$$W = \varrho a \int\limits_s v^2 ds \ , \tag{7.3.5$'$}$$

where s is a sphere in the wave zone. The solution of (7.3.5) for $l = 2$ in the near zone has the form

$$\tilde{\Phi} = A_1(1 - 6x + 12x^2) + A_2[-1 + 6x - 12x^2 + \mathrm{e}^{-1/x}(1 + 6x + 12x^2)] \ , \tag{7.3.6}$$

where A_1 and A_2 are arbitrary constants, and $x \equiv r/r_a$.

For $r \gg r_a$, in the wave zone we can neglect the second term in (7.3.5). Then the solution of (7.3.5), valid throughout the wave zone, takes the form

$$\tilde{\Phi} = \frac{Ba}{r\Omega}\mathrm{e}^{\mathrm{i}\Omega r/a}\left(1 - 3\mathrm{i}\frac{a}{\Omega r} - 3\frac{a^2}{(\Omega r)^2}\right) \ . \tag{7.3.7}$$

Here B is an arbitrary constant.

The second linearly independent solution is omitted in (7.3.7) because of the condition that there are no incident waves. In the case of a homogeneous atmosphere, $\tilde{\Phi}$ has the form (7.3.7) for all $r > R$.

Combining (7.3.6, 7), for $r_a \ll r \ll a/\Omega$ we obtain

$$\frac{A_2}{A_1} = \mathrm{i}\frac{32\,400}{(\Omega r_a/a)^5} \ ,$$

and hence $|A_2| \gg |A_1|$. Therefore in the near zone we neglect the term containing the factor A_1, and put

$$B = (\Omega r_a/a)^3 A_2/180 \ . \tag{7.3.8}$$

We adopt the notation

$$f(x) = 20[-24x + 6 + \mathrm{e}^{-1/x}(18 + 24x + 6/x + 1/x^2)] \ , \tag{7.3.9}$$

where $f(x)$ is the derivative with respect to x of the part of the expression (7.3.6) containing the factor A_2.

From the boundary condition (7.3.1), we have

$$\left.\frac{\partial\tilde{\Phi}}{\partial r}\right|_{r=R} \sim e^2 R\Omega \sin\alpha \quad .$$

Using (7.3.6), we find the constant A_2:

$$\frac{A_2 f(x_m)}{20 r_a} \sim e^2 \Omega R \sin\alpha \quad ,$$

and hence

$$A_2 \sim \frac{20 r_a R \Omega e^2 \sin\alpha}{f(x_m)} \quad , \quad x_m \equiv \frac{R}{r_a} \quad .$$

It is easy to calculate the flux of energy W in the wave zone: $W = 4\pi\varrho a B^2/15$. The final expression for the flux of acoustic energy generated in the presence of rotation of the magnetosphere can be represented, using (7.3.8), in the form

$$W = \left(\frac{4\pi e^4}{1215} \frac{R^{10}\Omega^8 \sin^2\alpha}{a^5}\right) \frac{1}{x_m^8 f^2(x_m)} \quad .$$

The expression in the brackets is equal to the emitted flux of acoustic energy in the case $p = \text{const}$. The factor $[x_m^8 f^2(x_m)]^{-1}$ is the ratio of the power of the acoustic energy generated in the presence of weak rotation of the magnetosphere in an isothermal atmosphere to the power of acoustic emission in a homogeneous ($p = \text{const}$) atmosphere for identical speeds of sound and identical densities for $r \to \infty$. For $x_m \to \infty$, the factor tends to unity, which is perfectly natural, since the density of an isothermal atmosphere tends exponentially fast to a constant value. For $x_m \to 0$, this factor attains large values, and this fact indicates that sound is generated more effectively in an isothermal atmosphere than in a homogeneous one.

In fact, according to the definition (7.3.9), with increasing x the function $f(x)$ falls off monotonically: $f(0) = 120$, $f(0.25) = 23.7$, and $f(x) \to 0$ as x^{-4} when $x \to \infty$, and, for example, for $x_m = 0.25$ the power of acoustic emission in a homogeneous atmosphere is less than in an isothermal one by a factor 117, i. e.,

$$[x_m^8 f^2(x_m)]^{-1} \approx 117 \quad .$$

7.3.2 Generation of Short Waves

We assume now that the angular velocity of rotation of the star satisfies the inequality $\Omega R/a > 1$. To ensure that the component of the velocity of its motion normal to the surface of the magnetosphere does not exceed the speed of sound, the following inequality must hold:

$$e^2 \Omega R \sin\alpha < a \quad .$$

In this case, the pulsations of the surface of the magnetosphere give a contribution to the acoustic emission of the same order as the contribution from the rotation of the magnetosphere as a solid body. An exact calculation in this case is extremely complicated. The order of magnitude of the power of the acoustic emission can be estimated by assuming that the boundary of the magnetosphere is rigid.

For short waves, the wave zone begins at once from the magnetosphere, and therefore we obtain the following expression for the emitted power of acoustic energy:

$$W = \frac{4\pi}{15} a\varrho e^4 R^4 \Omega^2 \sin^2 \alpha \quad .$$

It is of interest to note that in the case of an isothermal atmosphere the emitted power of acoustic energy according to this expression (for fixed α, Ω, a, and ϱ_∞) attains a minimum when the radius of the magnetosphere is one quarter of the radius r_a. Indeed, the expression $R^4 \exp(r_a/R)$ has a minimum at $R = r_a/4[W \sim R^4 \exp(r_a/R)$ if we take into account the fact that $\varrho = \varrho_\infty \exp(r_a/R)]$.

7.3.3 Inversion of Acoustic Waves

In the wave zone, each element of an acoustic wave moves approximately as a plane wave in the radial direction with phase velocity $a + v$, where v is the unperturbed radial velocity of the gas ($v < 0$ in the case of accretion). The motion of each element of the wave front is described by the equation of the characteristic, $\psi \equiv t - \int dr(a+v)^{-1} = \text{const.}$

The characteristic scale over which inversion of the wave takes place can be found by determining the point of intersection of the characteristics corresponding to the crest and the valley of the wave. We make use of the fact that up to the instant of inversion the flux of acoustic energy of the spherical waves is constant (see Sect. 7.1):

$$W = 4\pi(1+M)^2 \varrho a r^2 v'^2 \quad .$$

Here v' is the amplitude of a spherical wave propagating through an arbitrary spherically symmetric background, for which M is the Mach number, $M \equiv |v/a|$.

If at $r = r_0$ the amplitude of the perturbation was v_0', then from the condition of constancy of W we have

$$v' = \frac{(1+M_0) r_0 \sqrt{\varrho_0 a_0}}{(1+M) r \sqrt{\varrho a}} v_0' \quad .$$

In the acoustic wave, from the equation of motion we obtain, in the first approximation,

$$v' = \frac{2a'}{(\gamma - 1)} \quad .$$

Then the equation of motion of the crest can be written

$$t - \int_{r_0}^{r} \frac{dr}{a+v+a'+v'} = t - \int_{r_0}^{r} \frac{dr}{a+v} + b \int_{r_0}^{r} \frac{dr\sqrt{aW}}{r(a+v)^3\sqrt{\varrho}} = t_0 \ , \tag{7.3.10}$$

where

$$\sqrt{W} \equiv \sqrt{4\pi}\, v_0' r_0 \sqrt{\varrho_0 a_0}\,(1+M_0) \ , \quad b = \frac{(\gamma+1)}{(4\sqrt{\pi})} \ .$$

The equation of motion of the valley can be written by analogy, using the fact that it is emitted half a period earlier:

$$t - \int_{r_0}^{r} \frac{dr}{a+v} - b \int_{r_0}^{r} \frac{dr\sqrt{aW}}{r(a+v)^3\sqrt{\varrho}} = t_0 - \frac{\pi}{\Omega} \ . \tag{7.3.11}$$

Subtracting (7.3.11) from (7.3.10), for the determination of the point of intersection of the characteristics we obtain

$$b\Omega \int_{r_0}^{\tilde{r}} \frac{dr\sqrt{aW}}{r(a+v)^3\sqrt{\varrho}} = \frac{\pi}{2} \ , \tag{7.3.12}$$

or, introducing the notation σ for this integral, we have $\sigma = \pi/2$ for the point of intersection of the characteristics. Equation (7.3.12) is an algebraic equation for the inversion radius $\tilde{r}$. We shall consider the form of this equation for various models of the atmosphere.

For the homogeneous case in which $v = 0$, we obtain from (7.3.12) the well-known result [7.9]

$$\sigma = \Omega v_0'(\gamma+1) r_0 (\ln \tilde{r}/r_0)/(2a^2) = \pi/2 \ .$$

For an isothermal stationary atmosphere, $\gamma = 1$, $v = 0$, and $\varrho = \varrho_0 \exp(r_a/r)$. The inversion radius is determined by the condition

$$\sigma = \frac{\Omega\sqrt{W}}{2\sqrt{\pi\varrho_\infty a^5}} \int_{r_0}^{\tilde{r}} \frac{dr}{r} \mathrm{e}^{-r_a/2r} = \frac{\pi}{2} \ . \tag{7.3.13}$$

In comparison with the homogeneous case, inversion in an isothermal atmosphere takes place nearer the point of acoustic emission.

Stationary spherically symmetric accretion (the solution of *Bondi* [7.26]) is described by the expressions

$$a \sim r^{-1/2} \ , \quad v \sim r^{-1/2} \ , \quad \varrho \sim r^{-3/2} \ , \quad \gamma = \tfrac{5}{3} \ , \tag{7.3.14}$$

the radial velocity v is directed to the gravitating center, and the motion of the perfect gas is assumed to be isentropic. Then according to (7.3.12) we obtain

$$\sigma = \frac{\sqrt{W}}{3\sqrt{\pi}\, r_0^2 a_0^{5/2}(1+M)^3\sqrt{\varrho_0}}(\bar{r}^2 - r_0^2) = \frac{\pi}{2} \ . \tag{7.3.14'}$$

The foregoing expressions hold only for the wave zone. In the case of long-wave emission, we must interpret r_0 as the radius of the wave zone, and v_0' as the perturbation of the velocity at this radius.

7.3.4 Heating of the Plasma Due to Dissipation of Acoustic Waves

A more accurate analysis based on Sect. 7.1 for periodic waves shows that inversion of the acoustic perturbations begins at $\sigma = 1$, when on a discrete family of spheres the gradients of the perturbations of the velocity and density become infinite (the wave acquires a sawtooth profile), and not at $\sigma = \pi/2$, as given by (7.3.12).

For $\sigma > 2$, the flux of acoustic energy is dissipated according to the asymptotic law [7.10]

$$W = W_0(1+\sigma)^{-2} \ . \tag{7.3.15}$$

Equation (7.3.15) holds on the segment on which the sawtooth profile of the inverted wave has been completely formed. In the range $1 < \sigma < 2$, (7.3.15) is inaccurate. Equation (7.3.15) has the same form for both isothermal and adiabatic atmosphere. The only difference is that in the first case σ must be replaced by the expression (7.3.13), and in the second case by (7.3.14'). In the dissipation of a sawtooth profile, there are irreversible losses in shock waves (in the approximation of large acoustic Reynolds numbers).

For acoustic waves propagating through a spherically symmetric stationary background perturbed by the rotation of an oblate magnetosphere with angular velocity Ω around the z axis, the flux of outgoing energy and the z component of the flux of angular momentum are related by the equation

$$L_z = W/\Omega \ .$$

Therefore the dissipation of acoustic waves for $\sigma > 1$ leads to the transfer of angular momentum to the layers of the flowing gas, and they acquire nonradial components of velocity (the total flux of angular momentum must be conserved, so that if the acoustic waves vanish, a rotational nonwave motion of the gas must appear).

The stronger the accretion, the closer it presses the zone of heating of the plasma by the acoustic waves to the magnetosphere. Strong energy release inhibits the accretion.

For vanishingly small viscosity, falling off with increasing r, there is practically no laminar outflow of angular momentum from the system if the zone of acoustic dissipation is situated sufficiently close to the magnetosphere and the flux of acoustic energy W vanishes according to (7.3.15) sufficiently rapidly.

Owing to the strong differential rotation of the layers of the gas in the zone of heating, an instability develops. As a result of the chaotic motion of turbulent sections, angular momentum flows out from the zone of heating. The discharge of heat in this zone leads to a superadiabatic temperature gradient, and this in turn leads to turbulent convection in the gas, which, as estimates show, carries away heat to infinity much more effectively than the electron heat conduction.

The acoustic waves themselves do not carry away angular momentum to infinity, but serve to induce turbulent angular momentum, which retards the rotation of the neutron star.

7.3.5 Inclusion of Viscosity of the Gas in Calculating the Torque Acting on the Rotating Magnetosphere

The problem of calculating the torque acting on a nonsymmetric body rotating around a stationary axis in a viscous compressible fluid is one of the least studied problems in classical fluid mechanics.

We shall describe briefly the facts about the motion of a fluid around a rotating sphere which have been well studied experimentally and theoretically [7.27]. At small Reynolds numbers (steady flow), the motion of the fluid is described by the Kirchhoff solution (a generalization of this solution to the case of viscosity which varies with the radius in the presence of spherical accretion is given below). At Reynolds numbers $1000 < \mathrm{Re} < 40000$, a laminar boundary layer is formed around the sphere. In this case, the rotating sphere experiences a torque of the viscous friction given by $m \approx 13\varrho\Omega^2 R^5/\sqrt{\mathrm{Re}}$ (ϱ is the density of the gas, Ω is the angular velocity of the sphere, and R is its radius).

The rotating sphere acts as a fan: it sucks in gas from all directions and ejects it in the equatorial plane. In a thin boundary layer, the gas flows along spirals toward the equator. In the equatorial plane, aerial jets collide and are ejected in the form of twisted plaits. The volume discharge of "fanned" air is then $V \approx 3R^2\sqrt{\nu\Omega}$. At Reynolds numbers $\mathrm{Re} > 400\,000$, a turbulent boundary layer is formed around the sphere, and at the equator this layer has thickness of the order of the radius of the sphere. The volume discharge of the gas flowing to the sphere and ejected in the equatorial plane is of order $R^3\Omega/(\mathrm{Re})^{1/5}$. The torque acting on the sphere is then of order

$$m \sim 10^{-1}\varrho\Omega^2 R^5 \sqrt[5]{\mathrm{Re}} \ .$$

The sphere experiences the strongest drag in the case of rotation in the turbulent regime. In view of the local character of the boundary layer, the expressions for the retarding torque at large Reynolds numbers can be used to estimate the characteristic time of retardation of a neutron star with a strong magnetic field. At small Reynolds numbers, it is necessary to perform an analysis of the problem as a whole with allowance for the nonhomogeneity of the distribution of the parameters of the gas.

In this section, we give a combined analysis of viscous and acoustic effects accompanying the rotation of an ellipsoid with small eccentricity in a nonhomogeneous atmosphere around an arbitrary stationary z axis[6] at small Reynolds numbers. We begin with the Navier–Stokes equations with zero second viscosity:

$$\varrho \frac{dv_i}{dt} = -\frac{\partial p}{\partial x^i} + \frac{\partial}{\partial x^j}\left[\mu\left(\frac{\partial v_i}{\partial x^j} + \frac{\partial v_j}{\partial x^i} - \frac{2}{3}\delta_{ij}\,\mathrm{div}\,\boldsymbol{v}\right)\right] \quad . \tag{7.3.16}$$

All the perturbations and the Navier–Stokes equations for them on a spherically symmetric background can be divided naturally into two groups: even and odd, in accordance with the different behavior under inversion of the coordinates on the sphere $r = \mathrm{const}$. The velocity component tangential to the sphere is a vector with respect to coordinate transformations on the sphere. Any vector tangential to the sphere can be represented in terms of a scalar Φ (an even perturbation) and a pseudoscalar (an odd perturbation):

$$v'_A = \Phi_{,A} + \varepsilon_{AB} g^{BC} \psi_{,C} \quad ; \quad A, B, C = 1, 2 \quad ,$$

where g_{AB} and ε_{AB} are the covariant components of the metric and of the Levi–Civita tensor on the sphere of radius r, respectively. In spherical coordinates θ, φ, the metric on the sphere has the form $ds^2 = r^2(d\theta^2 \sin^2\theta\, d\varphi^2)$. Then

$$v_\theta = \Phi_{,\theta} + \psi_{,\varphi}/\sin\theta \quad , \quad v_\varphi = \Phi_{,\varphi} - \psi_{,\theta}\sin\theta \quad .$$

By virtue of the Navier–Stokes equations, the function Φ is related to the perturbations of the density and pressure, i. e., to the acoustic perturbations. In contrast, the function ψ is related to the perturbation of the component of the curl of the velocity $\omega_r = -\tilde{\Delta}\psi/(2r^2)$, where $\tilde{\Delta}$ is the Laplacian on the sphere of unit radius. The odd perturbations are not related to the perturbation of the density. A classical example of a solution for the odd perturbations is the Kirchhoff solution for the rotation of a sphere in a viscous homogeneous fluid.

From the Navier–Stokes equations (7.3.16), linearized around a spherically symmetric stationary solution $\varrho = \varrho(r)$, $v_r = r(r)$ in which $\mu = \mu(r)$, $4\pi\varrho v_r r^2 = -\dot{M} = \mathrm{const}$, it follows that

$$\begin{aligned}
\varrho\left(\frac{\partial v'_A}{\partial t} + v_r \frac{\partial v'_A}{\partial r}\right) = {} & -\frac{\partial p'}{\partial x^A} + \frac{\partial \mu}{\partial r}\left(\frac{\partial v'_A}{\partial r} + \frac{\partial v'_r}{\partial x^A} - \frac{2v'_A}{r}\right) \\
& + \mu\left[\frac{\partial}{\partial x^A}\left(\frac{\tilde{\Delta}\Phi}{r^2} + \frac{\partial^2 \Phi}{\partial r^2}\right)\right. \\
& \left. + \varepsilon_{AB} g^{BC} \frac{\partial}{\partial x^C}\left(\frac{\tilde{\Delta}\psi}{r^2} + \frac{\partial^2 \psi}{\partial r^2}\right)\right] \\
& + \mu\left(\frac{2}{r}\frac{\partial v'_r}{\partial x^A} + \frac{1}{3}\frac{\partial}{\partial x^A}\,\mathrm{div}\,\boldsymbol{v}'\right) \quad .
\end{aligned}$$

[6] The axis of rotation evolves slowly with time under the influence of the torques acting on the ellipsoid. However, this evolution can be neglected in the calculation of the torque themselves.

Here and in what follows, the perturbed quantities are indicated by primes. As a consequence of these equations, we have the following relations:

$$\varrho\left(\frac{\partial\Phi}{dt}+v_r\frac{\partial\Phi}{dr}\right)=-p'+\frac{\partial\mu}{\partial r}\left(\frac{\partial\Phi}{dr}+v_r'-2\frac{\Phi}{r}\right)$$
$$+\mu\left[\frac{\tilde{\Delta}\Phi}{r^2}+\frac{\partial^2\Phi}{\partial r^2}+2\frac{v_r'}{r}+\frac{1}{3}\operatorname{div}\boldsymbol{v}'\right] \ , \tag{7.3.17}$$

$$\varrho\left(\frac{\partial\psi}{\partial t}+v_r\frac{\partial\psi}{\partial r}\right)=\frac{\partial\mu}{\partial r}\left(\frac{\partial\psi}{\partial r}-\frac{2\psi}{r}\right)+\mu\left(\frac{\partial^2\psi}{\partial r^2}+\frac{\tilde{\Delta}\psi}{r^2}\right) \ . \tag{7.3.18}$$

Odd Perturbations. From (7.3.18), using the equation of continuity for the unperturbed solution $4\pi\varrho v_r r^2=-\dot{M}=\text{const}$, where $\dot{M}$ is the flux of mass flowing to the magnetosphere per unit time, we have

$$\varrho r^2\frac{\partial\psi}{\partial t}-\frac{\dot{M}}{4\pi}\frac{\partial\psi}{\partial r}=\frac{\partial}{\partial r}\left[\left(r^2\frac{\partial\psi}{\partial r}-2r\psi\right)\mu\right]+\mu(\tilde{\Delta}+2)\psi \ .$$

If we expand ψ in spherical harmonics $Y_l^m(\theta,\varphi)$, a solution which generalizes the Kirchhoff solution is obtained for $l=1$ and with the assumption of stationarity. The equation for ψ in this case can be integrated with respect to r. We shall denote the constant of integration by $\tilde{q}$. Then we obtain

$$\mu r^2\frac{\partial\psi}{\partial r}-2\mu r\psi+\frac{\dot{M}}{4\pi}\psi=\tilde{q} \ . \tag{7.3.18$'$}$$

Let us explain the physical meaning of the constant $\tilde{q}$. From the equations of motion of a fluid with a symmetric stress tensor it follows that

$$\frac{\partial}{\partial t}[\varrho(\boldsymbol{v}\times\boldsymbol{r})]=\frac{\partial}{\partial x^k}\{\varrho v^k(\boldsymbol{r}\times\boldsymbol{v})+\boldsymbol{p}^k\times\boldsymbol{r}\} \ , \tag{7.3.19}$$

where $\boldsymbol{p}^k$ is the vector whose components belong to row k of the stress tensor. The moment of the hydrodynamical forces acting on a body rotating in the fluid is

$$\int_S[\varrho(v_n-D_n)(\boldsymbol{r}\times\boldsymbol{v})+\boldsymbol{p}^n\times\boldsymbol{r}]\,dS=\frac{d\boldsymbol{m}}{dt} \ , \tag{7.3.20}$$

where $\boldsymbol{m}$ is the angular momentum of the rotating body, S is the surface of the body, and $\boldsymbol{n}$ is the outward normal to S.

The total angular momentum of the body and the fluid is

$$\boldsymbol{m}+\int\varrho(\boldsymbol{v}\times\boldsymbol{r})\,dV=\boldsymbol{M} \ ,$$

where the integration extends over the entire volume occupied by the fluid. It follows from (7.3.19,20) that

$$\frac{d}{dt}\boldsymbol{M} = \int\limits_{\Sigma} \{\varrho v^k(\boldsymbol{r} \times \boldsymbol{v}) + \boldsymbol{p}^k \times \boldsymbol{r}\} n_k \, d\Sigma \ , \tag{7.3.21}$$

where Σ is a surface which tends to infinity.

The rotating magnetosphere will experience a torque which effectively retards it only if angular momentum is carried away to infinity.

We choose Σ to be a sphere of radius r, since in the stationary case the expression (7.3.21) does not depend on the value of r, and for the gas we adopt the Navier–Stokes model with zero second viscosity:

$$p_{ij} = -p\delta_{ij} + \mu(v_{i,j} + v_{j,i} - \tfrac{2}{3}\delta_{ij} \operatorname{div} \boldsymbol{v}) \ .$$

Substituting into (7.3.21) the odd perturbations with $l = 1$, in the stationary case we readily obtain

$$\text{const} = L_z = \frac{dM_z}{dt} = \frac{8\pi}{3}\left(\frac{\dot{M}}{4\pi}\psi + \mu r^2 \frac{\partial \psi}{\partial r} - 2r\mu\psi\right) \ . \tag{7.3.21'}$$

For the odd perturbations with $l = 1$, we shall neglect the deviation of the magnetosphere from a sphere and write the condition of adhesion at $r = R$. Therefore the constant $\tilde{q}$ in (7.3.18') is determined by the equation

$$8\pi\tilde{q}/3 = L_z \ .$$

Equation (7.3.21′) can be integrated in explicit form. It is necessary here to use the boundary condition of adhesion on a sphere having radius R and rotating around the z axis with angular velocity Ω. For the viscosity we adopt a power-law dependence on the radius, $\mu = \mu_0(r/R)^{-\alpha}$, and we introduce the new variable $x \equiv r/R$ and the constants

$$\gamma = \frac{\dot{M}}{4\pi\mu_0 R} \ , \quad q = \frac{3}{8\pi\mu_0 R^3} L_z \ .$$

Then (7.3.21′) can be written as the equation

$$x^{4-\alpha}\frac{df}{dx} + \gamma f x^2 = q \ , \quad f = \frac{\psi}{r^2} \ . \tag{7.3.22}$$

with the boundary condition $f(1) = \Omega$. The properties of the solution of (7.3.22) for $0 < \alpha < 1$ and $\alpha > 1$ are very different. Physically, this is due to the fact that the condition of smallness of the Reynolds number throughout the region occupied by the gas may or may not be satisfied:

$$\text{Re} = \frac{\varrho|v_r|r}{\mu} = \frac{\dot{M}}{4\pi r\mu} = \frac{\dot{M}}{4\pi R\mu_0}\left(\frac{r}{R}\right)^{\alpha-1} = \gamma x^{\alpha-1} \ .$$

From the condition of smallness of the Reynolds number it follows that $\gamma < 1$. For $\alpha < 1$, the rotation of the sphere produces smooth motion of the external layers of the gas, and the angular momentum determined by the angular velocity

Ω will be carried away to infinity. For $\alpha > 1$, a viscous boundary layer is formed around the sphere. The Reynolds number grows as r increases, and therefore in this case the effects of viscosity become unimportant with increasing distance from the sphere. In an ideal gas, for barotropic motions the vortices are frozen into the fluid. They can be arbitrary in magnitude. For $\alpha > 1$, the flux of angular momentum is not determined by the rotation of the sphere and can be arbitrary.

An analysis of the solutions of (7.3.22) confirms what we have said above. For $\alpha < 1$ (7.3.22) has a unique particular solution which does not increase as $r \to \infty$. It has the form

$$f(x) = -q \exp\left(\frac{\gamma x^{\alpha-1}}{1-\alpha}\right) \int_x^\infty t^{\alpha-4} \exp\left(-\frac{\gamma t^{\alpha-1}}{1-\alpha}\right) dt \ . \tag{7.3.23}$$

From the boundary condition $f(1) = \Omega$ we obtain an expression which determines the flux of angular momentum carried away to infinity:

$$-\Omega = q \int_1^\infty t^{\alpha-4} \exp\left(-\frac{\gamma(t^{\alpha-1}-1)}{1-\alpha}\right) dt \ . \tag{7.3.24}$$

As an example, let us consider the case of an isothermal atmosphere in which $\mu = \text{const}$ and $\alpha = 0$. In this case, from (7.3.24) we obtain

$$-q = \frac{\Omega\gamma^3}{2[e^\gamma - 1 - \gamma - \gamma^2/2]} \ . \tag{7.3.25}$$

For $\gamma \to 0$ (the case of small accretion), we obtain by means of (7.3.25) the classical Kirchhoff result $L_z = -8\pi\mu R^3 \Omega$ [7.28]. The solutions for arbitrary $\alpha < 1$ have analogous properties. In the case of weak accretion ($\gamma \to 0$), for the torque acting on the sphere we obtain

$$L_z = \frac{8\pi}{3}\Omega R^3(\alpha - 3) \ .$$

In the case $\alpha > 1$, (7.3.23) does not hold, since all the solutions of (7.3.22) are physically meaningful. The solution of the homogeneous equation (7.3.22) with exponential damping at infinity gives the distribution of velocities in the boundary layer, and the general solution which takes into account the boundary condition $f(1) = \Omega$ has the form

$$f(x) = \exp\left(\frac{\gamma(x^{\alpha-1}-1)}{1-\alpha}\right)\left[\Omega + q\int_1^x t^{\alpha-4} \exp\left(\frac{\gamma t^{\alpha-1}}{\alpha-1}\right) dt\right] \ .$$

The constant q is determined by the flux of angular momentum frozen (in view of the small viscosity) in the barotropically accreting gas.

For $\gamma > 1$ and $\alpha > 1$, the characteristic depth of penetration Δr of the viscous forces is found from the relation $\Delta x = \Delta r/R \sim 1/\gamma$ to be

$$\Delta r \sim \frac{4\pi\mu_0 R^2}{\dot{M}} \quad .$$

The physical condition for the formation of a boundary layer is the smallness of the mean free path λ of the molecules in comparison with the thickness of the layer. From the elementary kinetic theory of a gas, we have $\mu_0 \sim \lambda a \varrho$. Therefore, from the condition $\lambda < 4\pi\mu_0 R^2/\dot{M} = \mu_0/\varrho v_r$ it follows that $v_r/a < 1$. Thus, in the case $\alpha > 1$ the viscous-fluid model is applicable only for subsonic accretion. In the case of supersonic accretion with $\alpha > 1$, it is necessary to consider the interaction of a rarefied gas with the magnetosphere in the framework of kinetic theory.

In the case of the Bondi *adiabatic atmosphere* (the exponent of the adiabat is equal to 5/3), μ falls off with increasing r as $r^{-5/2}$, i.e., $\alpha = 5/2$. Therefore in this case the model of retardation of a pulsar solely by smooth motions of a viscous gas is inapplicable.

Even (Acoustic) Perturbations in the Bondi Adiabatic Atmosphere in the Case of Subsonic Accretion. In what follows, we shall assume that the accretion is weak and shall always neglect the square of the Mach number in comparison with unity. We shall make use of the fact that for completely ionized hydrogen the viscosity depends on the temperature according to the law $\mu = 1.2 \times 10^{-16} T^{5/2}$ g/cm s, where the temperature is measured in degrees Kelvin. Therefore the dependence of μ on the radius is $\mu = \mu_* r^{-5/2}$ with $\mu_* = \text{const}$. We recall that in the Bondi atmosphere the density and the speed of sound depend on the radius as follows:

$$\varrho = \varrho_* r^{-3/2} \quad , \quad a^2 = a_*^2/r \quad , \quad a_*^2 = 2GM/3 \quad ,$$

where M is the mass of the star. We shall assume that the perturbations have a time dependence $\exp(\mathrm{i}\Omega t)$.

The Navier–Stokes equations (7.3.16) for the even perturbations in the case of the Bondi atmosphere for small Mach numbers take the form

$$\mathrm{i}\Omega v_r' = -\frac{\partial}{\partial r}\left(\frac{p'}{\varrho}\right) + \frac{\nu^*}{r}\left[\frac{\partial^2 v_r'}{\partial r^2} - \frac{5}{2r}\frac{\partial v_r'}{\partial r} + \frac{\tilde{\Delta} v_r'}{r^2} - \frac{2\tilde{\Delta}\Phi}{r^3}\right] \quad , \tag{7.3.26}$$

$$\mathrm{i}\Omega\Phi = -\frac{p'}{\varrho} + \frac{\nu^*}{r}\left[\frac{\partial^2 \Phi}{\partial r^2} - \frac{5}{2r}\frac{\partial \Phi}{\partial r} + \frac{(\tilde{\Delta}+5)\Phi}{r^2} - \frac{\mathrm{i}}{3}\Omega\frac{\varrho'}{\varrho}\right] \quad , \tag{7.3.27}$$

$$\nu_* = \mu_*/\varrho_* \quad .$$

Here we neglect the viscous dissipation and use the adiabatic dependence of the pressure on the density, taking into account the equation of equilibrium for the unperturbed solution. Equation (7.3.27) represents (7.3.17) for the case of the Bondi atmosphere with allowance for the perturbed equation of continuity, which has the form

$$\mathrm{i}\Omega\frac{\varrho'}{\varrho}+\frac{\partial v_r'}{\partial r}+\frac{1}{2}\frac{v_r'}{r}+\frac{\tilde{\Delta}\Phi}{r^2}=0\ . \tag{7.3.28}$$

We rewrite (7.3.26–28) by introducing the notation

$$x=\frac{2}{3}\sqrt{\frac{r^3\Omega}{\nu^*}}\ ,\quad \varepsilon=\frac{\Omega\nu^*}{a_*^2}\ ,$$

$$Q=\frac{a_*^2}{\sqrt{\Omega\nu^* r}}\frac{\varrho'}{\varrho}\ ,\quad v_r'=v\ ,\quad \chi=\frac{\Phi}{r}\ ,$$

and using the fact that $p'=a_*^2\varrho'/r$ and

$$\mathrm{i}v=-\frac{\partial}{\partial x}Q+\frac{1}{3}\frac{Q}{x}+\frac{\partial^2 v}{\partial x^2}-\frac{20}{9}\frac{v}{x^2}+\frac{4\mathrm{i}\varepsilon}{3}\frac{Q}{x}\ ; \tag{7.3.26'}$$

$$\mathrm{i}\chi=-\frac{2}{3}\frac{Q}{x}+\frac{\partial^2\chi}{\partial x^2}-\frac{14}{9}\frac{\chi}{x^2}-\frac{2}{9}\mathrm{i}\varepsilon\frac{Q}{x}\ ; \tag{7.3.27'}$$

$$\mathrm{i}\varepsilon Q+\frac{\partial v}{\partial x}+\frac{1}{3}\frac{v}{x}+\frac{2}{3}\frac{\tilde{\Delta}\chi}{x}=0\ . \tag{7.3.28'}$$

In what follows, we shall assume that $\varepsilon \ll 1$. Physically, this means that the characteristic scale $\sqrt{\nu/\Omega}$ (the penetration depth of tangential viscous perturbations) which arises in periodic motions of the viscous fluid is small in comparison with the wavelength a/Ω. In addition, we shall assume that $\sqrt{\nu/\Omega}$ is much larger than the radius R of the magnetosphere. It follows from this that

$$x_R \ll \varepsilon \ll 1\ ,\quad \text{where}\quad 3x_R=2\sqrt{R^3\Omega/\nu^*}\ .$$

In the zone of action of the viscous forces (in the zone near the magnetosphere), the effects of compressibility can be neglected.

In (7.3.26′–28′), in a first approximation, we can drop the terms containing the coefficient ε. In contrast, in the wave zone in (7.3.26–28) we can neglect the viscous forces, but we must take into account the compressibility. Actually, in the wave zone the scale of variation of the functions of x is of order $1/\sqrt{\varepsilon}$, and (7.3.26–28) can be rewritten in the form

$$\mathrm{i}v=-\frac{\partial}{\partial z}\tilde{Q}+\frac{1}{3}\frac{\tilde{Q}}{z}+\varepsilon\left(\frac{\partial^2}{\partial z^2}v-\frac{20}{9}\frac{v}{z^2}+\frac{4\mathrm{i}}{3}\frac{\tilde{Q}}{z}\right)\ ; \tag{7.3.26''}$$

$$\mathrm{i}\tilde{\chi}=-\frac{2\tilde{Q}}{3}+\varepsilon\left(\frac{\partial^2}{\partial z^2}\tilde{\chi}-\frac{14}{9z^2}\tilde{\chi}-\frac{2\mathrm{i}\tilde{Q}}{9z}\right)\ ; \tag{7.3.27''}$$

$$\mathrm{i}\tilde{Q}+\frac{\partial v}{\partial z}+\frac{v}{3z}+\frac{2}{3}\frac{\tilde{\Delta}\tilde{\chi}}{z}=0\ ; \tag{7.3.28''}$$

$$\tilde{Q}=\sqrt{\varepsilon}\,Q\ ,\quad z=\sqrt{\varepsilon}\,x\ ,\quad \tilde{\chi}=\sqrt{\varepsilon}\,\chi\ .$$

We first consider the behavior of the perturbations in the near zone. Substituting $\tilde{\Delta} = -6$ for the quadrupole acoustic waves and eliminating Q and χ, from the system (7.3.26′–28′) we obtain

$$\frac{1}{x^2}\left[\frac{9}{2}x^4\frac{d^4}{dx^4} + 18x^3\frac{d^3}{dx^3} - 12x^2\frac{d^2}{dx^2} + \frac{268}{9}\right] v$$
$$= -\mathrm{i}\left(\frac{9}{2}x^2\frac{d^2}{dx^2} + 9x\frac{d}{dx} - 11\right) v \ . \qquad (7.3.29)$$

For $x \ll 1$, the solution of (7.3.29) can be expanded in a series in powers of x:

$$v = \sum_{k=1}^{4} A_k x^n k(1 - \mathrm{i}a_k x^2 + \ldots) \ , \qquad (7.3.30)$$

where the A_k $(k = 1,2,3,4)$ are arbitrary constants. In (7.3.30), n_k $(k = 1,2,3,4)$ is one of the four roots of the equation

$$\Phi(n) \equiv n(n-1)[\tfrac{9}{2}n(n-1) - 21] + \tfrac{268}{9} = 0 \ , \qquad (7.3.31)$$

which is obtained by equating the left-hand side of (7.3.29) to zero.

The fact that the roots of (7.3.31) are complex has the consequence that the trajectories of the fluid particles near the magnetosphere are bounded. The appearance of mixing near the magnetosphere is a consequence of the variation of the coefficient of viscosity with the radius; the effect disappears in the case of constant viscosity (an isothermal atmosphere).

The roots of (7.3.31) can be represented in the form

$$n = (1 \pm a \pm \mathrm{i}b)/2 \ , \quad a \approx 3.28 \ ; \quad b \approx 0.66 \ . \qquad (7.3.32)$$

For $x \gg 1$ but $\sqrt{\varepsilon}\, x \ll 1$, the solution of (7.3.29) has the asymptotic behavior (C and D are arbitrary constants)

$$v = Ce^{-x\sqrt{\mathrm{i}}}\left(1 - \frac{19\sqrt{\mathrm{i}}}{3x} + \ldots\right) + Dx^{m_-}\left(1 + \frac{\Phi(m)\mathrm{i}}{9(1-m)x^2} + \ldots\right) \ . \qquad (7.3.33)$$

Here m_- is the negative root of the equation obtained by equating the right-hand side of (7.3.29) to zero:

$$9m(m+1) - 22 = 0 \ , \quad m \approx -2.13 \ . \qquad (7.3.33')$$

In (7.3.33) we have omitted two solutions of (7.3.29) which increase[7] as $x \to \infty$. The condition that the two increasing solutions of (7.3.29) are absent imposes two conditions on the four arbitrary constants A_k in the asymptotic

[7] One of them increases with x as an exponential $\exp(\sqrt{\mathrm{i}}\, x)$, and the other as a power x^{m_+}, where m_+ is the positive root of (7.3.33′).

behavior of (7.3.30) for v as $x \to 0$. We obtain two other conditions from the boundary conditions on the magnetosphere by assuming that the functions v'_r and χ are given at $r = R$, i.e., at $x = x_R \ll 1$: $v'_r = v_0$, $\chi = \chi_0$. We recall that for a weakly oblate impenetrable ellipsoid $v_0 \approx D_n \sim e^2 \Omega R \sin\alpha$.

Thus, on the magnetosphere $x = x_R$ we have the conditions

$$\begin{aligned} &a_1A_1 + a_2A_2 + a_3A_3 + a_4A_4 = 0 \ , \\ &b_1A_1 + b_2A_2 + b_3A_3 + b_4A_4 = 0 \ , \\ &A_1x_R^{n_1} + A_2x_R^{n_2} + A_3x_R^{n_3} + A_4x_R^{n_4} = v_0 \ , \\ &n_1A_1x_R^{n_1} + n_2A_2x_R^{n_2} + n_3A_3x_R^{n_3} + n_4A_4x_R^{n_4} = \chi_0 \ . \end{aligned} \tag{7.3.34}$$

The first two equations, which contain the constants a_k and b_k ($k = 1,2,3,4$) are the conditions for the absence of solutions which increase at infinity, and in the last two equations we have used the asymptotic behavior of (7.3.30) for $x_R \ll 1$. The coefficient D in (7.3.33) is a certain linear combination of the coefficients A_k:

$$D = c_1A_1 + c_2A_2 + c_3A_3 + c_4A_4 \ . \tag{7.3.35}$$

We now study the behavior of the solutions in the wave zone by means of the system (7.3.26″–28″). For this, we use the matching condition in the region $1 \ll x \ll 1\sqrt{\varepsilon}$, in which the exponentially small term in (7.3.33) can be neglected. In the wave zone, the analytic continuation of the solution (7.3.33) has the form

$$v'_r = D\frac{\mathrm{i}\sin b\pi\,\Gamma(1-b)}{1/6-b}\left(\frac{\Omega}{2a_*}\right)^b \delta^{m-}\zeta^{1/3}\frac{d}{d\zeta}\left[\zeta^{1/6}H_b\left(\frac{\Omega\zeta}{a_*}\right)\right] \ , \tag{7.3.36}$$

where $\zeta = 2r^{3/2}/3$, $b = \sqrt{96}/6 \approx 1.63$, $H_b(y)$ is a Hankel function, $\Gamma(x)$ is the Euler gamma function, and $\delta = (1-\mathrm{i})\sqrt{\Omega/2\nu^*}$.

The solution (7.3.36) is an exact solution of the system (7.3.26″–28″) for $\varepsilon = 0$, corresponding to a diverging quadrupole acoustic wave. Therefore the flux of energy (7.3.5′) carried away to infinity by the acoustic waves is given by the expression

$$W \sim \left(\frac{\Omega\nu_*}{a_*^2}\right)^{b+0.5} |D|^2\varrho_*a_* \ . \tag{7.3.37}$$

Thus, the problem of the flux of energy or angular momentum carried away to infinity by the acoustic waves reduces to the problem of determining the coefficient D.

According to the system (7.3.34) and the relations (7.3.32, 35), for $v_0 \gtrsim \chi_0$ we have

$$|D| \sim \max_k |A_k| \sim v_0x_R^{(a-1)/2} \approx v_0x_R^{1.14} \ . \tag{7.3.38}$$

From (7.3.37), using (7.3.38), we finally obtain

$$W \sim \varrho_* a_* v_0^2 (x_R)^{2.28} \left(\frac{\Omega \nu_*}{a_*^2}\right)^{2.13} = \varrho a R^2 v_0^2 \left(\frac{\Omega \nu}{a^2}\right)^{2.13} \left(\frac{R^2 \Omega}{\nu}\right)^{1.14} . \tag{7.3.39}$$

In (7.3.39–39″) the values of all the quantities are taken on the magnetosphere $r = R$.

In the case in which there is no accretion and the magnetosphere is impenetrable, we have $v_0 \sim e^2 \Omega R \sin\alpha$, and then from (7.3.39) we obtain

$$W \sim e^4 \Omega^2 R^4 \varrho a \sin^2\alpha \left(\frac{\Omega \nu}{a^2}\right)^{2.13} \left(\frac{R^2 \Omega}{\nu}\right)^{1.4} . \tag{7.3.39'}$$

We note that if no allowance is made for viscosity in the case of rotation of a weakly oblate ellipsoid as a solid body in an adiabatic atmosphere, the following expression is obtained for the output of energy radiated as a result of long acoustic waves:

$$W \approx \gamma e^4 \varrho \Omega^3 R^5 \left(\frac{\Omega R}{a}\right)^{2b} \sin^2\alpha (2^{2b+1} \sin^2\alpha + \cos^2\alpha) , \tag{7.3.39''}$$
$$b \approx 1.63 \ ; \quad \gamma = 16 \sin^2(b\pi) |\Gamma(1-b)|^2 / [15 \cdot 3^{2b+1} (\tfrac{1}{6} - b)^2] \approx 0.05 .$$

7.3.6 Supersonic Rotation of a Pulsar

When a pulsar rotates supersonically, $e^2 \Omega R \sin\alpha > a$, the surface of the magnetosphere is very different from its static shape. Indeed, otherwise, part of the surface of the static magnetosphere would have a velocity greater than the speed of sound with respect to the gas. At certain instants of time, the surface of the static magnetosphere would move away from the gas with velocity greater than the maximum velocity of the gas expanding into empty space, $v_{max} = 2a/(\gamma - 1) = 3a$ for $\gamma = 5/3$. The gas would break away from the surface of the magnetosphere, and it would not be possible to satisfy the boundary condition which requires that the pressure of the gas be equal to the pressure of the magnetic field on the magnetosphere.

Let us consider the figure formed by the static surface of the magnetosphere during a period $2\pi/\Omega$. We shall refer to it as an "apple". In the case of rapid rotation, the gas cannot penetrate deep inside the apple: the characteristic penetration depth is of order a/Ω. Therefore the shape of the magnetosphere in the case of rapid rotation will be approximately the same as the shape of the apple. The fields produced by the currents flowing along the surface of the rotating magnetosphere are extremely complex, and we shall not enter into their analysis. We note only that the main variation in the pressure of the magnetic field on the magnetosphere is produced by the rotation of the magnetic dipole. The pressure of the magnetic field on the magnetosphere is stationary in the rotating coordinate system, and it can therefore be expanded in spherical harmonics, with a dependence on the azimuthal angle φ and the time t which appears in the form of the expression $\varphi - \Omega t$. We shall assume that the variation of the pressure of the magnetic field is determined mainly by the rotation of the pulsar, and not by the currents flowing on the magnetosphere.

In the case of supersonic rotation, perturbations in the gas will be generated mainly by oscillations of the surface of the magnetosphere. The wave zone for acoustic waves begins immediately from the surface of the magnetosphere, each segment of which, independently of the others, generates a practically plane wave.

The speed of sound for $\gamma = 5/3$ is related to the pressure by the equation $a'/a = 5^{-1}p'/p$, where the primes designate the perturbed values of the corresponding quantities. We shall assume that each segment of the magnetosphere acts like a weightless bar to which a magnetic pressure, varying according to a harmonic law. is applied. The resulting motion of the bar in the gas causes the generation of waves, which are inverted at a certain distance from it. If the process of emission of acoustic waves is to be periodic, there must be a loss of heat, which is liberated in shock waves.

The perturbation of the velocity is related to the perturbation of the speed of sound by the law $v' = 3a'$. Therefore the power of the acoustic emission by the whole surface of the magnetosphere [see (7.3.5′)] is given by

$$W = \int a\varrho v'^2 ds \approx 0.36a^3\varrho \int \left(\frac{p'}{p}\right)^2 ds \sim \varrho a^3 R^2 \sin^2\alpha \quad .$$

Here all the quantities are evaluated on the magnetosphere, and it is assumed that the relative oscillations of the pressure on the magnetosphere are, on the average, of order $\sin\alpha$.

Owing to the dissipation of acoustic waves, the gas is heated in the vicinity of the magnetosphere. The instability of this equilibrium of the atmosphere leads to the appearance of turbulent thermal convection, in which the surplus of heat in comparison with adiabatic lamination is carried away to infinity.

7.3.7 A Model Problem of Stationary Accretion of a Gas with Index of the Adiabat $\gamma < 5/3$ in the Presence of Energy Release

We shall show that stationary spherical accretion in the presence of strong energy release is impossible if the mechanisms of cooling of the gas are sufficiently weak.

For this, we consider the following model problem. Suppose that in a unit volume of a perfect gas an amount of heat Q is released per unit time, where this quantity depends on the radius according to the law $Q = Ar^{-5/2}$ with $A = \text{const}$. This heat is transferred completely to the incident gas. We denote the expression $p\varrho^{-\gamma}$ by S. The function S depends only on the entropy S.

The equations of motion and the energy can be written as follows:

$$\frac{d}{dr}\frac{v^2}{2} = -\gamma S \varrho^{\gamma-2}\frac{d\varrho}{dr} - \frac{GM}{r^2} - \varrho^{\gamma-1}\frac{dS}{dr} \ , \tag{7.3.40}$$

$$\frac{\varrho^{\gamma-1}}{\gamma-1} v \frac{\partial S}{\partial r} = Q = Ar^{-5/2} \ . \tag{7.3.41}$$

We seek a particular solution of this system in the form

$$v = v_* r^{-1/2} \ , \quad \varrho = \varrho_* r^{-3/2} \ , \quad S = S_* r^{[3(\gamma-1)/2-1]} \ ,$$

where the asterisks indicate the corresponding constants.

Eliminating the entropy from this system of equations, we obtain

$$v_*^3 - 2GMv_* = 5(\gamma-1)[1-3(\gamma-1)/2]^{-1}A \ . \tag{7.3.42}$$

We introduce the notation

$$B \equiv 5(\gamma-1)[1-3(\gamma-1)/2]^{-1}A \ ,$$
$$v_* = 3Bx/2GM \ , \quad \alpha \equiv 27B^2/(2GM)^3 \ .$$

Then (7.3.42) can be written in the form

$$\alpha x^3 - 3x - 1 = 0 \ . \tag{7.3.43}$$

In the regime of accretion, the velocity of the gas is directed toward the center, so that $v_* < 0$ or $x < 0$. However, negative solutions of (7.3.43) exist only for $\alpha < 4$. Therefore, if the rate of energy release is sufficiently large ($\alpha > 4$), accretion of this type is impossible.

7.3.8 The Stability of Rotation of Pulsars in Close Binary Systems

In this section, we show that in the presence of accretion the rotation of a pulsar with a long-period variation of its rotational angular velocity is stable. This corresponds to a stable limit cycle in the corresponding dynamical system and is confirmed by the statistics of observations of pulsars with acceleration and retardation of the rotational angular velocity [7.1]. An important point in the discussion is the assumption that the pulsar is nonspherical and hence that there is a moment of the gravitational forces. In fact, prolonged accretion of matter in the region of the magnetic poles makes the ellipsoid of inertia of the pulsar triaxial in the general case (excluding the case in which the axis of rotation coincides with the axis of the magnetic dipole). An indirect theoretical argument in favor of asymmetry of the pulsar is the instability of the axially symmetric figures of equilibrium (Maclaurin spheroids) of a gravitating homogeneous fluid in the case of supercritical angular velocities of rotation, when the triaxial Jacobi ellipsoids are stable.

Thus, we consider a pulsar as a solid body moving along a circular orbit and simultaneously rotating around some axis. We apply the theorem on angular momentum to this body. In the coordinate system associated with the principal central axes of inertia, Euler's dynamical equations have the well-known form

$$\begin{aligned} A\frac{d\omega_x}{dt} + (C - B)\omega_y\omega_z &= \mathcal{M}_x \ , \\ B\frac{d\omega_y}{dt} + (A - C)\omega_x\Omega_z &= \mathcal{M}_y \ , \\ C\frac{d\omega_z}{dt} + (B - A)\omega_x\omega_y &= \mathcal{M}_z \ . \end{aligned} \tag{7.3.44}$$

Here ω_x, ω_y, and ω_z are the components of the instantaneous angular-velocity vector in the indicated rotating system, A, B, and C are the moments of inertia of the pulsar with respect to the principal axes of inertia, and $\mathcal{M}_x$, $\mathcal{M}_y$, and $\mathcal{M}_z$ are the projections of the moment of the external forces with respect to the center of mass of the pulsar in the same system of moving axes.

The moment of the external forces acting on the pulsar can be divided into three parts:

1) The stellar wind from the second component, which is partially trapped by the gravitational field of the pulsar and accretes onto it, imparts an angular momentum $\bar{\mathcal{M}}_{\text{ac}}$ per unit time to the latter.
2) The magnetosphere of the pulsar experiences retarding hydrodynamical forces $\bar{\mathcal{M}}_{\text{h}}$, which we studied in detail above.
3) Because the pulsar is nonspherical, it experiences a moment of the gravitational forces $\bar{\mathcal{M}}_{\text{gr}}$ (in the center-of-mass system of the pulsar).

We calculate the moment $\bar{\mathcal{M}}_{\text{gr}}$ under the natural assumption that the distance between the centers of mass of the pulsar and of the second component is much greater than their dimensions. The moment of the Newtonian force acting on an elementary mass $\varrho_1 d\bar{r}_1$ with radius vector $\bar{r}_1$ from the mass $\varrho_2 d\bar{r}_2$ with radius vector $\bar{R}_0 + \bar{r}_2$ is

$$G\varrho_1\varrho_2\left[\bar{r}_1 \times \nabla_1\left(\frac{1}{R_{12}}\right)\right] d\bar{r}_1 d\bar{r}_2 \ . \tag{7.3.45}$$

Hence $\bar{R}_0$ is the radius vector of the center of mass of the second component in the center-of-mass system of the pulsar, and $R_{12} \equiv |\bar{R}_0 + \bar{r}_2 - \bar{r}_1|$.

To find $\bar{\mathcal{M}}_{\text{gr}}$, the expression (7.3.45) must be integrated over the volume of both stars.

Expanding $1/R_{12}$ in a series in the small quantities $\bar{r}_1/R_0$ and $\bar{r}_2/R_0$, and including only the largest terms, after elementary calculations we obtain

$$4R_0^5\mathcal{M}_{\text{gr}\,k} = 3GM_2\varepsilon_{klm}I_{ln}X_mX_n \ .$$

In (7.3.45), $\mathcal{I}_{ln}$ is the moment of inertia tensor of the pulsar

$$\mathcal{I}_{ln} = \int \varrho_1(\delta_{ln}r_1^2 - x_l x_n)\, d\bar{r}_1 \ ;$$

ε_{kln} is Levi–Civita tensor in R_3; x_1, x_2, and x_3 are the components of $\bar{r}_1$; M_2 is the mass of the second component; and $X_1 \equiv X$, $X_2 \equiv Y$, $X_3 \equiv Z$. In the system of principal axes of inertia, we have

$$4R_0^5 \mathcal{M}_{\mathrm{gr}\,x} = 3GM_2(B-C)YZ \ ,$$
$$4R_0^5 \mathcal{M}_{\mathrm{gr}\,y} = 3GM_2(C-A)XZ \ ,$$
$$4R_0^5 \mathcal{M}_{\mathrm{gr}\,z} = 3GM_2(A-B)XY \ .$$

Let us consider the simplest case in which the axis of rotation of the pulsar is perpendicular to the plane of the circular orbit. We take the x axis along the axis of rotation, and the y and z axes in the plane of the orbit, in such a way that $B > C$. Then $\omega_y = \omega_z = \mathcal{M}_{\mathrm{gr}\,y} = \mathcal{M}_{\mathrm{gr}\,z} = 0$. Suppose that in the fixed system

$$\bar{R}_0(0, R_0 \cos\psi, R_0 \sin\psi) \ ,$$

where

$$\psi = \sqrt{G(M_1+M_2)/R_2^3}\, t$$

according to the well-known equations for circular motion.

Let φ be the angle of rotation of the pulsar around the axis of rotation: $\omega_x = d\varphi/dt$. Then for $\mathcal{M}_{\mathrm{gr}\,x}$ we have

$$\mathcal{M}_{\mathrm{gr}\,x} = a \sin 2(\varphi+\psi) \ , \quad a \equiv \frac{3GM_2(B-C)}{R_0^3} \ . \tag{7.3.46}$$

We now write (7.3.44) under the assumption that the gas jet accreting on the pulsar is symmetric with respect to the plane of the orbit and carries a time-independent angular momentum:

$$\bar{\mathcal{M}}_{\mathrm{ac}} = (m_{\mathrm{ac}}, 0, 0) \ , \quad m_{\mathrm{ac}} = \mathrm{const} \ .$$

According to the treatment of this section, the moment of the hydrodynamical forces can be taken in the form

$$\mathcal{M}_n = (m_n(2\omega)^n, 0, 0) \ , \quad m_n = \mathrm{const} \ , \quad n = \mathrm{const} \ ,$$

where ω is the angular velocity of rotation of the pulsar with respect to the second component: $\omega = \dot{\varphi} + \dot{\psi}$. We recall that $n = 1$ for slow (smooth) rotation [see (7.3.25)] and $n = 2$ for fast rotation in the turbulent regime (see Sect. 7.3.5). The deviation of the shape of the magnetosphere from the spherical shape leads to excitation of quadrupole acoustic waves and shock waves (see Sects. 7.3.1, 2, 5, 6). The exponent n then varies over a wide range. Therefore, using (7.3.46), the relations (7.3.44) reduce to

$$A\frac{d^2\varphi}{dt^2} = m_{\mathrm{ac}} - m_n\left(2\frac{d\varphi}{dt} + 2\frac{d\psi}{dt}\right)^n + a \sin 2(\varphi+\psi) \ . \tag{7.3.47}$$

We introduce the dimensionless time variable $\tau = \sqrt{A/2a}\,t$ and put $2(\varphi + \psi) \equiv \Phi + \pi$. Then from (7.3.47) we have

$$\frac{d^2\Phi}{d\tau^2} + \sin\Phi + \beta\left(\frac{d\Phi}{d\tau}\right)^n = \alpha \quad , \quad \alpha \equiv \frac{m_{ac}}{a} \quad , \quad \beta \equiv \frac{m_n}{a} \quad . \tag{7.3.48}$$

It is remarkable that (7.3.48) has the form of the equation for the oscillations of a mathematical pendulum in the presence of a constant moment in a medium with nonlinear resistance. Equation (7.3.48) on the phase cylinder $\Omega \equiv \dot{\Phi}$, Φ modulo 2π has the form

$$\Omega\frac{d\Omega}{d\Phi} + \beta\Omega^n = \alpha - \sin\Phi \quad . \tag{7.3.49}$$

For $\alpha > 1$, (7.3.49) has no singular points. We shall show that there exists a stable limit cycle for $\alpha > 1$, i.e., a stable periodic solution $\Omega = \Omega(\Phi)$. In fact, for sufficiently small ε and $\Omega = \varepsilon$ it follows from (7.3.49) that $d\Omega/d\Phi > 0$, i.e., all the trajectories enter the region $\Omega > 0$ as Φ increases. On the other hand, for $\Omega = \Omega_* + \varepsilon$, where $\Omega_* = \sqrt[n]{(\alpha+1)/\beta}$, all the trajectories also enter the region $(0, \Omega_*)$ as Φ increases, since for $\Omega = \Omega_* + \varepsilon$ we have $d\Omega/d\Phi < 0$. For the direction of winding of the cylinder by the integral curves of (7.3.49) to change inside the region $(0, \Omega_*)$, one of the following conditions must hold: either (a) there is a closed curve on which $d\Phi/d\Omega$ changes sign, becoming equal to zero on this curve, or (b) there is at least one closed curve onto which the integral curves wind indefinitely both "from above" (from the direction $\Omega > \Omega_*$) and "from below" (from the direction $\Omega < 0$. However, version (a) is impossible by virtue of the boundedness of $d\Omega/d\Phi$ in $[\varepsilon, \Omega_* + \varepsilon]$. Thus, we have proved that there exists at least one limit cycle.

We now consider the case $n = 2$, in which (7.3.49) admits an explicit solution:

$$\Omega^2 = \mathrm{const}\exp(-2\beta\Phi) + \frac{\alpha}{\beta} - \frac{4\beta}{4\beta^2+1}\sin\Phi + \frac{2}{4\beta^2+1}\cos\Phi \quad .$$

Obviously, here a limit cycle is given by the closed (on the cylinder) curve

$$\Omega^2 = \frac{\alpha}{\beta} - \frac{4\beta}{4\beta^2+1}\sin\Phi + \frac{2}{4\beta^2+1}\cos\Phi \quad . \tag{7.3.50}$$

The dimensionless period of variation τ of the angular velocity can be calculated from the formula

$$\tau = \int_0^{2\pi} \frac{d\Phi}{\Omega} \quad . \tag{7.3.51}$$

Substituting (7.3.50) into (7.3.51), we have

$$\tau = \frac{4\sqrt{\beta\sqrt{4\beta^2+1}}}{\sqrt{\alpha\sqrt{4\beta^2+1}+2\beta}}K(x) \quad , \quad x^2 \equiv \frac{4\beta}{\alpha\sqrt{4\beta^2+1}+2\beta} \quad . \tag{7.3.52}$$

Here $K(x)$ is the complete elliptic integral. For small friction $\beta \to 0$, the period of variation of the angular velocity is also small:

$$\tau \approx 2\pi\sqrt{\beta/\alpha} \ .$$

For large friction $\beta \to \infty$, the period τ according to (7.3.52) increases without limit:

$$\tau \approx 4\sqrt{\beta/(\alpha+1)}\,K\left(\sqrt{2/(\alpha+1)}\right) \ .$$

Thus, knowing from observations the minimum $\omega_{\min}$ and the maximum $\omega_{\max}$ of the rotational angular velocity of the pulsar, as well as the period of its variation T, we can use (7.3.50, 52) to calculate the parameters α and β. In fact, from (7.3.50) we have

$$\Omega^2_{\min} = \left(\alpha\sqrt{4\beta^2+1} - 2\beta\right) \Big/ \left(\beta\sqrt{4\beta^2+1}\right) \ ,$$

$$\Omega^2_{\max} = \left(\alpha\sqrt{4\beta^2+1} + 2\beta\right) \Big/ \left(\beta\sqrt{4\beta^2+1}\right) \ .$$

Then

$$\omega_{\min}T = \Omega_{\min}\tau = \varphi_1(\alpha,\beta) \ , \quad \omega_{\max}T = \Omega_{\max}\tau = \varphi_2(\alpha,\beta) \ .$$

By using the observations to calculate the parameters of the orbital motion, we can estimate the degree of deviation of the inertia tensor of the pulsar from sphericity [i.e., the parameter $(B-C)/A$]. In fact, according to the definition of τ, from (7.3.47) we have $\tau = \sqrt{A/2a}\,t$ and hence

$$\tau = T\sqrt{A/(B-C)}\sqrt{4R_0^3/3M_2G} \ . \qquad (7.3.53)$$

The period T_1 of the orbital circular motion is $2\pi\sqrt{R_0^3/G(M_1+M_2)}$. Therefore, knowing the ratio of the masses of the components in a close binary system and the period of the orbital motion, we can use (7.3.53) to calculate the parameter $(B-C)/A$, which (as was shown above) is determined by the observable quantities $\omega_{\min}$, $\omega_{\max}$, T, T_1, and M_1/M_2.

7.4 The Stability of Uniform Nonlinear Pulsations of Gravitating Gaseous Spheres

Among the various models of the pulsations of variable stars like the Cepheids, a special place is occupied by a solution with uniform nonlinear pulsations of gravitating gaseous spheres [7.15, 29, 30]. This is a physically simple model, which can be calculated analytically. It does not explain why the maximum luminosity of a star occurs not at the instant of greatest compression, but, as observations show, a quarter of a period later. Moreover, in the theory the asymmetry of the curve giving the periodic time dependence of the ray velocity for a star of hydrogen manifests itself only for large amplitudes of the pulsations, whereas observations show that for the majority of the Cepheids the degree of compression during the pulsations does not exceed 10–15 %. Nevertheless, this solution predicts quite accurately the period of the pulsations on the basis of the density of the star (if the Cepheid mass is determined on the basis of the empirical mass-luminosity law).

The methods of investigating the stability of gravitating objects arose in the theory of figures of equilibrium. The main results in this field apply to an incompressible fluid. Besides solutions describing the rotation of masses of an incompressible fluid as a solid body (the Maclaurin and Jacobi spheroids, the Laplace ring, etc.), studies have been made of problems involving a velocity field linear in the coordinates (the solutions of Dirichlet, Dedekind, Poincaré, and others; see the references in [7.31]).

Lyapunov showed that the only possible static equilibrium shape of a gravitating mass of incompressible fluid is a sphere. If we assume that in the conceivable varied states the body rotates as a solid object, we can regard it as a dynamical system. Then a given equilibrium of the body will be stable if in this state the energy of the body attains a minimum value for fixed values of the generalized momenta corresponding to the cyclic coordinates. This idea was used to study the stability of the linear series of Maclaurin and Jacobi (Poincarè, Darwin, Jeans, and others). In its general formulation, the problem of small oscillations was studied by Poincaré and Bryan. With inclusion of small corrections of the post-Newtonian theory in general relativity, the problem of the stability of classical figures of equilibrium was formulated and investigated by *Chandrasekhar* [7.32], who, in particular, studied the way in which the stability is influenced by a new phenomenon of general relativity: the emission of gravitational waves in the presence of oscillations.

Allowance for the compressibility of the gas leads to new effects characteristic of nonlinear nonstationary motions. Self-similar solutions and solutions with velocities linear in the coordinates were considered for the first time in gas dynamics by *Sedov* [7.15] and in magnetohydrodynamics by *Kulikovskii* and *Lyubimov* [7.33]. Nonradial oscillations of Emden spheres were investigated in [7.34].

In this section, we study the stability of the exact solution with a uniform density which varies periodically with the time, about which we spoke above. Below, we show that the equations for arbitrary small perturbations of this solution lead, after expansion of the perturbations in a certain system of eigenfunctions, to a system of ordinary differential equations with periodic coefficients.

The Lyapunov stability of the solutions of this system is related to the behavior of the characteristic exponents of the system, which depend parametrically on the amplitude of the pulsations of the background. It turns out that for each eigenvalue there exists a corresponding critical value of the amplitude of the pulsations of the background at which the system exhibits a parametric resonance of spherical acoustic waves. In the linear theory, the amplitudes of the spherical perturbations grow without limit for a supercritical amplitude of the pulsations of the background. The regime of nonlinear uniform pulsations with a temperature which decreases to zero at the boundary of the sphere is found to be unstable with respect to nonspherical perturbations with a sufficiently large number of spherical harmonics. A method is developed for constructing approximately the characteristic exponent corresponding to the unstable mode of oscillations. When there is thermal mixing of particles with different entropy as a consequence of rotational adiabatic motions of the gas, essentially nonlinear effects come into play, and these are not considered in this section.

Unperturbed Solution. We shall give a brief description of the unperturbed solution, referring to [7.15, 29] for further details. In the studied motions of a gravitating gas, the quantity r, the distance of a fluid particle from the center of symmetry at time t, is a linear function of its initial distance ξ: $r = \xi\mu^{-1}(t)$. The density of the gas during the pulsations remains uniform in the particles: $\varrho = \varrho_0\mu^3(t)$, where $\varrho_0 = \text{const}$. For adiabatic pulsations with index of the adiabat $\gamma > 4/3$, the entropy is variable in the particles, and the pressure is a function of ξ and $\mu(t)$:

$$p = p_0\varrho_0(R^2 - \xi^2)\mu^{3\gamma} \quad . \tag{7.4.1}$$

Here R is the value of the Lagrangian coordinate ξ at the boundary of the sphere.

After substituting these expressions into Euler's equations and using Poisson's equation for the gravitational potential, we obtain an equation for the function $\mu(t)$, a first integral of which has the form

$$\left(\frac{d\mu}{dt}\right)^2 = \mu^4\left(-\frac{4p_0}{3(\gamma-1)}\mu^{3(\gamma-1)} + \frac{8\pi G\varrho_0}{3}\mu + \chi\right) \quad , \tag{7.4.2}$$

where χ is an arbitrary constant of integration.

It can be shown [7.15, 29] that the right-hand side of (7.4.2) has two roots for values of the constant χ in the interval

$$0 > \chi > -\frac{8\pi}{9}\left(\frac{3\gamma-4}{\gamma-1}\right)\left(\frac{2\pi G\varrho_0}{3p_0}\right)^{1(3\gamma-4)} G\varrho_0 \quad .$$

The corresponding integral (7.4.2) characterizes the nonlinear periodic pulsations.

Derivation of Equations for Small Perturbations. We write the equations of gas dynamics with inclusion of the gravitational force in the variables t and $\xi(\xi_1, \xi_2, \xi_3)$:

$$\frac{\partial v}{dt} + \mu\left[\left(v + \frac{\xi}{\mu^2}\frac{d\mu}{dt}\right)\cdot\nabla\right]v = -\frac{\mu}{\varrho}\nabla p + \mu\nabla\Phi \quad , \tag{7.4.3}$$

$$\mu^2\nabla^2\Phi = -4\pi G\varrho \ , \tag{7.4.4}$$

$$\frac{\partial\varrho}{\partial t} + \mu\left[\left(v + \frac{\xi}{\mu^2}\frac{d\mu}{dt}\right)\cdot\nabla\right]\varrho + \mu\varrho\,\mathrm{div}\, v = 0 \ , \tag{7.4.5}$$

$$\frac{\partial}{\partial t}\left(\frac{p}{\varrho^\gamma}\right) + \mu\left[\left(v + \frac{\xi}{\mu^2}\frac{d\mu}{dt}\right)\cdot\nabla\right]\left(\frac{p}{\varrho^\gamma}\right) = 0 \ . \tag{7.4.6}$$

[The vector operations of differentiation in (7.4.3–6) are taken with respect to the variable ξ.]

We shall assume that the perturbations of v, p, and ϱ are small and linearize (7.4.3–6) around the solutions described earlier. Let $K = \delta\varrho/\varrho$ be the relative perturbation of the density. Taking the divergence of (7.4.3) and making use of (7.4.4, 5), we obtain

$$\begin{aligned}&\mu\frac{\partial}{\partial t}\left[\frac{1}{\mu^2}\frac{\partial}{\partial t}K\right] - \frac{\nabla^2\delta p}{\varrho_0\mu^2} - 6p_0\mu^{-2+3\gamma}K \\ &\quad - 2p_0\mu^{-2+3\gamma}(\xi\cdot\nabla)K - 4\pi G\varrho_0\mu^2 K = 0 \ .\end{aligned} \tag{7.4.7}$$

We replace δv by a new unknown function w according to the relation

$$\mu\delta v = \partial w/\partial t \ . \tag{7.4.8}$$

Then the first integral of (7.4.6) for small perturbations can be written in the form

$$\frac{\delta p}{\mu^{3\gamma}} - \gamma p_0\varrho_0(R^2 - \xi^2)K - 2p_0\varrho_0 w\cdot\xi = f(\xi) \ , \tag{7.4.9}$$

where $f(\xi)$ is an arbitrary function.

Since according to (7.4.8) the addition of an arbitrary vector function $\varphi(\xi)$ to w does not change δv, the function $f(\xi)$ can be made to vanish.

Using (7.4.9), from (7.4.7) we obtain

$$\begin{aligned}&\mu\frac{\partial}{\partial t}\left[\frac{1}{\mu^2}\frac{\partial}{\partial t}K\right] - \mu^{3\gamma-2}p_0\{\nabla^2[(R^2 - \xi^2)K] + 2K - 2Q \\ &\qquad + (2\xi\cdot\nabla + 4)(K + \mathrm{div}\, w)\} - 4\pi G\varrho_0\mu^2 K = 0 \ , \\ &Q \equiv \xi\cdot\mathrm{curl\,curl}\, w = \xi\cdot(\nabla\,\mathrm{div}\, w - \Delta w) \ .\end{aligned} \tag{7.4.10}$$

The first integral of the linearized equation (7.4.5) has the form

$$K + \mathrm{div}\, w = \psi(\xi) \ , \tag{7.4.11}$$

where $\psi(\xi)$ is an arbitrary function.

From (7.4.3) we obtain an equation for the function Q:

$$\begin{aligned}&\mu\frac{\partial}{\partial t}\left[\frac{1}{\mu^2}\frac{\partial}{\partial t}Q\right] = 2p_0\mu^{3\gamma-2}\Delta_1 K \ , \\ &\Delta_1 K \equiv \frac{1}{\sin\theta}\frac{\partial}{\partial\theta}\left(\sin\theta\frac{\partial}{\partial\theta}K\right) + \frac{1}{\sin^2\theta}\frac{\partial^2}{\partial\varphi^2}K \ .\end{aligned} \tag{7.4.12}$$

Without loss of generality, we set the function $\psi(\xi)$ in (7.4.11) equal to zero [for $\psi(\xi)\neq 0$, the system (7.4.10, 11) becomes nonhomogeneous].

We shall introduce the new variable τ and use a dot to denote differentiation with respect to τ: $\sqrt{2p_0}\,\mu^2 dt \equiv d\tau$. Then the homogeneous equations (7.4.10, 12) can be rewritten in the form

$$\mu^{5-3\gamma}\ddot{K} - \frac{1}{2}\gamma\nabla^2[(R^2-\xi^2)K] - K - \frac{2\pi G\varrho_0}{p_0}\mu^{4-3\gamma}K + Q = 0 \ , \quad (7.4.13)$$

$$\mu^{5-3\gamma}\ddot{Q} = \Delta_1 K \ . \quad (7.4.14)$$

We note that any perturbation, if it is not spherically symmetric, will necessarily be rotational by virtue of the nonuniformity of the entropy distribution in the unperturbed solution.

We represent the functions Q and K as expansions in spherical harmonics:

$$\begin{Bmatrix} Q \\ K \end{Bmatrix} = \sum_{l=0}^{\infty}\sum_{m=-1}^{l} \begin{Bmatrix} Q_{lm}(\xi,t) \\ K_{lm}(\xi,t) \end{Bmatrix} P_l^m(\cos\theta)\mathrm{e}^{im\varphi} \ . \quad (7.4.15)$$

[Here the $P_l^m(\cos\theta)$ are the associated Legendre polynomials.]

We consider in the interval $(0, R)$ the eigenfunctions of the operator D with eigenvalues $-E_n$:

$$D[y_n] \equiv \frac{1}{\xi^2}\frac{d}{d\xi}\left[\xi^2\frac{d}{d\xi}(R^2-\xi^2)y_n\right] - \frac{l(l+l)y_n(R^2-\xi^2)}{\xi^2} = -E_n y_n \ . \quad (7.4.16)$$

Below, we shall show that this operator has a discrete spectrum. We introduce the substitution

$$(R^2-\xi^2)y_n(\xi) = \xi^l q_n(x) \ , \quad x \equiv (2\xi^2/R^2) - 1 \quad (7.4.17)$$

and rewrite (7.4.16) in the form

$$(1-x^2)\frac{d^2q_n}{dx^2} + \frac{1}{2}(1-x)(2l+3)\frac{dq_n}{dx} + \frac{E_n}{4}q_n = 0 \ . \quad (7.4.18)$$

It follows from (7.4.17) that the functions q_n must vanish at $x = 1$. For arbitrary E_n, the solution of (7.4.18) can be expressed in terms of hypergeometric functions; however, it is only for certain discrete values of E_n that the solution of (7.4.18) vanishes at $x = 1$. This is the well-known system of Jacobi orthogonal polynomials [7.35]:

$$q_n = P_n^{-1,\,l+1/2}(x) = \sum_{m=0}^{n-1}\binom{n-1}{m}\binom{n+l+1/2}{n-m}(x-1)^{n+m}(x+1)^m \ . \quad (7.4.19)$$

Here $E_n = 4n(n+l+1/2)$.

We write down the first three eigenfunctions $y_n(\xi)$ with the normalization $y_n(R) = 1$:

$$\begin{aligned} y_1 &= \sigma^l \ , \quad y_2 = 2^{-2}\sigma^l[(2l+7)\sigma^2 - 2l - 3] \ , \\ y_3 &= 6^{-1}\sigma^l[(l^2+7l+69/4)\sigma^4 - \sigma^2(2l^2+11l+15) \\ &\quad + l^2 + 4l + 15/4] \ , \\ \sigma &\equiv \xi/R \ . \end{aligned} \qquad (7.4.19')$$

After expanding the functions $K_{lm}(\xi,t)$ and $Q_{lm}(\xi,t)$ in the complete system of polynomials $y_n(\xi)$,

$$K_{lm} = \sum K_n(t)y_n(\xi) \ , \quad Q_{lm} = \sum Q_n(t)y_n(\xi) \ ,$$

we readily obtain from (7.4.13, 14) a fourth-order ordinary differential equation for $Q_n(t)$:

$$\begin{aligned} &\mu^{5-3\gamma}[\mu^{5-3\gamma}\ddot{Q}_n]^{\cdot\cdot} + [2\gamma n(n+l+1/2) - 1 - 2\pi G\varrho_0\mu^{4-3\gamma}/p_0]\mu^{5-3\gamma}\ddot{Q}_n \\ &\quad - (l+1)lQ_n = 0 \ . \end{aligned} \qquad (7.4.20)$$

Analysis of the Stability in the Absence of Pulsations. Let us consider the stability of the equilibrium configuration when $\mu \equiv 1$ and $3p_0 = 2\pi G\varrho_0$. In this case, the solution of (7.4.16) can be sought in the form $Q = \text{const}\cdot\exp(\lambda\tau)$, and for λ we obtain the equation

$$\lambda^4 + [2\gamma n(n+l+1/2) - 4]\lambda^2 - l(l+1) = 0 \ . \qquad (7.4.21)$$

Two roots of this equation are conjugates and purely imaginary. They correspond to acoustic waves. The other two roots correspond to rotational perturbations, one of which increases exponentially.

The appearance of an instability can be understood as follows. Consider the stability in the gravitational field of a gas which is cooler "above" and warmer "below" (the gravitational forces are directed downward). For small adiabatic perturbations, a small particle from above which penetrates (slightly at first) into the lower "warm" layer will have density proportional to $(p_2/p_1)^{1/\gamma}$. But then the Archimedean force acting on the particle will be less than the weight of the particle, and the particle will be carried downward. Convection occurs when the lower "warm" layers begin to mix with the upper "cool" layers.

We note that from the form of the functions $y_n(\xi)$ for $n \sim 1$ and $l \gg 1$ [see (7.4.19)] it follows that these functions always differ little from zero, except in a small neighborhood of the boundary.

On the other hand, according to (7.4.21), λ_n has the largest rates of growth for highly nonspherical perturbations with $l \gg 1$ and $n \sim 1$, when $\lambda_n \sim \sqrt{l}$. Therefore the strongest thermal mixing occurs near the boundary of a gravitating static gaseous sphere.

Stability of the Nonlinear Pulsations with Respect to Spherical Perturbations. We now consider whether the nonlinear pulsations can prevent the indicated instability. Suppose that $\gamma = 5/3$. Then $\mu = 1 + A\cos\tau$, where $A = \text{const}$ is the relative amplitude of the pulsations, and (7.4.20) is reduced to the form

$$Q^{IV} + \left[\frac{10}{3}n\left(n + l + \frac{1}{2}\right) - 4 + \frac{3A\cos\tau}{1 + A\cos\tau}\right]\ddot{Q} - (l+1)lQ = 0\,. \qquad (7.4.22)$$

We first consider spherical perturbations, with respect to which the static configuration will be stable.

Suppose that $l = 0$ and $n = 2$ [the case with $n = 1$ and $l = 0$ corresponds to a change of the amplitude of the uniform oscillations, and one of the solutions of (7.4.13) with $Q = 0$ has the form $K = \text{const}\cdot\mu\, d\mu/d\tau$].

Following the general scheme for finding the characteristic Lyapunov exponents for a second-order equation with periodic coefficients [7.36], we shall seek a numerical solution of the equation with Poincaré's initial values.

$$\begin{gathered}\ddot{K} + \left(\frac{47}{3} - \frac{3}{1 + A\cos\tau}\right)K = 0\ , \\ K_{(1)}(0) = 1\ ,\quad \dot{K}_{(1)}(0) = 0\ ,\quad K_{(2)}(0) = 0\ ,\quad \dot{K}_{(2)}(0) = 1\ .\end{gathered} \qquad (7.4.23)$$

Then the characteristic exponents (more precisely, their exponentials) will be roots of the equation

$$s^2 - \alpha s + 1 = 0\ ,\quad (\alpha \equiv K_{(1)}(2\pi) + \dot{K}_2(2\pi))\ .$$

The numerical values of the coefficients α for various values of the amplitude A are as follows:

A	0.1	0.2	0.3	0.4	0.5	0.6	0.7	0.8	0.9
α	−1.88	−1.91	−1.95	−1.99	−1.99	−1.85	−1.23	−1.22	−17.65

Obviously, the case of parametric resonance corresponds to $\alpha = \pm 2$. If $|\alpha| < 2$, the waves vary as almost periodic functions without becoming stronger. On the other hand, if $|\alpha| > 2$, the waves become stronger after a period, and the system becomes unstable. The numerical calculation shows that for an amplitude $A < 0.815$ of the nonlinear oscillations the regime of pulsations is stable, while for $A > 0.815$ the regime is unstable with respect to proper acoustic spherical oscillations with $n = 2$.

Stability of the Nonlinear Pulsations with Respect to Nonspherical Perturbations. If $l \neq 0$, it can be shown that the solutions of (7.4.20) include four independent functions Q_i $(i = 1,2,3,4)$ such that $Q_i(\tau) = s_i Q_i(\tau + 2\pi)$ $(i = 1,2,3,4)$. Then the exponentials of the characteristic exponents can be found from the equation

$$\begin{vmatrix} y_1 - s & y_2 & y_3 & y_4 \\ \dot{y}_1 & \dot{y}_2 - s & \dot{y}_3 & \dot{y}_4 \\ \ddot{y}_1 & \ddot{y}_2 & \ddot{y}_s - s & \ddot{y}_4 \\ \dddot{y}_1 & \dddot{y}_2 & \dddot{y}_3 & \dddot{y}_4 - s \end{vmatrix} = 0\ . \qquad (7.4.24)$$

Here $y_i(\tau)$ are particular solutions of (7.4.22) which at $\tau = 0$ take the values

$$y_k^{(n-1)}(0) = \delta_k^n \quad , \quad k, n = 1, 2, 3, 4 \quad .$$

Thus, the problem reduces to numerical integration of (7.4.22) with the initial data (7.4.25), followed by solution of the algebraic equation (7.4.24).

If $l \gg 1$ and $n \sim 1$, there is a possible analytic approach for the determination of the characteristic exponent and the solution of (7.4.22), corresponding to an increasing mode. We introduce the notation $10n\ (n + l + 1/2)/3 - 4 \equiv N$. For $n \sim 1$ and $N \sim l \gg 1$, we shall seek a solution of (7.4.22) in the form

$$Q = \text{const} \cdot \exp(\lambda\tau) \left(1 + \sum_{m=1}^{\infty} \Phi_{(m)}(\tau)\right) \quad ,$$

where the $\Phi_{(m)}(\tau)$ are periodic functions of τ: $\Phi_{(m)}(\tau + 2\pi) = \Phi_{(m)}(\tau)$, with $|\Phi_{(m)}(\tau)| \sim l^{-m/2}$.

We shall also seek the exponent λ in the form of an expansion

$$\lambda = \lambda_n + \sum_{m=0}^{\infty} \kappa_m \quad , \quad \kappa_m \sim l^{-m/2} \quad .$$

Here λ_n is the positive root of (7.4.21). We shall choose the numbers κ_m in such a way that in the expressions for $\Phi_{(m)}$ there are no terms which rise linearly in τ. We shall obtain the desired equations for κ_m by integration, over a period, of the equations for $\Phi_{(m)}$, which have the form

$$\begin{aligned} &\lambda_n(4\lambda_n^2 + 2N)\dot{\Phi}_{(m)} + \kappa_m(4\lambda_n^3 + 2N\lambda_n) \\ &\quad = L_m(\Phi_{(1)}, \ldots, \Phi_{(m-1)}, \kappa_1, \ldots, \kappa_{m-1}, \tau) \quad , \end{aligned}$$

where L_m is a linear differential operator on the periodic functions $\Phi_{(1)}, \ldots,$ $\Phi_{(m-1)}$ with coefficients which are fourth-degree polynomials in $\kappa_1, \ldots, \kappa_{m-1}$ and periodic in τ. From the requirement of periodicity of $\Phi_{(m)}$ it follows that

$$\kappa_m \cdot 2\pi(4\lambda_n^3 + 2N\lambda_n) = \int_0^{2\pi} L_m \, d\tau \quad .$$

The calculations for κ_0, κ_1, and κ_2 give zero values for them, and

$$\kappa_3(4\lambda_n^2 + 2N)^3 = 9A^2\lambda_n(2\lambda_n^2 + 3n)\left[\frac{4(1 - \sqrt{1 - A^2})}{A^2\sqrt{1 - A^2}} - 1\right] \quad .$$

The equations for $\Phi_{(1)}$ and $\Phi_{(2)}$ have the form

$$(4\lambda_n^2 + 2N)\dot{\Phi}_{(1)} + \frac{3A\cos\tau}{1 + A\cos\tau}\lambda_n = 0 \quad ,$$

$$(4\lambda_n^3 + 2N\lambda_n)\dot{\Phi}_{(2)} + (6\lambda_n^2 + N)\ddot{\Phi}_{(1)} + \frac{3A\cos\tau}{1 + A\cos\tau}\Phi_{(1)}\lambda_n^2 = 0 \quad .$$

Thus, for large l the Lyapunov exponent λ for the increasing mode of perturbations in a pulsating sphere differs from the corresponding exponent λ_n for the static case by an amount of order $l^{-3/2}$.

In conclusion, we note that (as we found above) the exact solution with nonlinear pulsations of homogeneous gaseous spheres has an interesting type of convective instability, which increases strongly as the outer layers are approached. Emission from the inner layers (not taken into account in this section) inhibits thermal mixing, and, as a result, in the inner layers the regime of nonlinear pulsations can be stable. The indicated instability evidently occurs only near the surface, and, as a result, a "corona" of turbulent gas is formed around the pulsating sphere.

7.5 Nonlinear Transverse Oscillations at Resonance in a Layer of an Ideally Conducting Fluid in a Magnetic Field

In this section, we demonstrate the equivalence of the equations of one-dimensional motions of an isotropic, nonlinearly elastic body with plane waves and an ideally conducting compressible magnetizable fluid moving along the direction of the intensity vector of an external magnetic field, when the magnetic permeability of the fluid is an arbitrary function of the density and the length of the intensity vector of the magnetic field. For these cases of the motion of a continuous medium, we consider the essentially nonlinear problem of the transverse oscillations excited in an infinite layer of the medium by an external tangential force which acts periodically on one of the plane boundaries of the layer. We examine the behavior of forced oscillations at resonance when in an elastic body the speed of the longitudinal waves is much greater than the speed of the transverse waves, and in a fluid the speed of sound is much greater than the speed of the Alfvén waves. We establish relations between the amplitude of the driving force and the proximity of its frequency to the resonance frequency at which Alfvén discontinuities appear in the layer.

7.5.1 Magnetohydrodynamic Analogy of One-Dimensional Motions of a Nonlinearly Elastic Body with Plane Waves

a) Nonlinearly Elastic Layer. The equations of the dynamics of an isotropic elastic body for one-dimensional motions with planar symmetry can be written in the form

$$\varrho_0 \frac{\partial^2 w_1}{\partial t^2} = \frac{\partial}{\partial \xi} p_1 \ , \quad \varrho_0 \frac{\partial^2 w_i}{\partial t^2} = \frac{\partial}{\partial \xi} p_i \ , \quad i = 2,3 \ ,$$
$$p_1 = \varrho_0 \frac{\partial F}{\partial d} \ ; \quad p_i = \varrho_0 h_i \frac{\partial F}{\partial h^2/2} \ , \quad d \equiv \frac{\partial w_1}{\partial \xi} \ ; \tag{7.5.1}$$
$$h_i \equiv \frac{\partial w_i}{\partial \xi} \ , \quad i = 2,3 \ .$$

Here p_1 and p_i are the components of the stress tensor, $p_1 \equiv p_{11}$ and $p_i \equiv p_{1i}$; w_1, w_2, and w_3 are the components of the displacement vector of a Lagrangian particle in the Cartesian coordinate system x^1, x^2, x^3 of an observer; ξ is a Lagrangian coordinate which in an initial discharged state ($p_1 = p_{1i} = 0$ for $d = h_2 = h_3 = 0$) coincides with the Cartesian coordinate; ϱ_0 is the initial density; and $F = F(d, h^2/2)$ is the free-energy density, where $h^2 \equiv h_2^2 + h_3^2$ (we assume that the temperature is constant).

The relations (7.5.1) are obtained from the general equations for the dynamics of an elastic body in Piola–Kirchhof form [7.37], which in a Cartesian frame of reference take the form

$$\frac{\partial^2 w_i}{\partial t^2} = \frac{\partial}{\partial \xi^j}\left[\left(\delta_{ik} + \frac{\partial w_i}{\partial \xi^k}\right)\frac{\partial F}{\partial \varepsilon_{jk}}\right] \quad , \quad i,j,k = 1,2,3 \ ,$$

where ξ^1, ξ^2, and ξ^3 are the Lagrangian coordinates, and ε_{jk} are the components of the deformation tensor. In the derivation of (7.5.1), allowance has been made for the fact that the invariants of the deformation tensor in the one-dimensional case with plane waves can be expressed in terms of d and h^2, and the nonzero components ε_{jk} have the form

$$\varepsilon_{11} = d + (d^2 + h^2)/2 \ , \quad \varepsilon_{1k} = h_k/2 \ , \quad k = 2,3 \ .$$

In what follows, we shall confine ourselves to the case $h_3 = 0$. We introduce the notation $h_2 \equiv h$. Then from the system (7.5.1) for the displacements we obtain the following equations for the longitudinal, p_1, and shear, p_2, stresses:

$$\begin{aligned}
&\frac{\partial}{\partial t}\left[\frac{F_{hh}}{F_{dd}a_2^2}\frac{\partial p_1}{\partial t} - \frac{F_{dh}}{F_{dd}a_2^2}\frac{\partial p_2}{\partial t}\right] = \frac{\partial^2}{\partial \xi^2}p_1 \ , \\
&\frac{\partial}{\partial t}\left[-\frac{F_{dh}}{F_{dd}a_2^2}\frac{\partial p_1}{\partial t} + \frac{1}{a_2^2}\frac{\partial p_2}{\partial t}\right] = \frac{\partial^2}{\partial \xi^2}p_2 \ , \\
&F_{hh} \equiv \frac{\partial^2 F}{\partial h^2} \ , \quad F_{hd} \equiv \frac{\partial^2 F}{\partial h\, \partial d} \ , \quad F_{dd} \equiv \frac{\partial^2 F}{\partial d^2} \ , \\
&a_2^2 \equiv F_{hh} - \frac{(F_{hd})^2}{F_{dd}} \ .
\end{aligned} \qquad (7.5.2)$$

The equations of the characteristics of the systems (7.5.1,2) (for $h_3 = 0$) are

$$\left(\frac{\partial \xi}{\partial t}\right)^2 = \frac{1}{2}\left[F_{dd} + F_{hh} \pm \sqrt{(F_{dd} - F_{hh})^2 + 4F_{dh}^2}\right] \ , \qquad (7.5.3)$$

Here $d\xi/dt$ has the meaning of the speed of propagation of the wave fronts of the stresses in the particles.

b) Layer of Ideally Conducting Fluid. The pondermotive force in magnetizable media in the absence of polarization and volume charge is [7.37]

$$\frac{1}{c}[j \times B] + \frac{1}{8\pi}(B_k \nabla H^k - H_k \nabla B^k) \ .$$

Using this relation, it can be shown that one-dimensional motions with plane waves of a magnetizable and compressible ideally conducting fluid in an external magnetic field B_1 applied in the direction of wave propagation are described in the Lagrangian coordinates by the equations

$$\begin{aligned} \varrho_0 \frac{\partial^2 w_1}{\partial t^2} &= -\frac{\partial}{\partial \xi}\left(p + \frac{H_i B^i}{8\pi}\right) \ , \\ \varrho_0 \frac{\partial^2}{\partial t^2}\left[B_i\left(1 + \frac{\partial w_1}{\partial \xi}\right)\right] &= \frac{B_1^2}{4\pi\varrho_0}\frac{\partial^2 H_i}{\partial \xi^2} \ , \quad i = 2,3 \ . \end{aligned} \tag{7.5.4}$$

Here w_1 is the component of the displacement vector in the direction of the x^1 axis, and all the quantities depend on only x^1 and t; B_1, B_2, and B_3 are the components of the magnetic induction vector, and it follows from Maxwell's equations that $B_1 = \text{const}$; ϱ_0 is the unperturbed density of the fluid for $w_1 = 0$; and H_i are the components of the intensity vector of the magnetic field.

From the equation for heat flow in the reversible adiabatic case, it follows [7.37] that

$$du = \boldsymbol{H} \cdot d\left(\frac{\boldsymbol{B}}{4\pi\varrho}\right) - \left(p + \frac{\boldsymbol{H} \cdot \boldsymbol{B}}{8\pi}\right) d\frac{1}{\varrho} \ , \tag{7.5.5}$$

where u is the total density of the internal energy of the fluid and the magnetic field.

c) The Analogy. It follows from (7.5.4, 5) that if we introduce potentials w_i ($i = 2, 3$) for the new unknown functions $h_i \equiv B_i(1 + d)/B_1$ ($h_i = \partial w_i/\partial \xi$), the relations (7.5.4) for w_1, w_2, and w_3 will have the form (7.5.1) if the function u depends on the components of the vector $\boldsymbol{B}$ through its length.

In the absence of magnetization of the fluid, the system (7.5.4) becomes the system of equations for one-dimensional magnetohydrodynamics [7.33]. In this particular case, the function u takes the form

$$u = U(d) + \frac{B_1^2 h^2}{[8\pi\varrho_0(1 + d)]} \ ,$$

where $U(d)$ is the density of internal energy of internal energy of the fluid at constant entropy.

In our notation, the equations of the characteristics of the system (7.5.4) will be identical to (7.5.3).

The indicated magnetohydrodynamic analogy for nonlinear waves in an elastic body was first established in a paper by the present author [7.38][8]. (It is

[8] Of course, the analogy also holds in general, and not just for one-dimensional motions. Moreover, it is possible to establish an analogy between the model of an ideally conducting and magnetizable

(continued on p. 329)

clear that the analogy holds for waves in unbounded systems; for problems with boundary conditions, we must make a further analysis, as is done below in part (b) of Sect. 7.5.2.)

7.5.2 Derivation of Basis Equations for Oscillations Near Resonance

a) Oscillations in an Elastic Layer. Suppose that an elastic body is such that $F_{dd} \gg F_{dh} \gg F_{hh}$ in the body in the range of values of d and h characteristic of the problems under investigation. Then from (7.5.3) it follows that the speed of propagation a_1 of "fast" longitudinal waves is $\sqrt{F_{dd}}$, while the speed of slow "transverse" waves is determined by the last relation in (7.5.2), where $a_2^2 = dF_h(d(h), h)/dh$ along the curve $\partial F/\partial d = \text{const}$.

Let us consider a layer of such an elastic material of thickness L, resting without a break on an absolutely solid base with no tangential frictional forces; acting per unit area of the surface of the upper boundary of the layer, there is a periodic tangential force $A\sin\omega t$, as well as a constant normal load q_0, corresponding to the following boundary conditions for the system (7.5.1) for (7.5.2):

$$\begin{aligned} p_2 &= 0 \ , \quad w_1 = 0 \quad \text{for} \quad \xi = 0 \ , \\ p_1 &= q_0 \ , \quad p_2 = A\sin\omega t \quad \text{for} \quad \xi = L \ . \end{aligned} \tag{7.5.6}$$

We shall seek a periodic solution of this problem, for the realization of which the layer must not store energy, i.e., the work done by the external tangential force during a period must be equal to zero:

$$\int_{-\pi/\omega}^{\pi/\omega} A\sin\omega t(\partial w_2/\partial t)\,dt = 0 \ . \tag{7.5.7}$$

If the amplitude of the driving force is sufficiently small, the desired solution can be obtained from the solution of a linear problem with linearized boundary conditions. This will be a standing transverse wave

$$p_2 = A\sin\omega t\frac{\sin(\omega\xi/a_2)}{\sin(\omega L/a_2)} \ , \quad p_1 = q_0 \ . \tag{7.5.8}$$

However, as the frequency ω approaches the resonance frequency $\omega = \omega_* \equiv n\pi a_2/L$ $(n = 1, 2, \ldots)$, infinite stresses occur inside the layer. It follows from this that our boundary-value problem is essentially nonlinear near the resonance, even for a small amplitude of the driving force.

---

elastic ferromagnet (with internal-energy density depending arbitrarily on the components of the induction vector of the magnetic field, the deformation tensor, and the entropy) and the model of an anisotropic nonhomogeneous elastic body. This fact is essentially due to the integrals of the equation for the induction of a frozen magnetic field in the Lagrangian description of a continuous medium [see (5.4.2′)].

In view of the fact that the problem has a characteristic length L and a characteristic time $T = 2\pi/\omega$, with $L/T \approx a_2 \ll a_1$, we can seek a solution in the form of an expansion in the two small parameters F_{hd}/F_{dd} and F_{hh}/F_{dd}. In a first approximation, the longitudinal stress is constant, corresponding to the fact that no longitudinal waves are emitted. The relation between the longitudinal and transverse deformations is finite, owing to the equality $p_1 = p_1(d, h) = q_0 = \text{const}$. Expressing d in terms of h, we find $a_2 = a_2(h)$.

From (7.5.1) we obtain an equation for the shear displacement:

$$\frac{\partial^2 w_2}{\partial t^2} - a_2^2 \frac{\partial^2 w_2}{\partial \xi^2} = 0 \ . \tag{7.5.9}$$

From (7.5.2) we have an equation for the shear stress:

$$\frac{\partial}{\partial t}\left[\frac{1}{a_2^2}\frac{\partial}{\partial t}p_2\right] = \frac{\partial^2}{\partial \xi^2}p_2 \ .$$

Equation (7.5.9) can be written in two equivalent forms:

$$\left(\frac{\partial}{\partial t} \pm a_2 \frac{\partial}{\partial \xi}\right)\left[\frac{\partial w_2}{\partial t} \mp \int_0^h a_2(h)\, dh\right] = 0 \ .$$

This leads to the equations

$$\frac{\partial w_2}{\partial t} \mp \int_0^h a_2(h)\, dh = 2a\varphi^{\pm}(C_{\pm}) \ , \tag{7.5.10}$$

where $C_{\pm}$ can be found from the equation $\xi = \xi(t, C_{\pm})$, which determines the family of integral curves of the equation

$$d\xi/dt = \pm a_2(h) \ ,$$

and the constant a is determined below by (7.5.12). In view of the periodicity of the required solution, the functions φ^- and φ^+ must also be periodic. In order to eliminate the arbitrariness in choosing the parameters of the families of characteristics $C_{\pm}$, we shall set them equal to the times t at which the characteristics intersect the curve $\xi = 0$. From the conditions (7.5.6) at $\xi = 0$ it follows that

$$\varphi^-(C) = \varphi^+(C) = \varphi(C) \ . \tag{7.5.11}$$

We now study the case of a weakly nonlinear material whose free energy is an analytic function of the invariants of the deformation tensor J_1, J_2, and J_3. It can be shown that in this case the function $a_2(h)$ can be expanded in a series in even powers of h near the point $J_1 = J_2 = J_3 = 0$. We confine ourselves to the first two terms:

$$a_2(h) \approx a(1 + 3\alpha h^2) \ . \tag{7.5.12}$$

With accuracy up to the terms $\sim h^4$, from the equations of the characteristics we have

$$C_+ = \lambda + 3\alpha \int_{\lambda}^{(\lambda+\mu)/2} [\varphi(\lambda) - \varphi(2\tau - \lambda)]^2 \, d\tau \ ,$$

$$C_- = \mu - 3\alpha \int_{(\lambda+\mu)/2}^{\mu} [\varphi(\mu) - \varphi(2\tau - \mu)]^2 \, d\tau \ , \tag{7.5.13}$$

$$\lambda \equiv t - \xi/a \ , \quad \mu \equiv t + \xi/a \ .$$

From (7.5.10), using (7.5.11), we obtain

$$h + \alpha h^3 = \varphi(C_-) - \varphi(C_+) \ . \tag{7.5.14}$$

We find the function φ from the conditions (7.5.6) at $\xi = L$:

$$\varrho_0 a^2 (h + 2\alpha h^3) = A \sin \omega t \ .$$

Substituting here the solution (7.5.14) at $\xi = L$, we obtain for φ the functional equation

$$\varrho_0 a^2 [\varphi(C_-) - \varphi(C_+) + \alpha(\varphi(\lambda) - \varphi(\mu))^3] = A \sin \omega t \ . \tag{7.5.15}$$

For a small deviation of the frequency of the driving force from the resonance frequency, in the nonlinear terms of (7.5.15) we put $\omega = \omega_*$, $\lambda = t - n\pi/\omega$, $\mu = t + n\pi/\omega$, and we replace the difference $\varphi(C_-) - \varphi(C_+)$ by the expression

$$-\frac{2n\pi}{\omega} \frac{d\varphi(\lambda)}{dt} [3\alpha\varphi^2(\lambda) + 3\alpha\overline{\varphi^2} + (\omega_* - \omega)\omega_*^{-1}] \ . \tag{7.5.16}$$

Here we have made use of the condition and the notation

$$\int_{-\pi/\omega}^{+\pi/\omega} \varphi(t) \, dt = 0 \ , \quad \overline{\varphi^2} \equiv \frac{\omega}{2\pi} \int_{-\pi/\omega}^{\pi/\omega} \varphi^2(t) \, dt \tag{7.5.17}$$

and the periodicity of $\varphi(t)$. We shall neglect the last term on the left-hand side of (7.5.15), since it is proportional to $(\omega - \omega_*)^3$. Integrating (7.5.15) with respect to the time, we finally obtain

$$\varphi^3(t) + \varphi(t)(3\overline{\varphi^2} + \vartheta) + \nu \cos \omega t = 0 \ ,$$
$$\vartheta = (\omega_* - \omega)/\alpha\omega_* \ , \quad \nu = (-1)^n A / 2n\pi\alpha\varrho_0 a^2 \ . \tag{7.5.18}$$

b) Magnetohydrodynamic Oscillations in a Layer of a Weakly Compressible Fluid. We assume that the speed of sound in a weakly compressible fluid is much greater than the speed of Alfvén transverse waves a_2 [in the presence of magnetization, $a_2^2 \neq B_1^2/4\pi\varrho_0$].

The boundary conditions (7.5.6) in this case acquire a different interpretation: a layer of this fluid is enclosed between two infinitely conducting planes. One of the planes is fixed and inactive, and on the other there is a periodic current, and a constant normal load is applied to it. If the frequency ω of the current approaches the resonance frequency ω^*, then even for small amplitudes of the current the problem becomes nonlinear. With the method explained above, the determination of the solution in the case of weak nonlinearity again reduces to the analysis of the algebraic equation (7.5.18).

If on the boundary $\xi = L$ instead of the condition $p_1 = q_0$ we impose the condition $w_1 = 0$, then instead of (7.5.18) we obtain the equation

$$\varphi^3(t) + \varphi(t)(\overline{\varphi^2} + \vartheta) + \nu \cos \omega t = 0 \quad . \tag{7.5.19}$$

7.5.3 Investigation of Oscillations in an Elastic Layer

All the analysis which follows reduces to the determination of solutions of the system (7.5.17, 18) [the case of the system (7.5.17, 19) can be analyzed in a similar manner]. It is convenient to simplify (7.5.18) by means of the substitution $\varphi = \sqrt[3]{\nu}\, y(\tau)$, $\tau = \omega t$. From (7.5.17, 18) we obtain the system

$$\begin{aligned} &y^3(\tau) + 3y(\tau)(\overline{y^2} + \Omega) = \cos\tau \quad , \\ &2\pi\overline{y^2} = \int_{-\pi}^{\pi} y^2(\tau)\, d\tau \quad , \end{aligned} \tag{7.5.20}$$

$$\int_{-\pi}^{\pi} y(\tau)\, d\tau = 0 \quad , \quad \Omega \equiv \vartheta / 3\sqrt[3]{\nu} \quad . \tag{7.5.21}$$

Depending on the value of Ω, (7.5.20) can determine either (I) one real function for all values of τ or (II) three real functions for all values of τ, or (III) for some τ (7.5.20) has one real root, and for other τ it has three real roots. Therefore we divide the whole range of variation of Ω into three intervals.

I) We first consider the case in which $\Omega + \overline{y^2} \geq 0$. Making the substitution $y = 2\sqrt{\overline{y^2} + \Omega}\sinh\mu_1$, from (7.5.20) we obtain $\cos\tau = 2\sqrt{(\overline{y^2} + \Omega)^3}\sinh 3\mu_1$. Therefore the only real solution of the system (7.5.20) has the form

$$y(\tau) = 2\sqrt{\overline{y^2} + \Omega}\sinh\mu_1 \quad ; \quad \mu_1 \equiv \frac{1}{3}\sinh^{-1}\left(\frac{\cos\tau}{2\sqrt{(\overline{y^2} + \Omega)^3}}\right) \quad , \tag{7.5.22}$$

where $\overline{y^2}$ is found from the solution of the equation

$$\pi\overline{y^2} = 4(\overline{y^2} + \Omega) \int_0^{\pi} \sinh^2 \mu_1 d\tau \quad . \tag{7.5.23}$$

For $\Omega \gg 1$, the solution (7.5.22) becomes a linear solution:

$$y(\tau) = \frac{\cos\tau}{3\Omega} \quad , \quad \varphi(t) = -\frac{\nu\cos(\omega t)}{\vartheta} \quad . \tag{7.5.24}$$

For $\omega = \omega_*$ (the resonance frequency), $\Omega = 0$ and $\overline{y^2}$ satisfies the equation

$$\pi = 4 \int_0^{\pi} \sinh^2 \left\{ \frac{1}{3} \sinh^{-1} \left[\cos\tau/2 \left(\sqrt{\overline{y^2}} \right)^3 \right] \right\} d\tau \quad .$$

If $\Omega + \overline{y^2} = 0$, it follows directly from the system (7.5.20) that

$$y(\tau) = \sqrt[3]{\cos\tau} \quad , \quad \pi\overline{y^2} = \int_0^{\pi} \sqrt[3]{\cos^2\tau}\, d\tau = \sqrt{\pi}\, \frac{\Gamma(5/6)}{\Gamma(4/3)} \quad .$$

Here $\Gamma(x)$ is Euler's gamma function. Therefore the region $\Omega + \overline{y^2} \geq 0$ corresponds to the range of variation of Ω from $\Omega_1 \equiv -\Gamma(5/6)/(\sqrt{\pi}\,\Gamma(4/3))$ to $+\infty$ ($\Omega_1 \approx -0.75$).

II) Suppose now that $-(\Omega + \overline{y^2}) \geq 2^{-2/3}$. In this case, in (7.5.20) we make the substitution $y = -2\sqrt{-(\overline{y^2} + \Omega)} \sin\mu_2$, after which we obtain $2\sqrt{-(\overline{y^2} + \Omega^3} \times \sin 3\mu_2 = \cos\tau$. Therefore in case (II) (7.5.20) determines three continuous functions

$$y_k = -2\sqrt{-(\overline{y^2} + \Omega)} \sin\left(\mu_2 + \frac{2\pi k}{3}\right) \quad ,$$
$$\mu_2 = \frac{1}{3} \arcsin\left(\frac{\cos\tau}{2\sqrt{-(\overline{y^2} + \Omega)^3}} \right) \quad . \tag{7.5.25}$$

Of the three solutions $y_k(\tau)$, $k = 0, 1, 2$, only the solution $y_0(\tau)$ satisfies the condition (7.5.21). Therefore $\overline{y_0^2}$ satisfies the equation

$$\pi\overline{y_0^2} = -4(\overline{y_0^2} + \Omega) \int_0^{\pi} \sin^2 \mu_2\, d\tau \quad . \tag{7.5.26}$$

From smooth segments of the continuous solutions $y_1(\tau)$ and $y_2(\tau)$, we can construct discontinuous solutions $y_a(\tau)$ and $y_b(\tau)$ which ensure fulfillment of (7.5.7) and satisfy the condition (7.5.21):

$$y_a(\tau) = \begin{cases} y_1(\tau) \\ y_2(\tau) \end{cases} , \quad y_b(\tau) = \begin{cases} y_2(\tau) & \text{for} \quad -\pi/2 < \tau < \pi/2 \ , \\ y_1(\tau) & \text{for} \quad \pi/2 < \tau < 3\pi/2 \ . \end{cases}$$

However, for $\Omega \to -\infty$ the discontinuous solutions $y_a(\tau)$ and $y_b(\tau)$ [unlike the continuous solution $y_0(\tau)$] does not tend to the linear solution (7.5.24).

We now determine the boundary of the interval in Ω, $\Omega = \Omega_{\rm II}$, in which there is a solution in the form (7.5.25, 26). For $2\sqrt{-(\overline{y_0^2}+\Omega)^3} \to 1$, the smooth solution tends to the continuous periodic solution $y_0(\tau)$ having a discontinuity in the derivative at $\tau = \pi m$ for $m = 0, \pm 1, \ldots$ (Fig. 7.4):

$$y_0(\tau) = 2^{2/3}(-1)^N \sin\left(\frac{\tau}{3} - (2N+1)\frac{\pi}{6}\right) \quad .$$

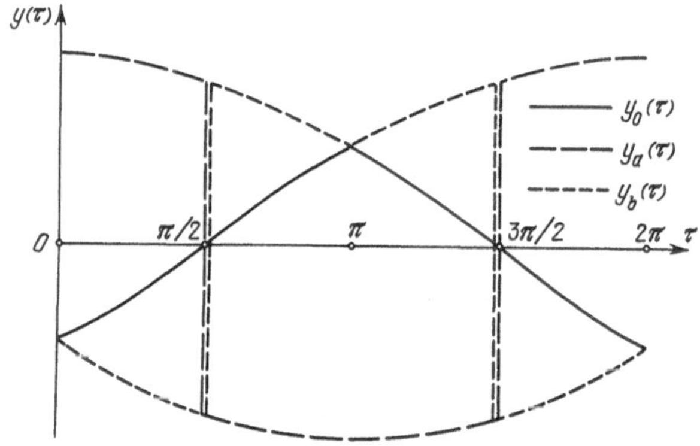

Fig. 7.4. Continuous solution $y_0(\tau)$ and discontinuous solutions $y_a(\tau)$ and $y_b(\tau)$ for $\Omega = \Omega_{\rm II}$

Here N is the integer part of τ/π. Calculating $\overline{y_0^2}$, we readily obtain $\overline{y_0^2} = 2^{1/3}(\pi - 3\sqrt{3}/2)/\pi$. Hence

$$\Omega_{11} = -2^{1/3}\left(\pi - 3\frac{\sqrt{3}}{2}\right) \Big/ \pi - 2^{-2/3} \approx -0.848 \quad .$$

III) Suppose now that Ω varies in the interval $(\Omega_{\rm II}, \Omega_{\rm I})$, which corresponds to the inequality $0 < -(\overline{y^2}+\Omega) < 2^{-2/3}$. In this interval, the system (7.5.20) determines a unique, but multiple-valued, function $y(\tau)$. Its form is shown in Fig. 7.5. Since $y(\tau)$ is multiple-valued in this range of variation of Ω, a continuous solution does not exist.

Let us denote $\arccos\left(2\sqrt{-(\overline{y^2}+\Omega)^3}\right)$ by δ. In the interval $(\Omega_{\rm II}, \Omega_{\rm I})$, the quantity δ is real and satisfies the inequalities $0 < \delta < \pi/2$. We shall distinguish two ranges of variation of τ: $|\cos\tau| > \cos\delta$ and $|\cos\tau| \le \cos\delta$.

For the ranges of τ in which $|\cos\tau| > \cos\delta$, we make the substitution

$$y(\tau) = 2\sqrt{-(\overline{y^2}+\Omega)}\,(\cosh\mu)\,\mathrm{sgn}\cos\tau \quad . \tag{7.5.27}$$

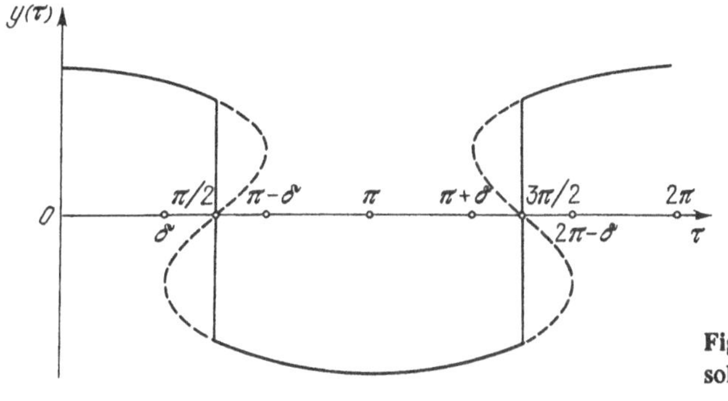

Fig. 7.5. Typical discontinuous solution for $\Omega \in (\Omega_{\rm II}, \Omega_{\rm I})$

Here $\operatorname{sgn} x = +1$ for $x > 0$, $\operatorname{sgn} x = -1$ for $x < 0$, and $\operatorname{sgn} x = 0$ for $x = 0$. Then the single-valued solution of (7.5.20) has the form (7.5.27), where $3\mu = \cosh^{-1}(|\cos\tau|/\cos\delta)$.

Suppose now that $|\cos\tau| \le \cos\delta$. In this case, we make the substitution

$$y = 2\sqrt{-(\overline{y^2} + \Omega)}\,(\operatorname{sgn}\cos\tau)\cos\tilde{\mu} \quad . \tag{7.5.28}$$

Then from (7.5.20) it follows that the discontinuous solution of (7.5.20) has the form (7.5.28), where

$$3\tilde{\mu} = \arccos(|\cos\tau|/\cos\delta) \quad .$$

For $\tau = \pi k + \pi/2$ ($k = 0, \pm 1, \ldots$), the solution (7.5.28) has a tangential discontinuity, since

$$y\left(\frac{\pi}{2} + \pi k - 0\right) = \sqrt{3}(-1)^k\sqrt{-(\Omega + \overline{y^2})} \quad ,$$

$$y\left(\frac{\pi}{2} + \pi k + 0\right) = -\sqrt{3}(-1)^k\sqrt{-(\Omega + \overline{y^2})} \quad .$$

For $|\cos\tau| = \cos\delta$, the solution (7.5.27) can be continued analytically to the solution (7.5.28).

In the interval $\Omega_{\rm II}, \Omega_{\rm I})$, we have for $\overline{y^2}$ the equation

$$\pi\overline{y^2} = -8(\overline{y^2} + \Omega)\left[\int\limits_0^{\delta} \cosh^2\mu\, d\tau + \int\limits_{\delta}^{\pi/2} \cos^2\tilde{\mu}\, d\tau\right] \quad .$$

Thus, our investigation leads to the interesting conclusion that when the dimensionless parameter $\Omega \equiv (\omega_* - \omega)(2n\pi\varrho a^2)^{2/3}/3\sqrt[3]{\alpha}\,\omega_* A^{2/3}$ varies in the interval $(\Omega_{\rm II}, \Omega_{\rm I})$ ($\Omega_{\rm II} \approx -0.848$, $\Omega_{\rm I} \approx -0.75$) forced transverse oscillations in a layer of weakly compressible fluid (Alfvén waves) must have tangential discontinuities.

For (7.5.19) (the case of a fixed boundary $\xi = L$), the values of $\Omega^*_{\rm II}$ and $\Omega^*_{\rm I}$ are as follows:

$$\Omega_{\mathrm{II}}^* = -\frac{1}{3\sqrt{\pi}}\frac{\Gamma(5/6)}{\Gamma(4/3)} \approx -0.25 \ ;$$

$$\Omega_{\mathrm{I}}^* = -2^{1/3}\left(\pi - 3\frac{\sqrt{3}}{2}\right) \Big/ 3\pi - 2^{-2/3} \approx -0.70 \ .$$

In the interval $(\Omega_{\mathrm{II}}^*, \Omega_{\mathrm{I}}^*)$, forced transverse oscillations in a layer with fixed boundaries have tangential discontinuities.

7.6 Excitation of Shock Waves in a Layer of an Ideally Conducting Gas at Resonance in a Magnetic Field

To study the effects of weak nonlinear interactions, it is of interest to consider resonators in which the dimensions of the system are multiples of the length of a periodically excited wave (the linear theory in this case gives an infinite amplitude). Gas-dynamical waves excited by periodic motion of a piston in a closed tube near resonance have been studied theoretically and experimentally [7.39–44].

Small nonlinear effects in a plasma accompanying the propagation of waves in extended systems have been described [7.45] in the framework of hydrodynamic and kinetic approaches[9]. In Sect. 7.5 we demonstrated the formation of tangential discontinuities at resonance of transverse oscillations in a layer, assuming that the speed of the longitudinal waves is much greater than the speed of the transverse waves.

In this section, we consider nonlinear oscillations of the density of the gas and of the magnetic field intensity in a layer of thickness L of an ideally conducting gas bounded by two stationary parallel planes in the case of a double resonance. A periodic electric current passes through one of the boundaries, and the other plate is assumed to be a dielectric. The external magnetic field is assumed to be perpendicular to the boundary planes. Near the resonance, the transverse oscillations cease to be standing waves and excite longitudinal acoustic waves. In the case of a double resonance, when the Alfvén and acoustic speeds are similar, the nonlinear interaction of the longitudinal and transverse waves is particularly pronounced. For this case, we examine below the evolution of the form of the periodic oscillations of the density and of the magnetic field intensity (in certain intervals with magnetohydrodynamic shock waves) as the frequency of the electric current exciting the oscillation varies.

---

[9] It is worthwhile to underline the remarkable contribution of english school of M.J. Lighthill and G.B. Whitham as well as french school of P. Germain (P. Bois, P. Gatignol, J.-P. Guiraud, G.A. Maugin, M. Roseau and others) in this topic for continuous media.

7.6.1 Formulation of the Problem

For one-dimensional motions with plane waves, the equations of magnetohydrodynamics take the form (see, for example, [7.33])

$$\varrho\left(\frac{\partial v_x}{\partial t}+v_x\frac{\partial v_x}{\partial x}\right)=-\frac{\partial}{\partial x}\left(p+\frac{H_\perp^2}{8\pi}\right)\ , \tag{7.6.1}$$

$$\varrho\left(\frac{\partial \boldsymbol{v}_\perp}{\partial t}+v_x\frac{\partial \boldsymbol{v}_\perp}{\partial x}\right)=\frac{H_x}{4\pi}\frac{\partial \boldsymbol{H}_\perp}{\partial x}\ , \tag{7.6.2}$$

$$\frac{\partial \boldsymbol{H}_\perp}{\partial t}+\frac{\partial}{\partial x}(\boldsymbol{H}_\perp v_x)=H_x\frac{\partial \boldsymbol{v}_\perp}{\partial x}\ , \tag{7.6.3}$$

$$\frac{\partial \varrho}{\partial t}+\frac{\partial}{\partial x}(\varrho v_x)=0\ ,\quad H_x=\mathrm{const}\ . \tag{7.6.4}$$

Here p is the pressure, ϱ is the mass density, v_x and H_x are, respectively, the components of the velocity and of the magnetic field intensity along the x axis, and $\boldsymbol{v}_\perp$ and $\boldsymbol{H}_\perp$ are the vector components of the vectors $\boldsymbol{v}$ and $\boldsymbol{H}$ perpendicular to the direction of propagation of the wave. We have already shown in Sect. 7.5 that [even if allowance is made for the reversible magnetization of the medium, $\mu=\mu(\varrho,H^2)$] the equations of magnetohydrodynamics (7.6.1–4) are identical, after a change of notation, to the equations of the one-dimensional dynamics of an isotropic nonlinearly elastic body.

The system written above has particular solutions in the form of simple waves: Alfvén, fast, and slow magnetohydrodynamic waves in the presence of a constant component of the transverse magnetic field [7.33]. We exclude rotational Alfvén waves from consideration, assuming that the oscillations of the magnetic field take place in a fixed plane passing through the direction of propagation of the wave. If the constant component $\boldsymbol{H}_\perp$ tends to zero, the speed of the fast magnetohydrodynamic wave tends to the larger of the speeds $a_{\mathrm{A}}^2=\sqrt{H_x^2/4\pi\varrho_0}$, $a_0=\sqrt{\partial p/\partial\varrho|_S}$, and the speed of the slow wave tends to the smaller of these speeds. To simplify the terminology in what follows, we shall speak of the interaction of the transverse and acoustic waves. Therefore the transverse waves will be either the fast or slow magnetohydrodynamic waves for $a_{\mathrm{A}}>a_0$ or $a_{\mathrm{A}}<a_0$, respectively, and will have a characteristic speed or propagation a_{A}, and the acoustic waves will have speed a_0. We rewrite the system (7.6.1–4) in terms of the Lagrangian coordinates t and ξ, where ξ coincides with the coordinate x in the absence of waves, when the density of the gas is ϱ_0. The equation of continuity then takes the form $\varrho(d\xi+dw)=\varrho_0\,d\xi$, where $w=x-\xi$. Eliminating $\boldsymbol{v}_\perp$ from (7.6.2, 3), from the system (7.6.1–4) we obtain

$$\begin{aligned}&\frac{\partial^2}{\partial t^2}\left[\left(1+\frac{\partial w}{\partial \xi}\right)\boldsymbol{H}_\perp\right]=a_{\mathrm{A}}^2\frac{\partial^2}{\partial \xi^2}\boldsymbol{H}_\perp\ ,\\ &\varrho_0\frac{\partial^2}{\partial t^2}w=-\frac{\partial}{\partial \xi}\left(p+\frac{H_\perp^2}{8\pi}\right)\ .\end{aligned} \tag{7.6.5}$$

Assuming that the motion of the gas is isentropic, for the dependence of the pressure on the specific volume $V \equiv \varrho^{-1}$ we use the expression which approximates the adiabat with accuracy up to terms of the second order inclusive:

$$p = p_0 - \left(\frac{a_0}{V_0}\right)^2 (V - V_0) + \frac{b_0^2}{2V_0^3}(V - V_0)^2, a_0, p_0, V_0, b_0 = \text{const} . \qquad (7.6.6)$$

The desired system, obtained from (7.6.5) using (7.6.6), has the form

$$\begin{aligned} &\frac{\partial^2}{\partial t^2} h - a_A^2 \frac{\partial^2}{\partial \xi^2} h = -\frac{\partial^2}{\partial t^2}\left(h \frac{\partial w}{\partial \xi}\right) , \quad h \equiv \frac{H_\perp}{H_x} , \\ &\frac{\partial^2}{\partial t^2} w - a_0^2 \frac{\partial^2}{\partial \xi^2} w = -\frac{1}{2}\frac{\partial}{\partial \xi}\left[b_0^2 \left(\frac{\partial w}{\partial \xi}\right)^2 + a_A^2 h^2\right] . \end{aligned} \qquad (7.6.7)$$

We shall seek periodic solutions of the system of equations (7.6.7) satisfying the boundary conditions

$$\begin{aligned} &h = 0 , \quad w = 0 \quad \text{for} \quad \xi = 0 ; \\ &h = h_0 \sin \omega t , \quad w = 0 \quad \text{for} \quad \xi = L . \end{aligned} \qquad (7.6.8)$$

The conditions (7.6.8) correspond to the fact that the boundaries of the layer are assumed to be stationary, a periodic electric current passes through the boundary $\xi = L$, and the boundary $\xi = 0$ is assumed to be a dielectric.

Far from resonance, both in the speed of sound and in the Alfvén speed, i.e., when $|2\omega L/a_0 - n\pi| \gg \varepsilon$, $|\omega L/a_A - m\pi| \gg \varepsilon$, where m and n are integers and $\varepsilon \ll 1$, the solution of (7.6.7) for a small amplitude of the current at the boundary $\xi = L$ with the boundary conditions (7.6.8) will be described by a standing transverse wave with amplitude of order ε, which leads to the appearance of gas-dynamical oscillations of order ε^2:

$$h = h_0 \lambda(\xi) \sin \frac{\omega t}{\lambda(L)} , \quad \lambda(\xi) \equiv \sin(\omega \xi / a_A) , \qquad (7.6.9)$$

$$\begin{aligned} w = {} & (16\omega\lambda^3(L))^{-1} h_0^2 a_A \{(a_A/a_0)^2 [\xi L^{-1} \lambda(2L) - \lambda(2\xi)] \\ & + a_A^2 (a_0^2 - a_A^2)^{-1} \cos 2\omega t [\lambda(2\xi) - \lambda(2L) \sin(2\omega\xi/a_0)/\sin(2\omega L/a_0)]\} . \end{aligned} \qquad (7.6.10)$$

In the neighborhood of a resonance in the Alfvén speed, $|\omega L/a_A - m\pi| \lesssim \varepsilon$, the expression (7.6.10) does not correctly describe the behavior of the oscillations, since the amplitudes of the oscillations of the magnetic field increase without limit according to the linear theory. An analogous situation occurs near a resonance in the speed of sound, i.e., for $|2\omega L/a_0 - n\pi| \lesssim \varepsilon$, since according to the equation of continuity the expression (7.6.10) then gives unbounded amplitudes of the oscillations of the density.

The point is that (although between the nodes of the standing wave the nonlinear terms are small, as before) near the nodes of the waves the nonlinear and the linear terms are of the same order of smallness. Near a resonance, fulfillment of the boundary condition at $\xi = L$ occurs near a node, since at resonance there is an integer number n of wavelengths in the layer. Therefore the form of the waves at resonance will be determined essentially by the nonlinear terms of the solution subject to the boundary conditions at $\xi = L$.

7.6.2 Investigation of the Form of the Oscillations Near a Resonance

We shall investigate resonant oscillations when the speeds of propagation of the acoustic and transverse waves are similar, i.e., we shall find the asymptotic behavior of the solution of the problem (7.6.7, 8) under the assumptions

$$h \sim \varepsilon \;, \quad a_0^{-1}|a_0 - a_{\mathrm{A}}| \lesssim \varepsilon \;, \quad |L\omega/a_0 - N\pi| \lesssim \varepsilon \;. \tag{7.6.11}$$

In the first approximation, the oscillations at resonance will be described by the expressions

$$w_1 = f(t+\xi/a_0) - f(t-\xi/a_0) \;, \quad h_1 = \varphi(t+\xi/a_{\mathrm{A}}) - \varphi(t-\xi/a_{\mathrm{A}}) \;. \tag{7.6.12}$$

We shall find the form of the functions f and φ by means of a nonlinear solution with boundary conditions at $\xi = L$. We substitute the expressions (7.6.12) into the right-hand side of (7.6.2) and find the next approximation for w and h. In calculating the values w_2 and h_2 of the second approximation on the boundary $\xi = L$ with accuracy up to infinitesimals of order higher than the second in ε, we replace $L\omega/a_0$ by $N\pi$ and a_{A} by a_0.

Suppose that, by definition, $L/a_0 = N\omega^{-1}\pi(1 + \Delta_f/2)$ and $L/a_{\mathrm{A}} = \omega^{-1}N\pi(1 + \Delta_\varphi/4)$, where $\Delta_f \sim \varepsilon$ and $\Delta_\varphi \sim \varepsilon$ in view of the conditions (7.6.11). Then because of the periodicity of φ and f with periods $2\pi/\omega$ and π/ω, respectively, for the values of w_1 and h_1 on the boundary $\xi = L$ we have

$$w_1 = \omega^{-1}N\pi\Delta_f f'(t + N\pi/\omega) \;, \quad h_1 = \Delta_\varphi N\pi\varphi'(t + N\pi/\omega)/2\omega \;, \tag{7.6.13}$$

where the primes indicate derivatives of the functions with respect to their arguments.

At $\xi = L$, the conditions $w_1 + w_2 = 0$ and $h_1 + h_2 = h_0 \sin\omega t$ must hold. From the first of these conditions, we obtain, using (7.6.13), the equation

$$2\Delta_f f'(t) + a_0[\varphi^2(t) + b_0^2(f'(t))^2/a_0^4 + A] = 0 \;. \tag{7.6.14}$$

We find the constant A by integrating the left-hand side of (7.6.14) over a period

$$\pi A/\omega = -\int_0^{\pi/\omega} [\varphi^2 + b_0^2 f'^2/a_0^4]\, dt \;.$$

From the boundary condition for h, we obtain

$$\Delta_\varphi\varphi(t) + a_0^{-1} f'(t)\varphi(t) + (-1)^N 2h_0 \cos(2\omega t/\pi N) = 0 \;. \tag{7.6.15}$$

We write $\Delta_\varphi - \Delta_f a_0^2/b_0^2 \equiv d$. For convenience in analyzing (7.6.14, 15) for $d \neq 0$ we introduce instead of f' and φ the new functions F and Φ, and instead of the given constants Δ_φ, Δ_f, h_0, and N the new constants Ω, y_0, and Σ, defined as follows:

$$\begin{aligned}
&F \equiv \frac{2f'}{a_0 d} + \Omega \;, \quad \Omega \equiv d^{-1}(\Delta_\varphi + \Delta_f a_0^2 b_0^{-2}) \;, \\
&\Phi \equiv \frac{2\varphi a_0}{b_0 d} \;, \quad y_0 \equiv \frac{4h_0 a_0}{N\pi b_0 d} \;, \\
&y \equiv (-1)^N y_0 \cos\omega t \;, \quad \tau \equiv \omega t \;, \\
&\Sigma \equiv a_0^2[-A + \Delta_f(a_0/b_0)^2](b_0 d)^{-2} \;.
\end{aligned} \tag{7.6.16}$$

(The case $d = 0$ is considered below.) Then from (7.6.14, 15) we obtain equations for Φ and F, and a relation between them:

$$(F^2 - 1)^2 + y^2 - 4\Sigma(F + 1)^2 = 0 \;, \quad (F + 1)\Phi + y = 0 \;, \tag{7.6.17}$$

$$(2\Phi + y)^2 + \Phi^4 - 4\Sigma\Phi^2 = 0 \;. \tag{7.6.17'}$$

For given Ω and y_0, the positive constant Σ can be found from the condition $\int_0^{2\pi}(F - \Omega)\, d\tau = 0$, which follows from the definition of F in (7.6.16) and the periodicity of f.

According to (7.6.17), we can easily construct the family of curves $F(y, \Sigma)$ for various values of the parameter Σ [Fig. 7.6(A)]. In the plane of F and y, on the two curves defined by the equation

$$(F^2 - 1)(F + 1)^2 - y^2 = 0 \;, \tag{7.6.18}$$

the derivatives dF/dy become infinite. [For fixed y, (7.6.18) has only two real roots.] The critical trajectory corresponding to $\Sigma = 1$,

$$(F + 1)^3(F - 3) + y^2 = 0 \;, \tag{7.6.19}$$

separates the closed curves with no self-intersections ($0 < \Sigma < 1$) from curves having the form of a nonsymmetric "figure of eight" with a point of self-intersection at $F = -1$. For these curves, $\Sigma > 1$. In Fig. 7.7(A) we show the qualitative character of the family of curves $\Phi = \Phi(y, \Sigma)$ defined by (7.6.17').

The parameter Ω characterizes the difference of the oscillation frequency from the resonance frequency. We shall find the relation between Ω and Σ by means of the condition

$$\int_0^{2\pi}(F - \Omega)\, d\tau = 0 \quad \text{or} \quad \frac{1}{\pi}\int_{-y_0}^{y_0} F(y, \Sigma)\frac{dy}{\sqrt{y_0^2 - y^2}} = \Omega \;.$$

In the (F, y) and (Φ, y) planes, we draw the lines $y = \pm y_0$ [according to (7.6.16), y_0 is proportional to the oscillation amplitude h_0]. These lines are tangential to two curves from the family (7.6.17) at the points $F = F_+$ and $F = F_-$, where F_+

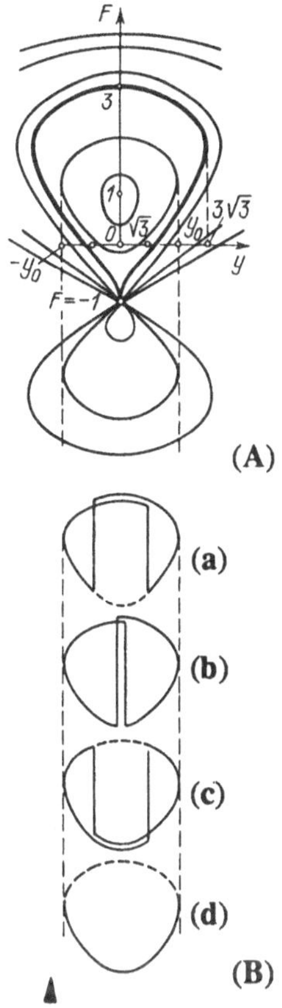

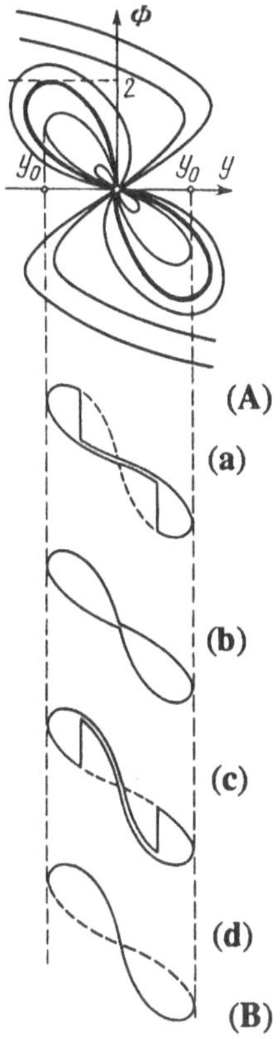

Fig. 7.6. (A) Family of curves $F = F(y, \Sigma)$ [see (7.6.17)]. (B) Evolution of the discontinuous solutions for F in the case $y_0 < \sqrt{27}$ for $\Omega \in (\Omega_3, \Omega_1)$

Fig. 7.7. (A) Family of curves $\Phi = \Phi(y, \Sigma)$. (B) Evolution of the discontinuous solutions for Φ corresponding to the solutions for F in Fig. 7.6

and F_- are the positive and negative roots of (7.6.18), respectively. We determine the corresponding values of Σ_+ and Σ_- by means of (7.6.17):

$$2\Sigma_+ = F_+^2 - F_+ \ , \quad 2\Sigma_- = F_-^2 - F_- \ .$$

Let us consider the change in the qualitative character of the oscillations as Ω varies from $+\infty$ to $-\infty$ under the condition $y_0 < \sqrt{27}$ ($y_0 = \sqrt{27}$ corresponds to the vertical tangent to the critical trajectory $\Sigma = 1$). In the interval $\Omega > \Omega_1$, the oscillations are smooth. In order to find the value of Ω_1 corresponding to Σ_+, we introduce the indefinite integral

$$\psi(F, F_\pm) \equiv 2 \int F \left(\frac{\partial \tau}{\partial F} \right)_\Sigma dF \ ,$$

where $F_\pm$ is equal to either F_+ or F_-.

For $\Sigma = \Sigma_+$, we have

$$d\tau|_\Sigma = -\frac{dy}{\sqrt{y_0^2 - y^2}} = \frac{2(F + F_+ - 1)\,dF}{\sqrt{(F + F_+)^2 + 2(F_+ - 1)}\sqrt{2(F_+^2 - F_+) - (F - 1)^2}} \ .$$

Then from the definition of ψ we have

$$\pi\Omega_1 = \psi\left(1 + \sqrt{2(F_+^2 - F_+)},\ F_+\right) - \psi(F_+, F_+) \ .$$

In the interval (Ω_3, Ω_1), continuous oscillations are impossible, and continuous motions reappear only when

$$\Omega < \Omega_3, -\pi\Omega_3 = \psi\left(1 - \sqrt{2(F_+^2 - F_+)},\ F_+\right) - \psi(F_+, F_+) \ .$$

In the interval (Ω_3, Ω_1), we shall construct discontinuous solutions on the basis of the following requirements: (1) in magnetohydrodynamics, discontinuities of rarefaction are impossible, since they correspond to transitions with a decrease of the entropy [7.33] without supply of heat; (2) the shock waves are weak, so that on the breaks we can neglect the discontinuity of the entropy. It follows from this that a break is admissible only from one branch to another branch of the curves $F(y, \Sigma)$ and $\Phi(y, \Sigma)$ for a fixed value of Σ (along the curves $\Sigma = \text{const}$ the value of the entropy does not change).

The oscillations of the density are described by the function

$$\varrho - \varrho_0 = -\varrho_0/a_0[f'(t + \xi/a_0) + f'(t - \xi/a_0)] \ .$$

This solution consists of two waves, traveling in opposite directions. It follows from the requirement (1) that the function f' on the cut must fall off with increasing t. Therefore as τ increases on the cut the function $F \equiv 2f'/a_0 d + \Omega$ must fall off for $d > 0$ and rise for $d < 0$, and, according to the corollary of the condition (2), the discontinuity connects different branches of the nonunique curve $F = F(y, \Sigma)$. In Fig. 7.6(B) we show the evolution of the discontinuous solutions for F over one period in the interval (Ω_3, Ω_1) for $y_0 < \sqrt{27}$, which are constructed by means of segments of the closed curve $F = F(y, \Sigma_+)$ with $\Sigma_+ < 1$. The direction of increase of τ for $d > 0$ corresponds to passage through the breaks with decreasing F, and for $d < 0$ the direction must be chosen so that F increases at the breaks. In Fig. 7.7(B) we show the corresponding discontinuous curves $\Phi(y)$ over a single period in τ. The discontinuous curves for Φ must be traversed in the direction consistent with the direction taken for the discontinuous curves for F. In the interval (Ω_3, Ω_1), the discontinuous solutions for the density are qualitatively similar to the near-resonance oscillations of a gas in a closed tube excited by periodic motion of a piston [7.39, 43]. The break in F becomes most intense for $\Omega = \Omega_2$, when there is a resonance in the speed of sound, $\Omega_2 = (\Omega_1 + \Omega_3)/2$ [see Fig. 7.6(b)]. Then the corresponding curve $\Phi(y)$ is continuous, with a break in the derivative at $y = 0$ [Fig. 7.7(b)]. In the interval

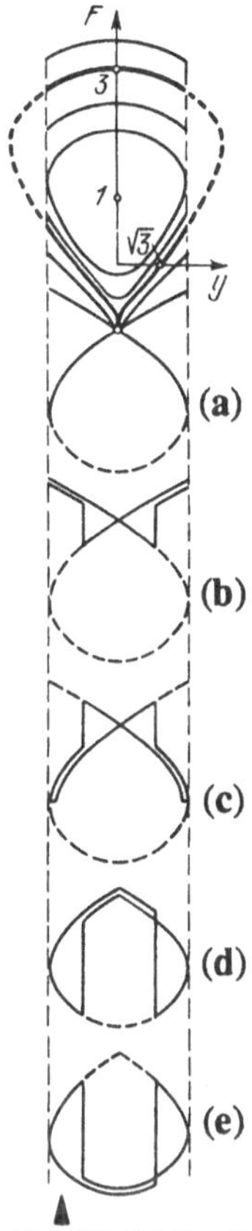

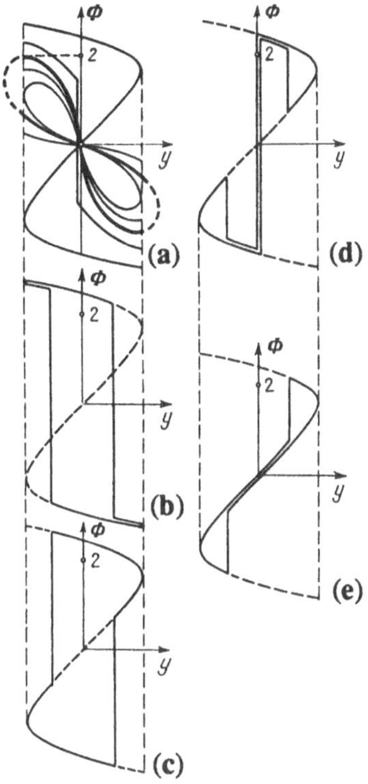

Fig. 7.8. (a) Continuous oscillations of F for $\Omega \in (\Omega_5, \Omega_3)$, $y_0 < \sqrt{27}$; **(b, c, d, e)** evolution of the discontinuous solutions for F for $\Omega \in (\Omega_9, \Omega_5)$

Fig. 7.9. Form of the oscillations of Φ corresponding to the oscillations of F in Fig. 7.8

$\Omega_4 < \Omega < \Omega_3$, the solutions are again smooth. The value Ω_4 corresponds to the value $\Sigma = 1$:

$$\Omega_4 = 4 \int_{-1}^{F_1} F \frac{\sqrt{(F+1)}(2-F)\,dF}{\sqrt{(3-F)}\sqrt{y_0^2 + (F+1)^3(F-3)}} \,.$$

Here F_1 is the smaller root of (7.6.19) for $y = y_0$. For $\Omega = \Omega_4$, the solutions for F and Φ have vertical tangents at $y = 0$ and are described by the part of the curve (7.6.19) for $y < y_0$. In the region $\Omega_5 < \Omega < \Omega_4$, the solutions for Φ have Alfvén breaks, where Φ changes sign without a change in its magnitude. At these points, the function F has a break in its derivative. Here $\pi\Omega_5 = \psi(-F_- - \sqrt{2(1-F_-)}, F_-) - \psi(-1, F_-)$.

In Figs. 7.8(a) and 7.9(a) we show the evolution of the continuous oscillations of F and Φ in the interval (Ω_5, Ω_3).

For $\Omega < \Omega_5$, the Alfvén breaks become weak magnetoacoustic shock waves, which will be present in the solution up to $\Omega = \Omega_9$, where $\pi\Omega_9 = \psi\left(1 - \sqrt{2(F_-^2 - F_-)}, F_-\right) - \psi(F_-, F_-)$. These discontinuous solutions are constructed from segments of the curves $F = F(y, \Sigma_-)$, $\Phi = \Phi(y, \Sigma_-)$. In the interval $\Omega_5 < \Omega < \Omega_9$, the evolution of the discontinuous solutions for the functions F and Φ is shown in Figs. 7.8 (b, c, d, e) and 7.9(b, c, d, e), respectively. The rule formulated above for the direction in which the discontinuous curves are traversed remains valid here. Besides weak magnetoacoustic shock waves, the oscillations of Φ have Alfvén breaks, and the oscillations of F have weak breaks in the interval (Ω_7, Ω_8), where $\pi\Omega_7 = \psi(F_-, F_-) - \psi(-1, F_-)$ and $2\Omega_8 = \Omega_7 + \Omega_9$. The oscillations of F and Φ have breaks of maximum intensity for $\Omega = \Omega_8$ and $\Omega = \Omega_6$, $2\Omega_6 = \Omega_5 + \Omega_7$. All these values of Ω can be expressed in terms of F_+ and F_- using Jacobi elliptic integrals. The parameters F_+ and F_- in turn can be expressed in terms of y_0 as the roots of the fourth-degree equation (7.6.17).

For $y_0 > \sqrt{27}$, the pattern of oscillations is somewhat different. In this case, Ω_1 can be expressed in terms of F_+, just as in the case $y_0 < \sqrt{27}$, and the expression for Ω_3 has the form $\psi(F_+, F_+) - \psi(-1, F_+)$. In the interval (Ω_2, Ω_3), the oscillations have not only magnetoacoustic breaks but also Alfvén breaks for Φ and weak breaks for F when $y = 0$. In Fig. 7.10(a, b, c) we show the evolution of the curves $F = F(y)$ for $\Omega_1 > \Omega > \Omega_3$. In the interval (Ω_3, Ω_9), the oscillations for $y_0 > \sqrt{27}$ are qualitatively similar to the oscillations for $y_0 < \sqrt{27}$ in the interval (Ω_4, Ω_9).

If the curves $F = F(y)$ and $\Phi = \Phi(y)$ represented as functions $F = F(\tau)$ and $\Phi = \Phi(\tau)$ for various values of Ω from Ω_1 to Ω_9, we obtain the picture of the evolution of the oscillations in the density and magnetic field intensity shown in Figs. 7.11(a, b) and 7.12(a, b). Figure 7.11(a, b) corresponds to $y_0 < \sqrt{27}$, and Fig. 7.12(a, b) corresponds to $y_0 > \sqrt{27}$ for $d > 0$. In order to obtain the picture of the evolution of the oscillations for $d < 0$, we must make the transformation $\tau' = -\tau$. We stress that in Figs. 7.11(a) and 7.12(a) the average value of F over a period is equal to Ω, and in Figs. 7.11(b) and 7.12(b) the average value of Φ over a period is equal to zero.

We now describe the form of the oscillations in the case $d = 0$. We introduce the new notation

$$F \equiv \left(\frac{y_0 a_0^3}{b_0}\right)^{-1/2} f' + \Delta \ , \quad \Phi \equiv \left(\frac{y_0 b_0}{a_0}\right)^{-1/2} \varphi(t) \ ,$$

$$\Delta \equiv \left(\frac{y_0 a_0}{b_0}\right)^{-1/2} \Delta\varphi \ .$$

Then (7.6.14, 15) can be rewritten in the form

$$F\Phi + \cos\tau = 0 \ , \quad \Phi^2 + F^2 = \Sigma \ , \quad \Sigma \equiv \left[\left(\frac{a_0 \Delta_f}{b_0}\right)^2 - A\right] y_0^{-1} \ . \quad (7.6.20)$$

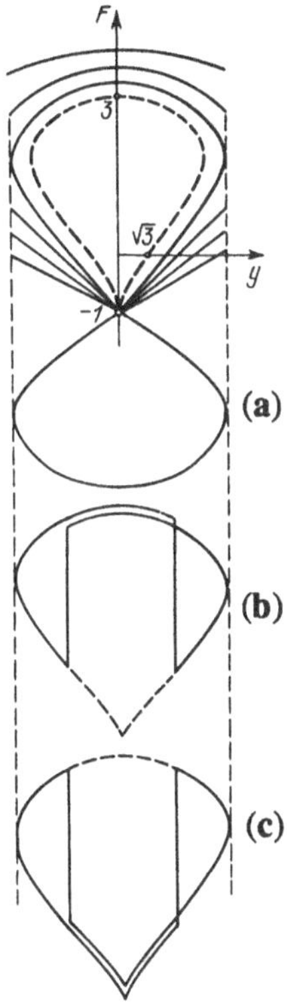

Fig. 7.10. Evolution of the form of the oscillations of F for $\Omega \in (\Omega_3, \Omega_1)$ in the case $y_0 > \sqrt{27}$

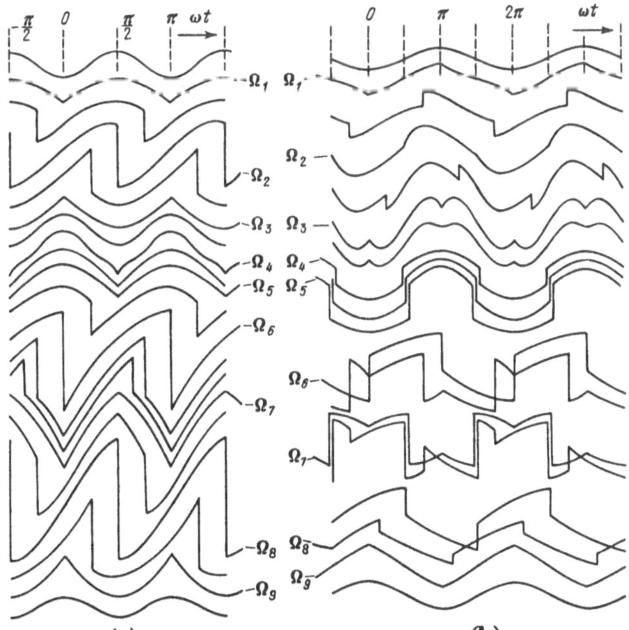

(a) (b)

Fig. 7.11. Evolution of the form of the oscillations of the density [curves (a)] and of the magnetic field intensity [curves (b)] as Ω varies for $y_0 < \sqrt{27}$

For a given value of Δ, the required value of Σ can be found from the requirement

$$\int_0^\pi (F - \Delta)\, d\tau = 0 \quad . \tag{7.6.21}$$

Qualitatively, the evolution of the oscillations in this case is similar to the evolution of the oscillations for $d > 0$ in the interval from Ω_9 to Ω_6, and then the picture goes in the opposite order (from Ω_6 to Ω_9). Therefore the entire range of variation of Δ can be divided into eight parts.

I) For $\pi\Delta > 4$ (i.e., in the case of continuous oscillations), from (7.6.20, 21) we readily obtain

$$\begin{aligned} 2\Phi &= \sqrt{\Sigma - 2\cos\tau} - \sqrt{\Sigma + 2\cos\tau} \quad , \\ 2F &= \sqrt{\Sigma - 2\cos\tau} + \sqrt{\Sigma + 2\cos\tau} \quad . \end{aligned} \tag{7.6.22}$$

According to (7.6.21), the constant Σ in (7.6.22) must be found from the equation

$$4E(\mu) = \pi\mu\Delta \quad , \tag{7.6.23}$$

where $E(\mu)$ is the complete elliptic integral of the first kind. Equation (7.6.23) has a unique solution for $\pi\Delta > 4$.

II) For $|\pi\Delta| < 4$, breaks occur for $\tau = -\theta_0 + \pi k$. We shall describe the solutions only in a half-period, since the magnetic field intensity in the other half-period has the opposite sign and the perturbation of the density has period π. For brevity, we introduce the notation

$$s \equiv \sin\tau/2 \quad , \qquad c \equiv \cos\tau/2 \quad .$$

In the interval $2\sqrt{2} < \pi\Delta < 4$, the solution has the form $\Phi = s - c$, $F = s + c$ for $-\theta_0 < \tau < \pi - \theta_0$. The point of the break θ_0 lies in the interval $0 < \theta_0 < \pi/2$, and $\pi\Delta = 4\cos\theta_0/2$.

III) For $4(\sqrt{2} - 1) < \pi\Delta < 2\sqrt{2}$, the point of the break θ_0 lies in the interval $\pi/2 < \theta_0 < \pi$, and $\pi\Delta = 4(\sqrt{2} - \sin(\theta_0/2))$. The solution of the system (7.6.20) has the form $\Phi = \pm(s - c)$, $F = \pm(s + c)$, where the upper sign is taken for the interval $\pi/2 > \tau > \pi - \theta_0$, and the lower sign for the interval $\pi - \theta_0 > \tau > -\pi/2$.

For $\pi\Delta = \pm 4(\sqrt{2} - 1)$, the solution of (7.6.20, 21) has Alfvén breaks for $\tau = \pi/2 + \pi k$ and weak breaks for $\tau = \pi k$ ($k = 0, \pm 1, \ldots$).

IV) For $0 < \pi\Delta < 4(\sqrt{2} - 1)$, the point of the break θ_0 lies in the interval $\pi/2 < \theta_0 < \pi$, and $\pi\Delta = 4[\sin(\theta_0/2) + \cos(\theta_0/2) - 1]$. The solution has the form $\Phi = s - c$, $F = s + c$ for $0 > \tau > -\theta_0$, and $\Phi = c - s$, $F = -s - c$ for $-\theta_0 > \tau > -\pi$. When $\Delta = 0$, the solution has the form $\Phi = s - c$, $F = s + c$ for $-\pi < \tau < 0$.

V) In the interval $-4(\sqrt{2} - 1) < \pi\Delta < 0$, the point of the break is determined by the equation $\pi\Delta = \pi(1 - \cos(\theta_0/2) - \sin(\theta_0/2))$ ($0 < \theta_0 < \pi/2$). Here the

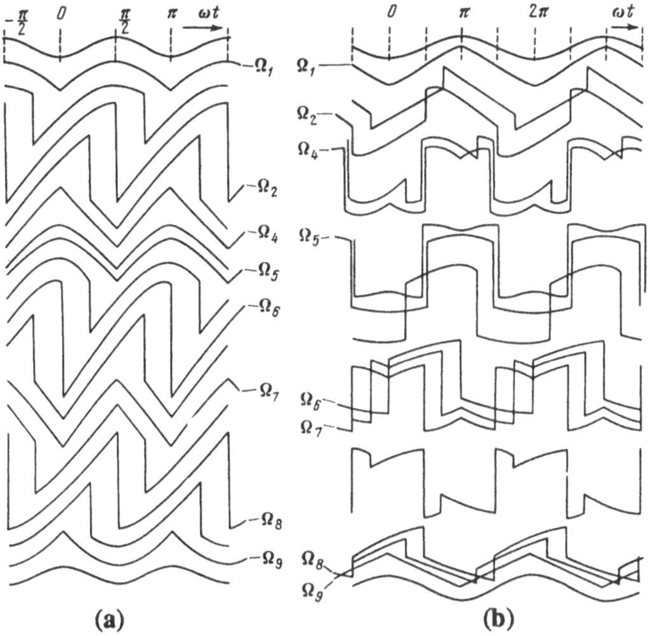

Fig. 7.12. Evolution of the form of the oscillations of the density [curves (a)] and of the magnetic field intensity [curves (b)] as Ω varies for $y_0 > \sqrt{27}$

solution is $\Phi = \pm(s - c)$, $F = \pm(s + c)$ (the upper sign for $-\pi < \tau < -\theta_0$, and the lower sign for $-\theta_0 < \tau < 0$).

VI) For $-2\sqrt{2} < \pi\Delta < -4(\sqrt{2} - 1)$, we have $\pi\Delta = -4(\sqrt{2} - \sin(\theta_0/2))$, and the point of the break θ_0 lies in the interval $0 < \theta_0 < \pi/2$. The solution has the form $\Phi = s + c$, $F = s - c$ for $-\theta_0 < \tau < \pi/2$, and $\Phi = -(s + c)$, $F = c - s$ for $\pi/2 < \tau < \pi - \theta_0$.

VII) For $-4 < \pi\Delta < -2\sqrt{2}$, for the determination of θ_0 we have $\pi\Delta = -4\sin\theta_0/2$, $\pi/2 < \theta_0 < \pi$. The solution here is $\Phi = s + c$, $F = s - c$ for $-\theta_0 < \tau < \pi - \theta_0$.

VIII) For $\pi\Delta < -4$, the solution again becomes continuous:

$$2\Phi = \sqrt{\Sigma + 2\cos\tau} - \sqrt{\Sigma - 2\cos\tau} \ ,$$

$$-2F = \sqrt{\Sigma + 2\cos\tau} + \sqrt{\Sigma - 2\cos\tau} \ ,$$

where Σ can be found by means of (7.6.23) with Δ replaced by $-\Delta$.

References

Chapter 1

1.1 N.H. Ibragimov: *Transformation Groups Applied to Mathematical Physics* (Reidel, Dordrecht 1985)

1.2 R. Courant: "Partielle Differentialgleichungen" (unpublished lecture notes, Göttingen 1932; Russian translation published as a book by Mir, Moscow 1964)

1.3 P. Günther, V. Wünsch: Math. Nachr. B **63**, 97 (1974)

1.4 E.T. Newman, R. Penrose: J. Math. Phys. **3**, 566 (1962); **4**, 998 (1963)

1.5 R.K. Sachs: Proc. R. Soc. London, Ser. A **264**, 309 (1961); **270**, 103 (1962)

1.6 G.A. Alekseev, V.I. Khlebnikov: "Application of the Newman-Penrose formalism in general relativity" [in Russian], Preprints Nos. 66, 67, Institute of Applied Mathematics, Moscow (1977)

1.7 V.P. Frolov: Tr. Fiz. Inst. Akad. Nauk SSSR **96**, 72 (1977)

1.8 F.I. Fedorov: In *Gravitation* [in Russian] (Naukova Dumka, Kiev 1972)

1.9 A.J. Janis, E.T. Newman: J. Math. Phys. **6**, 902 (1965)

1.10 A.Z. Petrov: *Einstein Spaces* (Pergamon Press, Oxford 1969)

1.11 A.Z. Petrov: In *Proceedings of the Second Soviet Gravitational Conference* [in Russian] (Tbilisi State University Press, Tbilisi 1967) p. 12

1.12 V.D. Zakharov: *Gravitational Waves in Einstein's Theory of Gravitation* [in Russian] (Nauka, Moscow 1972)

1.13 F. Pirani: Phys. Rev. **105**, 1089 (1957)

1.14 A.A. Logunov: *Lectures on the Theory of Relativity: Modern Analysis of the Problem* [in Russian] (Moscow State University Press, Moscow 1983)

1.15 L.I. Sedov: Prikl. Mat. Mekh. **36**, 3 (1972)

1.16 V.B. Braginskii: Znanie, Ser. Fiz., No. 1 (1977)

1.17 W.H. Press, K.S. Thorne: "Gravitational-wave astronomy", Preprint OAP-273, Caltech (1972)

1.18 K.S. Thorne: "Generation of gravitational waves", Preprints OAP-450, OAP-482, OAP-495, Caltech (1976–1977); preprints Nos. 662, 663, 664, C.N.S.R., Ithaca (1977)

1.19 K.S. Thorne, V.B. Braginsky: "Gravitational wave bursts from the nuclei of distant galaxies and quasars: proposals for detection using Doppler tracking of interplanetary spacecraft", Preprint OAP-426, Caltech (1975)

1.20 S.W. Hawking, G.F.R. Ellis: *The Large Scale Structure of Space-Time* (Cambridge University Press, Cambridge 1977)

1.21 G. Dautcourt: Bull. Acad. Pol. Sci. **20**, 417 (1972)

1.22 F.K. Stellmacher: Math. Ann. B **115**, 740 (1938)

1.23 N.R. Sibgatullin: "The problem of gravitational waves in the theory of gravitation" [in Russian], Scientific Report No. 1496, Institute of Mechanics, Moscow State University (1973)

1.24 A. Lichnerowitz: Colloq. Int. CNRS, No. 220, 47 (1974)

1.25 A. Papapetrou, H. Treder: Math. Nachr. B **20**, 59 (1959)

1.26 N.R. Sibgatullin: Zh. Eksp. Teor. Fiz. **66**, 1187 (1974) [English transl.: Sov. Phys. JETP **39**, 579 (1974)]

1.27 R.A. Penrose: In *Battelle Recontres: 1967 Lectures in Mathematics and Physics*, ed. by C.M. DeWitt, J.A. Wheeler (Benjamin, New York 1968) p. 121

1.28 I. Robinson, A. Trautman: Proc. R. Soc. London, Ser. A **265**, 463 (1962)
1.29 A. Trautman: C.R. Acad. Sci. **246**, 1500 (1957)
1.30 G. Dautcourt: J. Math. Phys. **3**, 908 (1962)
1.31 R.K. Sachs: J. Math. Phys. **3**, 908 (1962)
1.32 H. Bondi, F.A.E. Pirani, I. Robinson: Proc. R. Soc. London, Ser. A **251**, 519 (1959)
1.33 H. Takeno: Tensor **7**, No. 2, (1957); **8**, No. 1 (1958)
1.34 Yu. S. Vladimirov: Zh. Eksp. Teor. Fiz. **45**, 251 (1963) [English transl.: Sov. Phys. JETP **18**, 176 (1964)]
1.35 M.E. Gertsenshtein: Zh. Eksp. Teor. Fiz. **40**, 114 (1961) [English transl.: Sov. Phys. JETP **13**, 81 (1961)]
1.36 N.V. Mitskevich: *Physical Fields in General Relativity* [in Russian] (Nauka, Moscow 1969)
1.37 R.A. Isaacson: Phys. Rev. **166**, 1272 (1958)
1.38 N.R. Sibgatullin: Dokl. Akad. Nauk SSSR **200**, 308 (1971) [English transl.: Sov. Phys. Dokl. **16**, 697 (1972)]
1.39 G.V. Scrotsky: Dokl. Akad. Nauk SSSR, **114**, 73 (1957)
1.40 N.L. Balazs: Phys. Rev. **110**, 236 (1958)
1.41 Ya. B. Zel'dovich, I.D. Novikov: *Structure and Evolution of the Universe* [in Russian] (Nauka, Moscow 1975)
1.42 V.L. Ginzburg, V.N. Tsytovich: Zh. Eksp. Teor. Fiz. **65**, 1818 (1973) [English transl.: Sov. Phys. JETP **38**, 909 (1974)]
1.43 U.H. Gerlach: Phys. Rev. Lett. **32**, 1 (1974)
1.44 Tokuoka Tatsuo: J. Math. Phys. **15**, 1972 (1974)
1.45 Y. Choquet-Bruhat: Colloq. Int. CNRS, No. 220, 85 (1974)
1.46 W. Kinnersley: J. Math. Phys. **10**, 1195 (1969)
1.47 P.E. Krasnushkin: Dokl. Akad. Nauk SSSR **239**, 815 (1978) [English transl.: Sov. Phys. Dokl. **23**, 237 (1978)]
1.48 P.E. Krasnushkin, R.V. Khokhlov: Zh. Tekh. Fiz. **19**, 931 (1949)
1.49 V.I. Arnold: *Mathematical Methods of Classical Mechanics* (Springer Verlag, Berlin 1978)
1.50 V.P. Maslov: *Perturbation Theory and Asymptotic Methods* [in Russian] (Moscow State University Press, Moscow 1965)
1.51 D. Ludwig: Commun. Pure Appl. Math. **19**, 215 (1966)
1.52 V.R. Kaigorodov, A.B. Pestov: *Gravitation and Relativity*, Kazan **6**, 46 (1969)
1.53 O. Yu Dinariev; N.R. Sibgatullin: Zh. Eksp. Teor. Fiz. **72**, 1231 (1977) [English transl.: Sov. Phys. JETP **45**, 645 (1977)]
1.54 K.A. Khan, R. Penrose: Nature **229**, 185 (1971)
1.55 R.A. Penrose: Rev. Mod. Phys. **37**, 215 (1965)
1.56 P. Szekeres: J. Math. Phys. **13**, 286 (1972)
1.57 P. Bell, P. Szekeres: Gen. Relativ. Gravitat. **5**, 275 (1974)
1.58 J.B. Griffiths: Commun. Math. Phys. **28**, 295 (1972)
1.59 J.B. Griffiths: Ann. Phys. (N.Y.) **102**, 388 (1976)
1.60 Yu. G. Sbytov, Zh. Eksp. Teor. Fiz. **70**, 2001 (1976) [English transl.: Sov. Phys. JETP **43**, 1036 (1976)]
1.61 G.M. Bereshkov, A.I. Poltavtsev: Zh. Eksp. Teor. Fiz. **71**, 3 (1976) [English transl.: Sov. Phys. JETP **44**, 1 (1976)]
1.62 N.R. Sibgatullin, G.A. Alekseev: Zh. Eksp. Teor. Fiz. **67**, 1233 (1974) [English transl.: Sov. Phys. JETP **40**, 613 (1975)]
1.63 V.L. Berdichevskii: *Variational Principles of the Mechanics of Continuous Media* [in Russian] (Nauka, Moscow 1983)
1.64 M.V. Lur'e: Prikl. Mat. Mekh. **30**, 747 (1966)
1.65 L.I. Sedov: Usp. Mat. Nauk **20**, 120 (1965)
1.66 L.I. Sedov: Dokl. Akad. Nauk SSSR **164**, 519 (1965) [English transl.: Sov. Phys. Dokl. **10**, 824 (1966)]
1.67 L.I. Sedov: Prikl. Mat. Mekh. **47**, 180 (1983)

1.68 D. Ivanenko: Phys. Rep. **94**, 1 (1983)
1.69 A.A. Logunov, V.I. Denisov, A.A. Vlasov, M.A. Mestvirishvili, V.N. Folomeshkin: Teor. Mat. Fiz. **40**, 291 (1979)
1.70 L.I. Sedov: In *Irreversible Aspects of Continuum Mechanics and Transfer of Physical Characteristics in Moving Fluids* (Springer Verlag, New York 1968)
1.71 L.I. Sedov: *A Course in Continuum Mechanics* (Wolters-Noordhoff, Groningen 1971)
1.72 S. Chandrasekhar: *The Mathematical Theory of Black Holes* (Clarendon Press, Oxford 1983)
1.73 D.V. Gal'tsov: *Particles and Fields Near Black Holes* [in Russian] (Moscow State University Press, Moscow 1986)
1.74 I.D. Novikov, V.P. Frolov: *The Physics of Black Holes* [in Russian] (Nauka, Moscow 1986)

Additional References

1.75 R. Penrose, W. Rindler: *Spinor and Twistor Methods in Space-Time Geometry* (Cambridge University Press, 1988)
1.76 S. Chandrasekhar, B.C. Xanthopoulos: Proc. R. Soc. London, Ser. A **398**, 223 (1985); ibid. **402**, 37 (1985); **402**, 205 (1985); **403**, 189 (1986); **410**, 311 (1987); **420**, 93 (1988)
1.77 V. Ferrari, J. Ibanez: Gen. Relativ. Gravitat. **19**, 383 (1987)
1.78 M. Halilsoy: Phys. Rev. D **37**, 2121 (1988)
1.79 D. Kramer, H. Stephani, M. MacCallum, B. Herlt: *Exact Solutions of the Einstein Field Equations* (VEB Deutscher Verlag der Wiss. Berlin, 1980)
1.80 U. Yurtsever: Phys. Rev. D **38**, 1706 (1988)
1.81 J.B. Griffiths: *Colliding Waves in General Relativity* (Oxford University Press, 1990) (in press)
1.82 L.I. Sedov, A.G. Tsipkin: *Macroscopic Theories of Gravitation and Electromagnetism* (Nauka, Moscow 1988)

Chapter 2

2.1 R. Giacconi, H. Gursky, F.R. Paolini, B.B. Rossi: Phys. Rev. Lett. **9**, 439 (1962)
2.2 H. Gursky, E. Schreier: "The binary x-ray stars: the observational picture", Preprint No. 119, Center for Astrophysics, Texas (1974)
2.3 *X-Ray Binaries: Proceedings of the Symposium* (Washington 1976)
2.4 N.I. Shakura: *Neutron Stars and Black Holes in Binary Systems* [in Russian] (Zmanie, Moscow 1976)
2.5 F.J. Dyson: *Neutron Stars and Pulsars* (Academia Nazionale dei Lincei, Rome 1971)
2.6 E.A. Dibai, S.A. Kaplan: *Dimensions and Similarity of Astrophysical Quantities* [in Russian] (Nauka, Moscow 1976)
2.7 *Origin and Evolution of Galaxies and Stars* [in Russian] (Nauka, Moscow 1976)
2.8 R.A. Sunjaev, N.I. Shakura: In *Nonstationary Phenomena and Solar Evolution* [in Russian] (Nauka, Moscow 1974)
2.9 A.G.W. Cameron: In *The Crab Nebula*, IAU Symposium No. 46 (Dordrecht 1975)
2.10 C.W. Misner, K.S. Thorne, J.A. Wheeler: *Gravitation* (Freeman, San Francisco 1973)
2.11 L.I. Sedov: *Similarity and Dimensional Methods in Mechanics* (Academic Press, New York 1959)
2.12 I.S. Shklovsky: *Supernovae* (Wiley, London 1968)
2.13 K.S. Thorne: Usp. Fiz. Nauk **118**, 453 (1976) [Russian transl.]
2.14 A. Khundkhauzen: *Expansion of the Corona and the Solar Wind* [in Russian] (Mir, Moscow 1976)
2.15 S.J. Akasofu, S. Chapman: *Solar Terrestrial Physics* (Clarendon Press, Oxford 1972)
2.16 J.E. Pringle, M.J. Rees: Astron. Astrophys. **21**, 1 (1972)
2.17 E.P.T. Liang, R. Price: "Accretion diskcoronas and Cygnus X-1", Preprint, University of Utah (1976)
2.18 A.P. Lightman, D.M. Eardley: Astrophys. J. **187**, L1 (1974)

2.19 I.D. Novikov, K.S. Thorne: In *Black Holes* (Gordon and Breach, New York 1973)
2.20 A.H. Batten: *Binary and Multiple Systems of Stars* (Pergamon Press, New York 1973)
2.21 R.A. Sunjaev, N.I. Shakura: Pis'ma Astron. Zh. 1, 6 (1975) [English transl.: Sov. Astron. Lett. **1**, 2 (1975)]
2.22 V.F. Shvartsman: Astrofizika **6**, 309 (1970)
2.23 T. Damour: Doctoral Thesis, Université Pierre et Marie Curie, Paris (1979)
2.24 E.R. Harrison: Nature **264**, 525 (1976)
2.25 R. Ruffini, J.R. Wilson: Phys. Rev. D **12**, 2959 (1975)
2.26 S. Rosseland: Mon. Not. R. Astron. Soc. **841**, 720 (1924)
2.27 G.W. Gibbons: Commun. Math. Phys. **44**, 245 (1975)
2.28 R.K. Sachs: Proc. R. Soc. London, Ser. A **264**, 309 (1961); **270**, 103 (1962)
2.29 R.A. Penrose: In *Battelle Recontres: 1967 Lectures in Mathematics and Physics*, ed. by. C.M. DeWitt and J.A. Wheeler (Benjamin, New York 1968) p. 121
2.30 S.W. Hawking, G.F.R. Ellis: *The Large Scale Structure of Space-Time* (Cambridge University Press, Cambridge 1977)
2.31 S.W. Hawking: "Fundamental breakdown of physics in gravitational collapse", Preprint OAP-240, Caltech (1975)
2.32 J.W. Milnor, A. Wallace: *Differential Topology* [Russian transl.] (Mir, Moscow 1972)
2.33 P.K. Rashevskii: *Geometric Theory of Partial Differential Equations* [in Russian] (Gostekhizdat, Moscow 1947)
2.34 B. Carter: In *Black Holes* (Gordon and Breach, New York 1973)
2.35 A. Lichnerowitz: *Théories Relativistes de la Gravitation et de l'Electromagnétisme* (Masson, Paris 1955)
2.36 S.W. Hawking: In *Black Holes* (Gordon and Breach, New York 1973)
2.37 W. Israel: Phys. Rev. **164**, 1776 (1967); Commun. Math. Phys. **8**, 245 (1968)
2.38 A. Papapetrou: Ann. Inst. Henri Poincaré **4**, 83 (1966)
2.39 F. Ernst: Phys. Rev. **167**, 1175 (1968); J. Math. Phys. **15**, 1409 (1974)
2.40 W. Kinnersley, D.M. Chitre: J. Math. Phys. **18**, 1538 (1977); **19**, 2037 (1978)
2.41 F.J. Ernst: Phys. Rev. **167**, 1175 (1968); J. Math. Phys. **15**, 1409 (1974)
2.42 R.H. Boyer, R.W. Lindquist: J. Math. Phys. **8**, 265 (1967)
2.43 J.C. Graves, D.R. Brill: Phys. Rev. **120**, 1507 (1960)
2.44 I.D. Novikov, A.A. Starobinskii: "Quantum-electrodynamical effects inside a charged black hole and the problem of Cauchy horizons", Preprint No. 512, Institute for Space Research, USSR Academy of Sciences (1979)
2.45 D.S. Robinson: Phys. Rev. D **10**, 458 (1974)
2.46 D.S. Robinson: Phys. Rev. Lett. **34**, 905 (1975)
2.47 R.P. Kerr: Phys. Rev. Lett. **11**, 237 (1963)
2.48 J.M. Bardeen, B. Carter, S.W. Hawking: Commun. Math. Phys. **31**, 61 (1973)
2.49 S.W. Hawking: Phys. Rev. D **13**, 191 (1976)
2.50 B.J. Carr: Astrophys. J. **206**, 8 (1976)
2.51 B. Carter: Phys. Rev. Lett. **33**, 558 (1974)
2.52 S.W. Hawking: Commun. Math. Phys. **43**, 199 (1975)
2.53 D.N. Page: Phys. Rev. D **13**, 198 (1976)
2.54 D.N. Page: "Charged leptons from a nonrotating hole", Preprint OAP-490, Caltech (1977)
2.55 D.N. Page, S.W. Hawking: "Gamma rays from primordial black holes", Preprint OAP-430, Caltech (1975)
2.56 A.G. Doroshkevich, Ya. B. Zel'dovich, I.D. Novikov: Zh. Eksp. Teor. Fiz. **49**, 170 (1965) [English transl.: Sov. Phys. JETP **22**, 122 (1966)]
2.57 A.Z. Patashinskii, A.A. Khar'kov: Zh. Eksp. Teor. Fiz. **59**, 574 (1970) [English transl.: Sov. Phys. JETP **32**, 313 (1971)]
2.58 I. Piir: Izv. Akad. Nauk Est. SSR, Ser. Fiz. Mat. **20**, 253 (1971)
2.59 N.R. Sibgatullin, G.A. Alekseev: Zh. Eksp. Teor. Fiz. **67**, 1233 (1974) [English transl.: Sov. Phys. JETP **40**, 613 (1975)]

2.60 S. Chandrasekhar: Proc. R. Soc. London, Ser. A **343**, 289 (1975)
2.61 A. Papapetrou: J. Phys. A **8**, 313 (1975)
2.62 W.H. Press, S. Teukolsky: Astrophys. J. **185**, 649 (1973)
2.63 R. Price: Phys. Rev. D **5**, 2419 (1972)
2.64 T. Regge, J.A. Wheeler: Phys. Rev. **108**, 1063 (1957)
2.65 D. Christodoulou: Phys. Rev. Lett. **25**, 1596 (1970)
2.66 D. Christodoulou, R. Ruffini: Phys. Rev. **4**, 3552 (1971)
2.67 J.D. Bekenstein: Phys. Rev. D **5**, 1239 (1972)
2.68 J.D. Bekenstein: Phys. Rev. D **7**, 2333 (1973)
2.69 R.A. Penrose, R.M. Floyd: Nature **229**, 177 (1971)
2.70 W.H. Press: Astrophys. J. **175**, 245 (1972)
2.71 Ya. B. Zel'dovich, I.D. Novikov: *Structure and Evolution of the Universe* [in Russian] (Nauka, Moscow 1975)
2.72 A.A. Starobinskii: Zh. Eksp. Teor. Fiz. **64**, 48 (1973) [English transl.: Sov. Phys. JETP **37**, 28 (1973)]
2.73 A.A. Starobinskii, S.M. Churilov: Zh. Eksp. Teor. Fiz. **65**, 3 (1973) [English transl.: Sov. Phys. JETP **38**, 1 (1974)]
2.74 C.W. Misner: Bull. Am. Phys. Soc. **117**, 472 (1972)
2.75 S.W. Hawking: Phys. Rev. Lett. **26**, 1344 (1971)
2.76 K. Smarr, A. Cader, B. DeWitt, K. Eppley: Phys. Rev. D **14**, 24 (1976)
2.77 Plebanski, M. Demianski: Ann. Phys. (USA) **98**, 98 (1976)
2.78 A. Fischer, J. Marsden: Gen. Relativ. Gravitat. **5**, 89 (1974)

Chapter 3

3.1 N.R. Sibgatullin: Zh. Eksp. Teor. Fiz. **66**, 1187 (1974) [English transl.: Sov. Phys. JETP **39**, 579 (1974)]
3.2 R.P. Geroch: J. Math. Phys. **13**, 394 (1972)
3.3 W. Kinnersley, D.M. Chitre: J. Math. Phys. **18**, 1538 (1977); **19**, 2037 (1978)
3.4 J. Hauser, F.J. Ernst: J. Math. Phys. **19**, 1316 (1978)
3.5 J. Hauser, F.J Ernst: Phys. Rev. D **20**, 362, 1783 (1979)
3.6 G.A. Alekseev: Pis'ma Zh. Eksp. Teor. Fiz. **32**, 301 (1980) [English transl.: JETP Lett. **32**, 277 (1980)]
3.7 G. Neugebauer: Phys. Lett. **86A**, 91 (1983)
3.8 A.N. Golubyatnikov: Dokl. Akad. Nauk SSSR **192**, 55 (1970)
3.9 J.B. Griffiths: Commun. Math. Phys. **28**, 295 (1972)
3.10 J.B. Griffiths: Ann. Phys. (N.Y.) **102**, 388 (1976)
3.11 V.I. Repchenkov: Zh. Eksp. Teor. Fiz. **77**, 1233 (1979) [English transl.: Phys. JETP **50**, 621 (1979)]
3.12 M. Henneaux. Phys. Rev. D **21**, 857 (1980)
3.13 A.V. Bitsadze: *Some Classes of Partial Differential Equations* [in Russian] (Nauka, Moscow 1981)
3.14 L.I: Sedov: *Planar Problems of Hydrodynamics and Aerodynamics* [in Russian] (Nauka, Moscow 1966)
3.15 H. Sato, A. Tomimatzu: Phys. Rev. Lett. **29**, 19 (1972)
3.16 F.D. Gakhov: *Boundary Value Problems* (Addison-Wesley, Reading, Mass. 1966)
3.17 V.A. Belinskii, V.E. Zakharov: Zh. Eksp. Teor. Fiz. **77**, 3 (1979)
3.18 G. Springer: *Introduction to Riemann Surfaces* (Addison-Wesley, Reading, Mass. 1957)
3.19 N.R. Sibgatullin: Dokl. Akad. Nauk SSSR **291**, 302 (1986) [English transl.: Sov. Phys. Dokl. **31**, 863 (1986)]
3.20 V.E. Zakharov, A.B. Shabat: Funkts. Anal. Prilozh. **13**, No. 13, 13 (1979)
3.21 C.S. Gardner, J.M. Greene, M.D. Kruskal, R.M. Miura: Phys. Rev. Lett **19**, 1095 (1967)
3.22 P.D. Lax: Commun. Pure. Appl. Math. **21**, 467 (1968)

3.23 V.E. Zakharov, S.V. Manakov, S.P. Novikov, L.P. Pitaevskii: *Theory of Solitons* [in Russian] (Nauka, Moscow 1980)
3.24 L.D. Faddeev, L.A. Takhtajan: *Hamiltonian Methods in the Theory of Solitons* (Springer Verlag, Berlin 1987)

Additional References

3.25 D. Maison: J. Math. Phys. **20**, 871 (1979)
3.26 D. Kramer, G. Neugebauer: J. Phys. A: Math. Gen. **16**, 1927 (1983)
3.27 M.I. Ablowitz, H. Segur: *Solitons and Inverse Scattering Transforms* (Soc. for Industrial and Applied Math. USA, 1981)
3.28 A.C. Newell: *Solitons in Mathematics and Physics* (Soc. for Industrial and Applied Math. USA, 1985)
3.29 R.K. Dodd, J.C. Eilbeck, J.D. Gibbon, H.C. Morris: *Solitons and Nonlinear Wave Equations* (Academic, London 1984)
3.30 C.M. Cosgrove: J. Math. Phys. **23**, 615 (1982)
3.31 K. Ueno, Y. Nakamura: in *Nonlinear Integrable Systems*, ed. by M. Jimbo (World Scientific, Singapore 1981)

Chapter 4

4.1 J.D. Bekenstein: Phys. Rev. D **5**, 1239 (1972)
4.2 K.S. Thorne: "Nonspherical gravitational collapse: a short review", Preprint OAP-236, Caltech (1971)
4.3 M.A. Markov: Usp. Fiz. Nauk **111**, 3 (1973) [English transl.: Sov. Phys. Usp. **16**, 587 (1974)]
4.4 E.T. Newman, E. Couch, K. Chinnapared, A. Exton, A. Prakash, R. Torrence: J. Math. Phys. **6**, 918 (1965)
4.5 W. Israel: Phys. Rev. **164**, 1776 (1967); Commun. Math. Phys. **8**, 245 (1968)
4.6 T. Regge, J.A. Wheeler: Phys. Rev. **108**, 1063 (1957)
4.7 F.J. Zerilli: J. Math. Phys. **11**, 2203 (1970)
4.8 V.L. Ginzburg, L.M. Ozernoi: Zh. Eksp. Teor. Fiz. **47**, 1030 (1964) [English transl.: Sov. Phys. JETP **20**, 689 (1965)]
4.9 A.G. Doroshkevich, Ya. B. Zel'dovich, I.D. Novikov: Zh. Eksp. Teor. Fiz. **49**, 170 (1965) [English transl.: Sov. Phys. JETP **22**, 122 (1966)]
4.10 I. Piir: Izv. Akad. Nauk. Est. SSR, Ser. Fiz. Mat. **20**, 253 (1971)
4.11 R. Price: Phys. Rev. D **5**, 2419 (1972)
4.12 K.S. Thorne: Preprint OAP-236, Caltech (1971)
4.13 S. Chandrasekhar: *The Mathematical Theory of Black Holes* (Clarendon Press, Oxford 1983)
4.14 G.A. Alekseev, N.R. Sibgatullin: In *Problems of the Theory of Gravitation and Elementary Particles*, No. 5 [in Russian] (Atomizdat, Moscow 1974)
4.15 A.Z. Patashinskii, A.A. Khar'kov: Zh. Eksp. Teor. Fiz. **59**, 574 (1970) [English transl.: Sov. Phys. JETP **32**, 313 (1971)]
4.16 S. Chandrasekhar: Proc. R. Soc. London, Ser. A **343**, 289 (1975)
4.17 A. Papapetrou: J. Phys. A **8**, 313 (1975)
4.18 K. Tomita: Prog. Theor. Phys. **52**, No. 4 (1974)
4.19 G.A. Alekseev: In *Abstracts of Contributions to the Minsk Gravitational Conference* [in Russian] (Institute of Physics, Belorussian Academy of Sciences, 1976)
4.20 D.M. Chitre: Phys. Rev. D **11**, 760 (1975)
4.21 D.M. Chitre: R.H. Price, V.D. Sandberg: Phys. Rev. D **11**, 747 (1975)
4.22 R. Ruffini, J. Tiomno, C.V. Vishveshwara: Nuovo Cimento Lett. **3**, 211 (1972)
4.23 F.J. Zerilli: Phys. Rev. D **2**, 2141 (1970)
4.24 A.Z. Patashinskii, V.K. Pinus, A.A. Khar'kov: Zh. Eksp. Teor. Fiz. **66**, 393 (1974) [English transl.: Sov. Phys. JETP **39**, 187 (1974)]

4.25 J. Bičak: Gen. Relativ. Gravitat. **3**, 331 (1972)
4.26 N.R. Sibgatullin, G.A. Alekseev: Zh. Eksp. Teor. Fiz. **67**, 1233 (1974) [English transl.: Sov. Phys. JETP **40**, 613 (1975)]
4.27 N.R. Sibgatullin: Zh. Eksp. Teor. Fiz. **66**, 1187 (1974) [English transl.: Sov. Phys. JETP **39**, 579 (1974)]
4.28 A. Erdélyi (ed.): *Higher Transcendental Functions* [Bateman Manuscript Project], Vol. 1 (McGraw-Hill, New York 1953)
4.29 N.R. Sibgatullin: Dokl. Akad. Nauk SSSR **209**, 815 (1973) [English transl.: Sov. Phys. Dokl. **18**, 205 (1973)]
4.30 V.P. Maslov: *Perturbation Theory and Asymptotic Methods* [in Russian] (Moscow State University Press, Moscow 1965)
4.31 W.H. Press: Astrophys. J. **170**, L105 (1971)
4.32 G.A. Alekseev: Pis'ma Zh. Eksp. Teor. Fiz. **32**, 301 (1980) [English transl.: Lett. JETP **32**, 277 (1980)]
4.33 L.D. Landau, E.M. Lifshitz: *The Classical Theory of Fields* (Pergamon Press, Oxford 1975)
4.34 L.D. Faddeev: Tr. Mat. Inst. Akad. Nauk SSSR **73**, 314 (1964)
4.35 V. Monkrief: Phys. Rev. D **9**, 2707 (1974); ibid. **12**, 1526 (1975)

Chapter 5

5.1 V.G. Pisarenko: *Problems of Many-Body Relativistic Dynamics and Nonlinear Field Theory* [in Russian] (Naukova Dumka, Kiev 1964)
5.2 N.Ya. Vilenkin, Ya.A. Smorodinskii: Zh. Eksp. Teor. Fiz. **46**, 1793 (1964) [English transl.: Sov. Phys. JETP **19**, 1209 (1964)]
5.3 N.R. Sibgatullin: Dokl. Akad. Nauk SSSR **188**, 1224 (1969)
5.4 N.R. Sibgatullin: Dokl. Akad. Nauk SSSR **180**, 48 (1968) [English transl.: Sov. Phys. Dokl. **13**, 403 (1968)]
5.5 L.D. Landau, E.M. Lifshitz: *Fluid Mechanics* (Pergamon Press, Oxford 1969)
5.6 J.L. Synge: *The Relativistic Gas* (North-Holland, Amsterdam 1957)
5.7 H.A. Bethe: Ann. Rev. Nucl. Sci. **21**, 93 (1971)
5.8 W.D. Myers, W.J. Swiatecki: Nucl. Phys. **81**, 1 (1966)
5.9 F.J. Dyson: *Neutron Stars and Pulsars* (Accademia Nazionale dei Lincei, Rome 1971)
5.10 G.S. Saakyan: *Equilibrium Configurations of Degenerate Gaseous Masses* [in Russian] (Nauka, Moscow 1972)
5.11 V.M. Galitskii: Priroda (Moscow), No. 1, 76 (1976)
5.12 K.A. Brueckner: Phys. Rev. **97**, 1353 (1955)
5.13 A.G.W. Cameron: In *The Crab Nebula*, IAU Symposium No. 46 (Dordrecht 1975)
5.14 J. Goldstone: Proc. R. Soc. London, Ser. A **239**, 267 (1957)
5.15 V.A. Ambartsumyan, G.S. Saakyan: Astron. Zh. 37, 193 (1960)
5.16 S.Z. Belen'kii, L.D. Landau: Usp. Fiz. Nauk **56**, 309 (1955)
5.17 P.Ya Pomeranchuk: Dokl. Akad. Nauk SSSR **78**, 889 (1951)
5.18 I.M. Khalatnikov: Zh. Eksp. Teor. Fiz. **27**, 529 (1954)
5.19 E. Fermi: Phys. Rev. **81**, 683 (1951)
5.20 F.I. Frankl': Zh. Eksp. Teor. Fiz. **31**, 490 (1956) [English transl.: Sov. Phys. JETP **4**, 401 (1957)]
5.21 N. Sibgatullin: C.R. Acad. Sci., Ser. A **290**, 233 (1980)
5.22 Y. Choquet-Bruhat: J. Math. Pur. Appl. **48**, 117 (1969)
5.23 L.I. Sedov: *A Course in Continuum Mechanics* (Wolters-Noordhoff, Groningen 1971)
5.24 P.J.E. Peebles: Phys. Rev. D **1**, 39 (1970)
5.25 F.A. Baum, S.A. Kaplan, K.P. Stanyukovich: *Introduction to Cosmic Gas Dynamics* [in Russian] (Fizmatgiz, Moscow 1958)
5.26 V.A. Zhelnorovich: *Models of Material Continuous Media with Internal Electromagnetic and Mechanical Angular Momenta* [in Russian] (Moscow State University, Moscow 1980)

5.27 A. Lichnerowitz: *Relativistic Hydrodynamics and Magnetohydrodynamics* (Benjamin, New York 1967)
5.28 G.A. Maugin, A.C. Eringen: *Electromagnetics of Continua* (Springer, Berlin-Heidelberg 1989) Chap. 14
5.29 A.G. Kulikovskii, G.A. Lyubimov: *Magnetohydrodynamics* [in Russian] (Fizmatgiz, Moscow 1962)
5.30 R. Courant, K.O. Friedrichs: *Supersonic Flow and Shock Waves* (Interscience, New York 1948)
5.31 S.W. Hawking: Commun. Math. Phys. **43**, 199 (1975)
5.32 D.N. Page: Phys. Rev. D **13**, 198 (1976)
5.33 D.N. Page "Charged leptons from a nonrotating hole", Preprint OAP-490, Caltech (1977)
5.34 D.N. Page, S.W. Hawking: "Gamma rays from primordial black holes", Preprint OAP-430, Caltech (1975)
5.35 S.W. Hawking: "Fundamental breakdown of physics in gravitational collapse", Preprint OAP-240, Caltech (1975)
5.36 S.W. Hawking: Phys. Rev. D **13** , 191 (1976)
5.37 F.C. Michel: Astrophys. Space Sci. **15**, 153 (1972)
5.38 S.L. Shapiro: Astrophys. J. **180**, 531 (1971)
5.39 B. Carter, G.W. Gibbons, D.N.C. Lin, M.J. Perry: Astron. Astrophys. **52**, 427 (1976)
5.40 N.R. Sibgatullin: "Some problems of gas dynamics in the theory of relativity" [in Russian], Candidate's Dissertation, Moscow State University (1969)
5.41 N.R. Sibgatullin: Dokl. Akad. Nauk SSSR **187**, 531 (1969) [English transl.: Sov. Phys. Dokl. **14**, 619 (1970)]
5.42 V.A. Skripkin: Dokl. Akad. Nauk SSSR **136**, 971 (1961)
5.43 K.P. Stanyukovich: Zh. Eksp. Teor. Fiz. **43**, 199 (1962) [English transl.: Sov. Phys. JETP **16**, 142 (1966)]
5.44 H.H. Chiu: Phys. Fluids **16**, 825 (1973)
5.45 D.M. Eardley: Commun. Math. Phys. **37**, 287 (1974)
5.46 L.I. Sedov: *Similarity and Dimensional Methods in Mechanics* (Academic Press, New York 1959)
5.47 E. Milne: *Relativity, Gravitation and World Structure* (Oxford University Press, London 1935)

Chapter 6

6.1 P.J.E. Peebles: *Physical Cosmology* (Princeton University Press, Princeton, N.J. 1971)
6.2 R.H. Dicke, P.J.E. Peebles, P.G. Roll, D.T. Wilkinson: Astrophys. J. **142**, 414 (1965)
6.3 P.J.E. Peebles: Astrophys. J. **146**, 542 (1966)
6.4 T.A. Agekyan: *Stars, Galaxies, Metagalaxies* [in Russian] (Nauka, Moscow 1970)
6.5 E.A. Dibai, S.A. Kaplan; *Dimensions and Similarity of Astrophysical Quantities* [in Russian] (Nauka, Moscow 1976)
6.6 Ya. B. Zel'dovich, I.D. Novikov: *Structure and Evolution of the Universe* [in Russian] (Nauka, Moscow 1975)
6.7 *Origin and Evolution of Galaxies and Stars* [in Russian] (Nauka, Moscow 1976)
6.8 J.E. Gunn: "Observational tests in cosmology", Preprint OAP-464, Caltech (1976)
6.9 C.W. Misner, K.S. Thorne, J.A. Wheeler: *Gravitation* (Freeman, San Francisco 1973)
6.10 M. Rees, R. Ruffini, J.A. Wheeler: *Black Holes, Gravitational Waves, and Cosmology* (Gordon and Breach, New York 1974)
6.11 A.L. Zel'manov: In *Proc. of the 6th Meeting on Problems of Cosmology in June 1957* [in Russian] (USSR Academy of Sciences Press, Moscow 1959) p. 144
6.12 V.A. Belinskii, I.M. Khalatnikov: Zh. Eksp. Teor. Fiz. **57**, 2163 (1969) [English transl.: Sov. Phys. JETP **30**, 1174 (1970)]
6.13 S.P. Novikov: Zh. Eksp. Teor. Fiz. **62**, 1977 (1972) [English transl.: Sov. Phys. JETP **35**, 1031 (1972)]
6.14 C.W. Misner: Phys. Rev. **186**, 1319 (1969)

6.15 J. Silk: Astrophys. J. **151**, 459 (1968)
6.16 L.D. Landau, E.M. Lifshitz: *The Classical Theory of Fields* (Pergamon Press, Oxford 1975)
6.17 R.K. Sachs, A.M. Wolfe: Astrophys. J. **147**, 73 (1968)
6.18 A.G. Doroshkevich: Astrofizika **2**, 37 (1966)
6.19 N.R. Sibgatullin: "Some problems of gas dynamics in the theory of relativity" [in Russian], Candidate's Dissertation, Moscow State University (1969)
6.20 N.R. Sibgatullin: Dokl. Akad. Nauk SSSR **193**, 1267 (1970) [English transl.: Sov. Phys. Dokl. **15**, 739 (1971)]
6.21 A.G. Doroshkevich; Astrofizika **1**, 255 (1965)
6.22 I.S. Shikin: Dokl. Akad. Nauk SSSR **171**, 73 (1966) [English transl.: Sov. Phys. Dokl. **11**, 944 (1967)]
6.23 K.S. Jacobs: Astrophys. J. **155**, 379 (1969)
6.24 I.S. Shikin: Commun. Math. Phys. **26**, 24 (1972)
6.25 B.P. Gleizer, A.A. Ruzmaikin, D.D. Sokolov: In *Abstracts of Contributions to the Minsk Gravitational Conference* [in Russian] (Institute of Physics, Belorussian Academy of Sciences 1976)
6.26 H. Lamb: *Hydrodynamics* (Cambridge University Press, Cambridge 1932)
6.27 N.R. Sibgatullin: Dokl. Akad. Nauk SSSR **188**, 1224 (1969)
6.28 V.A. Skripkin: Astron. Zh. **38**, 192 (1961)
6.29 V.A. Skripkin: Dokl. Akad. Nauk SSSR **136**, 971 (1961)
6.30 V. Ts. Gurovich, K.P. Stanyukovich, O.V. Sharshekeev: Dokl. Akad. Nauk SSSR **165**, 510 (1965) [English transl.: Sov. Phys. Dokl. **10**, 1030 (1966)]
6.31 K.P. Stanyukovich: Zh. Eksp. Teor. Fiz. **66**, 826 (1974) [English transl.: Sov. Phys. JETP **39**, 399 (1974)]
6.32 V. Ts. Gurovich: Dokl. Akad. Nauk SSSR **169**, 62 (1966) [English transl.: Sov. Phys. Dokl. **11**, 569 (1967)]
6.33 T.V. Ruzmaikina, A.A. Ruzmaikin: Zh. Eksp. Teor. Fiz. **59**, 1576 (1970) [English transl.: Sov. Phys. JETP **32**, 862 (1971)]
6.34 O.I. Bogoyavlenskii: Zh. Eksp. Teor. Fiz. **73**, 1201 (1977) [English transl.: Sov. Phys. JETP **46**, 633 (1977)]
6.35 D.M. Eardley: Commun. Math. Phys. **37**, 287 (1974)
6.36 M.E. Cahill, A.H. Taub: Commun. Math. Phys. **21**, 1 (1971)
6.37 B.J. Carr, S.W. Hawking: Mon. Not. R. Astron. Soc. **168**, 399 (1974)
6.38 N.R. Sibgatullin, O.Yu. Dinariev: Zh. Eksp. Teor. Fiz. **73**, 1599 (1977) [English transl.: Sov. Phys. JETP **46**, 840 (1977)]
6.39 L.I. Sedov: *Similarity and Dimensional Methods in Mechanics* (Academic Press, New York 1959)
6.40 L.I. Sedov: Prikl. Mat. Mekh. **36**, 3 (1972)
6.41 N.R. Sibgatullin: Dokl. Akad. Nauk SSSR **187**, 531 (1969) [English transl.: Sov. Phys. Dokl. **14**, 619 (1970)]
6.42 A. Ori, T. Piran: Gen. Relativ. Gravitat. **20**, 7 (1988)

Chapter 7

7.1 *X-Ray Binaries: Proceedings of the Symposium* (Washington 1976)
7.2 R.A. Syunyaev, N.I. Shakura: In *Nonstationary Phenomena and Solar Evolution* [in Russian] (Nauka, Moscow 1974)
7.3 R. Giacconi, H. Gursky, F.R. Paolini, B.B. Rossi: Phys. Rev. Lett. **9**, 439 (1962)
7.4 H. Gursky, E. Schreier: "The binary x-ray stars: the observational picture", Preprint No. 119, Center for Astrophysics, Texas (1974)
7.5 S. Rappoport, P.S. Joss: "Accretion torques in x-ray pulsars", Preprint, Massachusetts Institute of Technology (1977)
7.6 J.E. Gunn, J.B. Ostriker: Astrophys. J. **160**, 979 (1970)

7.7 P. Goldreich, W.H. Julian: Astrophys. J. **157**, 869 (1969)
7.8 A.F. Illarionov, R.A. Sunjaev: Astron. Astrophys. **39**, 185 (1975)
7.9 L.D. Landau, E.M. Lifshitz: *Fluid Mechanics* (Pergamon Press, Oxford 1969)
7.10 O.V. Rudenko, S.P. Soluyan: *Theoretical Principles of Nonlinear Acoustics* [in Russian] (Nauka, Moscow 1975)
7.11 A.A. Eikhenval'd: Usp. Fiz. Nauk **14**, 552 (1934)
7.12 J.P. Guiraud: *La Théorie du Bang Supersonique d'après la Théorie Nonlinéaire des Ondes Courtes* (Publ. de l'O.N.E.R.A. 1967)
7.13 J.B. Keller: J. Appl. Phys. **25**, 588 (1954)
7.14 V.I. Arnold: *Mathematical Methods of Classical Mechanics* (Springer Verlag, Berlin 1978)
7.15 L.I. Sedov: *Similarity and Dimensional Methods in Mechanics* (Academic Press, New York 1959)
7.16 G.G. Chernyi: *Gas Flows with High Supersonic Velocity* [in Russian] (Fizmatgiz, Moscow 1959)
7.17 R.J. Slutz: J. Geophys. Res. **67**, 505 (1962)
7.18 J. Arons, S.M. Lea: Astrophys. J. **207**, 914 (1976)
7.19 J.D. Cole, J.H. Huth: Phys. Fluids **2**, 624 (1959)
7.20 J. Ehlers: In *Les Théories Relativistes de la Gravitation* (Colloques Int. C.N.R.S., Paris 1959)
7.21 V.M. Lipunov: Astron. Zh. **57**, 1253 (1980) [English transl.: Sov. Astron. **24**, 722 (1980)]
7.22 R.F. Elsner, F.K. Lamb: Nature **262**, 356 (1977)
7.23 L.I. Sedov: *Planar Problems of Hydrodynamics and Aerodynamics* [in Russian] (Nauka, Moscow 1966)
7.24 F.D. Gakhov: *Boundary Value Problems* (Addison-Wesley, Reading, Mass. 1966)
7.25 G.G. Tumashev, M.T. Nuzhin: *Inverse Boundary Value Problems and Their Applications* [in Russian] (Kazan State University Press, Kazan 1965)
7.26 H. Bondi: Mon. Not. R. Astron. Soc. **112**, 195 (1952)
7.27 O. Sawatski: Acta Mecan. **9**, 11 (1971)
7.28 N.A. Slezkin: *Dynamics of a Viscous Incompressible Fluid* [in Russian] (Gostekhizdat, Moscow 1955)
7.29 K.P. Stanyukovich: *Irregular Motions of a Continuous Medium* [in Russian] (Nauka, Moscow 1971)
7.30 S. Rosseland: *The Pulsation Theory of Variable Stars* (Clarendon Press, Oxford 1949)
7.31 H. Lamb: *Hydrodynamics*, (Cambridge University Press, Cambridge 1932)
7.32 S. Chandrasekhar: *Ellipsoidal Figures of Equilibrium* (Yale University Press, New Haven, Conn. 1969)
7.33 A.G. Kulikovskii, G.A. Lyubimov: *Magnetohydrodynamics* [in Russian] (Fizmatgiz, Moscow 1962)
7.34 A.B. Severnyi: Dokl. Akad. Nauk SSSR **46**, No. 2, 57 (1945)
7.35 A. Erdélyi (ed.): *Higher Transcendental Functions* [Bateman Manuscript Project], Vol. 1 (McGraw-Hill, New York 1953)
7.36 E. Kamke: *Gewöhnlich Differentialgleichungen* (Acad. Verl., Leipzig 1959)
7.37 L.I. Sedov: *A Course in Continuum Mechanics* (Wolters-Noordhoff, Groningen 1971)
7.38 N.R. Sibgatullin: Prikl. Mat. Mekh. **36**, 79 (1972)
7.39 Sh.U. Galiev, M.A. Il'gamov, A.V. Sadykov: MZhG, No. 2, 57 (1970)
7.40 L.P. Gor'kov: Inzh. Zh. **3**, No. 2, 246 (1963)
7.41 A.I. Gulyaev, V.M. Kuznetsov: Inzh. Zh. **3**, No. 2, 236 (1963)
7.42 R. Bechov: Phys. Fluids **1**, 3 (1958)
7.43 W. Chester: J. Fluid Mech. **18**, 1 (1964)
7.44 J. Jimenez: J. Fluid Mech. **59**, 1 (1973)
7.45 A.A. Galeev, R.Z. Sagdeev: Vopr. Teor. Plazmy, No. 7 (1973)

Additional References

7.46 G.B. Whitham: *Linear and Nonlinear Waves* (Wiley-Interscience New York 1974)
7.47 P. Germain: *Méthods asymptotiques en mécaniques des fluides* (Gordon and Breach, New York 1973)
7.48 P.A. Bois: Doctoral thesis, Paris VI (1979)
7.49 P. Gatignol: Doctoral thesis, Paris VI (1978)
7.50 H. Cabannes: Rech. Aéro, **49**, 11 (1956)
7.51 M. Roseau: *Asymptotic Wave Theory* (North-Holland, Amsterdam 1976)

Subject Index